Student's Solutions Manual

Basic Technical
Mathematics
and
Basic Technical
Mathematics with Calculus

Tenth Edition

Allyn J. Washington

Dutchess Community College

Prepared by John R. Martin

Tarrant County College, Northeast Campus

D1511379

PEARSON

Boston   Columbus   Indianapolis   New York   San Francisco   Upper Saddle River

Amsterdam   Cape Town   Dubai   London   Madrid   Milan   Munich   Paris   Montreal   Toronto

Delhi   Mexico City   Sao Paulo   Sydney   Hong Kong   Seoul   Singapore   Taipei   Tokyo

Editorial Director: Vern Anthony
Acquisitions Editor: Sara Eilert
Editorial Assistant: Doug Greive
Director of Marketing: David Gesell
Marketing Manager: Stacey Martinez
Marketing Assistant: Les Roberts
Senior Managing Editor: JoEllen Gohr

Senior Project Manager: Rex Davidson
Senior Operations Supervisor: Pat Tonneman
Cover Designer: Integra
Printer/Binder: Edwards Brothers Malloy
Composition: John R. Martin and John Garlow
Cover Printer: Edwards Brothers Malloy

10 9 8 7 6 5 4 3 2 1

ISBN-13: 978-0-13-325351-1
ISBN-10:   0-13-325351-1

CONTENTS

CHAPTER 1 BASIC ALGEBRAIC OPERATIONS

CHAPTER 2 GEOMETRY

CHAPTER 3   FUNCTIONS AND GRAPHS

CHAPTER 4   THE TRIGONOMETRIC FUNCTIONS

CHAPTER 5   SYSTEMS OF LINEAR EQUATIONS; DETERMINANTS

CHAPTER 6   FACTORING AND FRACTIONS

CHAPTER 7   QUADRATIC EQUATIONS

CHAPTER 8   TRIGONOMETRIC FUNCTIONS OF ANY ANGLE

CHAPTER 9 VECTORS AND OBLIQUE TRIANGLES

CHAPTER 10 GRAPHS OF THE TRIGONOMETRIC FUNCTIONS

CHAPTER 11 EXPONENTS AND RADICALS

CHAPTER 12 COMPLEX NUMBERS

CHAPTER 13 EXPONENTIAL AND LOGARITHMIC FUNCTIONS

CHAPTER 14 ADDITIONAL TYPES OF EQUATIONS AND SYSTEMS OF EQUATIONS

CHAPTER 15 EQUATIONS OF HIGHER DEGREE

CHAPTER 16 MATRICES

CHAPTER 17   INEQUALITIES

CHAPTER 18   VARIATION

CHAPTER 19   SEQUENCES AND THE BINOMIAL THEOREM

CHAPTER 20 ADDITIONAL TOPICS IN TRIGONOMETRY

CHAPTER 21 PLANE ANALYTIC GEOMETRY

CHAPTER 22   INTRODUCTION TO STATISTICS

CHAPTER 23   THE DERIVATIVE

CHAPTER 24   APPLICATIONS OF THE DERIVATIVE

CHAPTER 25   INTEGRATION

CHAPTER 26   APPLICATIONS OF INTEGRATION

CHAPTER 27   DIFFERENTIATION OF TRANSCENDENTAL FUNCTIONS

CHAPTER 28   METHODS OF INTEGRATION

CHAPTER 29　PARTIAL DERIVATIVES AND DOUBLE INTEGRALS

CHAPTER 30　EXPANSION OF FUNCTIONS IN SERIES

CHAPTER 31   DIFFERENTIAL EQUATIONS

Chapter 1

BASIC ALGEBRAIC OPERATIONS

1.1 Numbers

1. The numbers -3 and 14 are integers. They are also rational numbes since they can be written as $\dfrac{-3}{1}$ and $\dfrac{14}{1}$.

5. 3: integer, rational $\left(\dfrac{3}{1}\right)$, real

$\sqrt{-4}$: imaginary

$-\dfrac{\pi}{6}$: irrational, real

$\dfrac{1}{8}$: rational, real

9. $6 < 8$

13. $-|-3| = -3 \Rightarrow -4 < -3 = -|-3|$

17. The reciprocal of $3 = \dfrac{1}{3}$. The reciprocal of

$-\dfrac{4}{\sqrt{3}}$ is $\dfrac{1}{-\frac{4}{\sqrt{3}}} = -\dfrac{\sqrt{3}}{4}$.

The reciprocal of $\dfrac{y}{b}$ is $\dfrac{1}{\frac{y}{b}} = \dfrac{b}{y}$.

21. An absolute value is not always positive, $|0| = 0$ which is not positive.

25. Since $0.13 = \dfrac{13}{100} \cdot \dfrac{\frac{3}{13}}{\frac{3}{13}} = \dfrac{3}{\frac{300}{13}} = \dfrac{3}{23.0769} \cdots <$

$\dfrac{3}{23} = \dfrac{3}{23} \cdot \dfrac{\frac{14}{3}}{\frac{14}{3}} = \dfrac{14}{107.3} < \dfrac{14}{100} = 0.14$

$\dfrac{3}{23}$ is a rational number between 0.13 and 0.14 with numerator 3 and an integer, 23, in the denominator.

29. (a) $b - a; b > a$, positive integer

(b) $a - b; b > a$, negative integer

(c) $\dfrac{b-a}{b+a}$, positive rational number less than 1

33. (a) x is a positive number located to the right of 0.

(b) x is a negative number located to the left of -4.

37. $a + bj = a + b\sqrt{-1}$ which for $b = 0$ is $a + bj = a$, a real number. The complex number $a + bj$ is a real number for all values of a and $b = 0$.

41. $N = \dfrac{a \text{ bits}}{\text{byte}} \cdot \dfrac{1000 \text{ bytes}}{\text{kilobytes}} \cdot n \text{ kilobytes}$

$= 1000 \, an \text{ bits}$

1.2 Fundamental Operations of Algebra

1. $16 - 2 \times (-3) = 16 - (-6) = 22$

5. $8 + (-4) = 8 - 4 = 4$

9. $-19 - (-16) = -19 + 16 = -3$

13. $-7(-5) = 35$

17. $-2(4)(-5) = (-8)(-5) = 40$

21. $-9 - |2 - 10| = -9 - 1|-8| = -9 - 8 = -17$

25. $8 - 3(-4) = 8 - (-12) = 8 + 12 = 20$

29. $30(-6)(-2) \div (0 - 40) = 360 \div (-40) = -9$

33. $-7 - \dfrac{|-14|}{2(2-3)} - 3|6-8| = -7 - \dfrac{14}{2(-1)} - 3|-2|$

$$= -7 - \dfrac{14}{-2} - 3(2)$$
$$= -7 - (-7) - 6$$
$$= 0 - 6$$
$$= -6$$

37. $6(7) = 7(6)$ demonstrates the commutative law of multiplication.

41. $3 + (5 + 9) = (3 + 5) + 9$ demonstrates the associative law of addition.

45. $-a + (-b) = -a - b$ which is expression (d).

49. $\big|5 - (-2)\big| = \big||-5| - |-2|\big|$

53. The definition $|x| = \begin{cases} x, & \text{if } x \ge 0 \\ -x, & \text{if } x < 0 \end{cases}$

is correct because it is equivalent to the definition on page 3 where $|x|$ is defined as x, the number itself for $x = 0$ or $x > 0$ (positive) and $|x|$ is the corresponding positive number $(= -x)$ for $x < 0$ (negative).

57. $10 - (-15) = 10 + 15 = 25$

61. $\dfrac{-7 + (-3) + (2) + (3) + (1) + (-4) + (-6)}{7} = -2^\circ \text{C}$

65. $100 \text{ m} + 200 \text{ m} = 200 \text{ m} + 100 \text{ m}$ illustrates the commutative law of addition.

1.3 Calculators and Approximate Numbers

1. Yes, 0.390 has three significant digits since the 0 after the 9 is not needed to locate the decimal.

5. 8 cylinders is exact because they can be counted. 55 mi/h is approximate since it is measured.

9. 1 cm and 9 g are approximate

13. 6.80 has 3 significant digits; the zero indicates precision.
6.08 has 3 significant digits; the zero is not used for decimal location, and is not a place-holder only.
0.068 has 2 significant digits.

17. (a) 0.01 is more precise (more decimal places).
(b) 30.8 is more accurate (more significant digits).

21. (a) 0.004 is more precise (more decimal places).
(b) Both have the same accuracy

25. (a) $-50.893 = -50.9$ (3 significant digits)
(b) $-50.893 = -51.0$ (2 significant digits)

29. (a) $0.9449 = 0.945$ (3 significant digits)
(b) $0.9949 = 0.94$ (2 significant digits)

33. (a) Estimate: $1 \times 4 - 9 = -5$
(b) Calculator: $0.6572 \times 3.94 - 8.651 = -5.99$

37. (a) Estimate: $\dfrac{20 \times 0.02}{10 - 8} = 0.2$
(b) Calculator: $\dfrac{23.962 \times 0.01537}{10.965 - 8.249} = 0.1356$

41. $0.9788 + 14.9 = 15.8788$ since 4 is the number of decimal places in the least precise number.

45. 2.745 MHz and 2.755 MHz are the least possible and greatest possible frequencies respectively.

49. (a) $2.2 + 3.8 \times 4.5 = 19.3$

(b) $(2.2 + 3.8) \times 4.5 = 27$

53.
```
61812311→X
        61812311
11381216→Y
        11381216
(X-Y)/9
        5603455
```

57. (a) $\dfrac{1}{3} = 0.333\cdots$

(b) $\dfrac{5}{11} = 0.454545\cdots$

(c) $\dfrac{2}{5} = 0.4000\cdots$

0 is the repeating part

61. $1\,\text{MB} = 1{,}048{,}576 \text{ bytes}$

$256\,\text{MB} = 256 \times 1{,}048{,}576$

$= 268{,}435{,}456 \text{ bytes}$

1.4 Exponents

1. $\left(-x^3\right)^2 = \left[(-1)x^3\right]^2 = (-1)^2\left(x^3\right)^2 = x^{3(2)} = x^6$

5. $x^3 \cdot x^4 = x^{3+4} = x^7$

9. $\dfrac{m^5}{m^3} = m^{5-3} = m^2$

13. $\left(P^2\right)^4 = P^{2 \cdot 4} = P^8$

17. $\left(aT^2\right)^{30} = a^{30}T^{2(30)} = a^{30}T^{60}$

21. $\left(\dfrac{x^2}{-2}\right)^4 = \dfrac{x^{2 \cdot 4}}{2^4} = \dfrac{x^8}{16}$

25. $-3x^0 = (-3)(1) = -3$

29. $\dfrac{1}{R^{-2}} = R^2$

33. $\left(2v^2\right)^{-6} = \left(2^{-6}\right)\left(v^{-2 \cdot 6}\right) = \dfrac{1}{64v^{12}}$

37. $\dfrac{2v^4}{(2v)^4} = \dfrac{2v^4}{2^4 v^4} = \dfrac{1}{2^3} = \dfrac{1}{8}$

41. $\left(\pi^0 x^2 a^{-1}\right)^{-1} = \left(1 \cdot x^2 \cdot a^{-1}\right)^{-1} = x^{-2(1)}\left(a^{-1(-1)}\right)$

$= \dfrac{a}{x^2}$

45. $\left(\dfrac{4x^{-1}}{a^{-1}}\right)^{-3} = \dfrac{4^{-3}x^{-1(-3)}}{a^{-1(-3)}} = \dfrac{x^3}{4^3 a^3} = \dfrac{x^3}{64a^3}$

49. $7(-4) - (-5)^2 = -28 - 25 = -53$

53. $\dfrac{3.07\left(-|-1.86|\right)}{(-1.86)^4 + 1.596} = \dfrac{3.07(-1.86)}{1.86^4 + 1.596} = -0.421$

57. $\left(\dfrac{1}{x^{-1}}\right)^{-1} = \dfrac{1^{-1}}{x^{-1(-1)}} = \dfrac{1}{x} = \text{reciprocal of } x, \text{ yes.}$

61. $\left(x^a \cdot x^{-a}\right)^5 = \left(x^{a+(-a)}\right)^5 = \left(x^0\right)^5$

$= (1)^5, \quad x \neq 0$

$= 1$

65. $\pi\left(\dfrac{r}{2}\right)^3\left(\dfrac{4}{3\pi r^2}\right) = \pi \cdot \dfrac{r^3}{8} \cdot \dfrac{4}{3\pi r^2} = \dfrac{r}{6}$

69.

n	$\left(1+\tfrac{1}{n}\right)^n$
1	2
10	2.59374246
100	2.704813829
1000	2.716923932

73. $65.2\dfrac{\text{m}}{\text{s}} = 65.2\dfrac{\text{m}}{\text{s}}\left(\dfrac{3600\text{ s}}{\text{h}}\right)\left(\dfrac{100\text{ cm}}{\text{m}}\right)$

$\left(\dfrac{\text{in.}}{2.54\text{ cm}}\right)\left(\dfrac{\text{ft}}{12\text{ in.}}\right)$

$= 770{,}000\dfrac{\text{ft}}{\text{h}}$

1.5   Scientific Notation

1. $8.06 \times 10^3 = 8060$

5. $2.01 \times 10^{-3} = 0.00201$; move decimal point 3 places to the left by adding 2 zeros.

9. $1.86 \times 10 = 18.6$; move decimal point 1 place to the right.

13. $0.0087 = 8.7 \times 10^{-3}$; move decimal point to the right 3 places.

17. $1 = 1.0 \times 10^0$

21. $\dfrac{88,000}{0.0004} = \dfrac{8.8 \times 10^4}{4.0 \times 10^{-4}} = 2.2 \times 10^8$

25. $\left(1.2 \times 10^{29}\right)^3 = 1.2^3 \times 10^{29(3)}$
$$= 1.728 \times 10^{87}$$

29. $\dfrac{0.0732(6710)}{0.00134(0.0231)} = \dfrac{7.32 \times 10^{-2} \times 6.71 \times 10^3}{1.34 \times 10^{-3} \times 2.31 \times 10^{-2}}$
$$= 1.59 \times 10^7$$

33. $\dfrac{\left(3.69 \times 10^{-7}\right)\left(4.61 \times 10^{21}\right)}{0.0504} = 3.38 \times 10^{16}$

37. $0.000003 \text{ W} = 3 \times 10^{-6} \text{ W}$

41. $12,000,000,000 \text{ m}^2 = 1.2 \times 10^{10} \text{ m}^2$

45. (a) $2300 = 2.3 \times 10^3$
    (b) $0.23 = 230 \times 10^{-3}$
    (c) $23 = 23 \times 10^0$
    (d) $0.00023 = 230 \times 10^{-6}$

49. $1.4 \times 10^7 = 110 \, D_E$
$$D_E = 1.37 \times 10^5 \text{ m}$$

53. $\dfrac{4.57 \times 10^4}{1.86 \times 10^5} = 2.46 \times 10^{-1} \text{ s}$

57. $R = k / d^2$;
$R = 0.00000002196 \div 0.00007998^2$
$$= 2196 \times 10^8 \div \left(7.998 \times 10^{-5}\right)^2$$
$$= 3.433 \; \Omega$$

1.6   Roots and Radicals

1. $-\sqrt[3]{64} = -4$ since $(-4)^3 = -64$

5. $\sqrt{81} = 9$

9. $-\sqrt{49} = -7$

13. $\sqrt[3]{125} = 5$

17. $\left(\sqrt{5}\right)^2 = 5$

21. $\left(-\sqrt[4]{53}\right)^4 = 53$

25. $2\sqrt{84} = 2\sqrt{4 \cdot 21} = 2 \cdot 2\sqrt{21} = 4\sqrt{21}$

29. $\sqrt[3]{-8^2} = \sqrt[3]{-64} = -4$

33. $\sqrt{36 + 64} = \sqrt{100} = 10$

37. $\sqrt{85.4} = 9.24$

41. (a) $\sqrt{1296 + 2304} = \sqrt{3600} = 60.00$
    (b) $\sqrt{1296} = \sqrt{2304} = 36.00 + 48.00 = 84.00$

45. $\sqrt{24s} = \sqrt{(24)(150)} = \sqrt{3600} = 60$ mi/h

49. $d = \sqrt{w^2 + h^2} = \sqrt{52.3^2 + 29.3^2} = 59.9$ in.

53. No, it is not true if $a < 0$. It is always true that $\sqrt{a^2} = |a|$.

57. (a) imaginary (b) real

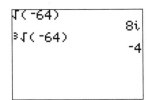

1.7 Addition and Subtraction of Algebraic Expressions

1. $3x + 2y - 5y = 3x - 3y$

5. $5x + 7x - 4x = 12x - 4x = 8x$

9. $2F - 2T - 2 + 3F - T = 5F - 3T - 2$

13. $s + 3a - 4 - 2 = s + (2a - 4) = 3s - 4$

17. $2 - 3 - (4 - 5a) = -1 - 4 + 5a = -5 + 5a = 5a - 5$

21. $-(t - 2u) + (3u - t) = -t + 2u + 3u - t = 5u - 2t$

25. $-7(6 - 3j) - 2(j + 4) = -42 + 21j - 2j - 8$
$$= 19j - 50$$

29. $2\left[4 - (t^2 - 5)\right] = 2\left[4 - t^2 + 5\right] = 2\left[9 - t^2\right] = 18 - 2t^2$

33. $aZ - \left[3 - (aZ + 4)\right] = aZ - \left[3 - aZ - 4\right]$
$$aZ - \left[-1 - aZ\right] = aZ + 1 + aZ$$
$$= 2aZ + 1$$

37. $5p - (q - 2p) - \left[3q - (p - q)\right]$
$$= 5p - q + 2p - \left[3q - p + q\right]$$
$$= 7p - q - \left[4q - p\right]$$
$$= 7p - q - 4q + p$$
$$= 8p - 5q$$

41. $5V^2 - \left(6 - (2V^2 + 3)\right) = 5V^2 - \left(6 - 2V^3 - 3\right)$
$$= 5V^2 - \left(3 - 2V^2\right)$$
$$= 5V^2 - 3 + 2V^2$$
$$= 7V^2 - 3$$

45. $-4\left[4R - 2.5(Z - 2R) - 1.5(2R - Z)\right]$
$$= -4\left[4R - 2.5Z + 5R - 3R + 1.5Z\right]$$
$$= -4\left[6R - Z\right]$$
$$= -24R + 4Z$$
$$= 4Z - 24R$$

49. $\left[\left(B + \dfrac{4}{3}\alpha\right) + 2\left(B - \dfrac{2}{3}\alpha\right)\right]$
$$-\left[\left(B + \dfrac{4}{3}\alpha\right) - \left(B - \dfrac{2}{3}\alpha\right)\right]$$
$$= \left[B + \dfrac{4}{3}\alpha + 2B - \dfrac{4}{3}\alpha\right] - \left[B + \dfrac{4}{3}\alpha - B + \dfrac{2}{3}\alpha\right]$$
$$= \left[3B\right] - \left[\dfrac{6}{3}\alpha\right] = 3B - 2\alpha$$

53. (a) $2x^2 - y + 2a + 3y - x^2 - b = x^2 + 2y + 2a - b$
(b) $2x^2 - y + 2a - (3y - x^2 - b)$
$$= 2x^2 - y + 2a - 3y + x^2 + b$$
$$= 3x^2 - 4y + 2a + b$$

57. $3 - 5x < 0 \Rightarrow |3 - 5x| = -(3 - 5x)$
$$5x - |3 - 5x| = 5x - \left(-(3 - 5x)\right)$$
$$= 5x + 3 - 5x$$
$$= 3$$

1.8   Multiplication of Algebraic Expressions

1. $2s^3\left(-st^4\right)^3\left(4s^2t\right)=2s^3\left(-s\right)^3\left(t^4\right)^3\left(4s^2t\right)$
$$=-2s^6t^{12}\left(4s^2t\right)$$
$$=-8s^8t^{13}$$

5. $\left(a^2\right)\left(ax\right)=a^{2+1}x=a^3x$

9. $\left(2ax^2\right)^2\left(-2ax\right)=4a^2x^4\left(-2ax\right)=-8a^3x^5$

13. $-3s\left(s^2-5t\right)=-3s\left(s^2\right)+3s\left(5t\right)=-3s^2+15st$

17. $3M\left(-M-N+2\right)=3M\left(-M\right)+3M\left(-N\right)+3M\left(2\right)$
$$=-3M^2-3MN+6M$$

21. $\left(x-3\right)\left(x+5\right)=x^2+5x-3x-15=x^2+2x-15$

25. $\left(2a-b\right)\left(-2b+3a\right)=6a^2-4ab-3ab+2b^2$
$$=6a^2-7ab+2b^2$$

29. $\left(x^2-1\right)\left(2x+5\right)=2x^3+5x^2-2x-5$

33. $2\left(a+1\right)\left(a-9\right)=2\left(a^2-8a-9\right)=2a^2-16a-18$

37. $2L\left(L+1\right)\left(4-L\right)=2L\left(4L-L^2+4-L\right)$
$$=2L\left(-L^2+3L+4\right)$$
$$=-2L^3+6L^2+8L$$

41. $\left(x_1+3x_2\right)^2=\left(x_1+3x_2\right)\left(x_1+3x_2\right)$
$$=x_1^2+3x_1x_2+3x_1x_2+9x_2^2$$
$$=x_1^2+6x_1x_2+9x_2^2$$

45. $2\left(x+8\right)^2=2\left(x^2+16x+64\right)=2x^2+32x+128$

49. $3T\left(T+2\right)\left(2T-1\right)=\left(3T^2+6T\right)\left(2T-1\right)$
$$=6T^3-3T^2+12T^2-6T$$
$$=6T^3+9T^2-6T$$

53. No, it should be
$$\left(x^2\right)\left(x^4\right)+\left(x^3\right)^5=x^{2+4}+x^{3(5)}$$
$$=x^6+x^{15}$$

57. $\left(x+y\right)^3=\left(x+y\right)\left(x+y\right)\left(x+y\right)$
$$=\left(x^2+2xy+y^2\right)\left(x+y\right)$$
$$=x^3+x^2y+2x^2y+2xy^2+xy^2+y^3$$
$$=x^3+3x^2y+3xy^2+y^3\neq x^3+y^3$$

61. $\left(2R-X\right)^2-\left(R^2+X^2\right)$
$$=4R^2-4RX+X^2-R^2-X^2$$
$$=3R^2-4RX$$

65. $\left(R_1+R_2\right)^2-2R_2\left(R_1+R_2\right)$
$$=R_1^2+2R_1R_2+R_2^2-2R_1R_2-2R_2^2$$
$$=R_1^2-R_2^2$$

1.9   Division of Algebraic Expressions

1. $\dfrac{-6a^2xy^2}{-2a^2xy^5} = 3y^{2-5} = 3y^{-3} = \dfrac{3}{y^3}$

5. $\dfrac{8x^3y^2}{-2xy} = -4x^2y$

9. $\dfrac{\left(15x^2\right)\left(4bx\right)\left(2y\right)}{30bxy} = 4x^2$

13. $\dfrac{3a^2x + 6xy}{3x} = \dfrac{3a^2x}{3x} + \dfrac{6xy}{3x} = a^2 + 2y$

17. $\dfrac{4pq^3 + 8p^2q^2 - 16pq^5}{4pq^2} = \dfrac{4pq^3}{4pq^2} + \dfrac{8p^2q^2}{4pq^2} - \dfrac{16pq^5}{4pq^2}$
$$= q + 2p - 4q^3$$

21. $\dfrac{-3ab^2 + 6ab^3 - 9a^2b^2}{-9a^2b^2} = \dfrac{3ab^2}{9a^2b^2} - \dfrac{6ab^3}{9a^2b^2} + \dfrac{9a^2b^2}{9a^2b^2}$
$$= \dfrac{1}{3a} - \dfrac{2b}{3a} + 1$$

25.
$$
\begin{array}{r}
2x+1 \\
x+3\overline{)2x^2 + 7x + 3} \\
\underline{2x^2 + 6x} \\
x + 3 \\
\underline{x + 3} \\
0
\end{array}
$$

29.
$$
\begin{array}{r}
4x^2 - x - 1 \\
2x-3\overline{)8x^3 - 14x^2 + \ x + 0} \\
\underline{8x^3 - 12x^2} \\
-2x^2 + \ x \\
\underline{-2x^2 + 3x} \\
-2x + 0 \\
\underline{-2x + 3} \\
-3
\end{array}
$$

33.
$$
\begin{array}{r}
x^2 + x - 6 \\
x+2\overline{)x^3 + 3x^2 - 4x - 12} \\
\underline{x^3 + 2x^2} \\
x^2 - 4x \\
\underline{x^2 + 2x} \\
-6x - 12 \\
\underline{-6x - 12} \\
0
\end{array}
$$

37.
$$
\begin{array}{r}
x^2 - 2x + 4 \\
x+2\overline{)x^3 + 0x^2 + 0x + 8} \\
\underline{x^3 + 2x^2} \\
-2x^2 + 0x \\
\underline{-2x^2 - 4x} \\
4x + 8 \\
\underline{4x + 8} \\
0
\end{array}
$$

41.
$$
\begin{array}{r}
x - y + z \\
x+y-z\overline{)x^2 + 0xy + 0xz - y^2 + 2yz - z^2} \\
\underline{x^2 + xy - xz} \\
-xy + xz - y^2 + 2yz - z^2 \\
\underline{-xy - y^2 + \ yz} \\
+xz + yz - z^2 \\
\underline{+xz + yz - z^2} \\
0
\end{array}
$$

$\dfrac{x^2 - y^2 + 2yz - z^2}{x + y - z} = x - y + z + \dfrac{0}{x + y - z}$

45.
$$
\begin{array}{r}
x^3 - x^2 + x - 1 \\
x+1\overline{)x^4 + 1} \\
\underline{x^4 + x^3} \\
-x^3 \\
\underline{-x^3 - x^2} \\
x^2 \\
\underline{x^2 + x} \\
-x + 1 \\
\underline{-x - 1} \\
2
\end{array}
$$

$\dfrac{x^4 + 1}{x + 1} = x^3 - x^2 + x - 1 + \dfrac{2}{x + 1} \neq x^3$

49. $\dfrac{8A^5 + 4A^3\mu^2 E^2 - A\mu^4 E^4}{8A^4}$

$= \dfrac{8A^5}{8A^4} + \dfrac{4A^3\mu^2 E^2}{8A^4} - \dfrac{A\mu^4 E^4}{8A^4}$

$= A + \dfrac{\mu^2 E^2}{2A} - \dfrac{\mu^4 E^4}{8A^3}$

53. $\left(\dfrac{s^2 - 2s - 2}{s^4 + 4}\right)^{-1} = \dfrac{s^4 + 4}{s^2 - 2s - 2}$

$$
\begin{array}{r}
s^2 + 2s + 6 \\
s^2 - 2s - 2\overline{)s^4 + 0s^3 + 0s^2 + 0s + 4} \\
\underline{s^4 - 2s^3 - 2s^2} \\
2s^3 + 2s^2 + 0s \\
\underline{2s^3 - 4s^2 - 4s} \\
6s^2 + 4s + 4 \\
\underline{6s^2 - 12s - 12} \\
16s + 16
\end{array}
$$

1.10   Solving Equations

1. (a) $x - 3 = -12$
$x - 3 + 3 = -12 + 3$
$x = -9$

(b) $x + 3 = -12$
$x + 3 - 3 = -12 - 3$
$x = -15$

(c) $\dfrac{x}{3} = -12$
$3\left(\dfrac{x}{3}\right) = 3(-12)$
$x = -36$

(d) $3x = -12$
$\dfrac{3x}{3} = \dfrac{-12}{3}$
$x = -4$

5. $x - 2 = 7$
$x = 7 + 2$
$x = 9$

9. $\dfrac{t}{2} = -5$
$t = 2(-5)$
$t = -10$

13. $3t + 5 = -4$
$3t = -4 - 5$
$3t = -9$
$t = -3$

17. $3x + 7 = x$
$3x - x = -7$
$2x = -7$
$x = \dfrac{-7}{2}$

21. $-(r - 4) = 6 + 2r$
$-r + 4 = 6 + 2r$
$-3r = 2$
$r = \dfrac{-2}{3}$

25. $0.1x - 0.5(x - 2) = 2$
$x - 5(x - 2) = 20$
$x - 5x + 10 = 20$
$-4x = 10$
$x = -2.5$

29. $\dfrac{4x - 2(x - 4)}{3} = 8$
$4x - 2(x - 4) = 24$
$4x - 2x + 8 = 24$
$2x = 16$
$x = 8$

33. $|2x - 3| = 5$
$2x - 3 = 5$ or $2x - 3 = -5$
$2x = 8$ $2x = -2$
$x = 4$ $x = -1$

37. $-0.24(C - 0.50) = 0.63$
$-0.24C + 0.12 = 0.63$
$-0.24C = 0.63 - 0.12$
$-0.24C = 0.51$
$C = -2.1$

41. $\dfrac{165}{223} = \dfrac{13V}{15}$
$\dfrac{15}{13}\left(\dfrac{165}{223}\right) = \dfrac{15}{13}\left(\dfrac{13V}{15}\right)$
$V = \dfrac{2475}{2899} = 0.85$

45.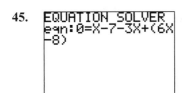

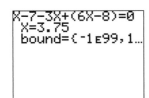

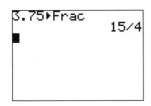

49. $1.1 = \dfrac{T - 76}{40}$

$44 = T - 76$

$T = 120°\,C$

53. $\dfrac{x}{5.5} = \dfrac{1250}{15}$

$x = \dfrac{1250(5.5)}{15}$

$x = 460\ \text{mi}$

1.11 Formulas and Literal Equations

1. $v - v_\circ = at$

$a = \dfrac{v - v_\circ}{t}$

5. $E = IR$

$\dfrac{E}{I} = \dfrac{IR}{I}$

$R = \dfrac{E}{I}$

9. $Q = SLd^2$

$L = \dfrac{Q}{Sd^2}$

13. $A = \dfrac{Rt}{PV}$

$Rt = APV$

$V = \dfrac{Rt}{AP}$

17. $T = \dfrac{c + d}{v}$

$Tv = c + d$

$d = Tv - c$

21. $a = \dfrac{2mg}{M + 2m}$

$aM + 2ma = 2mg$

$aM = 2mg - 2ma$

$M = \dfrac{2mg - 2ma}{a}$

25. $N = r(A - s)$

$N = rA - rs$

$rs = rA - N$

$s = \dfrac{rA - N}{r}$

29. $Q_1 = P(Q_2 - Q_1)$

$Q_1 = PQ_2 - PQ_1$

$Q_1(1 + P) = PQ_2$

$Q_2 = \dfrac{Q_1 + PQ_1}{P}$

33. $L = \pi(r_1 + r_2) + 2x_1 + x_2$

$L = \pi r_1 + \pi r_2 + 2x_1 + x_2$

$\pi r_1 = L - \pi r_2 - 2x_1 - x_2$

$r_1 = \dfrac{L - \pi r_2 - 2x_1 - x_2}{\pi}$

37. $C = \dfrac{2eAk_1k_2}{d(k_1 + k_2)}$

$Cd(k_1 + k_2) = 2eAk_1k_2$

$e = \dfrac{Cd(k_1 + k_2)}{2Ak_1k_2}$

41. $p(C - n) + n = A$

$pC - pn + n = A$

$n = \dfrac{A - pC}{1 - p}$

$n = \dfrac{13.0 - 0.25(15.0)}{1 - 0.25}$

$n = 12\ \text{L}$

45.
$$V_1 = \frac{VR_1}{R_1 + R_2}$$

$$R_1 + R_2 = \frac{VR_1}{V_1}$$

$$R_2 = \frac{VR_1}{V_1} - R_1 = \frac{12.0(3.56)}{6.30} - 3.56$$

$$R_2 = 3.22 \ \Omega$$

1.12   Applied Word Problems

1. x = number of 25 W lights

$37 - x$ = number of 40 W lights

$$25x + 40(37 - x) = 1000$$
$$25x + 1480 - 40 = 1000$$
$$-15x = -480$$
$$x = 32, \ 25 \text{ W lights}$$
$37 - x = 5, \ 40$ W lights

5. x = cost 6 years ago

$x + 5000$ = cost today

$$x + (x + 5000) = 64,000$$
$$2x + 5000 = 64,000$$
$$2x = 59,000$$
$$x = 29,500$$
$$x + 5000 = 34,500$$

$29,500 six years ago $34,500 is the cost today

9.   Let x = number of acres @ $200

$$200 \cdot x + 300 \cdot (140 - x) = 37,000$$
$$2x + 420 - 3x = 370$$
$$x = 50 \text{ acres @ } \$200$$
$$140 - x = 90 \text{ acres @ } \$300$$

13.   Let $x = 15$ m girders

$x - 4 = 18$ m girders

$$15x = 18(x - 4)$$
$$3x = 72$$
$$x = 24,$$

so there are twenty 18 m girders needed, from which L $= 20(18) = 360$ m for span.

17.   Let x = the main pipeline

$x + 2.6$ = the smaller pipeline

$$3(x + 2.6) + x = 35.4$$
$$3x + 7.8 + x = 35.4$$
$$4x = 27.6$$
$$x = 6.9 \text{ km for the main pipeline;}$$
$$9.5 \text{ km for the smaller pipeline}$$

21.   To completely pass each other, the rear of the 520 ft train must travel $(520 + 440)$ ft at a relative speed of 100 mph, from which

$$(520 + 440) \text{ ft} \cdot \left(\frac{\text{mi}}{5280 \text{ ft}}\right) = 100 \frac{\text{mi}}{\text{h}} \cdot t$$
$$t = 0.00\overline{18} \text{ h} = 6.\overline{54} \text{ s}$$

25. Let v = speed of French train

$$v\left(\frac{17}{60}\right) + (v - 8)\left(\frac{17}{60}\right) = 50 \Rightarrow 17v + 17v - 136$$
$$3000 \Rightarrow 34v = 3136 \Rightarrow v = 92.2$$
$$v - 8 = 84.2$$

The speed of the French train is 92.2 km/h. The speed of the English train is 84.2 km/h.

29. Assume the customer is located between A and B at a distance x from A.

$$3.40 + 0.0002x = 3.20 + 0.0002(228 - x)$$
$$x = 64 \text{ mi from A}$$

$\frac{15}{16} = 93.75\%$ gasoline $\left\{\begin{array}{c} x \\ \hline 100\% \text{ gasoline} \\ \hline 8 - x \\ \hline 75\% \text{ gasoline} \end{array}\right\}$ 8.0L

$$0.9375(8.0) = x + 0.75(8.0 - x) \text{ from which}$$
$$x = 6.0\text{L of gasoline added}$$

33. v = speed of car when it overtakes semitrailer

v = speed of car as it passes semitrailer

In passing, 25 m are coverfed in 10 s at a relative speed of $v - 70$. Changing 25 m to kilometers, 10 s to hours, and using $d = rt$ gives

$$\frac{25}{1000} = (v - 70)\left(\frac{10}{3600}\right) \text{ from which}$$

$$v = \frac{25(3600)}{1000(10)} + 70$$

$$v = 79 \text{ km/h}$$

Chapter 1 Review Exercises

1. $(-2) + (-5) - 3 = -7 - 3 = -10$

5. $-5 - |2(-6)| + \dfrac{-15}{3} = -5 - |-12| + (-5)$

$$= -5 - 12 - 5$$
$$= -17 - 5$$
$$= -22$$

9. $\sqrt{16} - \sqrt{64} = 4 - 8 = -4$

13. $\left(-2rt^2\right)^2 = 4r^2\left(t^2\right)^2 = 4r^2t^4$

17. $\dfrac{-16N^{-2}\left(NT^2\right)}{-2N^0T^{-1}} = 8N^{-2+1}T^{2-(-1)}$

$$= 8N^{-1}T^3$$
$$= \frac{8T^3}{N}$$

21. (a) 8000 has 1 significant digits

  (b) 8000 rounded to 2 significant digits is 8000

25. $37.3 - 16.92(1.067)^2 = 18.03676612$

on a calculator; 18.0

29. $a - 3ab - 2a + ab = a - 2a - 3ab + ab = -a - 2ab$

33. $(2x - 1)(5 + x) = 2x^2 + 10x - x - 5$

$$= 2x^2 + 9x - 5$$

37. $\dfrac{2h^3k^2 - 6h^4k^5}{2h^2k} = \dfrac{2h^3k^2}{2h^2k} - \dfrac{6h^4k^5}{2h^2k}$

$$= hk - 3h^2k^4$$

41. $2xy - \left\{3z - \left[5xy - (7z - 6xy)\right]\right\}$

$$= 2xy - \left\{3z - \left[5xy - 7z + 6xy\right]\right\}$$
$$= 2xy - \left\{3z - \left[11xy - 7z\right]\right\}$$
$$= 2xy - \left\{3z - 11xy + 7z\right\}$$
$$= 2xy - \left\{10 - 11xy\right\}$$
$$= 2xy - 10z + 11xy$$
$$= 13xy - 10z$$

45. $-3y(x - 4y)^2 = -3y\left(x^2 - 8xy + 16y^2\right)$

$$= -3x^2y + 24xy^2 - 48y^3$$

49. $\dfrac{12p^3q^2 - 4p^4q + 6pq^5}{2p^4q} = \dfrac{12p^3q^2}{2p^4q} - \dfrac{4p^4q}{2p^4q} + \dfrac{6pq^5}{2p^4q}$

$$= \frac{6q}{p} - 2 + \frac{3q^4}{p^3}$$

53.
$$
\require{enclose}
\begin{array}{r}
x^2 - 2x + 3 \\
3x - 1 \enclose{longdiv}{3x^3 - 7x^2 + 11x - 3} \\
\underline{3x^3 - x^2} \\
-6x^2 + 11x \\
\underline{-6x^2 + 2x} \\
9x - 3 \\
\underline{9x - 3}
\end{array}
$$

57. $-3\left\{(r + s - t) - 2\left[(3r - 2s) - (t - 2s)\right]\right\}$

$$= -3\left\{(r + s - t) - 2[3r - 2s - t + 2s]\right\}$$
$$= -3\left\{r + s - t - 2[3r - t]\right\}$$
$$= -3\left\{r + s - t - 6r + 2t\right\}$$
$$= -3\left\{-5r + s + t\right\}$$
$$= 15r - 3s - 3t$$

61. $3x + 1 = x - 8$

$$3x - x = -8 - 1$$
$$2x = -9$$
$$x = \frac{-9}{2}$$

65. $-6x + 5 = -3(x - 4)$

$$-6x + 5 = -3x + 12$$
$$3x = -7$$
$$x = \frac{-7}{3}$$

69. $3t - 2(7 - t) = 5(2t + 1)$

$$3t - 14 + 2t = 10t + 5$$
$$5t - 14 = 10t + 5$$
$$-5t = 19$$
$$t = \frac{-19}{5}$$

73. $60,000,000,000,000 = 6 \times 10^{13}$ bytes

77. 2.53×10^{13} mi $= 25,300,000,000,000$ mi

81. 1.5×10^{-1} Bq/L $= 0.15$ Bq/L

85. $P = \dfrac{\pi^2 E I}{L^2}$

$$E = \frac{P L^2}{\pi^2 I}$$

89. $d = (n - 1)A$

$$d = nA - A$$
$$na = d + A$$
$$n = \frac{d + A}{A}$$

93. $R = \dfrac{A(T_2 - T_1)}{H}$

$$RH = AT_2 - AT_1$$
$$AT_2 = RH + AT_1$$
$$T_2 = \frac{RH + AT_1}{A}$$

97. $\dfrac{5.25 \times 10^{13}}{6.4 \times 10^4} = 8.2 \times 10^8$

101. $\dfrac{R_1 R_2}{R_1 + R_2} = \dfrac{0.0275(0.0590)}{0.0275 + 0.0590} = 0.0188$ Ω

105. $4(t + h) - 2(t + h)^2 = 4t + 4h - 2(t^2 + 2th + h^2)$

$$= 4t + 4h - 2t^2 - 4th - 2h^2$$

109. $x - (3 - x) = 2x - 3$

$$x - 3 + x = 2x - 3$$
$$2x - 3 = 2x - 3, \text{the equation is an identity}$$

113. $|3 - x| + 7 = 2x$

$$|3 - x| = 2x - 7$$
$$3 - x = 2x - 7 \text{ or } 3 - x = -(2x - 7)$$
$$3x = 10 \text{ or } 3 - x = -2x + 7$$
$$x = \frac{10}{3} \text{ or } x = 4$$
$$x = \frac{10}{3} \text{ does not check}$$
$$3 - x = 3 - 4 = -1 < 0 \text{ and}$$
$$x = 4 \text{ checks}$$
$$x = 4 \text{ is the solution}$$

117. $\dfrac{8 \times 10^{-3}}{2 \times 10^4} = 4 \times 10^{-7}$

121. $x + x + 72 = 190 \Rightarrow 2x = 118 \Rightarrow x = \$59$

$$x + 72 = \$131$$

125. $2.4 \times 10^{-6} R + 2.4 \times 10^{-6}(R + 1200) = 12.0 \times 10^{-3}$

$$4.86 \times 10^{-6} R = 12.0 \times 10^{-3} - 1200 \times 2.4 \times 10^{-6}$$
$$R = 1900 \ \Omega$$
$$R + 1200 = 3100 \ \Omega$$

129. $17.4(t + 2) + 21.8t = 634 \Rightarrow t = \dfrac{634 - 34.8}{39.2}$

$$= 15.28571429,$$
$$\text{calculator}$$

Ships pass 15.3 h after the second ship starts.

133. $\dfrac{\text{square ft of tile}}{\text{square ft in house}} = 0.25$

$\dfrac{0.15(2200)}{2200 + x} = 0.25$

$x = 290 \text{ ft}^2 \text{ in kitchen and entry}$

Chapter 2

GEOMETRY

2.1 Lines and Angles

1. $\angle ABE = 90^\circ$

5. $\angle EBD$ and $\angle DBC$ are acute angles.

9. The complement of $\angle CBD = 65^\circ$ is 25°.

13. $\angle AOB = 90^\circ + 50^\circ = 140^\circ$

17. $\angle 1 = 180^\circ - 145^\circ = 35^\circ = \angle 2 = \angle 4$

21. $\angle 3 = 90^\circ - 62^\circ = 28^\circ$

25. $\angle DEB = 44^\circ$

29. $\dfrac{a}{4.75} = \dfrac{3.05}{3.20} \Rightarrow a = 4.75 \cdot \dfrac{3.05}{3.20} = 4.53$ m

33. $\measuredangle BHC = \measuredangle CGD = 25^\circ$

37. $\measuredangle GHA = \measuredangle HGE = 110^\circ$

41. $\angle BCD = 180^\circ - 47^\circ$
$\qquad = 133^\circ$

45. $\measuredangle 1 + \measuredangle 2 + \measuredangle 3 = 180^\circ$,
$\left(\measuredangle 1,\ \measuredangle 2,\ \text{and } \measuredangle 3 \text{ form a straight line} \right)$

2.2 Triangles

1. $\angle 5 = 45^\circ \Rightarrow \angle 3 = 45^\circ$
$\angle 2 = 180^\circ - 70^\circ - 45^\circ = 65^\circ$

5. $\angle A = 180^\circ - 84^\circ - 40^\circ = 56^\circ$

9. $A = \dfrac{1}{2}bh = \dfrac{1}{2}(7.6)(2.2) = 8.4 \text{ ft}^2$

13. $A = \dfrac{1}{2}bh = \dfrac{1}{2}(3.46)(2.55) = 4.41 \text{ ft}^2$

17. $p = 205 + 322 + 415$
$p = 942$ cm

21. $c = \sqrt{13.8^2 + 22.7^2} = 26.6$ ft

25. $\angle B = 90^\circ - 23^\circ = 67^\circ$

29.

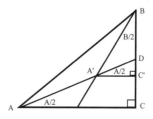

$\Delta ADC \sim \Delta A'DC' \Rightarrow \measuredangle DA'C' = A/2$
$\measuredangle$ between bisectors $= \measuredangle BA'D$
$\Delta BA'C',\ \dfrac{B}{2} + \left(\measuredangle BA'D + A/2 \right) = 90^\circ$
from which $\measuredangle BA'D = 90^\circ - \left(\dfrac{A}{2} + \dfrac{B}{2} \right)$
or $\measuredangle BA'D = 90^\circ - \left(\dfrac{A+B}{2} \right) = 90^\circ - \dfrac{90^\circ}{2} = 45^\circ$

33.

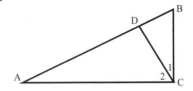

$$\angle A + \angle B = 90°$$
$$\angle 1 + \angle B = 90°$$
$$\Rightarrow \quad \angle A = \angle 1$$

redraw $\triangle BDC$ as

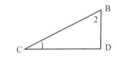

$$\angle 1 + \angle 2 = 90°$$
$$\angle 1 + \angle B = 90°$$
$$\Rightarrow \quad \angle 2 = \angle B$$

and $\triangle ADC$ as

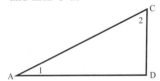

$\triangle BDC$ and $\triangle ADC$ are similar.

37. Since $\triangle MKL \sim \triangle MNO$; $KN = KM - MN$; $15 - 9 = 6$

$$= KM; \; \frac{KM}{MN} = \frac{LM}{MO}; \; \frac{6}{9} = \frac{LM}{12}; \; 9LM = 72; \; LM = 8$$

$$\frac{\text{street}}{4.0} = \frac{20.0}{1} \Rightarrow \text{street} = 4.0(20.0)$$

$$\text{ramp} = \sqrt{\left(4.0(20.0)\right)^2 + 4.0^2} = 80.09993758,$$

calculator

ramp = 80 in. (two significant digits)

41. $\angle = \dfrac{180° - 50°}{2} = 65°$

45. $A = \dfrac{1}{2}bh = \dfrac{1}{2}(8.0)(15) = 60 \text{ ft}^2$

49. $d = \sqrt{18^2 + 12^2 + 8^2} = 23 \text{ ft}$

53.

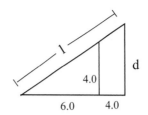

$$\frac{d}{8.0} = \frac{4.0}{6.0} \Rightarrow d = \frac{4.0(8.0)}{6.0}$$

$$l^2 = 8.0^2 + d^2 = 8.0^2 + \left(\frac{4.0(8.0)}{6.0}\right)^2$$

$$l = 9.6 \text{ ft}$$

2.3 Quadrilaterals

1.

b_2

b_1

trapezoid

5. $p = 4s = 4(65) = 260 \text{ m}$

9. $p = 2l + 2w = 2(3.7) + 2(2.7) = 12.8 \text{ m}$

13. $A = s^2 = 2.7^2 = 7.3 \text{ mm}^2$

17. $A = bh = 3.7(2.5) = 9.3 \text{ m}^2$

21. $p = 2b + 4a$

25. The parallelogram is a rectangle.

29.

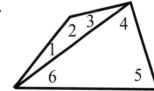

sum of interior angles

$= \measuredangle 1 + \measuredangle 2 + \measuredangle 3 + \measuredangle 4 + \measuredangle 5 + \measuredangle 6$

$= 180° + 180°$

$= 360°$

33. The diagonal always divides the rhombus into two congruent triangles. All outer sides are always equal.

37.

$w + 2.5 = 4w - 4.7$

$w = 2.4$ ft

$4w = 9.6$ ft

$l = \sqrt{130^2 - 70^2}$

$p = 2l + 2w$

$p = 2\sqrt{130^2 - 70^2} + 2(70)$

$p = 360$ yd

41.

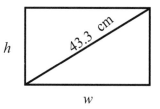

$\dfrac{w}{h} = 1.60 \Rightarrow w = 1.60h$

$43.3^2 = h^2 + w^2 = h^2 + (1.6h)^2$

$\qquad\qquad h = 22.9$ cm

$w = 1.60h = 36.7$ cm

45. $360°$. A diagonal divides a quadrilateral into two triangles, and the sum of the interior angles of each triangle is $180°$.

2.4   Circles

1. $\angle OAB + OBA + \angle AOB = 180°$

$\qquad \angle OAB + 90° + 72° = 180°$

$\qquad\qquad\qquad \angle OAB = 18°$

5. (a) AD is a secant line.

(b) AF is a tangent line.

9. $c = 2\pi r = 2\pi(275) = 1730$ ft

13. $A = \pi r^2 = \pi(0.0952^2) = 0.0285$ yd^2

17. $\angle CBT = 90° - \angle ABC = 90° - 65° = 25°$

21. ARC $BC = 2(60°) = 120°$

25. $022.5°\left(\dfrac{\pi}{180°}\right) = 0.393$ rad

29. $P = \dfrac{1}{4}(2\pi r) + 2r = \dfrac{\pi r}{2} + 2r$

33. All are on the same diameter.

37. $c = 2\pi r \Rightarrow \pi = \dfrac{c}{2r}$; $d = 2r \Rightarrow r = \dfrac{d}{2}$ from which

$\pi = \dfrac{c}{2 \cdot \frac{d}{2}}$

$\pi = \dfrac{c}{d} \cdot \pi$ is the ratio of the circumference to the diameter.

41.

$h = 11.5$ km

d

$r = 6378 - 11.5$ km

$r = 6378$ km

$$(6378 + 11.5)^2 = d^2 + 6378^2$$
$$d = 383 \text{ km}$$

45. $C = 2\pi r = 2\pi(3960) = 24{,}900$ mi

49. $c = 112; \; c = \pi d; \; d = c / \pi = 112 / \pi = 35.7$ in.

53. Length $= (2)\dfrac{3}{4}(2\pi)(5.5) + (4)(5.5) = 73.8$ in.

2.5 Measurement of Irregular Areas

1. The use of smaller intervals improves the approximation since the total omitted area or the total extra area is smaller.

5. $A_{\text{trap}} = \dfrac{2.0}{2}\left[0.0 + 2(6.4) + 2(7.4) + 2(7.0) + 2(6.1)\right]$

$\left[+2(5.2) + 2(5.0) + 2(5.1) + 0.0\right]$

$A_{\text{trap}} = 84.4 = 84 \text{ m}^2$ to two significant digits

9. $A_{\text{trap}} = \dfrac{0.5}{2}\left[0.6 + 2(2.2) + 2(4.7) + 2(3.1) + 2(3.6)\right]$

$\left[+2(1.6) + 2(2.2) + 2(1.5) + 0.8\right]$

$A_{\text{trap}} = 9.8 \text{ m}^2$

13. $A_{\text{trap}} = \dfrac{45}{2}\left[170 + 2(360) + 2(420) + 2(410)\right.$

$\left. + 2(390) + 2(350) + 2(330) + 2(290) + 230\right]$

$A_{\text{trap}} = 120{,}000 \text{ ft}^2$

17. A_{trap}

$= \dfrac{0.500}{2}\left[0.0 + 2(1.732) + 2(2.000) + 2(1.732)\right.$

$\left. + 0.0\right] = 2.73 \text{ in.}^2$

This value is less than 3.14 in.^2 because all of the trapezoids are inscribed.

2.6 Solid Geometric Figures

1. $V_1 = lwh_1, \; V_2 = (2l)(w)(2h) = 4lwh = 4V_1$

The volume is four times as much.

5. $V = e^3 = 7.15^3 = 366 \text{ ft}^3$

9. $V = \dfrac{4}{3}\pi r^3 = \dfrac{4}{3}\pi\left(0.877^3\right) = 2.83 \text{ yd}^3$

13. $V = \dfrac{1}{3}Bh = \dfrac{1}{3}\left(0.76^2\right)(1.30) = 0.25 \text{ in.}^3$

17. $V = \dfrac{1}{2}\left(\dfrac{4}{3}\pi r^3\right) = \dfrac{2}{3}\pi\left(\dfrac{0.83}{2}\right)^3 = 0.15 \text{ yd}^3$

21. $V = \dfrac{4}{3}\pi r^3 = \dfrac{4}{3}\pi\left(\dfrac{d}{2}\right)^3 = \dfrac{4}{3}\pi\dfrac{d^3}{8}$

$V = \dfrac{1}{6}\pi d^3$

25. $\dfrac{\text{final surface area}}{\text{original surface area}} = \dfrac{4\pi(2r)^2}{4\pi r^2} = \dfrac{4}{1}$

29. $V = \pi r^2 h = \pi(d/2)^2 h = \pi(4.0/2)^2 (3{,}960{,}000)$

$= 5.0 \times 10^7 \text{ ft}^3$ or 0.00034 mi^3

33. $V = \dfrac{1}{3}BH = \dfrac{1}{3}\left(250^2\right)(160) = 3{,}300{,}000 \text{ yd}^3$

37. $A = l^2 + \dfrac{1}{2}ps = 16^2 + \dfrac{1}{2}(4)(16)\sqrt{8^2 + 40^2}$

$\qquad = 1560 \text{ mm}^2$

41. $V = \text{cylinder} + \text{cone (top of rivet)}$

$\qquad = \pi r^2 h + \dfrac{1}{3}\pi r^2 h$

$\qquad = \pi(0.625/2)^2(2.75) + \dfrac{1}{3}\pi(1.25/2)^2(0.625)$

$\qquad = 1.10 \text{ in.}^3$

Chapter 2 Review Exercises

1. $\angle CGE = 180° - 148° = 32°$

5. $c = \sqrt{9^2 + 40^2} = 41$

9. $a = \sqrt{0.736^2 - 0.380^2} = 0.630$

13. $P = 3s = 3(8.5) = 25.5 \text{ mm}$

17. $C = \pi d = \pi(98.4) = 309 \text{ mm}$

21. $V = Bh = \dfrac{1}{2}(26.0)(34.0)(14.0) = 6190 \text{ cm}^3$

25. $A = 6e^2 = 6(0.520) = 1.62 \text{ m}^2$

29. $\angle BTA = \dfrac{50°}{2} = 25°$

33. $\angle ABE = 90° - 37° = 53°$

37. $P = b + \sqrt{b^2 + (2a^2)^2} + \dfrac{1}{2}\pi(2a) = b + \sqrt{b^2 + 4a^2} + \pi a$

41. A square is a rectangle with four equal sides and a rectangle is a parallelogram with perpendicular intersecting sides so a square is a parallelogram. A rhombus is a parallelogram with four equal sides and since a square is a parallelogram, a square is a rhombus.

45.

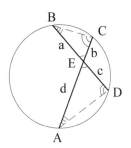

$\angle BEC = \angle AED$, vertical $\angle$'s.

$\angle BCA = \angle ADB$, both are inscribed in $\overset{\frown}{AB}$

$\angle CBE = \angle CAD$, both are inscribed in $\overset{\frown}{CD}$

which shows $\triangle AED \sim \triangle BEC \Rightarrow \dfrac{a}{d} = \dfrac{b}{c}$

49. $L = \sqrt{0.48^2 + 7.8^2} = 7.8 \text{ m}$

53. $\dfrac{AB}{13} = \dfrac{14}{18}$

$AB = 10 \text{ m}$

57. The longest distance in inches between points on the photograph is,

$\sqrt{8.00^2 + 10.0^2} = 12.8 \text{ in.}$ from which

$\dfrac{x}{12.8} = \dfrac{18,450}{1}$

$x = (12.8)(18,450) \text{ in.} \left(\dfrac{1 \text{ ft}}{12 \text{ in.}}\right)\left(\dfrac{\text{mi}}{5280 \text{ ft}}\right)$

$x = 3.73 \text{ mi}$

61. $A = (4.0)(8.0) - 2\pi \cdot \dfrac{1.0^2}{4} = 30 \text{ ft}^2$

65. $V = \pi r^2 h = \pi\left(\dfrac{4.3}{2}\right)^2 (13) = 190 \text{ m}^3$

69.

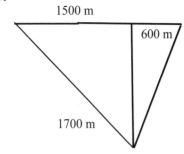

1500 m

600 m

1700 m

$$d^2 = 1700^2 - 1500^2 + 600^2$$
$$d = 1000 \text{ m}$$

73. $\dfrac{w}{h} = \dfrac{16}{9} \Rightarrow w = \dfrac{16h}{9}$

$152^2 = w^2 + h^2 = \left(\dfrac{16h}{9}\right)^2 + h^2 \Rightarrow h = 74.5 \text{ cm}$

$w = \dfrac{16h}{9} = 132 \text{ cm}$

77. Label the vertices of the pentagon ABCDE. The area is the sum of the areas of three triangles, one with sides 921, 1490, and 1490 and two with sides 921, 921, and 1490. The semi-perimeters are given by

$s_1 = \dfrac{921 + 921 + 1490}{2} = 1666$ and

$s_2 = \dfrac{921 + 1490 + 1490}{2} = 1950.5.$

$A = 2\sqrt{1666(1666 - 921)(1666 - 921)(1666 - 1490)}$

$\qquad + \sqrt{1950.5(1950.5 - 1490)(1950.5 - 1490)}$

$\qquad + \sqrt{(1950.5 - 921)}$

$\quad = 1,460,000 \text{ ft}^2$

Chapter 3

FUNCTIONS AND GRAPHS

3.1   Introduction to Functions

1. $f(x) = 3x - 7$

$f(-2) = 3(-2) - 7 = -13$

5.   (a) $A(r) = \pi r^2$

(b) $A(d) = \pi \left(\dfrac{d}{2}\right)^2 = \dfrac{1}{4}\pi d^2$

9.   From geometry, $A = s^2$; $= 2s^2 = d^2$

$$A = \dfrac{d^2}{2}$$

$$d = \sqrt{2A}$$

13. $f(x) = 2x + 1$; $f(1) = 2(1) + 1 = 3$;

$f(-1) = 2(-1) + 1 = -1$

17. $\phi(x) = \dfrac{6 - x^2}{2x}$

$\phi(2\pi) = \dfrac{6 - (2\pi)^2}{2(2\pi)} = \dfrac{6 - 4\pi^2}{4\pi}$

$\phi(2\pi) = \dfrac{3 - 2\pi^2}{2\pi}$

$\phi(-2) = \dfrac{6 - (-2)^2}{2 \cdot (-2)} = \dfrac{2}{-4} = -\dfrac{1}{2}$

21. $K(s) = 3s^2 - s + 6$

$K(-s + 2) = 3(-s + 2)^2 - (-s + 2) + 6$

$= 3s^2 - 11s + 16$

$K(-s) + 2 = 3(-s)^2 - (-s) + 6 + 2$

$= 3s^2 + s + 8$

25. $f(x) = 5x^2 - 3x$; $f(3.86) = 5 \cdot 3.86^2 - 3 \cdot 3.86 = 62.9$

$f(-6.92) = 5 \cdot (-6.92)^2 - 3 \cdot (-6.92) = 260$

29. $f(x) = x^2 + 2$; square x and add 2 to the result.

33.   Take 3 times the sum of twice the independent variable and 5, then subtract 1.

37. $A = 5e^2$

$f(e) = 5e^2$

41. $s = f(t) = 17.5 - 4.9t^2$; $f(12) = 17.5 - 4.9 \cdot (1.2)^2$

$s = 10.4$ m

45. $L = 1.2Q^2 + 1.5Q$

$= 1.2(4.50)^2 + 1.5(4.50)$

$= 31$ lb/in.^{2}

49.   (a) $f\{f(x)\}$ means "function of the function of x."
To evaluate, replace x in the function by the function itself.

(b) $f(x) = 2x^2$

$f(f(x)) = f(2x^2) = 2(2x^2)^2$

$f(f(x)) = 8x^4$

3.2   More About Functions

1. $f(x) = -x^2 + 2$ is defined for all real values of x; the domain is all real numbers. Since $-x^2 + 2 \le 2$ the range is all real numbers $f(x) \le 2$.

5. The domain and range of $f(x) = x + 5$ is all real numbers.

9. The domain of $f(s) = \dfrac{2}{s^2}$ is all real numbers except zero since it gives a division by zero. The range is all positive real numbers because $\dfrac{2}{s^2}$ is always positive.

13. The domain of $y = |x - 3|$ is all real numbers and the range is all nonnegative numbers, $y \ge 0$.

17. $f(D) = \dfrac{D^2 + 8D - 8}{D(D-2) + 4(D-2)}$

$= \dfrac{D^2 + 8D - 8}{(D-2)(D+4)}$ since division by zero is undefined, the domain must be restricted to exclude any value for which $D - 2$, $D + 4$ are equal to zero. In this case, $D \ne 2, -4$. So the domain is the set of all real numbers except $2, -4$.

21. $F(t) = 3t - t^2$ for $t \le 2$

$F(1) = 3 \cdot 1 - 1^2 = 2$

$F(2) = 3 \cdot 2 - 2^2 = 2;$

$F(3) =$ does not exist.

25. $d(t) = 55(2) + 40t = 110 + 40t$

29. $m(h) = \begin{cases} 110 \text{ for } h \le 1000 \\ 110 + 0.5(h - 1000) \text{ for } h > 1000 \end{cases}$

33. (a) $0.1x + 0.4y = 1200 \Rightarrow y(x) = \dfrac{1200 - 0.1x}{0.4}$

(b) $y(400) = \dfrac{1200 - 0.1 \cdot 400}{0.4} = 2900$ L

37. For the square, $p = x$, side $= \dfrac{x}{4}$;

$A_{\text{square}} = \left(\dfrac{x}{4}\right)^2 = \dfrac{x^2}{16} = \dfrac{p^2}{16}$

For the circle, $c = 60 - x = 2\pi r$;

$r = \dfrac{60 - x}{2\pi} = \dfrac{60 - p}{2\pi}$

$A_{\text{circle}} = \pi r^2 = \dfrac{\pi(60 - p)^2}{2\pi}$. Thus, the total

$A = \dfrac{p^2}{16} + \dfrac{(60 - p)^2}{4\pi}$

41. $s = f(t) = \dfrac{300}{t - 3}$

Domain: $t > 3$

Range: $s > 0$ (upper limit depends on truck)

45. From Ex. 21,

$m = f(h) = 110 + 0.5(h - 1000) = 0.5h - 390$

$m = \begin{cases} 0.5h - 390 & \text{for} \quad h > 1000 \\ 110 & \text{for} \quad 0 \le h \le 1000 \end{cases}$

49. $f(x + 2) = |x|$

$f(0) = f(-2 + 2) = |-2| = 2$

3.3   Rectangular Coordinates

1. $A(-1, -2), (4, -2), C(4, 1)$

D, the fourth vertex has coordinataes $(-1, 1)$.

5.

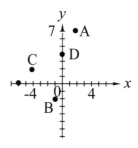

9. Rectangle

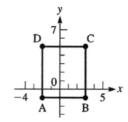

13. In order for the *x*-axis to be the perpendicular disector of the line segment join *P* and *Q*, *Q* must be $(3, -2)$.

17. All points $(x, 3)$, where *x* is any real number, are points on a line parallel to the *x*-axis, 3 units above it.

21. Abscissas are *x*-coordinates; thus the abscissa of all points on the *y*-axis is zero.

25. All points which lie to the left of a line that is parallel to the *y*-axis, one unit to the left have $x < -1$.

29. $xy = 0$ for $x = 0 (y\text{-axis})$ or $y = 0 (x\text{-axis})$
    $xy = 0$ on either axis.

33. (a) $d = 3 - (-5) = 8$
       (b) $d = 4 - (-2) = 6$

3.4   The Graph of a Function

1. $f(x) = 3x + 5$

x	y
−3	−4
−2	−1
−1	2
0	5
1	8

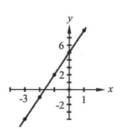

5. $y = 3x$

x	y
−1	−3
0	0
1	3

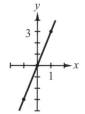

9. $s = 7 - 2t$

s	t
−1	9
0	7
1	5

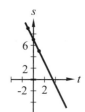

13. $y = x^2$

x	y
−2	4
−1	1
0	0
1	1
2	4

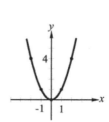

17. $y = \frac{1}{2}x^2 + 2$

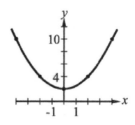

x	y
-4	10
-2	4
0	2
2	4
4	10

21. $y = x^2 - 3x + 1$

x	y
3	1
2	-1
1.5	-1.25
1	-1
0	1

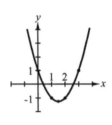

25. $y = x^3 - x^2$

x	y
-2	-12
-1	-2
0	0
2/3	-4/27
1	0
2	4

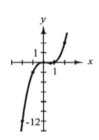

29. $P = \dfrac{8}{V} + 3$

V	P
-3	1/3
-2	-1
-1	-5
1	11
2	7
3	17/3

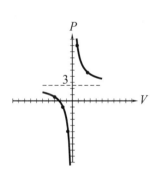

33. $y = \sqrt{9x}$

x	y
0	0
1	3
4	6
9	9
16	12

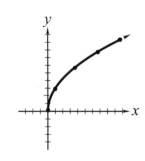

37. $n = 0.40m$

m	n
10	4
50	20
80	32

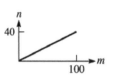

41. $H = 240I^2$

I	H
0	0
0.2	9.6
0.4	38.4
0.6	86.4
0.8	153.6

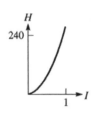

45. $P = 0.004v^3$

v	P
0	0
5	0.5
10	4.0
15	13.5
20	32.0

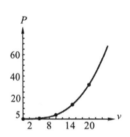

49. $l + w + h = l + w + 2w = 45$

$$l = 45 - 3w$$
$$V = lwh = (45 - 3w) \cdot w \cdot 2w$$
$$V = f(w) = 90w^2 - 6w^3$$

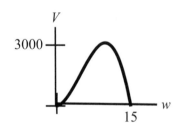

53. $N = \sqrt{n^2 - 1.69}$

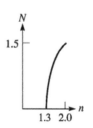

n	N
1.3	0
1.5	0.748
1.7	1.095
2.0	1.520

x	y
−3	−1
−2	0
−1	1
0	2
1	3
2	undefined

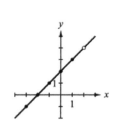

57. $(1, 2)$ on the graph means $f(1) = 2$ which says nothing about $f(2)$ which may or may not equal 1.

69. No. Some vertical lines will intercept the graph at multiple points.

61.

x	y
−2	2
−1	1
0	0
1	1
2	2

$y = x$ is the same as $y = |x|$ for $x \geq 0$.
$y = |x|$ is the same as $y = -x$ for $x < 0$.

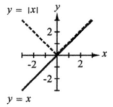

For negative values of x, $y = |x|$ becomes $y = -x$.

3.5 Graphs on the Graphing Calculator

1. $x^2 + 2x = 1 \Rightarrow x^2 + 2x - 1 = 0$. Graph $y = x^2 + 2x - 1$.

x	y
−4	7
−3	2
−2	−1
−1	−2
0	−1
1	2
2	7

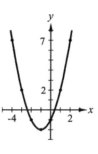

$x = -2.4$, $x = 0.4$

65. (a) $y = x + 2$

x	y
−3	−1
−2	0
−1	1
0	2
1	3
2	4

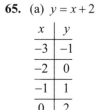

(b) $y = \dfrac{x^2 - 4}{x - 2}$

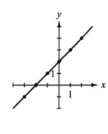

5. $y = 3x - 1$

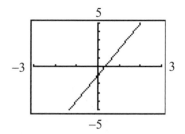

9. $y = 6 - \dfrac{x^3}{2}$

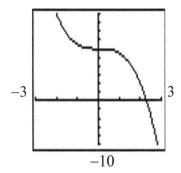

13. $y = \dfrac{2x}{x-2}$

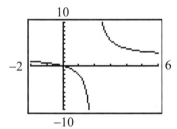

17. $y = 3 + \dfrac{2}{x}$

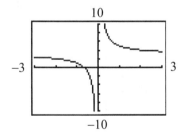

21. Graph $y = x^2 + x - 5$ and use the zero feature to solve.

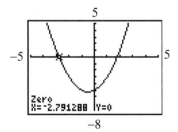

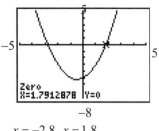

$x = -2.8, \; x = 1.8$

25. Graph $y_1 = x^3 - 4x^2 - 6$ and use the zero feature to solve.

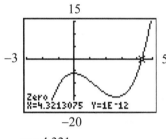

$x = 4.321$

29. $\dfrac{1}{x^2+1} = 0$, graph $y = \dfrac{1}{x^2+1}$. Since the graph does not cross the x-axis, the equation $\dfrac{1}{x^2+1} = 0$ has no solution.

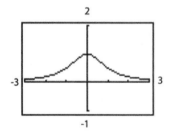

33. Graph $y = \dfrac{x^2}{x+1}$ and use the minimum and maximum feature to find the range.

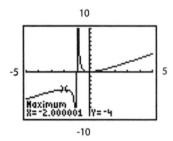

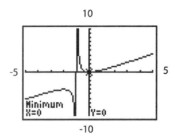

Range: $y \leq -4$ or $y \geq 0$

37. Graph $y = f(x) = \dfrac{x^2 + 8x - 8}{x(x-2) + 4(x-2)}$

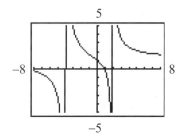

From the graph the range is all real numbers.

41. function: $y = \sqrt{x}$

function shifted right 3: $y = \sqrt{x - 3}$

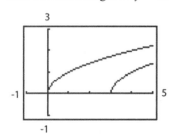

45. function: $y = \sqrt{2x + 1}$

function shifted up 1, left 1:

$y = \sqrt{2(x+1) + 1} + 1$

$y = \sqrt{2x + 3} + 1$

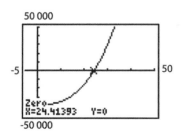

49.

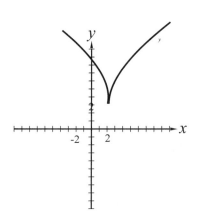

53. $e^3 + (e+5)^3 = 40{,}000 \Rightarrow e^3 + (e+5)^3 - 40{,}000 = 0$,

graph $y = x^3 + (x+5)^3 - 40{,}000$ and use the zero

feature to solve.

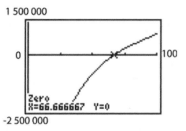

$e = 24.4$, $e + 5 = 29.4$

The edges of the coolers are 24.4 cm and 29.4 cm.

57. Graph $y = 9x^3 - 2400x^2 + 240{,}000x - 8{,}000{,}000$

and use the zero feature to solve.

The light sources are 67 ft apart.

61. $s = \sqrt{t - 4t^2}$, graph $y = \sqrt{x - 4x^2}$ and use the maximum feature to solve.

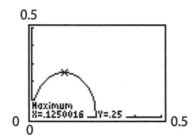

0.5

Maximum
X=.1250016 Y=.25

0 0.5
0

The maximum cutting speed is 0.25 ft/min.

65. Using the curve with a positive c, the curve with a negative c is reflected about the x-axis.

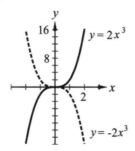

$y = 2x^3$

$y = -2x^3$

3.6   Graphs of Functions Defined by Tables of Data

1.

Month	Temp
J	6
F	7
M	12
A	18
M	24
J	28
J	31
A	29
S	26
O	19
N	13
D	7

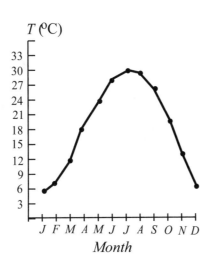

T (°C)

J F M A M J J A S O N D
Month

5.

Dist. (cm)	M. ind. (H)
0.0	0.77
2.0	0.75
4.0	0.61
6.0	0.49
8.0	0.38
10.0	0.25
12.0	0.17

M. ind. (H)

0.8
0.6
0.4
0.2

Distance (cm)
2 4 6 8 10 12

9. (a) Reading from the graph, $T = 132°\text{C}$ for $t = 4.3$ min.

(b) Reading from the graph, $t = 0.7$ min for $T = 145.0°\text{C}$.

13. $2\left[1.2\begin{bmatrix}8.0 & 0.38 \\ 9.2 & ? \\ 10.0 & 0.25\end{bmatrix}x\right] - 0.13$

$\dfrac{1.2}{2} = \dfrac{x}{-0.13}$, $x = -0.78$

Therefore, $M = 0.38 - 0.078 = 0.30$ H

17.

Height (ft)	Rate (ft³ / s)
0	0
1.0	10
2.0	15
4.0	22
6.0	27
8.0	31
12	35

(a) For $R = 20$ ft³ / s, $H = 3.4$ ft

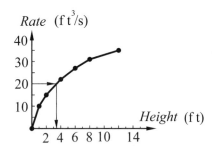

(b) For $H = 2.5$ ft, $R = 17$ ft^3 / s

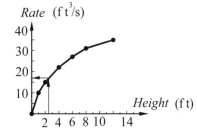

$$21. \quad 10 \begin{bmatrix} 6 \begin{bmatrix} 30 & 0.30 \\ 46 & ? \\ 40 & 0.37 \end{bmatrix} x \end{bmatrix} 0.07$$

$\dfrac{6}{10} = \dfrac{x}{0.07}$, $x = 0.042$

Therefore, $f = 0.30 + 0.042 = 0.34$

25. The graph is extended using a straight line segment.

$T \approx 130.3°\,\text{C}$ for $t = 5.3$ min

Chapter 3 Review Exercises

1. $A = \pi r^2 = \pi \left(2t\right)^2$

$A = 4\pi t^2$

5. $f\left(x\right) = 7x - 5$

$f\left(3\right) = 7\left(3\right) - 5 = 21 - 5 = 16$

$f\left(-6\right) = 7\left(-6\right) - 5 = -42 - 5 = -47$

9. $F\left(x\right) = x^3 + 2x^2 - 3x$

$F\left(3 + h\right) - F\left(3\right) = \left(3 + h\right)^3 + 2\left(3 + h\right)^2 - 3\left(3 + h\right)$

$\qquad - \left(3^3 + 2\left(3^2\right) - 3\left(3\right)\right)$

$\qquad = 27 + 27h + 9h^2 + h^3 + 18 + 12h$

$\qquad + 2h^2 - 9 - 3h - 27 - 18 + 9$

$\qquad = h^3 + 11h^2 + 36h$

13. $f\left(x\right) = 8.07 - 2x$

$f\left(5.87\right) = 8.07 - 2 \cdot 5.87 = -3.67$

$f\left(-4.29\right) = 8.07 - 2\left(-4.29\right) = 16.65 \approx 16.7$

17. The domain of $f\left(x\right) = x^4 + 1$ is $-\infty < x < \infty$.

The range is $y \geq 1$.

21. $f\left(n\right) = 1 + \dfrac{2}{\left(n - 5\right)^2}$ has domain $n \neq 5$ and

range $f\left(n\right) > 1$.

25. $s = 4t - t^2$

t	s
-1	-5
0	0
1	3
2	4
3	3
4	0
5	-5

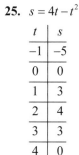

29. The graph of

$A = 6 - s^4$ is

A	s
-2	-10
-1	5
0	6
1	5
0	6

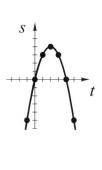

33. Graph $y_1 = 7x - 3$ and use the zero feature to solve.

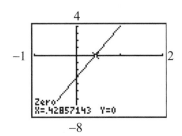

37. $x^3 - x^2 = 2 - x \Rightarrow x^3 - x^2 + x - 2 = 0$

Graph $y_1 = x^3 - x^2 + x - 2$ and use the zero feature to solve.

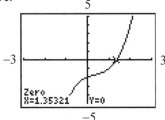

41. Graph $y_1 = x^4 - 5x^2$ and use the minimum feature, then from the graph the range is $y \geq -6.25$.

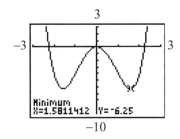

45. $A(a, b) = A(2, -3)$ is in QIV which $B(b, a)$ $= B(-3, 2)$ is in QII.

49. $\left| \dfrac{y}{x} \right| > 0$ for $\dfrac{y}{x} \neq 0 \Rightarrow$ all (x, y)

not on x-axis or y-axis.

53. $y = \sqrt{x - 1}$ shifted left 2 and up 1 is

$y = \sqrt{(x + 2) - 1} + 1$ $y = \sqrt{x + 1} + 1$

57. $(2, 3), (5, -1), (-1, 3)$ do not lie on a straight line. However, they do all lie on a parabola of the form.

$f(x) = ax^2 + bx + c, \ f(2) = 4a + 2b + c = 3$

$f(5) = 25a + 5b + c = -1, \ f(-1) = a - b + c = 3$

The solutions are $a = -\dfrac{2}{9}, \ b = \dfrac{2}{9}, \ c = \dfrac{31}{9}$. The 3

points lie on the graph of $f(x) = \dfrac{-2x^2}{9} + \dfrac{2x}{9} + \dfrac{31}{9}$.

61. $I = f(m) = 12.5\sqrt{1 + 0.5m^2}$

$f(0.55) = 12.5\sqrt{1 + 0.5(0.55)^2} \approx 13.4$

65. $T = f(t) = 28.0 + 0.15t, \ 0 \leq t \leq 30$

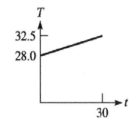

69. $L = 2\pi r + 12$

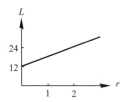

73. $P = f(i) = 1.5 \times 10^{-6} i^3 - 0.77, \ 80 \leq i \leq 140$

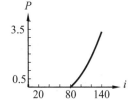

77. $d = f(D)$

81. $d = f(t) = 250 - 60y, \ 0 \leq t \leq 2;$

$d = f(t) = 130 - 40(t-2), \ 2 < t < 5.25$

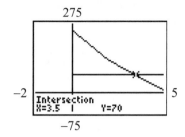

85. $v = 7.6x - 2.1x^2, \ 0 \leq x \leq 1.75$

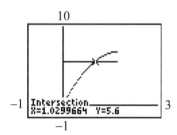

89. $T = \dfrac{4t^2}{t+2} - 20, \ t \geq 0$

$0 = \dfrac{4t^2}{t+2} - 20 \Rightarrow 4t^2 - 20(t+2) = 0$

$$4t^2 - 20t - 40 = 0$$

$$t^2 - 5t - 10 = 0$$

Graph $y_1 = x^2 - 5x - 10$ for $x \geq 0$ and use the zero feature to solve.

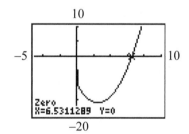

Chapter 4

THE TRIGONOMETRIC FUNCTIONS

4.1 Angles

1. $145.6° + 2(360°) = 865.6°$

5.

9. positive: $45° + 360° = 405°$

negative: $45° - 360° = -315°$

13. positive: $430°30' - 360° = 70°30'$

negative: $70°30' - 360° = -289°30'$

17. To change 0.265 rad to degrees multiply by $\dfrac{180}{\pi}$,

$0.265 \text{ rad} \left(\dfrac{180°}{\pi \text{ rad}} \right) \approx 15.18°$

21. $0.329 \text{ rad} \approx 18.85°$

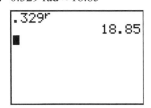

25. $56.0° = 0.977$ rad to three significant digits

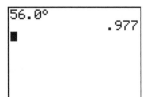

29. $47° + 0.5° \cdot \dfrac{60'}{1°} = 47° + 30' = 47°30'$

33. $15°12' = 15° + 12' \cdot \dfrac{1°}{60'} = 15.2°$

37. Angle in standard position terminal side passing through $(4, 2)$.

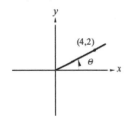

41. Angle in standard position terminal side passing through $(-7, 5)$.

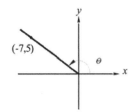

45. $31°$, QI; $310°$, QIV

49. 1 rad, QI; 2 rad, QII

53. $21°42'36''$

$= 21° + 42' \cdot \dfrac{1°}{60'} + 36'' \cdot \dfrac{1°}{60''} \cdot \dfrac{1°}{60'} = 21.710°$

4.2 Defining the Trigonometric Functions

1. $r = \sqrt{4^2 + 3^2} = \sqrt{25} = 5$

$\sin \theta = \dfrac{3}{5}$

$\cos \theta = \dfrac{4}{5}$

$\tan \theta = \dfrac{3}{4}$

$\csc \theta = \dfrac{5}{3}$

$\sec \theta = \dfrac{5}{4}$

$\cot \theta = \dfrac{4}{3}$

5. $r = \sqrt{15^2 + 8^2} = \sqrt{289} = 17$

$\sin \theta = \dfrac{y}{r} = \dfrac{8}{17}$

$\cos \theta = \dfrac{x}{r} = \dfrac{15}{17}$

$\tan \theta = \dfrac{y}{x} = \dfrac{8}{15}$

$\csc \theta = \dfrac{r}{y} = \dfrac{17}{8}$

$\sec \theta = \dfrac{r}{x} = \dfrac{17}{15}$

$\cot \theta = \dfrac{x}{y} = \dfrac{15}{8}$

9. $r = \sqrt{1 + 15} = \sqrt{16} = 4$

$\sin \theta = \dfrac{y}{r} = \dfrac{\sqrt{15}}{4}$

$\cos \theta = \dfrac{x}{r} = \dfrac{1}{4}$

$\tan \theta = \dfrac{y}{x} = \sqrt{15}$

$\csc \theta = \dfrac{r}{y} = \dfrac{4}{\sqrt{15}}$

$\sec \theta = \dfrac{r}{x} = 4$

$\cot \theta = \dfrac{x}{y} = \dfrac{1}{\sqrt{15}}$

13. $r = \sqrt{50^2 + 20^2} = \sqrt{2900} = 10\sqrt{29}$

$\sin \theta = \dfrac{y}{r} = \dfrac{20}{10\sqrt{29}} = \dfrac{2}{\sqrt{29}}$

$\cos \theta = \dfrac{x}{r} = \dfrac{50}{10\sqrt{29}} = \dfrac{5}{\sqrt{29}}$

$\tan \theta = \dfrac{y}{x} = \dfrac{20}{50} = \dfrac{2}{5}$

$\csc \theta = \dfrac{r}{y} = \dfrac{10\sqrt{29}}{20} = \dfrac{\sqrt{29}}{2}$

$\sec \theta = \dfrac{r}{x} = \dfrac{10\sqrt{29}}{50} = \dfrac{\sqrt{29}}{5}$

$\cot \theta = \dfrac{x}{y} = \dfrac{50}{20} = \dfrac{5}{2}$

17. $\cos \theta = \dfrac{12}{13} \Rightarrow x = 12$ and $r = 13$ with θ in QL

$r^2 = x^2 + y^2 \Rightarrow 169 = 144 + y^2 \Rightarrow y^2 = 25$

$\qquad\qquad\qquad\qquad\qquad\qquad\qquad y = 5$

$\sin \theta = \dfrac{y}{r} = \dfrac{5}{13}$, $\cot \theta = \dfrac{x}{y} = \dfrac{12}{5}$.

21. $\sin \theta = 0.750 \Rightarrow y = 0.750$ and $r = 1$ with θ in QI.

$r^2 = x^2 + y^2 = 1^2$

$\quad x^2 + 0.750^2 = 1$

$\quad x = \sqrt{0.4375}$

$\cot \theta = \dfrac{x}{y} = \dfrac{\sqrt{0.4375}}{0.750} = 0.882.$

$\csc \theta = \dfrac{r}{y} = \dfrac{1}{0.750} = 1.33.$

25. For $(3, 4)$, $r = 5$,

$\sin \theta = \dfrac{y}{r} = \dfrac{4}{5}$ and $\tan \theta = \dfrac{y}{x} = \dfrac{4}{3}$.

For $(6, 8)$, $r = 10$,

$\sin \theta = \dfrac{y}{r} = \dfrac{8}{10} = \dfrac{4}{5}$ and $\tan \theta = \dfrac{y}{x} = \dfrac{8}{6} = \dfrac{4}{3}$.

For $(4.5, 6)$, $r = 7.5$,

$\sin \theta = \dfrac{y}{r} = \dfrac{6}{7.5} = \dfrac{4}{5}$ and $\tan \theta = \dfrac{y}{x} = \dfrac{6}{4.5} = \dfrac{4}{3}$.

29. tan, cot, sec, csc can have value 1.1.

33. $\sin^2 \theta + \cos^2 \theta = \left(\dfrac{3}{5}\right)^2 + \left(\dfrac{4}{5}\right)^2$

$$= \dfrac{9}{25} = \dfrac{16}{25}$$

$$= \dfrac{25}{25} = 1$$

37. $\tan \theta = \dfrac{y}{x} = \dfrac{6}{-2} = \dfrac{4}{x+1}$

$$6x + 6 = -8$$

$$6x = -14$$

$$x = -\dfrac{7}{3}$$

4.3   Values of the Trigonometric Functions

1. $\sin \theta = 0.3527$

$\theta = 20.65°$

5. Answers may vary in exercises 5, 6, 7, 8. One set of measurements gives $x = 7.6$ and $y = 6.5$.

$\sin 40° = \dfrac{6.5}{10} = 0.65$

$\cos 40° = \dfrac{7.6}{10} = 0.76$

$\tan 40° = \dfrac{6.5}{7.6} = 0.86$

$\csc 40° = \dfrac{10}{6.5} = 1.54$

$\sec 40° = \dfrac{10}{7.6} = 1.32$

$\cot 40° = \dfrac{7.6}{6.5} = 1.17$

9. $\sin 22.4° = 0.381$

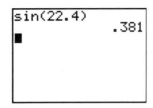

13. $\cos 15.71° = 0.9626$

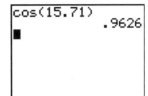

17. $\cot 67.78° = 0.4085$

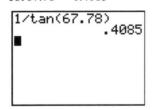

21. $\csc 0.49° = 116.9$

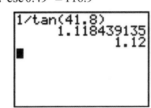

25. $\cos \theta = 0.3261$

$\theta = 70.97°$

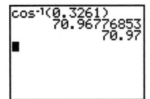

29. $\tan\theta = 0.207$

$\theta = 11.7^\circ$

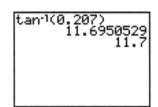

33. $\csc\theta = 1.245$

$\theta = 53.44^\circ$

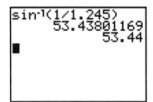

37. $\sec\theta = 0.305 \Rightarrow \cos\theta = \dfrac{1}{0.305} > 1,$

which has no solution

41. $\dfrac{\sin 43.7^\circ}{\cos 43.7^\circ} = \tan 43.7^\circ$

$= 0.96$

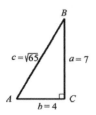

45. y is always less than r.

49. θ acute $\Leftrightarrow 0^\circ < \theta < 90^\circ$

$x + y > r$

$\dfrac{x}{r} + \dfrac{y}{r} > 1$

$\cos\theta + \sin\theta > 1$

53. $\sec\theta = 1.3698$

$\tan\theta = \tan\left(\cos^{-1}\left(1.3698^{-1}\right)\right)$

$\tan\theta = 0.93614$

57. $l = a\left(\sec\theta + \csc\theta\right)$

$= 28.0\left(\sec 34.5^\circ + \csc 34.5^\circ\right)$

$= 83.4$ cm

4.4   The Right Triangle

1.

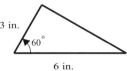

$\sin A = \dfrac{7}{\sqrt{65}} = 0.868$

$\cos A = \dfrac{4}{\sqrt{65}} = 0.496$

$\tan A = \dfrac{7}{4} = 1.75$

$\sin B = \dfrac{4}{\sqrt{65}} = 0.496$

$\cos B = \dfrac{4}{\sqrt{65}} = 0.868$

$\tan B = \dfrac{4}{7} = 0.571$

5. A 60° angle between sides of 3 in. and 6 in. determines the unique triangle shown in the figure below.

9. $\sin 77.8^\circ = \dfrac{6700}{c} \Rightarrow c = 6850$

$\angle B = 90^\circ - 77.8^\circ = 12.2^\circ$

$\tan 77.8^\circ = \dfrac{6700}{b} \Rightarrow b = 1450$

13. $\angle A = 90° - 32.1° = 57.9°,$

$$\sin 32.1° = \frac{b}{238} \Rightarrow b = 126$$

$$\cos 32.1° = \frac{a}{238} \Rightarrow a = 202$$

17. $\angle B = 90° - 32.1° = 57.9°,$

$$\sin 32.1° = \frac{a}{56.85} \Rightarrow a = 30.21$$

$$\cos 32.1° = \frac{b}{56.85} \Rightarrow b = 48.16$$

21. $\angle A = 90° - 37.5° = 52.5°,$

$$\tan 37.5° = \frac{b}{0.862} \Rightarrow b = 0.661$$

$$\cos 37.5° = \frac{0.862}{c} \Rightarrow c = 1.09$$

25. $\tan A = \frac{591.87}{264.93} \Rightarrow A = 65.89°;$

$$\tan B = \frac{264.93}{591.87} \Rightarrow B = 24.11°$$

$$c = \sqrt{264.93^2 + 591.87^2} = 648.46$$

29. $A = 90.0° - 9.56° = 80.44°$

$$\sin B = \frac{b}{c}$$

$$\Rightarrow b = 0.0973 \sin 9.56° = 0.0162$$

33. $\sin 61.7° = \frac{3.92}{x} \Rightarrow x = \frac{3.92}{\sin 61.7°} = 4.45$

37.

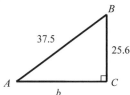

$$\sin A = \frac{25.6}{37.5}$$

$$A = 43.1° < B = 90° - 43.1° = 47.9°$$

41. $\angle B = 90° - \angle A,$

$$A = \frac{a}{c} \Rightarrow a = c \sin A,$$

$$\cos A = \frac{b}{c} \Rightarrow b = c \cos A$$

45. $\tan 30° = \frac{6}{r} \Rightarrow r = 6\sqrt{3}$

$$A = \pi r^2 = \pi \left(6\sqrt{3}\right)^2$$

$$A = 108\pi$$

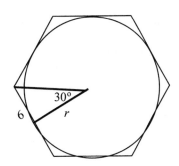

4.5 Applications of Right Triangles

1. $90° - 62.1° = 27.9°$

$$d = 2090 \text{ ft}$$

5. $\tan \theta = \frac{12.0}{85.0}$

$$\theta = 8°$$

9.

$$\sin 6° = \frac{2.65}{d} \Rightarrow d = \frac{2.65}{\sin 6°}$$

$$\sin 3° = \frac{2.65}{x} \Rightarrow x = \frac{2.65}{\sin 3°}$$

$$x - d = 25.3 \text{ ft longer}$$

13. $\cos 76.67° = \dfrac{196.0}{h}$

$\qquad h = \dfrac{196.0}{\cos 76.67°} = 850.1 \text{ cm}$

17. $\tan \alpha = \dfrac{6.75}{15.5}; \ \alpha = \tan^{-1}\left(\dfrac{6.75}{15.5}\right) = 23.5°$

21. $\theta = \tan^{-1}\dfrac{6.0}{100} = 3.4°$

25. Each angle of the pentagon is $\dfrac{360°}{5} = 72°$. Radii

drawn from the center of the pentagon (which is also the center of the circle) through adjacent vertices of the pentagon outward to the fence form an isosceles triangle with base 92.5 and equal sides x. A $\perp$ bisector from the center of the pentagon to

the base this isosceles triangle forms a right triangle with hypotenuse x and base 46.25. The

base angle of this right triangle is $\dfrac{180° - 72°}{2} = 54°$.

Thus,

$\cos(54°) = \dfrac{46.25}{x} \Rightarrow x = \dfrac{46.25}{\cos 54°}$

and $C = 2\pi(x + 25)$

$\qquad C = 2\pi\left(\dfrac{46.25}{\cos 54°} + 25\right) = 651 \text{ ft}$

29.

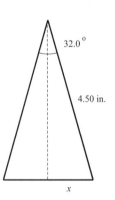

$\sin 16.0° = \dfrac{x}{4.50}$

$\qquad x = 1.24$

distance between ends $= 2x = 2.48$ in.

33. difference in angles $= \tan^{-1}\dfrac{65.3}{224} - \tan^{-1}\dfrac{65.3}{302} = 4.1°$

37. $\qquad \sin D = \dfrac{5380}{6680}$

$\qquad D = \sin^{-1}\left(\dfrac{5380}{6680}\right) = 53.7°$

$\qquad \tan 53.7° = \dfrac{2860}{h}$

$\qquad\qquad h = \dfrac{2860}{\tan 53.7°} = 2100 \text{ ft}$

41. $\tan \dfrac{\theta}{2} = \dfrac{\frac{d}{2}}{x}$

$\qquad \dfrac{d}{2} = x \tan \dfrac{\theta}{2}$

$\qquad d = 2x \tan \dfrac{\theta}{2}$

Chapter 4 Review Exercises

1. $17.0^\circ + 360.0^\circ = 377.0^\circ$, $17.0^\circ - 360.0^\circ = -343.0^\circ$

5. $31^\circ + 54'\left(\dfrac{1^\circ}{60'}\right) = 31.9^\circ$

9. $17.5^\circ = 17^\circ + 0.5^\circ\left(\dfrac{60'}{1^\circ}\right) = 17^\circ 30'$

13. $r = \sqrt{x^2 + y^2} = \sqrt{24^2 + 7^2} = \sqrt{625} = 25$

$\sin\theta = \dfrac{y}{r} = \dfrac{7}{25}$, $\qquad \csc\theta = \dfrac{r}{y} = \dfrac{25}{7}$

$\cos\theta = \dfrac{x}{r} = \dfrac{24}{25}$, $\qquad \sec\theta = \dfrac{r}{x} = \dfrac{25}{24}$

$\tan\theta = \dfrac{y}{x} = \dfrac{7}{24}$, $\qquad \cot\theta = \dfrac{x}{y} = \dfrac{24}{7}$

17. $r^2 = x^2 + y^2 \Rightarrow 13^2 = x^2 + 5^2 \Rightarrow x = 12$

$\cos\theta = \dfrac{x}{r} = \dfrac{12}{13} = 0.923$

$\cot\theta = \dfrac{x}{y} = \dfrac{12}{5} = 2.40$

21. $\sin 72.1^\circ = 0.952$ from the calculator

25. $\sec 18.4^\circ = 1.05$ from the calculator

29. $\cos\theta = 0.950 \Rightarrow \theta = \cos^{-1}(0.950)$

$\theta = 18.2^\circ$ from the calculator

33. $\csc\theta = 4.713 \Rightarrow \dfrac{1}{\sin\theta} = 4.13 \Rightarrow \sin\theta = \dfrac{1}{4.713}$

$\theta = \sin^{-1}\dfrac{1}{4.713} = 12.25^\circ$ from the calculator

37. $\cot\theta = \dfrac{1}{\tan\theta} = 7.117$

$\tan\theta = \dfrac{1}{7.117}$

$\theta = \tan^{-1}\dfrac{1}{7.117}$

$\theta = 8.00^\circ$ from the calculator

41. $\angle B = 90.0^\circ - 17.0^\circ = 73.0^\circ$

$\tan 17.0^\circ = \dfrac{a}{6.00} \Rightarrow a = 1.83$

$\cos 17.0^\circ = \dfrac{6.00}{c} \Rightarrow c = \dfrac{6.00}{\cos 17.0^\circ} = 6.27$

45. $\angle B = 90.0^\circ - 37.5^\circ = 52.5^\circ$

$\tan 37.5^\circ = \dfrac{12.0}{b} \Rightarrow b = 15.6$

$\sin 37.5^\circ = \dfrac{12.0}{c} \Rightarrow c = \dfrac{12.0}{\sin 37.5^\circ} = 19.7$

49. $\angle B = 90.0^\circ - 49.67^\circ = 40.33^\circ$

$\sin 49.67^\circ = \dfrac{a}{0.8253} \Rightarrow a = 0.6292$

$\cos 49.67^\circ = \dfrac{b}{0.8253} \Rightarrow b = 0.5341$

53.

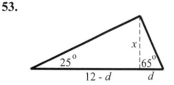

$90^\circ - 25^\circ = 65^\circ$

$\tan 65^\circ = \dfrac{x}{d} \Rightarrow d = \dfrac{x}{\tan 65^\circ}$

$\tan 25^\circ = \dfrac{x}{12 - d} = \dfrac{x}{12 - \frac{x}{\tan 65^\circ}}$

$x = \left(12 - \dfrac{x}{12 - d}\right)\tan 25^\circ$

$x = 12\tan 25^\circ - x\cdot\dfrac{\tan 25^\circ}{\tan 65^\circ}$

$x\left(1 + \dfrac{\tan 25^\circ}{\tan 65^\circ}\right) = 12\tan 25^\circ$

$x = \dfrac{12\tan 25^\circ}{1 + \frac{\tan 25^\circ}{\tan 65^\circ}}$

$x = 4.6$

57.

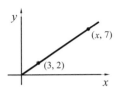

$$\tan \theta = \frac{2}{3} = \frac{7}{x}$$

$$2x = 21$$

$$x = 10.5$$

61.

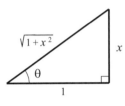

$$\tan \theta = x = \frac{x}{1}$$

$$\csc \theta = \frac{\sqrt{x^2 + 1}}{x}$$

65.

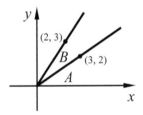

$$\tan A = \frac{2}{3} \Rightarrow A = 33.7°$$

$$\tan (A + B) = \frac{3}{2} \Rightarrow A + B = 56.3°$$

$$B = 56.3° - A = 22.6°$$

69. $e = E \cos \alpha \Rightarrow 56.9 = 339 \cos \alpha \Rightarrow \alpha$

$$= \cos^{-1} \frac{56.9}{339} = 80.3°$$

73. (a) A triangle with angle θ included between sides a and b, the base, has an altitude of $a \sin \theta$. The area, A, is $A = \frac{1}{2} \cdot b \cdot a \sin \theta$.

(b) The area of the tract is $A = \frac{1}{2} \cdot 31.96 \cdot 47.25$

$\sin 64.09° = 679.2 \text{ m}^2$.

77.

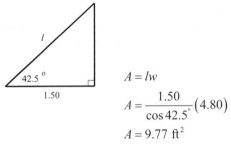

$$A = lw$$

$$A = \frac{1.50}{\cos 42.5°} (4.80)$$

$$A = 9.77 \text{ ft}^2$$

81. $d = a + b = \frac{1.85}{\tan 28.3°} + \frac{1.85}{\tan \left(90.0° - 28.3°\right)} = 4.43 \text{ m}$

85. Let x = length of window through which sun does not shine.

$$\tan 65° = \frac{x + 2.5}{2.0} \Rightarrow x = 2.0 \tan 65° - 2.5$$

$$\text{percent window shaded} = \frac{2.0 \tan 65° - 2.5}{3.2} = 56\%$$

89.

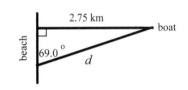

$$\sin 69° = \frac{2.75}{d} \Rightarrow d = \frac{2.75}{\sin 69°}$$

$$d = 37.5t \Rightarrow t = \frac{2.75}{37.5 \sin 69.0°} \cdot \frac{60 \text{ min}}{1 \text{ h}}$$

$$t = 4.71 \text{ min}$$

93. $\theta = \tan^{-1} \frac{1.25}{2.10} = 30.8°$

97.

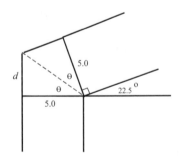

$$2\theta + 90^\circ + 22.5^\circ = 180^\circ$$

$$\theta = \frac{67.5^\circ}{2}$$

$$\tan\frac{67.5^\circ}{2} = \frac{d}{5.0}$$

$$d = 5.0\tan\frac{67.5^\circ}{2}$$

$$l = 5.0 + 65.0 + 5.0\tan\frac{67.5^\circ}{2} = 73.3 \text{ cm}$$

Chapter 5

Systems of Linear Equations: Determinants

5.1 Linear Equations

1. $x - \dfrac{y}{6} + z - 4w = 7 \Rightarrow x - \dfrac{1}{6}y + z - 4w = 7$ is linear.

5. The coordinates of the point $(3, 1)$ do satisfy the equation since $2(3) + 3(1) = 6 + 3 = 9$.

The coordinates of the point $(5, 1/3)$ do not satisfy

the equation since $2(5) + 3\left(\dfrac{1}{3}\right) = 10 + 1 = 11 \neq 9$.

9. $3(2) - 2y = 12;\ 2y = 6 - 12 = -6;\ y = -3$

$3(-3) - 2y = 12;\ 2y = -9 - 12 = -21;\ y = -\dfrac{21}{2}$

13. $24(2/3) - 9y = 16;\ 9y = 16 - 16 = 0;\ y = 0$

$24(-1/2) - 9y = 16;\ 9y = -12 - 16 = -28;$

$y = (-28/9)$

17. If the values $A = -2$ and $B = 1$ satisfy both equations, they are a solution.

$-2 + 5(1) = 3 \neq 7;\ 3(-2) - 4(1) = -10 \neq 4$

Since the given values do not satisfy both equations they are not a solution.

21. If the values $x = 0.6$ and $y = -0.2$ satisfy both equations, they are a solution.

$3(0.6) - 2(-0.2) = 2.2;\ 5(0.6) - 0.2 = 2.8$

Therefore the given values are a solution.

25. $(2 - x)(y + 2) = x(6 - y) \Rightarrow$

$8x - 2y = 4,$ linear

$8(2) - 2(6) = 4$

$x = 2,\ y = 6$ is a solution

29. If $p = 260$ mi/h and $w = 40$ mi/h, then

$p + w = 260 + 40 = 300$ mi/h

$p - w = 260 - 40 = 220$ mi/h

The speeds are 260 mi/h and 40 mi/h.

5.2 Graphs of Linear Functions

1. $(-1, -2), (3, -1)$

$m = \dfrac{-1 - (-2)}{3 - (-1)} = \dfrac{1}{4}$. The line rises 1 unit for each

4 units in going from left to right.

5. By taking $(3, 8)$ as (x_2, y_2) and $(1, 0)$ as (x_1, y_1)

$m = \dfrac{8 - 0}{3 - 1} = \dfrac{8}{2} = 4$

9. By taking $(-2, -5)$ as (x_2, y_2) and $(5, -3)$ as (x_1, y_1)

$m = \dfrac{-5 - (-3)}{-2 - 5} = \dfrac{-5 + 3}{-7} = \dfrac{2}{7}$

13. $m = 2, (0, -1)$

Plot the y-intercept point $(0, -1)$. Since the slope is 2/1, from this point, go over 1 unit and up 2 units, and plot a second point. Sketch the line between the 2 points.

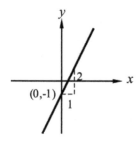

17. $m = \dfrac{1}{2}, (0, 0)$

Plot the y-intercept point $(0, 0)$. Since the slope is 1/2, from this point, go over 2 units and up 1 unit, and plot a second point. Sketch the line between the 2 points.

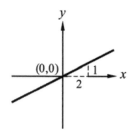

21. $m = -2x + 1$, $m = -2$, $b = 1$

Plot the y-intercept point $(0, 1)$. Since the slope is $-2/1$, from this point, go over 1 unit and down 2 units, and plot a second point. Sketch the line between the 2 points.

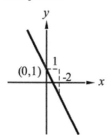

25. $5x - 2y = 40 \Rightarrow y = \dfrac{5}{2}x - 20$, $m = \dfrac{5}{2}$, $b = -20$

Plot the y-intercept point $(0, 20)$. Since the slope is 5/2, from this point, go over 2 units and up 5 units, and plot a second point. Sketch the line between these 2 point.

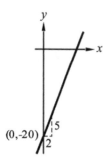

29. $x + 2y = 4\big|_{x=0} \Rightarrow 0 + 2y = 4 \Rightarrow y\text{-int} = 2$

$x + 2y = 4\big|_{y=0} \Rightarrow x + 2 \cdot 0 = 4 \Rightarrow x\text{-int} = 4$

Plot the y-intercept point $(4, 0)$ and the y-intercept point $(0, 2)$. Sketch the line between these 2 pts. A third point is found as a check. Let $x = -2$, $-2 + 2y = 4$, $2y = 6$, $y = 3$. Therefore the point $(-2, 3)$ should lie on the line.

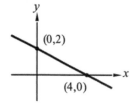

33. $y = 3x + 6\big|_{x=0} \Rightarrow y = 3 \cdot 0 + 6 \Rightarrow y\text{-int} = 6$

$y = 3x + 6\big|_{y=0} \Rightarrow 0 = 3x + 6 \Rightarrow x\text{-int} = -2$

Plot the x-intercept point $(-2, 0)$ and the y-intercept point $(0, 6)$. Sketch the line between these 2 pts. A third point is found as a check. Let $x = 1$, $y = 3(1) + 6 = 9$. Therefore the point $(1, 9)$ should lie on the line.

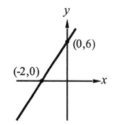

37. $3x - 2y = 6$

$x = 0, \Rightarrow y = -3,$ y-int: $(0, -3)$

$y = 0, \Rightarrow x = 2,$ x-int: $(2, 0)$

41. No

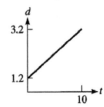

$$m\left[(1, -2), (3, -3)\right] = \frac{-3 - (-2)}{3 - 1} = -\frac{1}{2}$$

$$m\left[(5, -4), (7, -6)\right] = \frac{-6 - (-4)}{7 - 5} = -1$$

45. $d = 0.2l + 1.2$

The d-intercept is $(0, 1.2)$. Since the slope is 2/10, from the d-intercept go over 10 units and up 2 units the point $(10, 3.2)$ and plot a second point. Sketch the line between these 2 points.

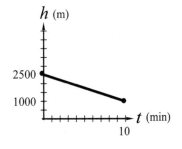

49. $h(t) = 2500 - 150t$

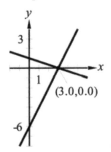

5.3 Solving Systems of Two Linear Equations in Two Unknowns Graphically

1. $2x + 5y = 10.$ Let $x = 0,$ $y = 0$ to find intercepts of $(0, 2), (5, 0)$. A third point is $\left(1, \frac{8}{5}\right)$.

$3x + y = 6.$ Let $x = 0,$ $y = 0$ to find intercepts of $(0, 6), (2, 0)$. A third point is $(1, 3)$. From the graph the solution is approximately $x = 1.5,$ $y = 1.4$.

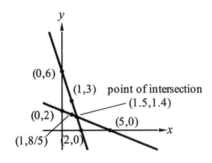

5. $y = 2x - 6;$ $y = -\left(\frac{1}{3}\right)x + 1$

The slope of the first line is 2, and the y-intercept is -6. The slope of the second line is $-1/3$ and the y-intercept is 1. From the graph, the point of intersection is $(3.0, 0.0)$. Therefore, the solution of the system of equations is $x = 3.0,$ $y = 0.0$.

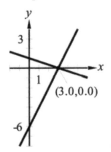

9. $2x - 5y = 10; \; 3x + 4y = -12$

The intercepts of the first line are $(5, 0), (0, -2)$.

A third point is $\left(1, -\frac{8}{5}\right)$. The intercepts of the

second line are $(-4, 0), (0, -3)$. A third point is

$\left(\frac{4}{3}, 4\right)$. From the graph, the point of intersection is

$x = -0.9, \; y = -2.3$.

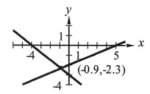

13. $y = -x + 3; \; y = -2x + 3$

The intercepts of the first line are $(3, 0), (0, 3)$.

A third point is $(2, 1)$. The intercepts of the

second line are $\left(\frac{3}{2}, 0\right), (0, 3)$. A third point is

$(1, 1)$. From the graph, the point of intersection is

$(0.0, 3.0)$. The solution of the system of equations

is $x = 0.0, \; y = 3.0$.

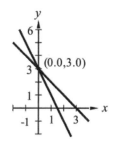

17. $-2r_1 + 2r_2 = 7; \; 4r_1 - 2r_2 = 1$

The intercepts of the first line are $\left(-\frac{7}{2}, 0\right), \left(0, \frac{7}{2}\right)$.

A third point is $\left(-2, \frac{3}{2}\right)$. The intercepts of the

second line are $\left(\frac{1}{4}, 0\right), \left(0, -\frac{1}{2}\right)$. A third point is

$\left(1, \frac{3}{2}\right)$. From the graph, the point of intersection

is $(4.0, 7.5)$, and the solution of the system of

equations is $r_1 = 4.0, \; r_2 = 7.5$.

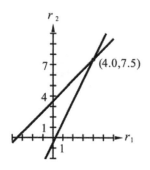

21. $x = 4y + 2 \Leftrightarrow y = \dfrac{x - 2}{4}$ and

$3y = 2x + 3 \Leftrightarrow y = \dfrac{2x + 3}{3}$

On a graphing calculator let $y_1 = \dfrac{x - 2}{4}$ and

$y_2 = \dfrac{2x + 3}{3}$. Using the intersect feature, the

point of intersection is $(-3.6, -1.4)$, and the

solution of the system of equations is $x = -3.600$,

$y = -1.400$.

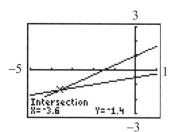

25. $x - 5y = 10 \Leftrightarrow y = \dfrac{x - 10}{5}$ and

$2x - 10y = 20 \Leftrightarrow y = \dfrac{x - 10}{5}$

On the graphing calculator let $y_1 = \frac{x-10}{5}$ and

$y_2 = \frac{x-10}{5}$. From the graph the lines are the

same. The system is dependent.

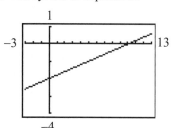

29. $5x - y = 3 \Leftrightarrow y = 5x - 3$ and

$$4x = 2y - 3 \Leftrightarrow y = \frac{4x + 3}{2}$$

On a graphing calculator let $y_1 = 5x - 3$ and $y_2 = \frac{4x+3}{2}$. Using the intersect feature, the point of intersection is $(1.5, 4.5)$, and the solution to the system of equations is $x = 1.500$, $y = 4.500$.

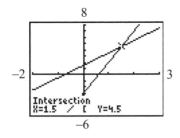

33. $y = 3x - 7$

$y = mx + b$

will have no solution for $m = 3$, $b \neq -7$.

37. $(p + w)(3) = 780 \Rightarrow p + w = 260$

$(p + 1.5w)(2.5) = 700 \Rightarrow p + 1.5w = 280$

$p = 220$ km/h, speed of plane in still air

$w = 40$ km/h, speed of wind

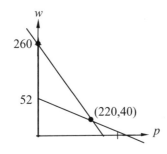

5.4 Solving Systems of Two Linear Equations in Two Unknowns Algebraically

1. (1) $x - 3y = 6 \Rightarrow x = 3y + 6$

(2) $2x - 3y = 3$

$2(3y + 6) - 3y = 3$ substitute x from (1) into (2)

$6y + 12 - 3y = 3$

$3y = -9$

$y = -3$

$x - 3(3) = 6$ substitute -3 for y in (1)

$x + 9 = 6$ $x = -3$

5. (1) $x = y + 3$

(2) $x - 2y = 5$

$(y + 3) - 2y = 5$ substitute x from (1) into (2)

$-y = 2$

$y = -2$ substitute -2 for y in (1)

$x = -2 + 3 = 1$

9. (1) $x + y = -5$, $y = -x - 5$

(2) $2x - y = 2$

$2x - (-x - 5) = 2$ substitute y from (1) into (2)

$3x = -3$

$x = -1$

$-1 + y = -5$ substitute -1 for x in (1)

$y = -4$

13. (1) $33x + 2y = 34 \Rightarrow y = -\frac{33}{2}x + 17$

(2) $40y = 9x + 11$

$40\left(-\frac{33}{2}x + 17\right) = 9x + 11$ substitute y from (1) in (2)

$-660x + 680 = 9x + 11$

$-669x = -669$

$x = 1$

$33(1) + 2y = 34$ substitute 1 for x in (1)

$2y = 1$

$y = \frac{1}{2}$

17. (1) $2x - 3y = 4$

(2) $\underline{2x + y = -4}$

$-4y = 8$ subtract (1) and (2)

$y = -2$

$2x - 3(-2) = 4$ substitute for x in (1)

$2x + 6 = 4$

$2x = -2$

$x = -1$

21. (1) $v + 2t = 7$

(2) $2v + 4t = 9$

(3) $2v + 4t = 14$ multiply (1) by 2

(4) $\underline{2v + 4t = 9}$ recopy (2)

$0 = 5$ inconsistent

25. (1) $2x - y = 5, y = 2x - 5$

(2) $6x + 2y = -5$

$6x + 2(2x - 5) = -5$ substitute y from (1) into (2)

$10x = 5$

$x = \dfrac{1}{2}$

$y = 2\left(\dfrac{1}{2}\right) - 5 = -4$ substitute $\dfrac{1}{2}$ for x in (1)

$x = \dfrac{1}{2}$, from (1) $y = 2x - 5 = 2 \cdot \dfrac{1}{2} - 5 = 1 - 5 = -4$

$\left(\dfrac{1}{2}, -4\right)$

29. (1) $15x + 10y = 11 \Rightarrow 75x + 50y = 55$

(2) $20x - 25y = 7 \Rightarrow \underline{40x - 50y = 14}$ add

$115x = 69$

$x = \dfrac{3}{5}$

$15\left(\dfrac{3}{5}\right) + 10y = 11$ substitute $\frac{3}{5}$ for x in (1)

$10y = 2$

$y = \dfrac{1}{5}$

33. (1) $44A = 1 - 15B \Rightarrow 44A + 15B = 1$

(2) $5B = 22 + 7A \Rightarrow 7A - 5B = -22$

(1) $44A + 15B = 1$

(2) $\underline{21A - 15B = -66}$ multiply (2) by 3

$65A = -65$

$A = -1$

$5B = 22 + 7(-1)$ substitute (-1) for A in (2)

$5B = 15$

$B = 3$

37. $0.3x - 0.7y = 0.4 \qquad 3x - 7y = 4 \qquad 15x - 35y = 30$

$0.2x + 0.5y = 0.7 \qquad 2x + 5y = 7 \qquad \underline{14x + 35y = 49}$

$29x = 69$

$x = \dfrac{69}{29}$

$2\left(\dfrac{69}{29}\right) + 5y = 7$

$5y = \dfrac{65}{29}$

$y = \dfrac{13}{29}$

41. $f(x) = ax + b$

$f(2) = 2a + b = 1$

$f(-1) = -a + b = -5$

from which $a = 2$, $b = -3$

45. $V_1 + V_2 = 15$

$\underline{V_1 - V_2 = 3}$

$2V_1 = 18$

$V_1 = 9\,V$

$9 + V_2 = 15 \Rightarrow V_2 = 6$, the solution is $V_1 = 9$ and $V_2 = 6\,V$

49. x = number of regular email messages

y = number of spam messages

$x + y = 78$ $x + 4x - 2 = 78$

     $y = 4x - 2$ $5x = 80$

                           $x = 16$ reg. messages

$y = 4(16) - 2 = 62$ spam messages

53.   (1) $2000t_1 = 3200t_2$

      (2) $t_1 - 12 = t_2$

$2000t_1 = 3200(t_1 - 12) \Rightarrow 2000t_1$

$ = 3200t_1 - 38,400 \Rightarrow -1200t_1$

$ = -38,400$

$t_1 = 32\ s$ and $t_2 = 32 - 12 = 20\ s$

57. $A = B + 2000$

$B + 0.01A = 0.99A + 1000 \Rightarrow$

$A - B = 2000$

$0.98A - B = -1000 \Rightarrow$

$A = 150,000$ votes, $B = 148,000$ votes

61. Let $x =$ the amount of sales; $x + 8000$

           $=$ this month.

      $x =$ last month

so $x + 8000 + x = 4000 - 2x$

      $2x + 8000 = 4000 - 2x$

$0 = 4000$ inconsistent; incorrect conclusion or error in sales figures.

65.   When $x = -3$, $\left(x, \frac{1}{3}x - 3\right) = \left(-3, \frac{1}{3}(-3) - 3\right)$

           $= (-3, -4)$ when

$x = 9$, $\left(x, \frac{1}{3}x - 3\right) = \left(9, \frac{1}{3}(9) - 3\right) = (9, 0)$

5.5   Solving Systems of Two Linear Equations in Two Unknowns by Determinants

1. $\begin{vmatrix} 4 & -6 \\ 3 & 17 \end{vmatrix} = 4(17) - 3(-6) = 68 + 18 = 86$

5. $\begin{vmatrix} 2 & 4 \\ 3 & 1 \end{vmatrix} = (2)(1) - (3)(4) = 2 - 12 = -10$

9. $\begin{vmatrix} 27 & -10 \\ 0 & 12 \end{vmatrix} = (27)(12) - (0)(-10) = 324$

13. $\begin{vmatrix} 0.75 & -1.32 \\ 0.15 & 1.18 \end{vmatrix} = 0.75(1.18) - (0.15)(-1.32)$

                       $= 0.885 + 0.198 = 1.083$

17. $\begin{vmatrix} 2 & a-1 \\ a+2 & a \end{vmatrix} = 2a - (a^2 + a - 2) = -a^2 + a + 2$

21. $2x - 3y = 4$

     $2x + y = -4$

$$x = \frac{\begin{vmatrix} 4 & -3 \\ -4 & 1 \end{vmatrix}}{\begin{vmatrix} 2 & -3 \\ 2 & 1 \end{vmatrix}} = \frac{4(1) - 4(-3)}{2(1) - 2(-3)} = \frac{4 - 12}{2 + 6} = \frac{-8}{8} = -1$$

$$y = \dfrac{\begin{vmatrix} 2 & 4 \\ 2 & -4 \end{vmatrix}}{\begin{vmatrix} 2 & -3 \\ 2 & 1 \end{vmatrix}} = \dfrac{2(-4) - 2(4)}{2(1) - 2(-3)} = \dfrac{-8 - 8}{2 + 6}$$

$$= \dfrac{-16}{8} = -2$$

25. $v + 2t = 7$

$2v + 4t = 9$

$$v = \dfrac{\begin{vmatrix} 7 & 2 \\ 9 & 4 \end{vmatrix}}{\begin{vmatrix} 1 & 2 \\ 2 & 4 \end{vmatrix}} = \dfrac{7(4) - 9(2)}{1(4) - 2(2)} = \dfrac{28 - 18}{4 - 4} = \dfrac{10}{0}$$

inconsistent

29. $x = \dfrac{\begin{vmatrix} 0.4 & -0.7 \\ 0.7 & 0.5 \end{vmatrix}}{\begin{vmatrix} 0.3 & -0.7 \\ 0.2 & 0.5 \end{vmatrix}} = \dfrac{0.69}{0.29} = 2$, one significant digit

$y = \dfrac{\begin{vmatrix} 0.3 & 0.4 \\ 0.2 & 0.7 \end{vmatrix}}{0.29} = \dfrac{0.13}{0.29} = 0.4$, one significant digit

33. $301x - 529y = 1520$

$385x - 741 = 2540$

$$x = \dfrac{\begin{vmatrix} 1520 & -529 \\ 2540 & -741 \end{vmatrix}}{\begin{vmatrix} 301 & -529 \\ 385 & -741 \end{vmatrix}} = -11.2$$

$$y = \dfrac{\begin{vmatrix} 301 & 1520 \\ 385 & 2540 \end{vmatrix}}{\begin{vmatrix} 301 & -529 \\ 385 & -741 \end{vmatrix}} = -9.26$$

37. For $c = d = 0$, $\begin{vmatrix} a & b \\ c & d \end{vmatrix} = \begin{vmatrix} a & b \\ 0 & 0 \end{vmatrix} = a(0) - 0(b) = 0$

41. Rewrite the system with both equations in standard form.

$F_1 + F_2 = 21$

$2F_1 - 5F_2 = 0$

$$F_1 = \dfrac{\begin{vmatrix} 21 & 1 \\ 0 & -5 \end{vmatrix}}{\begin{vmatrix} 1 & 1 \\ 2 & -5 \end{vmatrix}} = 15 \text{ lb}$$

$$F_2 = \dfrac{\begin{vmatrix} 1 & 21 \\ 2 & 0 \end{vmatrix}}{\begin{vmatrix} 1 & 1 \\ 2 & -5 \end{vmatrix}} = 6 \text{ lb}$$

45. x = number of three bedroom homes

y = number of four bedroom hoimes

$25{,}000x + 35{,}000y = 6{,}800{,}000$

$2000x + 3000\ \ y = \ \ 560{,}000$

from which

$y = 80$ four bedroom homes

$x = 160$ three bedroom homes

49. $p = 2(1.62w) + 2w = 4.20$

$w = 0.801526716$

$1.62w = l = 1.298473282$

dimensions: $l = 1.30$ m, $w = 0.802$ m

53. Convert 24 minutes to hours.

24 min = 0.4 h

$t_2 = t_1 - 0.4 \Leftrightarrow t_1 - t_2 = 0.4$

$42t_1 = 50t_2 \Leftrightarrow 42t_1 - 50t_2 = 0$

$$t_1 = \dfrac{\begin{vmatrix} 0.4 & -1 \\ 0 & -50 \end{vmatrix}}{\begin{vmatrix} 1 & -1 \\ 42 & -50 \end{vmatrix}} = 2.5 \text{ h}$$

$$t_2 = \dfrac{\begin{vmatrix} 1 & 0.4 \\ 42 & 0 \end{vmatrix}}{\begin{vmatrix} 1 & -1 \\ 42 & -50 \end{vmatrix}} = 2.1 \text{ h}$$

5.6 Solving Systems of Three Linear Equations in Three Unknowns Algebraically

1. (1) $4x + y + 3z = 1$
    (2) $2x - 2y + 6z = 12$
    (3) $\underline{-6x + 3y + 12z = -14}$
    (4) $\underline{8x + 2y + 6z = 2}$ (1) multiplied by 2
    (2) $\underline{2x - 2y + 6z = 12}$ add
    (5) $10x\qquad + 12z = 14$
    (6) $12x + 3y + 9z = 3$ (1) multiplied by 3
    (3) $\underline{-6x + 3y + 12z = -14}$ subtract
    (7) $18x\qquad - 3z = 17$
    (8) $72x\qquad - 12z = 68$
    (5) $\underline{10x\qquad + 12z = 14}$ add
    (9) $82x\qquad = 82$
    (10) $\qquad x = 1$
    (11) $18(1) - 3z = 17$ substituting $x = 1$ into (7)
    (12) $-3z = -1$
    (13) $z = \dfrac{1}{3}$
    (14) $4(1) + y + 3\left(\dfrac{1}{3}\right) = 1$ substitute $x = 1$ and $z = \dfrac{1}{3}$ into (1)
    (15) $4 + y + 1 = 1$
    (16) $y = -4$

Thus, the solution is $x = 1$, $y = -4$, $z = \dfrac{1}{3}$

5. (1) $2x + 3y + z = 2$
    (2) $-x + 2y + 3z = -1$
    (3) $-3x - 3y + z = 0$
    (4) $-2x + 4y + 6z = -2$ multiply (2) by 2
    (5) $7y + 7z = 0$ add (1) and (4)
    (6) $y = -z$
    (7) $-x + 2(-z) + 3z = -1$ substitute (5) in (2)
    (8) $-x - z = -1$
    (9) $-3x - 3(-z) + z = 0$ substitute (5) in (3)
    (10) $-3x + 4z = 0$
    $\qquad 3x = 4z$

5. (1) $2x + 3y + x = 2$

    (2) $-x + 2y + 3z = -1$

    (3) $-3x - 3y + z = 0$

    (4) $-2x + 4y + 6z = -2$ multiply (2) by 2

    (5) $7y + 7z = 0$ add (1) and (4)

    (6) $y = -z$

    (7) $-x + 2(-z) + 3z = -1$ substitute (5) in (2)

    (8) $-x - z = -1$

    (9) $-3x - 3(-z) + z = 0$ substitute (5) in (3)

    (10) $-3x + 4z = 0$

$$3x = 4z$$

9. (1) $2x - 2y + 3z = 5$

    (2) $2x + y - 2z = -1$

    (3) $4x - y - 3z = 0$

    (4) $-3y + 5z = 6$ subtract (1) and (2)

    (5) $4x + 2y - 4z = -2$ multiply (2) by 2

    (6) $-3y + z = 2$ subtract (5) and (3)

$$4z = 4$$
$$z = 1$$

    (10) $2x - 2\left(-\dfrac{1}{3}\right) + 3(1) = 5$ substitute (9), (6) into (1)

$$2x + \frac{2}{3} + 3 = 5$$

$$2x = \frac{15}{3} - \frac{11}{3}$$

$$2x = \frac{4}{3}$$

$$x = \frac{2}{3}$$

(7) $2x - 2y + 3(1) = 5$ substitute (6) in (1)

$$2x - 2y = 2$$

(8) $2x + y - 2(1) = -1$ substitute (6) in (2)

$$2x + y = 1$$

(9) $-3y = 1$ subtract (7) and (8)

$$y = -\frac{1}{3}$$

The solution is $x = 2/3, y = -1/3, z = 1$.

13. (1) $10x + 15y - 25z = 35$

    (2) $40x - 30y - 20z = 10$

    (3) $16x - 2y + 8z = 6$

    (4) $20x + 30y - 50z = 70$ multiply (1) by 2

    (2) $\underline{40x - 30y - 20z = 10}$ add

    (5) $60x - 70z = 80$

    (2) $40x - 30y - 20z = 10$

    (6) $\underline{-240x + 30y - 120z = -90}$ multiply (3) by -15, add

(7) $\quad -200x \qquad -140z = -80$

(8) $\quad \underline{-120x \qquad +140z = -160}$ multiply (5) by -2, add

$\qquad -320x \qquad\qquad = -240$

$$x = \frac{3}{4}$$

(5) $\quad 60\left(\dfrac{3}{4}\right) - 70z = 80 \Rightarrow z = -\dfrac{1}{2}$

(1) $\quad 10\left(\dfrac{3}{4}\right) + 15y - 25\left(-\dfrac{1}{2}\right) = 35 \Rightarrow y = 1$

The solution is $x = \dfrac{3}{4}$, $y = 1$, $z = -\dfrac{1}{2}$.

17. $\qquad\qquad Ax + By + Cz = D$

(1) $\quad 2A + 4B + 4C = 12$

(2) $\quad 3A - 2B + 8C = 12$

(3) $\quad -A + 8B + 6C = 12$

(4) $\quad -2A + 16B + 12C = 24$ multiply (3) by 2

(1) $\quad \underline{2A + 4B + 4C = 12}$ add

(5) $\qquad\quad 20B + 16C = 36 \Rightarrow 5B + 4C = 9$

(6) $\quad -3A + 24B + 18C = 36$

(2) $\quad \underline{3A - 2B + 8C = 12}$ add

(7) $\qquad\qquad 22B + 26C = 48 \Rightarrow 11B + 13C = 24$

(8) $\qquad\qquad 55B + 44C = 99$ multiply (5) by 11

(9) $\qquad \underline{-55B - 65C = -21}$ multiply (7) by -5, add

$\qquad\qquad\qquad -21C = -21$

$\qquad\qquad\qquad\quad C = 1$

(5) $5B + 4(1) = 9 \Rightarrow B = 1$

(1) $2A + 4(1) + 4(1) = 12 \Rightarrow A = 2$

The constants are $A = 2$, $B = 1$, $C = 1$ and the equation is $2x + y + z = 12$.

21. (1) $P + M + I = 1150$, solved for I, (4) $I = 1150 - M - P$

(2) $P = 4I - 100$ with I from (4), (5) $P = 4(1150 - M - P) - 100$

(3) $P = 6M + 50$. Substituting P from (3) into (5) gives

(6) $6M + 50 = 4\big(1150 - M - (6M + 50)\big) - 100 \Rightarrow 6M + 50 = 4(1150 - M - 6M - 50) - 100$

$6M + 50 = 4600 - 29M - 200 - 100$

$34M = 4250 \Rightarrow M = 125$, from (3) $P = 6 \cdot 125 + 50 = 800$.

From (1) $800 + 125 + I = 1150 \Rightarrow I = 225$.

$P = 800$ h, $M = 125$ h, $I = 225$ h is the solution.

25. $(A+B)+(90°-A)+(180°-3B)=180°$

$$-2B=-90°$$

$$B=45°$$

$A+(A+B)=90°$

$2A+45°=90°$

$2A=45°$

$A=22.5°$

$C=A+2B=22.5+2(45°)$

$C=112.5°$

29. $MA + MS + PhD = 420$

$MA = MS + PhD + 100$

$MS = 3PhD \qquad \Rightarrow$

$MA + MS + PhD = 420$

$MA - MS - PhD = 100$

$MS - 3PhD = 0 \qquad \Rightarrow$

$MA = 260, \ MS = 120, \ PhD = 40$

33. $(1)\ x-2y-3z=2 \Rightarrow x=2y+3z+2$ substituted in (2) and (3) gives

$(2)\ x-4y-13z=14 \Rightarrow 2y+3z+2-4y-13z=14 \Rightarrow -2y-10z=12 \Rightarrow$

$(4)\ y+5z=-6$

$(3)\ -3x+5y+4z=0 \Rightarrow -3(2y+3z+2)+5y+4z=0 \Rightarrow y+5z=-6$ which is (4).

Thus, $y=-6-5z$ and from $(1)\ x-2(-6-5z)-3z=2 \Rightarrow x=-7z-10$. The

solution is $x=-7z-10,\ y=-5z-6,\ z=z$. Letting $z=0,\ x=-10,\ y=-6,\ z=0$.

5.7 Solving Systems of Three Linear Equations in Three Unknowns by Determinants

1.
$$\begin{vmatrix} -2 & 3 & -1 \\ 1 & 5 & 4 \\ 2 & -1 & 5 \end{vmatrix} \begin{matrix} -2 & 3 \\ 1 & 5 \\ 2 & -1 \end{matrix}$$

$$= -2(5)(5) + 3(4)(2) + (-1)(1)(-1)$$
$$- 2(5)(-1) - (-1)(4)(-2) - 5(1)(3)| \quad = -38$$

5.
$$\begin{vmatrix} 8 & 9 & -6 \\ -3 & 7 & 2 \\ 4 & -2 & 5 \end{vmatrix} \begin{matrix} 8 & 9 \\ -3 & 7 \\ 4 & -2 \end{matrix}$$

$$= 280 - (-32) + 72 - (-135) + (-36) - (-168)$$
$$= 651$$

9.
$$\begin{vmatrix} 4 & -3 & -11 \\ -9 & 2 & -2 \\ 0 & 1 & -5 \end{vmatrix} \begin{matrix} 4 & -3 \\ -9 & 2 \\ 0 & 1 \end{matrix}$$
$$= -40 - (-8) + 0 - (-135) + 99 - 0 = 202$$

13.
$$\begin{vmatrix} 0.1 & -0.2 & 0 \\ -0.5 & 1 & 0.4 \\ -2 & 0.8 & 2 \end{vmatrix} \begin{matrix} 0.1 & -0.2 \\ -0.5 & 1 \\ -2 & 0.8 \end{matrix}$$
$$= 0.2 - 0.032 + 0.16 - 0.2 + 0 - 0 = 0.128$$

17.
$$x = \frac{\begin{vmatrix} 2 & 1 & 1 \\ 1 & 0 & -1 \\ 1 & 1 & 0 \end{vmatrix} \begin{matrix} 2 & 1 \\ 1 & 0 \\ 1 & 1 \end{matrix}}{\begin{vmatrix} 1 & 1 & 1 \\ 1 & 0 & -1 \\ 1 & 1 & 0 \end{vmatrix} \begin{matrix} 1 & 1 \\ 1 & 0 \\ 1 & 1 \end{matrix}}$$

$$= \frac{0 - (-2) + (-1) - 0 + 1 - 0}{0 - (-1) + (-1) - 0 + 1 - 0} = \frac{2}{1} = 2$$

$$y = \frac{\begin{vmatrix} 1 & 2 & 1 \\ 1 & 1 & -1 \\ 1 & 1 & 0 \end{vmatrix} \begin{matrix} 1 & 2 \\ 1 & 1 \\ 1 & 1 \end{matrix}}{1}$$

$$= \frac{0 - (-1) + (-2) - 0 + 1 - 1}{1} = \frac{-1}{1} = -1$$

$$z = \frac{\begin{vmatrix} 1 & 1 & 2 \\ 1 & 0 & 1 \\ 1 & 1 & 1 \end{vmatrix} \begin{matrix} 1 & 1 \\ 1 & 0 \\ 1 & 1 \end{matrix}}{1}$$

$$= \frac{0 - 1 + 1 - 1 + 2 - 0}{1} = \frac{1}{1} = 1$$

21.
$$l = \frac{\begin{vmatrix} 6 & 6 & -3 \\ -3 & -7 & -2 \\ 1 & 1 & -7 \end{vmatrix} \begin{matrix} 6 & 6 \\ -3 & -7 \\ 1 & 1 \end{matrix}}{\begin{vmatrix} 5 & 6 & -3 \\ 4 & -7 & -2 \\ 3 & 1 & -7 \end{vmatrix} \begin{matrix} 5 & 6 \\ 4 & -7 \\ 3 & 1 \end{matrix}}$$

$$= \frac{294 + 12 - 12 - 126 + 9 - 21}{245 + 10 - 36 + 168 - 12 - 63}$$
$$= \frac{156}{312} = \frac{1}{2}$$

$$w = \frac{\begin{vmatrix} 5 & 6 & -3 \\ 4 & -3 & -2 \\ 3 & 1 & -7 \end{vmatrix} \begin{matrix} 5 & 6 \\ 4 & -3 \\ 3 & 1 \end{matrix}}{312}$$

$$= \frac{105 + 10 - 36 + 168 - 12 - 27}{312}$$
$$= \frac{208}{312} = \frac{2}{3}$$

$$h = \frac{\begin{vmatrix} 5 & 6 & -3 \\ 4 & -3 & -2 \\ 3 & 1 & -7 \end{vmatrix} \begin{matrix} 5 & 6 \\ 4 & -3 \\ 3 & 1 \end{matrix}}{312}$$

$$= \frac{-35 + 15 - 54 - 224 + 24 + 126}{312}$$

$$= \frac{52}{312} = \frac{1}{6}$$

25.

$$x = \frac{\begin{vmatrix} 6 & -7 & 3 \\ 1 & 3 & 6 \\ 5 & -5 & 2 \end{vmatrix} \begin{matrix} 6 & -7 \\ 1 & 3 \\ 5 & -5 \end{matrix}}{\begin{vmatrix} 3 & -7 & 3 \\ 3 & 3 & 6 \\ 5 & -5 & 2 \end{vmatrix} \begin{matrix} 3 & -7 \\ 3 & 3 \\ 5 & -5 \end{matrix}}$$

$$= \frac{36 + 180 - 210 + 14 - 15 - 45}{18 + 90 - 210 + 42 - 45 - 45}$$

$$= \frac{-40}{-150} = \frac{4}{15}$$

$$y = \frac{\begin{vmatrix} 3 & 6 & 3 \\ 3 & 1 & 6 \\ 5 & 5 & 2 \end{vmatrix} \begin{matrix} 3 & 6 \\ 3 & 1 \\ 5 & 5 \end{matrix}}{-150}$$

$$= \frac{6 - 90 + 180 - 36 + 45 - 15}{-150}$$

$$= \frac{90}{-150} = -\frac{3}{5}$$

$$z = \frac{\begin{vmatrix} 3 & -7 & 6 \\ 3 & 3 & 1 \\ 5 & -5 & 5 \end{vmatrix} \begin{matrix} 3 & -7 \\ 3 & 3 \\ 5 & -5 \end{matrix}}{-150}$$

$$= \frac{45 + 15 - 35 + 105 - 90 - 90}{-150}$$

$$= \frac{-50}{-150} = \frac{1}{3}$$

29.

$$x = \frac{\begin{vmatrix} 10.5 & 4.5 & -7.5 \\ 1.2 & -3.6 & -2.4 \\ 1.5 & -0.5 & 2.0 \end{vmatrix} \begin{matrix} 10.5 & 4.5 \\ 1.2 & -3.6 \\ 1.5 & -0.5 \end{matrix}}{\begin{vmatrix} 3.0 & 4.5 & -7.5 \\ 4.8 & -3.6 & -2.4 \\ 4.0 & -0.5 & 2.0 \end{vmatrix} \begin{matrix} 3.0 & 4.5 \\ 4.8 & -3.6 \\ 4.0 & -0.5 \end{matrix}}$$

$$= \frac{-75.6 - 12.6 - 16.2 - 10.8 + 4.5 - 40.5}{-21.6 - 3.6 - 43.2 - 43.2 + 18 - 108}$$

$$= \frac{-151.2}{-201.6}$$

$$= \frac{3}{4}$$

$$y = \frac{\begin{vmatrix} 3.0 & 10.5 & -7.5 \\ 4.8 & 1.2 & -2.4 \\ 4.0 & 1.5 & 2.0 \end{vmatrix} \begin{matrix} 3.0 & 10.5 \\ 4.8 & 1.2 \\ 4.0 & 1.5 \end{matrix}}{-201.6}$$

$$= \frac{7.2 + 10.8 - 100.8 - 100.8 - 54 + 36 - 201.6}{-201.6}$$

$$= 1$$

$$z = \frac{\begin{vmatrix} 3.0 & 4.5 & 10.5 \\ 4.8 & -3.6 & 1.2 \\ 4.0 & -0.5 & 1.5 \end{vmatrix} \begin{matrix} 3.0 & 4.5 \\ 4.8 & -3.6 \\ 4.0 & -0.5 \end{matrix}}{-201.6}$$

$$= \frac{-16.2 + 1.8 + 21.6 - 32.4 - 25.2 + 151.2}{-201.6}$$

$$= \frac{100.8}{-201.6} = -\frac{1}{2}$$

33. $\begin{vmatrix} 4 & 2 & 1 \\ 7 & 8 & 6 \\ 7 & 9 & 8 \end{vmatrix} = 19$, using calculator

the value does not change.

37. $s_o + 2v_o + 2a = 20$
$s_o + 4v_o + 8a = 54$
$s_o + 6v_o + 18a = 104$

$$s_o = \frac{\begin{vmatrix} 20 & 2 & 2 \\ 54 & 4 & 8 \\ 104 & 6 & 18 \end{vmatrix}}{\begin{vmatrix} 1 & 2 & 2 \\ 1 & 4 & 8 \\ 1 & 6 & 18 \end{vmatrix}} = \frac{16}{8} = 2 \text{ ft}$$

$$v_o = \frac{\begin{vmatrix} 1 & 20 & 2 \\ 1 & 54 & 8 \\ 1 & 104 & 18 \end{vmatrix}}{8} = \frac{40}{8} = 5 \text{ ft/s}$$

$$a = \frac{\begin{vmatrix} 1 & 2 & 20 \\ 1 & 4 & 54 \\ 1 & 6 & 104 \end{vmatrix}}{8} = \frac{32}{8} = 4 \text{ ft/s}^2$$

41. $V = f(T) = a + bT + cT^2$

$a + b(2.0) + c(2.0)^2 = 6.4$

$a + b(4.0) + v(4.0)^2 = 8.6$

$a + b(6.0) + c(6.0)^2 = 11.6$

$$a = \frac{\begin{vmatrix} 6.4 & 2.0 & 2.0^2 \\ 8.6 & 4.0 & 4.0^2 \\ 11.6 & 6.0 & 6.0^2 \end{vmatrix}}{\begin{vmatrix} 1 & 2.0 & 2.0^2 \\ 1 & 4.0 & 4.0^2 \\ 1 & 6.0 & 6.0^2 \end{vmatrix}} = 5.0 \text{ using calculator}$$

$$b = \frac{\begin{vmatrix} 1 & 6.4 & 2.0^2 \\ 1 & 8.6 & 4.0^2 \\ 1 & 11.6 & 6.0^2 \end{vmatrix}}{\begin{vmatrix} 1 & 2.0 & 2.0^2 \\ 1 & 4.0 & 4.0^2 \\ 1 & 6.0 & 6.0^2 \end{vmatrix}} = 0.50 \text{ using calculator}$$

$$c = \frac{\begin{vmatrix} 1 & 2.0 & 6.4 \\ 1 & 4.0 & 8.6 \\ 1 & 6.0 & 11.6 \end{vmatrix}}{\begin{vmatrix} 1 & 2.0 & 2.0^2 \\ 1 & 4.0 & 4.0^2 \\ 1 & 6.0 & 6.0^2 \end{vmatrix}} = 0.10 \text{ using calculator}$$

$V = f(T) = 5.0 + 0.50T + 0.10T^2$

45. $v_j = 12v_c; \ 12v_c - v_j = 0$

$v_c = v_t + 15; \ v_c - v_t = 15$

$1.10v_c + 1.95v_j + 0.52v_t = 1140$

$$v_c = \frac{\begin{vmatrix} 0 & -1 & 0 \\ 15 & 0 & -1 \\ 1140 & 1.95 & 0.52 \end{vmatrix} \begin{matrix} 0 & -1 \\ 15 & 0 \\ 1140 & 1.95 \end{matrix}}{\begin{vmatrix} 12 & -1 & 0 \\ 1 & 0 & -1 \\ 1.10 & 1.95 & 0.52 \end{vmatrix} \begin{matrix} 12 & -1 \\ 1 & 0 \\ 1.10 & 1.95 \end{matrix}}$$

$$= \frac{0 - 0 + 1140 + 7.8 + 0 - 0}{0 + 23.4 + 1.10 + 0.52 + 0 - 0}$$

$$= \frac{1147.8}{25.02} = 45.9$$

$$v_j = \frac{\begin{vmatrix} 12 & 0 & 0 \\ 1 & 15 & -1 \\ 1.10 & 1140 & 0.52 \end{vmatrix} \begin{matrix} 12 & 0 \\ 1 & 15 \\ 1.10 & 1140 \end{matrix}}{25.02}$$

$$= \frac{93.6 + 13\,680 + 0 - 0 + 0 - 0}{25.02}$$

$$= \frac{13\,773.6}{25.02} = 551$$

$$v_t = \frac{\begin{vmatrix} 12 & -1 & 0 \\ 1 & 0 & 15 \\ 1.10 & 1.95 & 1140 \end{vmatrix} \begin{matrix} 12 & -1 \\ 1 & 0 \\ 1.10 & 1.95 \end{matrix}}{25.02}$$

$$= \frac{0 - 351 - 16.5 + 1140 + 0 - 0}{25.02}$$

$$= \frac{772.5}{25.02} = 30.9$$

The average speeds for the trip were 30.9 mi/h, 45.9 mi/h, 551 mi/h.

Chapter 5   Review Exercises

1. $\begin{vmatrix} -2 & 5 \\ 3 & 1 \end{vmatrix} = (-2)(1) - (3)(5) = -2 - 15 = -17$

5. $m = \dfrac{y_2 - y_1}{x_2 - x_1} = \dfrac{-8 - 0}{4 - 2} = \dfrac{-8}{2} = -4$

9. Comparing $y = -2x + 4$ to $y = mx + b$
gives a slope of -2 and y-intercept of 4.

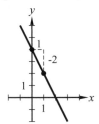

13.

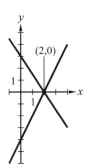

17. $7x = 2y + 14 \Rightarrow y = \dfrac{7x - 14}{2}$

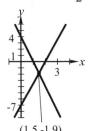

21. (1) $x + 2y = 5 \Rightarrow x = 5 - 2y$ which substitutes
into (2) $x + 3y = 7$ to give $5 - 2y + 3y = 7 \Rightarrow$
$y = 2$. From (1) $x = 5 - 2 \cdot 2 = 1$. The solution
is (1, 2).

25. $i = \dfrac{\begin{vmatrix} 29 & -27 \\ 69 & 33 \end{vmatrix}}{\begin{vmatrix} 10 & -27 \\ 40 & 33 \end{vmatrix}} = \dfrac{2820}{1410} = 2$

$v = \dfrac{\begin{vmatrix} 10 & 29 \\ 40 & 69 \end{vmatrix}}{1410} = \dfrac{-470}{1410} = -\dfrac{1}{3}$

The solution is $i = 2$, $v = -\dfrac{1}{3}$.

29. (1) $90x - 110y = 40 \Rightarrow x = \dfrac{11y + 4}{9}$ which
substitutes into (2) $30x - 15y = 25$ to give
$30\dfrac{11y + 4}{9} - 15y = 25 \Rightarrow y = \dfrac{7}{13}$. From (1)
$x = \dfrac{11 \cdot \frac{7}{13} + 4}{9} = \dfrac{43}{39}$. The solution is $\left(\dfrac{43}{39}, \dfrac{7}{13}\right)$.

33. $4x + 3y = -4 \Rightarrow 4x + 3y = -4$
$y = 2x - 3 \qquad\qquad 2x - y = 3$

$x = \dfrac{\begin{vmatrix} -4 & 3 \\ 3 & -1 \end{vmatrix}}{\begin{vmatrix} 4 & 3 \\ 2 & -1 \end{vmatrix}} = \dfrac{-5}{-10} = \dfrac{1}{2}$

$y = \dfrac{\begin{vmatrix} 4 & -4 \\ 2 & 3 \end{vmatrix}}{-10} = \dfrac{20}{-10} = -2$

37. $7x = 2y - 6 \Rightarrow 7x - 2y = -6$
$7y = 12 - 4x \Rightarrow 4x + 7y = 12$

$x = \dfrac{\begin{vmatrix} -6 & -2 \\ 12 & 7 \end{vmatrix}}{\begin{vmatrix} 7 & -2 \\ 4 & 7 \end{vmatrix}} = \dfrac{-18}{57} = -\dfrac{6}{19}$

$y = \dfrac{\begin{vmatrix} 7 & -6 \\ 4 & 12 \end{vmatrix}}{57} = \dfrac{108}{57} = \dfrac{36}{19}$

41. Exercise 33 is most easily solved by substitution
because the second equation is already solved
for y.

45. $\begin{vmatrix} 4 & -1 & 8 \\ -1 & 6 & -2 \\ 2 & 1 & -1 \end{vmatrix}\begin{matrix} 4 & -1 \\ -1 & 6 \\ 2 & 1 \end{matrix}$

$= 4(6)(-1) + (-1)(-2)(2) + 8(-1)(1) - 2(6)(8)$
$- 1(-2)(4) - (-1)(-1)(-1) = -115$

49. (1) $2x + y + z = 4$

(2) $\underline{x - 2y - z = 3} \quad$ add

(4) $3x - y \quad\quad = 7$

(1) $4x + 2y + 2z = 8$

(3) $\underline{3x + 3y - 2z = 1} \quad$ add

(5) $7x + 5y \quad\quad = 9$

(4) $\underline{15x - 5y = 35} \quad$ add

$\quad 22x \quad\quad = 44$

$\quad\quad x = 2$

(4) $3x - y = 7 \Rightarrow 3(2) - y = 7 \Rightarrow y = -1$

(1) $2x + y + z = 4 \Rightarrow 2(2) - 1 + z = 4 \Rightarrow z = 1$

53. Multiply both sides of all three equations by 10 to clear decimals.

(1) $36x + 52y - 10z = -22$ solve for z:

(4) $z = \dfrac{36x + 52y + 22}{10}$

(2) $32x - 48y + 39z = 81$

(3) $64x + 41y + 23z = 51$

(5) $32x - 48y + 39 \cdot \dfrac{36x + 52y + 22}{10} = 81$

(2) with z from (4) which simplifies to

(5) $1724x + 1548y = -48 \Rightarrow y = \dfrac{-48 - 1724x}{1548}$

(6) $64x + 41y + 23 \cdot \dfrac{36x + 52y + 22}{10} = 51$

(3) with z from (4) which simplifies to

(6) $1468x + 1606 \cdot \dfrac{-48 - 1724x}{1548} = 4 \Rightarrow$

$x = -0.1678084952.$ From (5) $y = \dfrac{-48 - 1724x}{1548}$

$= 0.1558797453.$ From (4)

$z = \dfrac{36x + 52y + 22}{10} = 2.406464093.$ The solution is

$(-0.17, 0.16, 2.4).$

57. $r = \dfrac{\begin{vmatrix} 8 & 1 & 2 \\ 5 & -2 & -4 \\ -3 & 3 & 4 \end{vmatrix}}{\begin{vmatrix} 2 & 1 & 2 \\ 3 & -2 & -4 \\ -2 & 3 & 4 \end{vmatrix}} = \dfrac{42}{14} = 3$

$s = \dfrac{\begin{vmatrix} 2 & 8 & 2 \\ 3 & 5 & -4 \\ -2 & -3 & 4 \end{vmatrix}}{14} = \dfrac{-14}{14} = -1$

$t = \dfrac{\begin{vmatrix} 2 & 1 & 8 \\ 3 & -2 & 5 \\ -2 & 3 & -3 \end{vmatrix}}{14} = \dfrac{21}{14} = \dfrac{3}{2}$

61. $\begin{vmatrix} 2 & 5 \\ 1 & x \end{vmatrix} = 3 \Rightarrow 2x - 5 = 3 \Rightarrow 2x = 8 \Rightarrow x = 4$

65. (1) $\dfrac{1}{x} - \dfrac{1}{y} = \dfrac{1}{2} \Rightarrow u - v = \dfrac{1}{2}$

(2) $\dfrac{1}{x} + \dfrac{1}{y} = \dfrac{1}{4} \Rightarrow u + v = \dfrac{1}{4}$

Adding (1), (2) $2u = \dfrac{3}{4} \Rightarrow u = \dfrac{3}{8} \Rightarrow x = \dfrac{8}{3}$, from (2)

$v = \dfrac{1}{4} - \dfrac{3}{8} = -\dfrac{1}{8}$, $y = -8$. $y = -8$. The solution is

$\left(\dfrac{8}{3}, -8 \right)$.

69. (1) $3x - ky = 6$

(2) $x + 2y = 2$. Multiplying (2) by 3 gives
$3x + 6y = 6$ which is (1) with $k = -6$. A k-value
of -6 makes the system dependent.

73. $F_1 = \dfrac{\begin{vmatrix} 280 & 2.0 & 0 \\ 0 & 0 & -1 \\ 600 & -4.0 & 0 \end{vmatrix}}{\begin{vmatrix} 1 & 2.0 & 0 \\ 0.87 & 0 & -1 \\ 3.0 & -4.0 & 0 \end{vmatrix}} = \dfrac{-2320}{-10} = 232 = 230 \text{ lb}$

$$F_2 = \dfrac{\begin{vmatrix} 1 & 280 & 0 \\ 0.87 & 0 & -1 \\ 3.0 & 600 & 0 \end{vmatrix}}{-10} = \dfrac{-240}{-10} = 24 \text{ lb}$$

$$F_3 = \dfrac{\begin{vmatrix} 1 & 2.0 & 280 \\ 0.87 & 0 & 0 \\ 3.0 & -4 & 600 \end{vmatrix}}{-10} = \dfrac{-2018.4}{-10} = 201.84 = 200 \text{ lb}$$

77. $I = 0.105$

$I = 2400 + 0.045$

$0.105 = 2400 + 0.045 \Rightarrow S = 40,000$

$I = 0.01(40,000) = 4000$

Both plans produce an income of \$4000 for sales of \$40,000.

81. $T = \dfrac{a}{x+100} + b$

$\dfrac{a}{0+100} + b = 14$

$\dfrac{a}{900+100} + b = 10$

$a = \dfrac{\begin{vmatrix} 14 & 1 \\ 10 & 1 \end{vmatrix}}{\begin{vmatrix} \frac{1}{100} & 1 \\ \frac{1}{1000} & 1 \end{vmatrix}} = 440 \text{ m} \cdot {}^\circ\text{C} \text{ using calculator}$

$b = \dfrac{\begin{vmatrix} \frac{1}{100} & 14 \\ \frac{1}{1000} & 10 \end{vmatrix}}{\begin{vmatrix} \frac{1}{100} & 1 \\ \frac{1}{1000} & 1 \end{vmatrix}} = 9.6\,^\circ\text{C} \text{ using calculator}$

85. $800x + 1100y = 49,200$

$900x + 1250y = 55,600 \Rightarrow$

$x = 34, \ y = 20$

89. $I^2 R = P$

$1.0^2 R_1 + 3.0^2 R_2 = 14.0$

$3.0^2 R_1 + 1.0^2 R_2 = 6.0$

$$R_1 = \dfrac{\begin{vmatrix} 14.0 & 3.0^2 \\ 6.0 & 1.0^2 \end{vmatrix}}{\begin{vmatrix} 1.0^2 & 3.0^2 \\ 3.0^2 & 1.0^2 \end{vmatrix}} = 0.50 \ \Omega \text{ using calculator}$$

$$R_2 = \dfrac{\begin{vmatrix} 1.0^2 & 14.0 \\ 3.0^2 & 6.0 \end{vmatrix}}{\begin{vmatrix} 1.0^2 & 3.0^2 \\ 3.0^2 & 1.0^2 \end{vmatrix}} = 1.5 \ \Omega \text{ using calculator}$$

93. (1) $A + B + C = 180$

(2) $A = 2B - 55 \Rightarrow 2B = A + 55$

   substitute into (1)

(3) $C = B - 25$

(4) $A + B + B - 25 = 180 \Rightarrow$

   $\underline{2B = -A + 205}$ add

   $4B = 260 \Rightarrow B = 65^\circ$

(3) $C = 65 - 25 \Rightarrow C = 40^\circ$

(1) $A + 65^\circ + 40^\circ = 180 \Rightarrow A = 75^\circ$

97. x = weight of gold in air

y = weight of silver in air

$x + y = 6.0 \Rightarrow y = 6.0 - x$

$0.947x + 0.9y = 5.6 \Rightarrow 0.947x + 0.9(6.0 - x) = 5.6$

$x = 4.3 \text{ N}$

$y = 6.0 - 4.3 = 1.7 \text{ N}$

Chapter 6

FACTORING AND FRACTIONS

6.1 Special Products

1. $(3r-2s)(3r+2s)=(3r)^2-(2s)^2$
$$=9r^2-4s^2$$

5. $40(x-y)=40x-40y$

9. $(T+6)(T-6)=T^2-6^2=T^2-36$

13. $(4x-5y)(4x+5y)=(4x)^2-(5y)^2$
$$=16x^2-25y^2$$

17. $(5f+4)^2=(5f)^2+2(5f)(4)+4^2$
$$=25f^2+40f+16$$

21. $(L^2-1)^2=(L^2)^2-2\cdot L^2\cdot 1+1^2$
$$=L^4-2L^2+1$$

25. $(0.6s-t)^2=(0.6s)^2-2(0.6s)(t)+t^2$
$$=0.36s^2-1.2st+t^2$$

29. $(3+C^2)(6+C^2)=18+(3C^2+6C^2)+(C^2)^2$
$$=18+9C^2+C^4$$

33. $(10v-3)(4v+15)=40v^2+138v-45$

37. $2(x-2)(x+2)=2(x^2-4)=2x^2-8$

41. $6a(x+2b)^2=6a(x^2+4bx+4b^2)$
$$=6ax^2+24abx+24ab^2$$

45. $\left[(2R+3r)(2R-3r)\right]^2=\left[4R^2-9r^2\right]^2$
$$=16R^4-72R^2r^2+81r^4$$

49. $\left[3-(x+y)^2\right]=9-6(x+y)+(x+y)^2$
$$=9-6x-6y+x^2+2xy+y^2$$

53. $(3L+7R)^3$
$$=(3L)^3+3(3L)^2(7R)+3(3L)(7R)^2+(7R)^3$$
$$=27L^3+189L^2R+441LR^2+343R^3$$

57. $(x+2)(x^2-2x+4)=x^3-2x^2+4x+2x^2-4x+8$
$$=x^3+8$$

61. $(x+y)^2(x-y)^2=(x^2+2x+y^2)(x^2-2xy+y^2)$
$$=x^4-2x^3y+x^2y^2+2x^3y-4x^2y^2$$
$$+2xy^3+x^2y^2-2xy^3+y^4$$
$$=x^4-2x^2y^2+y^4$$

65. $4(p+DA)^2=4(p^2+2pDA+D^2A^2)$
$$=4p^2+8pDA+4D^2A^2$$

69. $\dfrac{L}{6}(x-a)^3=\dfrac{L}{6}\left[x^3-3x^2a+3xa-a^3\right]$
$$=\dfrac{L}{6}x^3-\dfrac{L}{2}ax^2+\dfrac{L}{2}a^2x-\dfrac{L}{6}a^3$$

73. $(49)(51)=(50-1)(50+1)=50^2-1^2$
$$=2500-1$$
$$=2499$$

77. $(2x-1)^3(1-2x)=-(2x-1)^3(2x-1)$
$$=-(2x-1)^4$$
$$=-(1-2x)^4\neq(1-2x)^4$$

81. (a) $A=(x+y)^2=x^2+2xy+y^2$
(b) $A=x^2+xy+xy+y^2=x^2+2xy+y^2$

6.2   Factoring: Common Factor and Difference of Squares

1. $4ax^2 - 2ax = 2ax(2x-1)$

5. $6x + 6y = 6(x+y)$
(6 is a common monomial factor, c.m.f.)

9. $3x^2 - 9x = 3x(x-3)$ (3x is a c.m.f.)

13. $288n^2 + 24n = 24n(12n+1)$ (24n is a c.m.f.)

17. $3ab^2 - 6ab + 12ab^3 = 3ab(b-2+4b^2)$
(3ab is a c.m.f.)

21. $2a^2 - 2b^2 + 4c^2 - 6d^2 = 2(a^2 - b^2 + 2c^2 - 3d^2)$
(2 is a c.m.f.)

25. $100 - 9A^2 = (10 - 3A)(10 + 3A)$
(because $-30A + 30A = 0A = 0$)

29. $162s^2 - 50t^2 = 2(81s^2 - 25t^2)$
$\qquad = 2(9s - 5t)(9s + 5t)$
(because $-45st + 45st = 0st = 0$)

33. $(x+y)^2 - 9 = (x+y-3)(x+y+3)$

37. $300x^2 - 2700z^2 = 300(x^2 - 9z^2)$
$\qquad = 300(x-3z)(x+3z)$

41. $x^4 - 16 = (x^2 - 4)(x^2 + 4)$
$\qquad = (x-2)(x+2)(x^2+4)$

45. Solve $2a - b = ab + 3$ for a.
$2a - ab = b + 3$
$a(-b) = b + 3$
$a = \dfrac{b+3}{2-b}$

49. $(x + 2k)(x - 2) = x^2 3x - 4k$
$x^2 - 2x + 2kx - 4k = x^2 + 3x - 4k$
$2kx = 5x$
$k = \dfrac{5}{2}$

53. $a^2 + ax - ab - bx = (a^2 + ax) - (ab + bx)$
$\qquad = a(a+x) - b(a+x)$
$\qquad = (a+x)(a-b)$
$\qquad = (a-b)(a+x)$

57. $x^2 - y^2 + x - y = (x^2 - y^2) + (x - y)$
$\qquad = (x+y)(x-y) + (x-y)$
$\qquad = (x+y+1)(x-y)$
$\qquad = (x-y)(x+y+1)$

61. $n^2 + n = n(n+1)$, the product of two consecutive integers of which one must be even. Therefore, the product is even.

65. $81s - s^3 = s(81 - s^2) = s(9-s)(9+s)$

69.

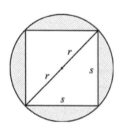

$s^2 + s^2 = (2r)^2$
$2s^2 = 4r^2$
$s^2 = 2r^2$ = area of square
Area left = Area of circle − Area of square
$= \pi r^2 - 2r^2$
$= r^2(\pi - 2)$

73.
$$i_1 R_1 = (i_2 - i_1) R_2 = i_2 R_2 - i_1 R_2$$
$$i_1 R_1 + i_1 R_2 = i_2 R_2$$
$$i_1 (R_1 + R_2) = i_2 R_2$$
$$i_1 = \frac{i_2 R_2}{R_1 + R_2}$$

77.
$$ER = A t T_0 - A a t T_1$$
$$= t A (T_0 - T_1)$$
$$t = \frac{ER}{A(T_0 - T_1)}$$

6.3   Factoring Trinomials

1. $x^2 + 4x + 3 = (x + 3)(x + 1)$

5. $2x^2 + 6x - 36 = 2(x^2 + 3x - 18) = 2(x + 6)(x - 3)$

9. $s^2 - s - 42 = (s - 7)(s + 6)$
(because $-s = -7s + 6s$)

13. $x^2 + 2x + 1 = (x + 1)(x + 1) = (x + 1)^2$
(because $2x = x + x$)

17. $3x^2 - 5x - 2 = (3x + 1)(x - 2)$
(because $-5x = -6x + x$)

21. $2s^2 + 13s + 11 = (2s + 11)(s + 1)$
(because $13s = 2s + 11s$)

25. $2t^2 + 7t - 15 = (2t - 3)(t + 5)$
(because $7t = 10t - 3t$)

29. $4x^2 - 3x - 7 = (4x - 7)(x + 1)$
(because $-3x = 4x - 7x$)

33. $12m^2 + 60m + 75 = 3(2m + 5)(2m + 5) = 3(2m + 5)^2$
(because $20m = 10m + 10m$)

37. $9t^2 - 15t + 4 = (3t - 4)(3t - 1)$
(because $-15t = -3t - 12t$)

41. $4p^2 - 25pq + 6q^2 = (4p - q)(p - 6q)$
(because $-25pq = -24pq - pq$)

45. $2x^2 - 14x + 12 = 2(x^2 - 7x + 6) = 2(x - 1)(x - 6)$
(because $-7x = -6x - x$)

49. $ax^3 + 4a^2x^2 - 12a^3x = ax(x^2 + 4ax - 12a^2)$
$$= ax(x + 6a)(x - 2a)$$

53. $25a^2 - 25x^2 - 10xy - y^2$
$$= 25a^2 - (25x^2 + 10xy + y^2)$$
$$= 25a^2 - (5x + y)^2$$
$$= (5a + 5x + y)(5a - 5x - y)$$

57. $16t^2 - 80t + 64 = 16t^2(t^2 - 5t + 4)$
$$= 16(t - 4)(t - 1)$$

61. $3Q^2 + Q - 30 = (3Q + 10)(Q - 3)$

65. $wx^4 - 5wLx^3 + 6wL^2x^2 = wx^2(x^2 - 5Lx + 6L^2)$
$$= wx^2(x - 3L)(x - 2L)$$

69. Write $4x^2 + 4x - k$ as
$(2x)^2 + 2(2x)(1) + (-k)$ and comparing to
$(2x)^2 + 2(2x)(1) + 1^2$, which is a perfect square,
gives $-k = 1^2 \Rightarrow k = -1$ from which
$4x^2 + 4x - k$ becomes $4x^2 + 4x + 1$ which factors
as $4x^2 + 4x + 1 = (2x + 1)^2$

73. $x^4 + 4 = x^4 + 4 - 4x^2 + 4x^2$

$\qquad = x^4 + 4x^2 + 4 - 4x^2$

$\qquad = \left(x^2 + 2\right)^2 - 4x^2$

$\qquad = \left(x + 2\right)^2 - \left(2x\right)^2$

$\qquad = \left(x^2 + 2 - 2x\right)\left(x^2 + 2 + 2x\right)$

$\qquad = \left(x^2 - 2x + 2\right)\left(x^2 + 2x + 2\right)$

37. $x^6 - y^6 = \left(x^2\right)^2 - \left(y^3\right)^2$

$\qquad = \left(x^2 - y^2\right)\left(x^4 + x^2 y^2 + y^4\right)$

$\qquad = (x + y)(x - y)\left(x^4 + x^2 y^2 + y^4\right)$

6.5 Equivalent Fractions

1. $\dfrac{18abc^6}{24ab^2 c^5} = \dfrac{6abc^5 \left(3c\right)}{6abc^5 \left(4b\right)} = \dfrac{3c}{4b}$

6.4 The Sum and Difference of Cubes

1. $x^3 - 8 = x^3 - 2^3 = \left(x - 2\right)\left(x^2 + 2x + 2^2\right)$

$\qquad = \left(x - 2\right)\left(x^2 + 2x + 4\right)$

5. $8 - t^3 = 2^3 - t^3 = \left(2 - t\right)\left(4 + 2t + t^2\right)$

9. $4x^3 + 32 = 4\left(x^3 + 8\right) = 4\left(x + 2\right)\left(x^2 - 2x + 4\right)$

13. $54x^3 y - 6x^3 y^4 = 6x^3 y \left(9 - y^3\right)$

17. $3a^6 - 3a^2 = 3a^2 \left(a^4 - 1\right)$

$\qquad = 3a^2 \left(a^2 - 1\right)\left(a^2 + 1\right)$

$\qquad = 3a^2 \left(a + 1\right)\left(a - 1\right)\left(a^2 + 1\right)$

21. $27L^6 + 216L^3 = 27L^3 \left(L^3 + 8\right)$

$\qquad = 27L^3 \left(L + 2\right)\left(L^2 - 2L + 4\right)$

25. $64 - x^6 = 4^3 - \left(x^2\right)^3 = \left(4 - x^2\right)\left(16 + 4x^2 + x^4\right)$

$\qquad = (2 + x)(2 - x)\left(16 + 4x^2 + x^4\right)$

29. $32x - 4x^4 = 4x\left(8 - x^3\right) = 4x\left(2^3 - x^3\right)$

$\qquad = 4x\left(2 - x\right)\left(4 + 2x + x^2\right)$

33. $QH^4 + Q^4 H = QH\left(H^3 + Q^3\right)$

$\qquad = QH\left(H + Q\right)\left(H^2 - HQ + Q^2\right)$

5. $\dfrac{2}{3} \cdot \dfrac{7}{7} = \dfrac{14}{21}$

9. $\dfrac{2}{\left(x + 3\right)} \cdot \dfrac{\left(x - 2\right)}{\left(x - 2\right)} = \dfrac{2\left(x - 2\right)}{\left(x + 3\right)\left(x - 2\right)} = \dfrac{2x - 4}{x^2 + x - 6}$

13. $\dfrac{28}{44} = \dfrac{\frac{28}{4}}{\frac{44}{4}} = \dfrac{7}{11}$

17. $\dfrac{2\left(R - 1\right)}{\left(R - 1\right)\left(R + 1\right)} = \dfrac{\frac{2(R-1)}{(R-1)}}{\frac{(R-1)(R+1)}{(R-1)}} = \dfrac{2}{R + 1}$

21. $\dfrac{A}{6y^2} = \dfrac{3x}{2y} \cdot \dfrac{3y}{3y} = \dfrac{9xy}{6y^2} \Rightarrow A = 9xy$

25. $\dfrac{A}{x^2 - 1} = \dfrac{2x^3 + 2x}{x^4 - 1} = \dfrac{2x\left(x^2 + 1\right)}{\left(x^2 + 1\right)\left(x^2 - 1\right)}$

$\qquad A = \dfrac{2x}{x^2 - 1} = 2x$

29. $\dfrac{2a}{8a} = \dfrac{2a}{2a \cdot 4} = \dfrac{1}{4}$

33. $\dfrac{b + a}{5a^2 + 5ab} = \dfrac{\left(a + b\right)}{5a\left(a + b\right)} = \dfrac{1}{5a}$

37. $\dfrac{4x^2+1}{4x^2-1}=\dfrac{4x^2+1}{(2x-1)(2x+1)}$

Since no cancellations can be made the fraction cannot be reduced.

41. $\dfrac{3+2y}{4y^3+6y^2}=\dfrac{(2y+3)}{2y^2(2y+3)}=\dfrac{1}{2y^2}$

45. $\dfrac{2w^4\,5w^2-3}{w^4+11w^2+24}=\dfrac{(2w^2-1)(w^2+3)}{(w^2+8)(w^2+3)}=\dfrac{2w^2-1}{w^2+8}$

49. $\dfrac{N^4-16}{8N-16}=\dfrac{(N^2+4)(N^2-4)}{8(N-2)}$

$=\dfrac{(N^2+4)(N+2)(N-2)}{8(N-2)}$

$=\dfrac{(N^2+4)(N+2)}{8}$

53. $\dfrac{(x-1)(3+x)}{(3-x)(1-x)}=\dfrac{(x-1)(3+x)}{-(3-x)(x-1)}=\dfrac{3+x}{-(3-x)}=\dfrac{x+3}{x-3}$

57. $\dfrac{n^3+n^2-n-1}{n^3-n^2-n+1}=\dfrac{n^3-1+n^2-n}{n^2+1-n^2-n}$

$=\dfrac{(n-1)(n^2+n+1)+n(n-1)}{(n+1)(n^2-n+1)-n(n+1)}$

$=\dfrac{(n-1)(n^2+n+1+n)}{(n+1)(n^2-n+1-n)}$

$=\dfrac{(n-1)(n^2+2n+1)}{(n+1)(n^2-2n+1)}$

$=\dfrac{(n-1)(n+1)(n+1)}{(n+1)(n-1)(n-1)}$

$=\dfrac{n+1}{n-1}$

61. $\dfrac{x^3+y^3}{2x+2y}=\dfrac{(x+y)(x^2-xy+y^2)}{2(x+y)}=\dfrac{x^2-xy+y^2}{2}$

65. (a) $\dfrac{x^2(x+2)}{x^2+4}$ will not reduce further since x^2+4 does not factor.

(b) $\dfrac{x^4+4x^2}{x^4-16}=\dfrac{x^2(x^2+4)}{(x^2+4)(x^2-4)}=\dfrac{x^2}{x^2-4}$

$=\dfrac{x^2}{(x+2)(x-2)}$

69. If $3-x<0\Rightarrow x>3$

$x^2>9$

$x^2-9>0$

$\dfrac{x^2-9}{3-x}<0$

If $3-x<0$ is $\dfrac{x^2-9}{3-x}>0$? No.

73. $\dfrac{mu^2-mv^2}{mu-mv}=\dfrac{m(u^2-v^2)}{m(u-v)}$

$=\dfrac{(u-v)(u+v)}{(u-v)}=u+v$

6.6   Multiplication and Division of Fractions

1. $\dfrac{4x+6y}{(x-y)^2}\times\dfrac{(x^2-y^2)}{6x+9y}=\dfrac{2(2x+3y)(x+y)(x-y)}{(x-y)(x-y)\cdot 3(2x+3y)}$

$=\dfrac{2(x+y)}{3(x-y)}$

5. $\dfrac{3}{8}\times\dfrac{2}{7}=\dfrac{3}{4}\times\dfrac{1}{7}=\dfrac{3}{28}$

(divide out a common factor of 2)

9. $\dfrac{2}{9}\div\dfrac{4}{7}=\dfrac{2}{9}\times\dfrac{7}{4}=\dfrac{1}{9}\times\dfrac{7}{2}=\dfrac{7}{18}$

(divide out a common factor of 2)

13. $\dfrac{4x+12}{5} \times \dfrac{15t}{3x+9} = \dfrac{4(x+3)}{5} \times \dfrac{5(3t)}{3(x+3)} = 4t$

(divide out common factors of $15(x+3)$)

17. $\dfrac{2a+8}{15} \div \dfrac{a^2+8a+16}{125} = \dfrac{2(a+4)}{3 \times 5} \times \dfrac{5 \times 5 \times 5}{(a+4)(a+4)}$

$= \dfrac{50}{3(a+4)}$

(divide out a common factor of $5(a+4)$)

21. $\dfrac{3ax^2-9ax}{10x^2+5x} \times \dfrac{2x^2+x}{a^2x-3a^2} = \dfrac{3ax(x-3)}{5x(2x+1)} \times \dfrac{x(2x+1)}{a^2(x-3)}$

$= \dfrac{3x}{5a}$

(divide out a common factor of $ax(2x+1)(x-3)$)

25. $\dfrac{\frac{x^2+ax}{2b-cx}}{\frac{a^2+2ax+x^2}{2bx-cx^2}} = \dfrac{x(x+a)}{(2b-cx)} \times \dfrac{x(2b-cx)}{(a+x)(a+x)} = \dfrac{x^2}{a+x}$

(divide out a common factor of $(a+x)(2b-cx)$)

29. $\dfrac{x^2-6x+5}{4x^2-17x-15} \times \dfrac{6x+21}{2x^2+5x-7}$

$= \dfrac{(x-5)(x-1)}{(4x+3)(x-5)} \times \dfrac{3(2x+7)}{(2x+7)(x-1)}$

$= \dfrac{3}{4x+3}$

(divide out common factor $(x-5)(x-1)(2x+7)$)

33. $\dfrac{7x^2}{3a} \div \left(\dfrac{a}{x} \times \dfrac{a^2x}{x^2} \right) = \dfrac{7x^2}{3a} \div \dfrac{a^3}{x^2} = \dfrac{7x^2}{3a} \times \dfrac{x^2}{a^3} = \dfrac{7x^4}{3a^4}$

(divide out a common factor of x)

37. $\dfrac{x^3-y^3}{2x^2-2y^2} \times \dfrac{y^2+2xy+x^2}{x^2+xy+y^2}$

$= \dfrac{(x-y)(x^2+xy+y^2)}{2(x-y)(x+y)} \times \dfrac{(x+y)(x+y)}{(x^2+xy+y^2)} = \dfrac{x+y}{2}$

(divide out common factors of

$(x-y), (x+y), (x^2+xy+y^2)$)

41. $\dfrac{x}{2x+4} \times \dfrac{x^2-4}{3x^2} = \dfrac{x(x+2)(x-2)}{2(x+2)(3x^2)} = \dfrac{x-2}{6x}$

45. $\dfrac{2\pi}{\lambda}\left(\dfrac{a+b}{2ab} \right)\left(\dfrac{ab\lambda}{2a+2b} \right) = \dfrac{2\pi}{\lambda}\left(\dfrac{a+b}{2ab} \times \dfrac{\lambda ab}{2(a+b)} \right)$

$= \dfrac{2\pi}{\lambda} \times \dfrac{\lambda}{4} = \dfrac{\pi}{2}$

(divide out common factors $ab, 2, (a+b)$ and λ)

6.7 Addition and Subtraction of Fractions

1. $4a^2b = 2 \cdot 2 \cdot a \cdot a \cdot b$

$6ab^3 = 2 \cdot 3 \cdot a \cdot b \cdot b \cdot b$

$4a^2b^2 = 2 \cdot 2 \cdot a \cdot a \cdot b \cdot b$

$L.C.D. = 2^2 \cdot 3 \cdot a^2 \cdot b^3 = 12a^2b^3$

5. $\dfrac{3}{5} + \dfrac{6}{5} = \dfrac{3+6}{5} = \dfrac{9}{5}$

9. $\dfrac{1}{2} + \dfrac{3}{4} = \dfrac{2}{4} + \dfrac{3}{4} = \dfrac{2+3}{4} = \dfrac{5}{4}$

13. $\dfrac{a}{x} - \dfrac{b}{x^2} = \dfrac{ax}{x^2} - \dfrac{b}{x^2} = \dfrac{ax-b}{x^2}$

17. $\dfrac{2}{5a} + \dfrac{1}{a} - \dfrac{a}{10} = \dfrac{4}{10a} + \dfrac{10}{10a} - \dfrac{a^2}{10a} = \dfrac{4+10-a^2}{10a}$

$= \dfrac{14-a^2}{10a}$

21. $L.C.D. = 2(2x-1)$

$\dfrac{3}{2x-1} + \dfrac{1}{4x-2} = \dfrac{3}{(2x-1)} \times \dfrac{2}{2} + \dfrac{1}{2(2x-1)}$

$= \dfrac{6+1}{2(2x-1)} = \dfrac{7}{2(2x-1)}$

25. $L.C.D. = 4(s-3)$

$\dfrac{s}{2s-6} + \dfrac{1}{4} - \dfrac{3s}{4s-12} = \dfrac{s}{2(s-3)} \times \dfrac{2}{2} + \dfrac{1}{4} \times \dfrac{(s-3)}{(s-3)}$

$- \dfrac{3s}{4(s-3)} = \dfrac{2s+(s-3)-3s}{4(s-3)} = \dfrac{-3}{4(s-3)}$

29. $L.C.D. = (x-4)(x-4) = (x-4)^2$

$$\frac{3}{x^2-8x+16} - \frac{2}{4-x}$$

$$= \frac{3}{(x-4)(x-4)} + \frac{2}{(x-4)} \times \frac{(x-4)}{(x-4)}$$

$$= \frac{3+2(x-4)}{(x-4)(x-4)} = \frac{3+2x-8}{(x-4)(x-4)}$$

$$= \frac{2x-5}{(x-4)(x-4)} = \frac{2x-5}{(x-4)^2}$$

33. $L.C.D. = (3x-1)(x-4)$

$$\frac{x-1}{3x^2-13x+4} - \frac{3x+1}{4-x}$$

$$= \frac{(x-1)}{(3x-1)(x-4)} + \frac{(3x+1)}{(x-4)} \times \frac{(3x-1)}{(3x-1)}$$

$$= \frac{x-1+9x^2-1}{(3x-1)(x-4)} = \frac{9x^2+x-2}{(3x-1)(x-4)}$$

$$= \frac{t^3+6t^2+9t-2t^3+2t^2+12t-t^3-5t^2-6t}{(t+3)^2(t-3)(t+2)}$$

$$= \frac{-2t^3+3t^2+15t}{(t+3)^2(t-3)(t+2)} = \frac{-t(2t^2-3t-15)}{(t+3)^2(t-3)(t+2)}$$

37. $L.C.D. = (w+1)(w^2-w+1)$

$$\frac{1}{w^3+1} + \frac{1}{w+1} - 2$$

$$= \frac{1}{(w+1)(w^2-x+1)} + \frac{(w^2-w+1)}{(w+1)(w^2-w+1)}$$

$$- \frac{2(w+1)(w^2-w+1)}{(w+1)(w^2-w+1)}$$

$$= \frac{1+w^2-w+1-2(w+1)(w^2-w+1)}{(w+1)(w^2-w+1)}$$

$$= \frac{w^2-w+2-2w^2-2}{(w+1)(w^2-w+1)}$$

$$= \frac{-2w^3+w^2-w}{(w+1)(w^2-w+1)}$$

41. $\dfrac{\frac{x}{y}-\frac{y}{x}}{1+\frac{y}{x}} \times \dfrac{xy}{xy} = \dfrac{x^2-y^2}{xy+y^2} = \dfrac{(x+y)(x-y)}{y(x+y)} = \dfrac{x-y}{y}$

45. $f(x) = \dfrac{x}{x+1}, \quad f(x+h)-f(x)$

$$= \frac{x+h}{x+h+1} - \frac{x}{x+1}$$

$$f(x+h)-f(x) = \frac{(x+h)}{(x+h+1)} \times \frac{(x+1)}{(x+1)} \times \frac{(x+h+1)}{(x+h+1)}$$

$$= \frac{x^2+x+hx+h-x^2-xh-x}{(x+1)(x+h+1)}$$

$$= \frac{h}{(x+1)(x+h+1)}$$

49. $\tan\theta \times \cot\theta (\sin\theta)^2 = \dfrac{y}{x} \times \dfrac{x}{y} + \left(\dfrac{y}{r}\right)^2 - \dfrac{x}{r}$

$$= 1 + \frac{y^2}{r^2} - \frac{x}{r} = \frac{r^2+y^2-rx}{r^2}$$

53. $f(x) = x - \dfrac{2}{x}$

$$f(a+1) = a+1 - \frac{2}{a+1} = \frac{(a+1)^2-2}{a+1}$$

$$= \frac{a^2+2a+1-2}{a+1}$$

$$= \frac{a^2+2a-1}{a+1}$$

57. $\dfrac{y^2-x^2}{y^2+x^2} = \dfrac{\left(\frac{mn}{m-n}\right)^2 - \left(\frac{mn}{m+n}\right)^2}{\left(\frac{mn}{m-n}\right)^2 + \left(\frac{mn}{m+n}\right)^2} \times \dfrac{(m-n)^2(m+n)^2}{(m-n)^2(m+n)^2}$

$$= \frac{m^2n^2(m+n)^2 - m^2n^2(m-n)^2}{m^2n^2(m+n)^2 + m^2n^2(m-n)^2}$$

$$= \frac{m^2n^2(m^2+2mn+n^2-m^2+2mn-n^2)}{m^2n^2(m^2+2mn+n^2+m^2-2mn+n^2)}$$

$$= \frac{4mn}{2m^2+2n^2}$$

$$= \frac{2mn}{m^2+n^2}$$

61. $\dfrac{2n^2-n-4}{2n^2+2n-4}+\dfrac{1}{n-1}=\dfrac{2n^2-n-4+2(n+2)}{2(n-1)(n+2)}$

$\qquad\qquad\qquad\qquad =\dfrac{2n^2-n-4+2n+4}{2(n-1)(n+2)}$

$\qquad\qquad\qquad\qquad =\dfrac{2n^2+n}{2(n-1)(n+2)}$

$\qquad\qquad\qquad\qquad =\dfrac{n(2n+1)}{2(n-1)(n+2)}$

65. $\dfrac{\frac{L}{C}+\frac{R}{sC}}{sL+R+\frac{1}{sC}}=\dfrac{\frac{Ls+R}{sC}}{\frac{(sL+R)sC+1}{sC}}=\dfrac{\frac{lS+R}{sC}}{\frac{s^2CL+CRs+1}{sC}}$

$\qquad\qquad =\dfrac{LsR}{Cs}\times\dfrac{Cs}{CLs^2+CRs+1}$

$\qquad\qquad =\dfrac{Ls+R}{CLs^2+CRs+1}$

69. $5=(v-w)t_1\Rightarrow t_1=\dfrac{5}{v-w}$

$\qquad 5=(v+w)t_2\Rightarrow t_2=\dfrac{5}{v+w}$

$\qquad t_1+t_2=t=\dfrac{5}{v-w}+\dfrac{5}{v+w}$

$\qquad\qquad t=\dfrac{10v}{(v+w)(v-w)}$

6.8 Equations Involving Fractions

1. $\dfrac{x}{2}-\dfrac{1}{b}=\dfrac{x}{2b}$

$\qquad xb-2=x$

$\qquad x(b-1)=2$

$\qquad\quad x=\dfrac{2}{b-1}$

5. $\dfrac{x}{2}+6=2x$

$\qquad x+12=4x$

$\qquad\quad 3x=12$

$\qquad\quad\; x=4$

9. $1-\dfrac{t-5}{6}=\dfrac{3}{4}$

Multiply both sides by the L.C.D. $=12$

$12-2(t-5)=9$

$12-2t+10=9$

$\qquad\quad 2t=13$

$\qquad\quad\; t=\dfrac{13}{2}$

13. $\dfrac{3}{T}+2=\dfrac{5}{3}$

Multiply both sides by the L.C.D. $=3T$

$9+6T=5T$

$\qquad T=-9$

17. $\dfrac{2y}{y-1}=5$

Multiply both sides by the L.C.D. $=y-1$

$2y=5y-5$

$3y=5$

$\;\; y=\dfrac{5}{3}$

21. $\dfrac{5}{2x+4}+\dfrac{3}{6x+12}=2$

Multiply both sides by the L.C.D. $=6(x+2)$

$\dfrac{5}{2(x+2)}+\dfrac{3}{6(x+2)}=2$

$\qquad\qquad 15+3=2\times6(x+2)$

$\qquad\qquad\quad 18=12x+24$

$\qquad\qquad\; 12x=-6$

$\qquad\qquad\quad\; x=-\dfrac{1}{2}$

25. $\dfrac{1}{4x}+\dfrac{3}{2x}=\dfrac{2}{x+1}$

Multiply both sides by L.C.D. $=4x(x+1)$

$(x+1)+6(x+1)=2\times4x$

$\qquad x+1+6x+6=8x$

$\qquad\qquad 7x+7=8x$

$\qquad\qquad\qquad x=7$

29. $\dfrac{1}{x^2-x}-\dfrac{1}{x}=\dfrac{1}{x-1}$

Multiply both sides by L.C.D. $=x(x-1)$

$$\dfrac{1}{x(x-1)}-\dfrac{1}{x}=\dfrac{1}{(x-1)}$$

$$1-(x-1)=x$$

$$1-x+1=x$$

$$2x=2$$

$$x=1,\text{ no solution}$$

33. $2-\dfrac{1}{b}+\dfrac{3}{c}=0$, for c

Multiply both sides by L.C.D. $=bc$

$$2bc-c+3b=0$$

$$c(2b-1)=-3b$$

$$c=\dfrac{3b}{1-2b}$$

37. $\dfrac{s-s_\circ}{t}=\dfrac{v+v_\circ}{2}$ for v

Multiply both sides by L.C.D. $=2t$

$$2(s-s_0)=t(v+v_0)$$

$$2(s-s_0)=tv+tv_0$$

$$2(s-s_0)-tv_0=tv$$

$$v=\dfrac{2(s-s_0)-tv_0}{t}$$

41. $z=\dfrac{1}{g_m}-\dfrac{jX}{g_mR}$ for R

$$g_mRz=R-jX$$

$$g_mRz-R=-jX$$

$$R(g_mz-1)=-jX$$

$$R=\dfrac{jX}{1-g_mz}$$

45. $\dfrac{1}{C}=\dfrac{1}{C_2}+\dfrac{1}{C_1+C_3}$

$$C_2(C_1+C_3)=C(C_1+C_3)+CC_2$$

$$C_1C_2+C_2C_3=CC_1+CC_3+CC_2$$

$$C_1(C_2-C)=CC_3+CC_2-C_2C_3$$

$$C_1=\dfrac{CC_3+CC_2-C_2C_3}{C_2-C}$$

49. $\dfrac{1}{5.0}\times t+\dfrac{1}{8.0}\times t=1$

$$8.0t+5.0t=40$$

$$13t=40$$

$$t=\dfrac{40}{13}$$

$$t=3.1\text{ h}$$

53. $d=2.0t_1$ for trip up

$d=2.2t_2$ for trip down

$$t_1+t_2+90=5.0(60)$$

$$\dfrac{d}{2.0}+\dfrac{d}{2.2}+90=5(60)$$

$$d=220\text{ m}$$

57. $0.75d=2t,\quad 0.25d=1\cdot t_2$

$$v=\dfrac{d}{t_1+t_2}=\dfrac{d}{\frac{0.75d}{21}+0.25d}$$

$$v=\text{mach }1.6$$

61. $\dfrac{V}{R_1}+\dfrac{V}{R_2}=i$

$$\dfrac{V}{2.7}+\dfrac{V}{6.0}=1.2$$

$$V=2.2\text{ V}$$

Chapter 6 Review Exercises

1. $3a(4x+5a)=12ax+15a^2$

5. $(2a+1)^2=4a^2+4a+1$

9. $(2x+5)(x-9)=2x^2-13x-45$

13. $3s+9t=3(s+3t)$

17. $W^2b^{x+2}-144b^x=b^x(W^2b^2-144)$
$$=b^x(Wb+12)(Wb-12)$$

21. $4(36t^2-24t+4)=4(3t-1)(3t-1)=4(3t-1)^2$

25. $x^2+x-56=(x+8)(x-7)$

29. $2k^2-k-36=(2k-9)(k+4)$

33. $10b^2+23b-5=(5b-1)(2b+5)$

37. $250-16y^6=2(125-8y^6)$
$$=2(5^3-(2y^2)^3)$$
$$=2(5-2y^2)(25+10y^2+4y^4)$$

41. $ab^2-3b^2+a-3=b^2(a-3)+(a-3)$
$$=(a-3)(b^2+1)$$

45. $\dfrac{48ax^3y^6}{9a^3xy^6}=\dfrac{16x^2}{3a^2}$

49. $\dfrac{4x+4y}{35x^2}\dfrac{28x}{x^2-y^2}=\dfrac{4(x+y)}{35x^2}\dfrac{28x}{(x+y)(x-y)}$
$$=\dfrac{16}{5x(x-y)}$$

53. $\dfrac{\frac{3x}{7x^2+13x-3}}{\frac{6x^2}{x^2+4x+4}}=\dfrac{3x}{(7x-1)(x+2)}\dfrac{(x+2)(x+2)}{3(2x^2)}$
$$=\dfrac{x+2}{2x(7x-1)}$$

57. $\dfrac{4}{9x}-\dfrac{5}{12x^2}=\dfrac{4}{9x}\dfrac{4x}{4x}-\dfrac{5}{12x^2}\dfrac{3}{3}=\dfrac{16x-15}{36x^2}$

61. $\dfrac{a+1}{a+2}-\dfrac{a+3}{a}=\dfrac{(a+1)}{(a+2)}\dfrac{a}{a}-\dfrac{(a+3)}{a}\dfrac{(a+2)}{(a+2)}$
$$=\dfrac{a(a+1)-(a+3)(a+2)}{a(a+2)}$$
$$=\dfrac{a^2+a-a^2-5a-6}{a(a+2)}$$
$$=\dfrac{-4a-6}{a(a+2)}=\dfrac{-2(a+3)}{a(a+2)}$$

65. $\dfrac{3x}{2x^2-2}-\dfrac{2}{4x^2-5x+1}=\dfrac{3x}{2(x+1)(x-1)}\times\dfrac{(4x-1)}{(4x-1)}$
$$-\dfrac{2}{(4x-1)(x-1)}\times\dfrac{2(x+1)}{2(x+1)}$$
$$=\dfrac{3x(4x-1)-4(x+1)}{2(4x-1)(x+1)(x-1)}=\dfrac{12x^2-3x-4x-4}{2(4x-1)(x+1)(x-1)}$$
$$=\dfrac{12x^2-7x-4}{2(4x-1)(x+1)(x-1)}$$

69. $\dfrac{6x^2-7x-3}{4x^2-8x+3}=\dfrac{(2x-3)(3x+1)}{(2x-1)(2x-3)}=\dfrac{3x+1}{2x-1}$

Graph $y_1=\dfrac{6x^2-7x-3}{4x^2-8x+3}$

and $y_2=\dfrac{3x+1}{2x-1}$. The graphs are the same.

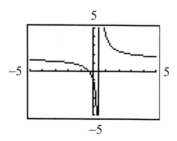

73. $x^2-5=(x+\sqrt{5})(x-\sqrt{5})$

77. $\dfrac{x}{2} - 3 = \dfrac{x-10}{4}$

$2x - 12 = x - 10$

$x = 2$

81. $\dfrac{2x}{2x^2 - 5x} - \dfrac{3}{x} = \dfrac{1}{4x - 10}$

$\dfrac{2x}{x(2x-5)} - \dfrac{3}{x} = \dfrac{1}{2(2x-5)}$

$4x - 6(2x-5) = x$

$4x - 12x + 30 = x$

$9x = 30$

$x = \dfrac{10}{3}$

85. (a) Changing an odd number of signs changes the sign of the fraction.

(b) Changing an even number of signs leaves the sign of the fraction unchanged.

89. $Pb(L+b)(L-b) = Pb(L^2 - b^2) = PbL^2 - Pb^3$

93. $cT_2 - cT_1 + RT_2 - RT_1 = c(T_2 - T_1) + R(T_2 - T_1)$

$= (T_2 - T_1)(c + R)$

97. $(n+1)^3 (2n+1)^3$

$= (n^3 + 3n^2 + 3n + 1)(8n^3 + 12n^2 + 6n + 1)$

$= 8n^6 + 12n^5 + 6n^4 + n^3 + 24n^5 + 36n^4 + 18n^3 + 3n^2$

$\quad + 24n^4 + 36n^3 + 18n^2 + 3n + 8n^3 + 12n^2 + 6n + 1$

$= 8n^6 + 36n^5 + 66n^4 + 63n^3 + 33n^2 + 9n + 1$

101. Increase in volume $= (x+4)^3 - x^3$

$= x^3 + 12x^2 + 48x + 64 - x^2$

$= 12x^2 + 48x + 64$

$= 4(3x^2 + 12x + 16)$

105. $\dfrac{\frac{\pi ka}{2}(R^4 - r^4)}{\pi ka(R^2 - r^2)} = \dfrac{1(R^2 + r^2)(R^2 - r^2)}{2(R^2 - r^2)} = \dfrac{R^2 + r^2}{2}$

109. $\dfrac{4k-1}{4k-4} + \dfrac{1}{2k} = \dfrac{(4k-1)}{4(k-1)} \times \dfrac{k}{k} + \dfrac{1}{2k} \times \dfrac{2(k-1)}{2(k-1)}$

$= \dfrac{4k^2 - k + 2k - 2}{4k(k-1)} = \dfrac{4k^2 + k - 2}{4k(k-1)}$

113. $\dfrac{\frac{u^2 2g}{}-x}{\frac{1}{2gc^2} - \frac{u^2}{2g} + x} \times \dfrac{2gc^2}{2gc^2} = \dfrac{u^2 c^2 - 2gc^2 x}{1 - u^2 c^2 + 2gc^2 x}$

117. $R = \dfrac{wL}{H(w+L)}$

$RHw + RHL = wL$

$wL - RHL = RHw$

$L(w - RH) = RHw$

$L = \dfrac{RHw}{w - RH}$

121. $s^2 + \dfrac{cs}{m} + \dfrac{kL^2}{mb^2} = 0$

$s^2 mb^2 + csb^2 + kL^2 = 0$

$csb^2 = -s^2 mb^2 - kL^2$

$c = \dfrac{-s^2 mb^2 - kL^2}{sb^2}$

125. $\dfrac{1}{4}t + \dfrac{1}{24}t = 1$

$\dfrac{7}{24}t = 1$

$t = 3.4 \text{ h}$

129. $d = \dfrac{w_a}{w_a - w_w} = \dfrac{1.097 w_w}{1.097 w_w - w_w} = \dfrac{1.097}{1.097 - 1} = 11.3$

133. $\dfrac{(1 + \frac{1}{s})(1 + \frac{1}{s/2})}{3 + \frac{1}{s} + \frac{1}{s/2}}$

When you "cancel" the basic operation being performed is division.

$= \dfrac{(\frac{s+1}{s})(\frac{s/2+1}{s/2})}{(3 + \frac{1}{s} + \frac{1}{s/2})} \times \dfrac{s(s/2)}{s(s/2)}$

$= \dfrac{(s+1)(s/2+1)}{3s(s/2) + \frac{s}{2} + s} \times \dfrac{2}{2}$

$= \dfrac{(s+1)(s+2)}{3s^2 + s + 2s} = \dfrac{(s+1)(s+2)}{3s(s+1)}$

$= \dfrac{s+2}{3s}$

Chapter 7

Quadratic Equations

7.1   Quadratic Equations; Solution By Factoring

1. $3x^2 + 7x + 2 = 0$

$(3x+1)(x+2) = 0$ factor

$3x+1 = 0$ or $x+2 = 0$

$3x = -1$ $x = -2$

$x = -\dfrac{1}{3}$

The roots are $x = -\dfrac{1}{3}$ and $x = -2$.

5. $x^2 = (x+2)^2$

$x^2 = x^2 + 4x + 4$

$4x + 4 = 0$, no x^2 term, not quadratic

9. $x^2 - 4 = 0$

$(x+2)(x-2) = 0$

$x+2 = 0$ or $x-2 = 0$

$x = -2$ $x = 2$

13. $x^2 - 8x - 9 = 0$

$(x-9)(x+1) = 0$

$x-9 = 0$ or $x+1 = 0$

$x = 9$ $x = -1$

17. $40x - 16x^2 = 0$

$2x^2 - 5x = 0$

$x(2x-5) = 0$

$x = 0$ or $2x-5 = 0$

$2x = 5$

$x = \dfrac{5}{2}$

21. $3x^2 - 13x + 4 = 0$

$(3x-1)(x-4) = 0$

$3x-1 = 0$ or $x-4 = 0$

$3x = 1$ $x = 4$

$x = \dfrac{1}{3}$

25. $6x^2 = 13x - 6$

$6x^2 - 13x + 6 = 0$

$(3x-2)(2x-3) = 0$

$3x-2 = 0$ or $2x-3 = 0$

$3x = 2$ $2x = 3$

$x = \dfrac{2}{3}$ $x = \dfrac{3}{2}$

29. $6y^2 + by = 2b^2$

$6y^2 + by - 2b^2 = 0$

$(2y-b)(3y+2b) = 0$

$2y-b = 0$ or $3y+2b = 0$

$y = \dfrac{b}{2}$ $y = \dfrac{-2b}{3}$

33. $(x+2)^3 = x^3 + 8$

$x^3 + 6x^2 + 12x + 8 = x^3 + 8$

$6x^2 + 12x = 0$

$6x(x+2) = 0$

$6x = 0$ or $x+2 = 0$

$x = 0$ $x = -2$

37. $x^2 + 2ax = b^2 - a^2$

$x^2 + 2ax + (a^2 - b^2) = 0$

$(x+(a+b))(x+(a-b)) = 0$

$x+a+b = 0$ or $x+a-b = 0$

$x = -a-b$ $x = b-a$

41. $12x^2 - 64x + 64 = 0$

$3x^2 - 16x + 16 = 0$

$(3x - 4)(x - 4) = 0$

$x = \dfrac{4}{3}$ reject since $x > 2$,

$x = 4$ is the solution.

45. $P = 4h^2 - 48h + 744$

$P = 664 = 4h^2 - 48h + 774$

$4h^2 - 48h + 80 = 0$

$4\left(h^2 - 12h + 20\right) = 0$

$4(h - 10)(h - 2) = 0$

$h - 2 = 0$ or $h = 2$

$h - 10 = 0$ $h = 10$

The power is 664 MW at 2 a.m. and 10 a.m.

49. $x^3 - x = 0$

$x\left(x^2 - 1\right) = 0$

$x(x + 1)(x - 1) = 0$

$x = 0$ or $x + 1 = 0$ or $x - 1 = 0$

$x = -1$ $x = 1$

The three roots are $-1, 0, 1$.

53. $\dfrac{1}{2x} - \dfrac{3}{4} = \dfrac{1}{2x + 3}$

$4(2x + 3) - 3(2x)(2x + 3) = 2x(4)$

$8x + 12 - 12x^2 - 18x = 8x$

$-12x^2 - 18x + 12 = 0$

$-6(2x - 1)(x + 2) = 0$

$2x - 1 = 0$ or $x + 2 = 0$

$2x = 1$ $x = -2$

$x = \dfrac{1}{2}$

57. (1) $d = v_1 t_1$, going

     (2) $d = v_2 t_2$, returning

     (3) $2d = 120 \Rightarrow d = 60$

     (4) $t_1 + t_2 = 3.5 \Rightarrow t_1 = 3.5 - t_2$

     (5) $v_1 + 10 = v_2$

     add (1) and (2) $2d = 120 = v_1 t_1 + v_2 t_2$

$120 = v_1\left(3.5 - t_2\right) + \left(v_1 + 10\right)t_2$

$120 = v_1\left(3.5 - \dfrac{d}{v_2}\right) + \left(v_1 + 10\right) \times \dfrac{d}{v_2}$

$120 = v_1\left(3.5 - \dfrac{60}{v_1 + 10}\right) + \left(v_1 + 10\right) \times \dfrac{60}{\left(v_1 + 10\right)}$

$120 = 3.5v_1 - \dfrac{60v_1}{v_1 + 10} + 60$

$120\left(v_1 + 10\right) = 3.5v_1\left(v_1 + 10\right) - 60v_1 + 60\left(v_1 + 10\right)$

$120v_1 + 1200 = 3.5v_1^2 + 35v_1 - 60v_1 + 60v_1 + 600$

$3.5v_1^2 - 85v_1 - 600 = 0$ from which

$v_1 = 30$ km/h

$v_2 = v_1 + 10 = 40$ km/h

7.2   Completing the Square

1. $x^2 + 6x - 8 = 0$

$x^2 + 6x = 8$

$x^2 + 6x + 9 = 8 + 9$

$(x + 3)^2 = 17$

$x + 3 = \pm\sqrt{17}$

$x = -3 \pm \sqrt{17}$

5. $x^2 = 7$

$\sqrt{x^2} = \pm\sqrt{7}$

$x = \sqrt{7}$ or $x = -\sqrt{7}$

9. $(y + 3)^2 = 7$

$\sqrt{(y + 3)^2} = \pm\sqrt{7}$

$y + 3 = \pm\sqrt{7}$

$y = -3 \pm \sqrt{7}$

13. $D^2 + 3D + 2 = 0$

$D^2 + 3D = -2$

$D^2 + 3D + \dfrac{9}{4} = -2 + \dfrac{9}{4}$

$\left(D + \dfrac{3}{2}\right)^2 = \dfrac{1}{4}$

$$D + \frac{3}{2} = -\frac{1}{2} \quad \text{or} \quad D + \frac{3}{2} = \frac{1}{2}$$

$$D = -\frac{4}{2} \qquad\qquad D = -\frac{2}{2}$$

$$= -2 \qquad\qquad\quad = -1$$

17.
$$v(v+2) = 15$$
$$v^2 + 2v = 15$$
$$v^2 + 2v - 15 = 0$$
$$v^2 + 2v = 15$$
$$v^2 + 2v + 1 = 15 + 1$$
$$(v+1)^2 = 16$$
$$v + 1 = \pm 4$$
$$v = -4 - 1 = -5$$
$$v = 4 - 1 = 3$$

21.
$$3y^2 = 3y + 2$$
$$3y^2 - 3y - 2 = 0$$
$$y^2 - y = \frac{2}{3}$$
$$y^2 - y + \frac{1}{4} = \frac{2}{3} + \frac{1}{4}$$
$$\left(y - \frac{1}{2}\right)^2 = \frac{11}{12}$$
$$y - \frac{1}{2} = \pm\sqrt{\frac{11}{12}}$$
$$y = \frac{1}{2} \pm \frac{1}{2}\sqrt{\frac{11}{3}} = \frac{1}{2} \pm \frac{1}{2}\frac{\sqrt{33}}{3}$$
$$= \frac{1}{2} \pm \frac{1}{6}\sqrt{33} = \frac{1}{6}\left(3 \pm \sqrt{33}\right)$$

25.
$$10T - 5T^2 = 4$$
$$5T^2 - 10T + 4 = 0$$
$$5\left(T^2 - 2T + 1\right) = -4 + 5$$
$$5(T-1)^2 = 1$$
$$(T-1)^2 = \frac{1}{5}$$
$$T = 1 \pm \frac{\sqrt{5}}{5}$$

29.
$$x^2 + 2bx + c = 0$$
$$x^2 + 2bx = -c$$
$$x^2 + 2bx + b^2 = -c + b^2$$
$$(x+b)^2 = -c + b$$
$$x + b = \pm\sqrt{b^2 - c}$$
$$x = -b \pm \sqrt{b^2 - c}$$

33.

$$34^2 = (x+14)^2 + x^2$$
$$1156 = x^2 + 28x + 196 + x^2$$
$$2x^2 + 28x - 960 = 0$$
$$x^2 + 14x - 480 = 0$$
$$(x-16)(x+30) = 0$$
$$x - 16 = 0 \quad \text{or} \quad x + 30 = 0$$
$$x = 16 \qquad\qquad x = -30, \text{ reject}$$

The camera is 16 ft above the ATM.

7.3   The Quadratic Formula

1. $x^2 + 5x + 6 = 0; \; a = 1, \; b = 5, \; c = 6$

$$x = \frac{-5 \pm \sqrt{5^2 - 4(1)(6)}}{2(1)} = \frac{-5 \pm \sqrt{1}}{2} = \frac{-5 \pm 1}{2}$$

$$x = \frac{-5+1}{2} = -2 \quad \text{or} \quad x = \frac{-5-1}{2} = -3$$

5. $x^2 + 2x - 8 = 0; \; a = 1, \; b = 2, \; c = -8$

$$x = \frac{-2 \pm \sqrt{2^2 - 4(1)(-8)}}{2(1)} = \frac{-2 \pm \sqrt{36}}{2}$$

$$= \frac{-2 \pm 6}{2}$$

$$x = 2 \quad \text{or} \quad x = -4$$

9. $x^2 - 4x + 2 = 0;\ a = 1,\ b = -4,\ c = 2$

$$x = \frac{-(-4) \pm \sqrt{(-4)^2 - 4(1)(2)}}{(2)}$$

$$= \frac{4 \pm \sqrt{8}}{2}$$

$$= \frac{4 \pm 2\sqrt{2}}{2}$$

$$= 2 \pm \sqrt{2}$$

13. $8s^2 + 20s = 12$

$$2s^2 + 5s = 3$$

$$2s^2 + 5s - 3 = 0;\ a = 2,\ b = 5,\ c = -3$$

$$s = \frac{-5 \pm \sqrt{5^2 - 4(2)(-3)}}{2(2)}$$

$$= \frac{-5 \pm \sqrt{49}}{4}$$

$$= \frac{-5 \pm 7}{4}$$

$$s = \frac{1}{2}\ \text{ or }\ s = -3$$

17. $z + 2 = 2z^2$

$$2z^2 - z - 2 = 0;\ a = 2,\ b = -1,\ c = -2$$

$$z = \frac{-(-1) \pm \sqrt{(-1)^2 - 4(2)(-2)}}{2(2)}$$

$$= \frac{1 \pm \sqrt{17}}{4}$$

21. $8t^2 + 61t = -120$

$$8t^2 + 61t + 120 = 0;\ a = 8,\ b = 61,\ c = 120$$

$$t = \frac{-61 \pm \sqrt{61^2 - 4(8)(120)}}{2(8)}$$

$$= \frac{-61 \pm \sqrt{-119}}{16}$$

25. $25y^2 - 121 = 0;\ a = 25,\ b = 0,\ c = -121$

$$y = \frac{-0 \pm \sqrt{0^2 - 4(25)(-121)}}{2(25)}$$

$$= \frac{\pm\sqrt{12{,}100}}{50}$$

$$= \frac{\pm 110}{50}$$

$$= \frac{\pm 100}{50} = \pm\frac{11}{5}$$

29. $x^2 - 0.20x - 0.40 = 0;\ a = 1,\ b = -0.20,\ c = -0.40$

$$x = \frac{-(-0.20) \pm \sqrt{(-0.20)^2 - 4(1)(-0.40)}}{2(1)}$$

$$= \frac{0.2 \pm \sqrt{1.64}}{2} = -0.54\ \text{ or }\ x = 0.74$$

33. $x^2 + 2cx - 1 = 0$

$$x = \frac{-2c \pm \sqrt{(2c)^2 - 4(1)(-1)}}{2(1)}$$

$$= \frac{-2c \pm \sqrt{4c^2 + 4}}{2}$$

$$= \frac{-2c \pm 2\sqrt{c^2 + 1}}{2}$$

$$= -c \pm \sqrt{c^2 + 1}$$

37. $2x^2 - 7x = -8$

$$2x^2 - 7x + 8 = 0;\ a = 2;\ b = -7;\ c = 8$$

$$D = \sqrt{(-7)^2 - 4(2)(8)} = \sqrt{-15},$$

unequal imaginary roots

41. $x^2 + 4x + k = 0$ will have a double root if

$$b^2 - 4ac = 0 \Rightarrow 4^2 - 4(1)(k) = 0$$

$$k = 4$$

45. $ax^2 + bx + c = 0$ has solution $\dfrac{-b \pm \sqrt{b^2 - 4ac}}{2a}$

$ax^2 - bx - c = 0$ has solution $\dfrac{b \pm \sqrt{b^2 - 4ac}}{2a}$

$$-\frac{-b \pm \sqrt{b^2 - 4ac}}{2a} = \frac{b \mp \sqrt{b^2 - 4ac}}{2a} = \frac{b \pm \sqrt{b^2 - 4ac}}{2a}$$

The solutions are opposites.

49. $8x^2 - 15Lx + 6L^2 = 0$

$$x = \frac{-(-15L) \pm \sqrt{(-15L)^2 - 4(8)(6L^2)}}{2(8)}$$

$$= \frac{\left(15 \pm \sqrt{33}\right)L}{16}$$

$$x = \frac{\left(15 - \sqrt{33}\right)L}{16} \text{ since } x < L.$$

53. $\dfrac{l}{w} = \dfrac{l+w}{l} \Rightarrow l^2 = lw + w^2$

$l^2 - wl - w^2 = 0;\ a = 1,\ b = -w,\ c = -w^2$

$$l = \frac{-(-w) \pm \sqrt{(-w)^2 - 4(1)(-w^2)}}{2(1)}$$

$$= \frac{w + w\sqrt{5}}{2} \text{ where } + \text{ is chosen to make } l > 0$$

$$\frac{l}{w} = \frac{1 + \sqrt{5}}{2} = 1.618$$

$$h = \frac{8r \pm \sqrt{64r^2 - 4(4)(b^2)}}{2(4)}$$

$$= \frac{8r \pm \sqrt{64r^2 - 16b^2}}{8}$$

$$= \frac{8r \pm 4\sqrt{4r^2 - b^2}}{8}$$

$$= \frac{2r \pm \sqrt{4r^2 - b^2}}{2}$$

57.

$0.8(33.8)(27.3) = (33.8 - 2x)(27.3 - 2x)$

$4x^2 - 122.2x + 184.548 = 0;\ a = 4,\ b = -122.2$

$c = 184.548$

$$x = \frac{-(-122.2) \pm \sqrt{(-122.2)^2 - 4(-4)(184.548)}}{2(4)}$$

$x = 1.6$ cm

61. $\quad v = $ truck speed

$v + 15 = $ car speed

From $d = vt$

$77 = vt,$ car

$$77 = v\left(t + \frac{18}{60}\right), \text{ truck}$$

$$77 = v\left(\frac{77}{v + 15} + \frac{18}{60}\right)$$

$$77(v + 15) = 77v + \frac{18}{60}v(v + 15)$$

$$\frac{18}{60}v^2 + \frac{18}{4}v - 1155 = 0;\ a = \frac{18}{60},\ b = \frac{18}{4},\ c = -1155$$

$$v = \frac{-\frac{18}{4} \pm \sqrt{\left(\frac{18}{4}\right)^2 - 4\left(\frac{18}{60}\right)(-1155)}}{2\left(\frac{18}{60}\right)}$$

$v = -70,\ 55$

The truck speed is 55 mi/h and the car speed is 70 mi/h.

7.4 The Graph of the Quadratic Function

1. $y = 2x^2 + 8x + 6$; $a = 2$, $b = 8$, $c = 6$

x-coordinate of vertex $= \dfrac{-b}{2a}$

$\qquad\qquad = \dfrac{-8}{2(2)} = -2$

y-coordinate of vertex $= 2(-2)^2 + 8(-2) + 6$

$\qquad\qquad = -2$

The vertex is $(-2, -2)$ and since $a > 0$, it is a minimum. Since $c = 6$, the y-intercept is $(0, 6)$ and the check is:

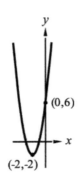

$(0,6)$

x

$(-2,-2)$

5. $y = -3x^2 + 10x - 4$; $a = -3$, $b = 10$.

This means that the x-coordinate of the extreme is

$$\frac{-b}{2a} = \frac{-10}{2(-3)} = \frac{10}{6} = \frac{5}{3}$$

and the y-coordinate is

$$y = -3\left(\frac{5}{3}\right)^2 + 10\left(\frac{5}{3}\right) - 4 = \frac{13}{3}.$$

Thus the extreme point is $\left(\dfrac{5}{3}, \dfrac{13}{3}\right)$.

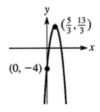

$\left(\frac{5}{3}, \frac{13}{3}\right)$

x

$(0, -4)$

Since $a < 0$, it is a maximum point.
Since $c = -4$, the y-intercept is $(0, -4)$. Use the maximum point $\left(\frac{5}{3}, \frac{13}{3}\right)$, and the y-intercept $(0, -4)$, and the fact that the graph is a parabola, to sketch the graph.

9. $y = x^2 - 4 = x^2 + 0x - 4$; $a = 1$, $b = 0$, $c = -4$

The x-coordinate of the extreme point is

$\dfrac{-b}{2a} = \dfrac{-0}{2(1)} = 0$, and the y-coordinate is

$y = 0^2 - 4 = -4$.

The extreme point is $(0, -4)$.

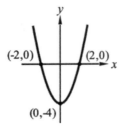

$(-2,0)$ $(2,0)$ x

$(0,-4)$

Since $a > 0$, it is a minimum point.
Since $c = -4$, the y-intercept is $(0, -4)$.
$x^2 - 4 = 0$, $x^2 = 4$, $x = \pm 2$ are the x-intercepts.
Use the minimum points and intercepts to sketch the graph.

13. $y = 2x^2 + 3 = 2x^2 + 0x + 3$; $a = 2$, $b = 0$, $c = 3$

The x-coordinate of the extreme point is

$\dfrac{-b}{2a} = \dfrac{-0}{2(2)} = 0$, and the y-coordinate is

$y = 2(0)^2 + 3 = 3$.

The extreme point is $(0, 3)$. Since $a > 0$ it is a minimum point.

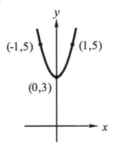

$(-1,5)$ $(1,5)$

$(0,3)$

x

Since $c = 3$, the y-intercept is $(0, 3)$ there are no x-intercepts, $b^2 - 4ac = -24$. $(-1, 5)$ and $(1, 5)$ are on the graph. Use the three points to sketch the graph.

17. $2x^2 - 3 = 0$. Graph $y = 2x^2 - 3$ on a graphing calculator and use the zero feature to find the roots. $x = -1.2$ and $x = 1.2$.

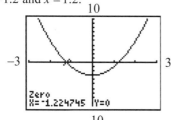

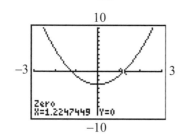

21. $x(2x - 1) = -3$.

Graph $y_1 = x(2x - 1) + 3$ and use the zero feature.

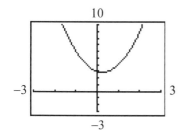

As the graph shows there are no real solutions.

25. (a) $y = x^2$ (b) $y = x^2 + 3$ (c) $y = x^2 - 3$

The parabola $y = x^2 + 3$ is shifted up $+3$ units (minimum point $(0, 3)$).

The parabola $y = x^2 - 3$ is shifted down -3 units (minimum point $(0, -3)$).

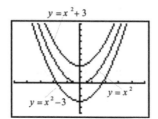

29. (a) $y = x^2$ (b) $y = 3x^2$ (c) $y = \frac{1}{3}x^2$

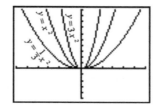

The graph of $y = 3x^2$ is the graph of $y = x^2$ narrowed. The graph of $y = \frac{1}{3}x^2$ is the graph of $y = x^2$ broadened.

33. $y = 2x^2 - 4x - c$ will have two real roots if

$$(-4)^2 - 4(2)(-c) \geq 0$$
$$16 + 8c \geq 0$$
$$c \geq -2$$

-2 is the smallest integral value of c such that $y = 2x^2 - 4x - c$ has two real roots.

37. $d = 2t^2 - 16t + 47$; $a = 2$, $b = -16$, $c = 47$

$$x\text{-coordinate of vertex} = \frac{-b}{2a} = \frac{-(-16)}{2(2)} = 4$$

$$y\text{-coordinate of vertex} = 2(4)^2 - 16(4) + 47 = 15$$

$$y\text{-intercept} = (0, 47)$$

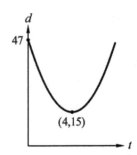

41. The maximum point is $\left(\dfrac{25}{3}, 208 \right)$.

45. $\pi (r+1)^2 = 96.0$

$(r+1)^2 = \dfrac{96.0}{\pi}$

$r+1 = \sqrt{\dfrac{96.0}{\pi}}$

$r = \sqrt{\dfrac{96.0}{\pi}} - 1$

$r = 4.53$ cm

Chapter 7 Review Exercises

1. $x^2 + 3x - 4 = 0$

$(x+4)(x-1) = 0$

$x+4 = 0$ or $x-1 = 0$

$x = -4$ $\qquad$ $x = 1$

5. $3x^2 + 11x = 4$

$3x^2 + 11x - 4 = 0$

$(3x-1)(x+4) = 0$

$3x-1 = 0$ or $x+4 = 0$

$3x = 1$ $\qquad$ $x = -4$

$x = \dfrac{1}{3}$

9. $6s^2 = 25s$

$6s^2 - 25s = 0$

$s(6s - 25) = 0$

$s = 0$ or $6s - 25 = 0$

$\qquad\qquad$ $6s = 25$

$\qquad\qquad$ $s = \dfrac{25}{6}$

13. $x^2 - x - 110 = 0$

$x = \dfrac{-(-1) \pm \sqrt{(-1)^2 - 4(1)(-110)}}{2(1)}$

$= \dfrac{\pm 21}{2}$

$x = -10$ or $x = 11$

17. $2x^2 - x = 36$

$2x^2 - x - 36 = 0$

$x = \dfrac{-(-1) \pm \sqrt{(-1)^2 - 4(2)(-36)}}{2(2)}$

$= \dfrac{1 \pm \sqrt{289}}{4}$

$= \dfrac{1 \pm 17}{4}$

$x = \dfrac{9}{2}$ or $x = -4$

21. $2.1x^2 + 2.3x + 5.5 = 0$

$x = \dfrac{-2.3 \pm \sqrt{2.3^2 - 4(2.1)(5.5)}}{2(2.1)}$

$= \dfrac{-2.3 \pm \sqrt{-40.91}}{4.2}$

25. $x^2 + 4x - 4 = 0$

$x = \dfrac{-4 \pm \sqrt{4^2 - 4(1)(-4)}}{2(1)}$

$= \dfrac{-4 \pm \sqrt{32}}{2}$

$= \dfrac{-4 \pm 4\sqrt{2}}{2}$

$x = -2 \pm 2\sqrt{2}$

29. $4v^2 = v + 5$

$4v^2 - v - 5 = 0$

$(v+1)(4v-5) = 0$

$v+1 = 0$ or $4v-5 = 0$

$v = -1$ $\qquad$ $4v = 5$

$\qquad\qquad$ $v = \dfrac{5}{4}$

33. $a^2x^2 + 2ax + 2 = 0$

$x = \dfrac{-2a \pm \sqrt{(2a)^2 - 4(a^2)(2)}}{2(a^2)}$

$x = \dfrac{-2a \pm \sqrt{-4a^2}}{2a^2}$ and for $a > 0$,

$x = \dfrac{-2a \pm 2a\sqrt{-1}}{2a^2}, \quad x = \dfrac{-1 \pm \sqrt{-1}}{a}$

37. $x^2 - x - 30 = 0$

$x^2 - x + \dfrac{1}{4} = 30 + \dfrac{1}{4} = \dfrac{121}{4}$

$\left(x - \dfrac{1}{2}\right)^2 = \dfrac{121}{4}$

$x - \dfrac{1}{2} = \dfrac{\pm 11}{2}$

$x = \dfrac{1}{2} + \dfrac{\pm 11}{2} = -5, 6$

41.
$$\frac{x-4}{x-1} = \frac{2}{x}$$
$$x(x-4) = 2(x-1)$$
$$x^2 - 4x = 2x - 2$$
$$x^2 - 6x + 2 = 0$$
$$x = \frac{-(-6) \pm \sqrt{(6)^2 - 4(1)(2)}}{2(1)}$$
$$= \frac{6 \pm \sqrt{28}}{2}$$
$$= \frac{6 \pm \sqrt{4 \times 7}}{2}$$
$$= \frac{6 \pm \sqrt{7}}{2}$$
$$x = 3 \pm \sqrt{7}$$

45. $y = 2x^2 - x - 1$; $a = 2$, $b = -1$, $c = -1$

$c = -1 \Rightarrow y\text{-intercept} = -1$

$2x^2 - x - 1 = 0 \Rightarrow x = -\frac{1}{2}$, $x = 1$, the x-intercepts

$x \text{ vertex} = \frac{-b}{2a} = \frac{-(-1)}{2(2)} = \frac{1}{4}$

$y \text{ vertex} = 2\left(\frac{1}{4}\right)^2 - \left(\frac{1}{4}\right) - 1 = -\frac{9}{8}$

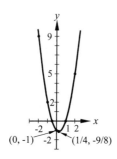

49. Graph $y_1 = 2x^2 + x - 4$ and use the zero feature.

$x = -1.7, 1.2$

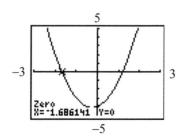

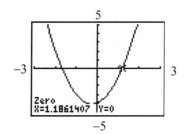

53. The roots are equally spaced on either side of $x = -1$.

$$\frac{x_1 + x_2}{2} = -1$$
$$x_1 + x_2 = -2$$
$$2 + x_2 = -2$$
$$x_2 = -4 \text{ is the other solution.}$$

57. $12x^2 - 80x + 96 = 0 \Rightarrow 3x^2 - 20x + 24 = 0$

$$x = \frac{-(-20) \pm \sqrt{(-20)^2 - 4(3)(24)}}{2(3)} = \frac{10 \pm 2\sqrt{7}}{3}$$

$$x = \frac{10 - 2\sqrt{7}}{3} = 1.569$$

61. $T^2 + 520T - 5300 = 0$

$(T - 10)(T + 530) = 0$

$T - 10 = 0$

$T = 10$

$212 - 10 = 202° \text{F}$

65.
$$\frac{n^2}{500,000} = 144 - \frac{n}{500}$$
$$n^2 + 1000n - 72,000,000 = 0$$
$$(n + 9000)(n - 8000) = 0$$
$$n - 8000 = 0 \quad \text{or} \quad n + 9000 = 0$$
$$n = 8000 \qquad\qquad n = -9000$$
$$\text{reject since } n > 0$$

69. $p_2 = p_1 + rp_1(1 - p_1)$

$p_2 = p_1 + rp_1 - rp_1^2$.

$rp_1^2 - (r + 1)p_1 + p_2 = 0$; $a = r$, $b = -(r + 1)$,

$c = p_2$

73.
$$V = e^3$$
$$V - 29 = (e - 0.10)^3$$
$$= e^3 - 3e^2(0.10) + 3e(0.10)^2 - (0.10)^3$$
$$e^3 - 29 = e^3 - 3(0.10)e^2 + 3(0.10)^2 e - (0.10)^3$$
$$3(0.10)e^2 - 3(0.10)^2 e - 29 + (0.10)^3 = 0$$
from which $e = 9.9$ cm

77. $V = 4.00(16.0)(12.0) = 768$
$$0.9(768) = 4.00(16.0 - x)(12.0 - x)$$
$$\Rightarrow x = 0.703$$
$$16.0 - 0.703 = 15.297$$
$$12.0 - 0.703 = 11.297$$
new dimensions: $l = 15.3$ cm, $w = 11.3$ cm

81. Suppose n poles are placed along the road for a distance of 1 km and x is the distance, in km, between the poles, then $n \times x = 1$. Increasing the distance between the poles to $x + 0.01$ and decreasing the number of poles to $n - 5$ gives $(n - 5)(x + 0.01) = 1$. Substitution gives

$$(n - 5)\left(\frac{1}{n} + 0.01\right) = 1$$
$$1 + 0.01n - \frac{5}{n} - 0.05 = 1$$
$$0.01n^2 - 0.05n - 5 = 0$$
$$n^2 - 5n - 500 = 0$$
$$(n + 20)(n - 25) = 0$$
$$n + 20 = 0 \quad \text{or} \quad n - 25 = 0$$
$$n = -20 \qquad n = 25$$

There are 25 poles being placed each kilometer.

85. Multiply both sides by $2R(R+1)$ to obtain
$$R(R+1) = 2(R+1) + 2R \text{ which simplifies to}$$
$$R^2 - 00.562 \text{ which is rejected since } R > 0 \text{ and}$$
$$R = 3.56$$

Chapter 8

TRIGONOMETRIC FUNCTIONS OF ANY ANGLE

8.1 Signs of the Trigonometric Functions

1. (a) $\sin\left(150^\circ + 90^\circ\right)$ is $-$

   $\cos\left(290^\circ + 90^\circ\right)$ is $+$

   $\tan\left(190^\circ + 90^\circ\right)$ is $-$

   $\cot\left(260^\circ + 90^\circ\right)$ is $-$

   $\sec\left(350^\circ + 90^\circ\right)$ is $+$

   $\csc\left(100^\circ + 90^\circ\right)$ is $-$

   (b) $\sin\left(300^\circ + 90^\circ\right)$ is $+$

   $\cos\left(150^\circ + 90^\circ\right)$ is $-$

   $\tan\left(100^\circ + 90^\circ\right)$ is $+$

   $\cot\left(300^\circ + 90^\circ\right)$ is $+$

   $\sec\left(200^\circ + 90^\circ\right)$ is $+$

   $\csc\left(250^\circ + 90^\circ\right)$ is $-$

5. $\csc 98^\circ$ is positive since 98° is in QII, where $\csc\theta$ is positive.

   $\cot 82^\circ$ is positive since 82° is in QI, where $\cot\theta$ is positive.

9. $\cos 348^\circ$ is positive since 348° is in QIII, where $\cos\theta$ is positive.

   $\csc 238^\circ$ is negative since 238° is in QIII, where $\csc\theta$ is negative.

13. $\cot -2^\circ$ is negative since -2° is in QIV, where $\cot\theta$ is negative.

   $\cos 710^\circ = \cos\left(710^\circ - 360^\circ\right) = \cos 350^\circ$ which is positive since 350° is in QIV, where $\cos\theta$ is positive.

17. $\left(-2, -3\right)$, $x = -2$, $y = -3$, $r = \sqrt{x^2 + y^2} = \sqrt{13}$

   $\sin\theta = \dfrac{y}{r} = \dfrac{-3}{\sqrt{13}}$

   $\cos\theta = \dfrac{x}{r} = \dfrac{-2}{\sqrt{13}}$

   $\tan\theta = \dfrac{y}{x} = \dfrac{-3}{-2} = \dfrac{3}{2}$

   $\csc\theta = \dfrac{r}{y} = \dfrac{\sqrt{13}}{-3}$

   $\sec\theta = \dfrac{r}{x} = \dfrac{\sqrt{13}}{-2}$

   $\cot\theta = \dfrac{x}{y} = \dfrac{-2}{-3} = \dfrac{2}{3}$

21. $\left(20, -8\right)$; $x = 20$, $y = -8$, $r = \sqrt{x^2 + y^2} = 4\sqrt{29}$

   $\sin\theta = \dfrac{y}{r} = \dfrac{-8}{4\sqrt{29}} = \dfrac{-2}{\sqrt{29}}$

   $\cos\theta = \dfrac{x}{r} = \dfrac{20}{4\sqrt{29}} = \dfrac{5}{\sqrt{29}}$

   $\tan\theta = \dfrac{y}{x} = \dfrac{-8}{20} = -\dfrac{2}{5}$

   $\csc\theta = \dfrac{r}{y} = \dfrac{4\sqrt{29}}{-8} = -\dfrac{\sqrt{29}}{2}$

   $\csc\theta = \dfrac{r}{y} = \dfrac{4\sqrt{29}}{-8} = -\dfrac{\sqrt{29}}{2}$

   $\sec\theta = \dfrac{r}{x} = \dfrac{4\sqrt{29}}{20} = \dfrac{\sqrt{29}}{5}$

   $\cot\theta = \dfrac{x}{y} = \dfrac{20}{-8} = -\dfrac{5}{2}$

25. $\tan\theta = 1.500 \Rightarrow$ QI, QIII

29. $\sin \theta$ is positive and $\cos \theta$ is negative

$\sin \theta$ is positive in QI and QII

$\cos \theta$ is negative in QII and QIII. The

terminal side of θ is in QII.

33. $\csc \theta$ is negative and $\tan \theta$ is negative

$\csc \theta$ is negative in QIII and QIV

$\tan \theta$ is negative in QII and QIV. The

terminal side of θ is in QIV.

37. $\sin \theta$ is positive and $\cot \theta$ is negative

$\sin \theta$ is positive in QI and QII

$\cot \theta$ is negative in QII and QIV. The

terminal side of θ is in QII.

41. For (x, y) in QIV, $x(+)$ and $y(-) \Rightarrow \dfrac{y}{x} = \dfrac{(-)}{(+)} = (-)$

8.2   Trigonometric Functions of Any Angle

1. $\sin 200° = -\sin 20° = -0.3420$

$\tan 150° = -\tan 30° = -0.5774$

$\cos 265° = -\cos 85° = -0.0872$

$\cot 300° = -\cot 60° = -0.5774$

$\sec 344° = \sec 16° = 1.040$

$\sin 397° = \sin 37° = 0.6018$

5. $\sin 160° = \sin \left(180° - 160°\right) = \sin 20°$

$\cos 220° = \cos \left(180° + 40°\right) = -\cos 40°$

9.

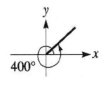

$$\sec 400° = \sec\left(360° + 40°\right)$$
$$= \sec 40°$$

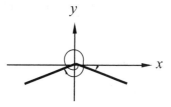

$$\sin\left(-520°\right) = \sin\left(200°\right) = -\sin 20°$$

13. $\cos 106.3° = -\cos 73.7° = -0.281$

17. $\tan\left(-91.5°\right) = \tan 88.5° = 38.12$

21. $\sin 310.36° = -0.7620$

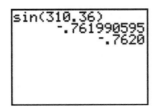

25. $\cos\left(-72.61°\right) = 0.2989$

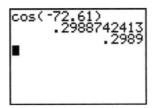

29. $\cos \theta = 0.4003;\ \theta_{ref} = \cos^{-1} 0.4003 = 66.4°$

Since $\cos \theta$ is positive, θ is in QI or QIV.

Therefore, $\theta = 66.40°$

or $\theta = 293.60°$

$0° \le \theta < 360°$

33. $\sin \theta = 0.870;\ \theta_{ref} = \sin^{-1} 0.870 = 60.4°$

Since $0° \le \theta < 360°$, $\cos\theta < 0$, θ is in QII and

$\theta = 119.5°$.

37. $\csc \theta = -1.366; \; \theta = \sin^{-1}\left(\dfrac{1}{-1.366}\right) = -47.1°$

Since $0° \le \theta < 360°$, $\cos\theta > 0$, θ is in QI or QIV

$\theta = 312.9°$.

41. $\cos 60° + \cos 70° + \cos 110°$

$= \cos 60° + \cos 70° - \cos 70°$

$= \dfrac{1}{2}$

45. $\sin \theta = -0.5736; \; \theta_{ref} = \sin^{-1}(-0.5736) = -35°$

Since $\cos \theta > 0$, θ is in QIV.

$\theta = 325° + k \times 360°$ where $k = 0, \pm 1, \pm 2, \cdots$

$\tan \theta = -0.7003$

49. $\sin 90° = 1, \; 2\sin 45° = 2 \times \dfrac{\sqrt{2}}{2} = \sqrt{2}$

$\qquad 1 < \sqrt{2}$

$\sin 90° < 2\sin 45°$

53. $\theta = 195°$ has $\theta_{ref} = 15° \Rightarrow \cos 195° = -\cos 15°$

$\qquad\qquad\qquad\qquad\qquad\quad = -\sin 75°$

$\qquad\qquad\qquad\qquad\qquad\quad = -0.9659$

57. $A + B + C = 180° \Rightarrow A = 180° - (B+C)$

$\tan A + \tan (B+C)$

$= \tan\big(180° - (B+C)\big) + \tan (B+C)$

$= -\tan (B+C) + \tan (B+C)$

$= 0$

61. $y \sin \alpha = x \sin \beta; \; x = 6.78, \; \alpha = 31.3°, \; \beta = 104.7°$

$y = \dfrac{x \sin \beta}{\sin \alpha}$

$= \dfrac{(6.78)\sin 104.7°}{\sin 31.3°}$

$= \dfrac{(6.78)(0.9673)}{(0.5195)}$

$= 12.6$ in.

8.3   Radians

1. $2.80 = \left(\dfrac{180°}{\pi}\right)(2.80) = 160°$

5. $15° = \dfrac{\pi}{180}(15) = \dfrac{\pi}{12}$

$150° = \dfrac{\pi}{180}(150) = \dfrac{5\pi}{6}$

9. $210° = \dfrac{\pi}{180}(210) = \dfrac{7\pi}{6}$

$117° = \dfrac{\pi}{180}(117) = \dfrac{13\pi}{20}$

13. $\dfrac{2\pi}{5} = \dfrac{180°}{\pi} \cdot \left(\dfrac{2\pi}{5}\right) = 72°$

$\dfrac{3\pi}{2} = \dfrac{180°}{\pi} \cdot \left(\dfrac{3\pi}{2}\right) = 270°$

17. $\dfrac{7\pi}{18} = \dfrac{180°}{\pi} \cdot \left(\dfrac{7\pi}{18}\right) = 70°$

$\dfrac{5\pi}{60} = \dfrac{180°}{\pi} \cdot \left(\dfrac{5\pi}{6}\right) = 150°$

21. $23.0° = \dfrac{\pi}{180}(23.0°) = 0.401$

25. $-333.5° = \dfrac{\pi}{180}(-333.5) = -5.821$

29. $0.750 = \dfrac{180°}{\pi} \cdot (0.750) = 43.0°$

33. $12.4 = \dfrac{180°}{\pi} \cdot (12.4) = 710°$

37. $\sin \dfrac{\pi}{4} = \sin\left[\left(\dfrac{\pi}{4}\right)\left(\dfrac{180}{\pi}\right)\right] = \sin 45° = 0.7071$

41. $\cos \dfrac{5\pi}{6} = \cos\left[\left(\dfrac{5\pi}{6}\right)\left(\dfrac{180}{\pi}\right)\right] = \cos 150° = -0.8660$

45. $\tan 0.7359 = 0.9056$

49. $\sec 2.07 = \dfrac{1}{\cos 2.07} = -2.1$

53. $\sin \theta = 0.3090,\ \theta = 0.3141$
$\sin \theta$ is positive, θ is in QI and QII.
QI, $\theta = 0.3141$
$QII,\ \theta = \pi - 0.3141 = 2.827$

57. $\cos \theta = 0.6742,\ \theta = 0.8309$
$\cos \theta$ is positive, θ is in QI and QIV.
QI, $\theta = 0.8309$
$QIV,\ \theta = 2\pi - 0.8309 = 5.452$

61. $\theta = \dfrac{s}{r} = \dfrac{15\ \text{cm}}{12\ \text{cm}} = \dfrac{5}{4}$

65. For $0 < \theta < \dfrac{\pi}{2},\ \dfrac{\pi}{2} < \dfrac{\pi}{2} + \theta < \pi \Rightarrow \dfrac{\pi}{2} + \theta$ is in QII.

$\dfrac{\pi}{2} + \theta$ has $\theta_{\text{ref}} = \dfrac{\pi}{2} - \theta \Rightarrow \tan\left(\dfrac{\pi}{2} + \theta\right)$

$$= -\tan\left(\theta - \dfrac{\pi}{2}\right) = -\cot \theta$$

69. $1.75(2\pi) = \dfrac{7\pi}{2}\ \text{rad} = 11.0\ \text{rad}$

73. $h = 1200 \tan \dfrac{5t}{3t+10}\ t < 10\ \text{s}.$ Let $t = 8.0\ \text{s}$

$h = 1200 \tan \dfrac{5(8.0)}{3(8.0)+10} = 1200 \tan \dfrac{40.0}{34.0}$

$= 1200 \tan 2.403 = 2900\ \text{m}$

8.4   Applications of Radian Measure

1. $s = \left(\dfrac{\pi}{4}\right)(3.00) = 2.36\ \text{in.}$

5. $s = r\theta = (3.30)\left(\dfrac{\pi}{3}\right) = 3.46\ \text{in.}$

9. $\theta = \dfrac{s}{r} = \dfrac{0.3913}{0.9449} = 0.4141 = 23.73°$

$A = \dfrac{1}{2}\theta r^2 = \dfrac{1}{2}(0.4141)(0.9449)^2 = 0.1849\ \text{mi}^2$

13. $A = \dfrac{1}{2}r^2\theta \Rightarrow r = \sqrt{\dfrac{2A}{\theta}} = \sqrt{\dfrac{2(0.0119)}{326°\,\frac{\pi}{180}}}$

$r = 0.0647\ \text{ft}$

17. $r = \dfrac{s}{\theta} = \dfrac{0.203}{\frac{3}{4}(2\pi)}$

$r = 0.0431\ \text{mi}$

21. From $\theta = wt$,

hour hand: $\theta = \dfrac{2\pi}{12}t$

minute hand: $\theta + \pi = \dfrac{2\pi}{1}t,\ t$ in hours

$\dfrac{\pi}{6}t + \pi = 2\pi t \Rightarrow t = \dfrac{6}{11}\ \text{hour} = 32.73\ \text{minutes}$

$t = 32\ \text{minutes}\ 44\ \text{seconds}$

at 32 minutes and 44 seconds after noon the hour
and minute hands will be at $180°$.

25. $w = \dfrac{\theta}{t} = \dfrac{\pi}{6.0}\ \text{rad/s} = 0.52\ \text{rad/s}$

29.

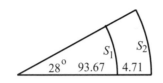

From $S = r\theta$

$$S_1 = 93.67(28.0°)\frac{\pi}{180°}$$

$$S_1 = 45.78$$

$$S_2 = (93.67 + 4.71)(28.0°)\frac{\pi}{180°}$$

$$S_2 = 48.08$$

$$S_2 - S_1 = 2.30 \text{ ft. Outer rail is 2.30 ft longer.}$$

33. $V = At = \left[\frac{1}{2}r_1^2\theta - \frac{1}{2}r_2^2\theta\right]t = \frac{1}{2}\theta\left(r_1^2 - r_2^2\right)t$

$$V = \frac{1}{2}(15.6°)\frac{\pi}{180°}\left((285+15.2)^2 - 285^2\right)(0.305)$$

$$V = 369 \text{ m}^3$$

37. $v = rw = 8.5(20)(2\pi) = 1070 \text{ ft/min}$

41. $v = \frac{2\pi r}{t} \cdot \frac{2\pi(3,500,000 \text{ mi})}{2.88da} \cdot \frac{da}{24 \text{ h}}$

$$v = 3.18 \times 10^5 \text{ mi/h}$$

45.

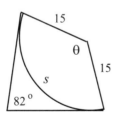

$$82.0° + 2(90°) + \theta = 360°$$

$$\theta = 98.0°$$

$$s = r\theta$$

$$s = 15.0(98.0°)\frac{\pi}{180°}$$

$$s = 25.7 \text{ ft}$$

49. $w = \frac{v}{r} = \frac{\frac{1}{4}(6.5)}{3.75} = 0.433 \text{ rad/s}$

53. $\theta = wt = 2400 \cdot \frac{\text{r}}{\text{min}} \cdot \frac{2\pi \text{ rad}}{\text{r}} \cdot \frac{\text{min}}{60 \text{ s}}(1 \text{ s})$

$$\theta = 80\pi \text{ rad} = 250 \text{ rad}$$

57. $x = \sqrt{1.10^2 - 0.74^2} = 0.81;$

$$\theta = \sin^{-1}\left(\frac{0.74}{1.10}\right) = 42.3°;$$

$$2\theta = 84.6° \frac{\pi}{180°} = 1.477$$

$$A_{\text{sector}} = \frac{1}{2}(1.10)^2(1.477) = 0.89$$

$$A_{\text{triangle}} = \frac{1}{2}(1.48)^2(0.81) = 0.60$$

$$A_{\text{segment}} = 0.89 - 0.60 = 0.29$$

$$A_{\text{circle}} = \pi(1.10)^2 = 3.80$$

$$A_{\text{tank}} = 3.80 - 0.29 = 3.51$$

$$A = 3.51 \times 4.25 = 14.9 \text{ m}^3$$

61. $0.2'' = \frac{0.2°}{3600}$

$$= (5.556 \times 10)° \frac{\pi}{180°}$$

$$= 9.696 \times 10^{-7} \text{ rad}$$

$$12.5 \text{ light years} = (12.5)(5.88 \times 10^{12})\text{min}$$

$$\tan 0.2'' = \frac{x}{12.5 \times 5.88 \times 10^{12}}$$

$$x = (12.5)(5.88 \times 10^{12})(9.696 \times 10^{-5})$$

$$= 7.13 \times 10^7 \text{ mi}$$

Chapter 8 Review Exercises

1. $r = \sqrt{6^2 + 8^2} = 10$ for $(6, 8)$

$\sin\theta = \dfrac{y}{r} = \dfrac{8}{10} = \dfrac{4}{5}$

$\cos\theta = \dfrac{x}{r} = \dfrac{6}{10} = \dfrac{3}{5}$

$\tan\theta = \dfrac{y}{x} = \dfrac{8}{6} = \dfrac{4}{3}$

$\csc\theta = \dfrac{r}{y} = \dfrac{5}{4}$

$\sec\theta = \dfrac{r}{x} = \dfrac{5}{3}$

$\cot\theta = \dfrac{x}{y} = \dfrac{3}{4}$

5. $\cos 132° = -\cos(180° - 132°) = -\cos 48°$

$\tan 194° = \tan(194° - 180°) = \tan 14°$

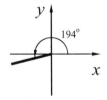

9. $40° \cdot \dfrac{\pi}{180°} = \dfrac{2\pi}{9}$

$153° \cdot \dfrac{\pi}{180°} = \dfrac{17\pi}{20}$

13. $\dfrac{7\pi}{5} \cdot \dfrac{180°}{\pi} = 252°$; $\dfrac{13\pi}{18} \cdot \dfrac{180°}{\pi} = 130°$

17. $0.560 \cdot \dfrac{180°}{\pi} = 32.1°$

21. $102° \cdot \dfrac{\pi}{180°} = 1.78$

25. $262.05° \cdot \dfrac{\pi}{180°} = 4.5736$

29. $\cos 265.5° = -0.0785$

33. $\csc 247.82° = -1.080$

37. $\tan 301.4° = -1.64$

41. $\sin \dfrac{9\pi}{4} = -0.5878$

45. $\sin 0.5906 = 0.5569$

49. $\tan\theta = 0.1817, 0 \le \theta < 360°$

$\theta = \tan^{-1}(0.1817) = 10.30°$ in QI

$\theta = 180° + 10.3° = 190.30°$ in QIII

53. $\cos\theta = 0.8387, 0 \le \theta < 2\pi$

$\theta = \cos^{-1}(0.8387) = 0.5759$ in QI

$\theta = 2\pi - 0.5759 = 5.707$ in QIV

57. $\cos\theta = -0.7222,\ \sin\theta < 0$ for $0° \le \theta < 360° \Rightarrow \theta$ in QIII

$\theta = \cos^{-1}(-0.7222) = 136.2364165$ from

calculator reference angle $= 180° - \theta = 43.76°$

QIII angle $= 180° +$ reference angle $= 223.76°$.

61. $r = \dfrac{s}{\theta} = \dfrac{20.3 \text{ in.}}{107.5° \cdot \dfrac{\pi}{180°}} = 10.8 \text{ in.}$

65. $s = r\theta = r\dfrac{2A}{r^2} = \dfrac{2A}{r}$

$s = \dfrac{2(32.8)}{4.62} = 14.2 \text{ m}$

69. $\tan 200° + 2 \cot 110° + \tan \left(-160°\right)$

$= \tan 20° + 2\left(-\cot 70°\right) + \tan 20°$

$= 2 \tan 20° - 2 \tan 20°$

$= 0$

73. $A = A_{\text{triangle}} + A_{\text{sector}} = \dfrac{1}{2}bh + \dfrac{1}{2}r^2\theta$

$A = \dfrac{1}{2}\left(40.0\right)\left(30.0\right) + \dfrac{1}{2}\left(40.0^2 + 30.0^2\right) \cdot 20° \dfrac{\pi}{180°}$

$A = 1040 \text{ m}^2$

(b) perimeter $= 40.0 + 30.0 + 50.0\left(1 + 20° \dfrac{\pi}{180°}\right)$

$= 137.5 \text{ m}$

77. $\theta = \dfrac{s}{r} = \dfrac{6.60}{8.25} = 0.800 = 45.8°$

81. velocity $= \dfrac{\text{distance}}{\text{time}} = \dfrac{2\pi\left(240,000\right) \text{ mi}}{28 \text{ day}} \cdot \dfrac{\text{day}}{24 \text{ h}}$

velocity $= 2200 \text{ mi/h}$

85. velocity $= \dfrac{\text{distance}}{\text{time}} = \dfrac{\pi\left(25.0\right)}{\frac{1}{60}} = 4710 \text{ cm/s}$

89. $v = rw = \dfrac{7.20}{2}\left(80,000 \text{ r/min}\right)\dfrac{2\pi}{\text{r}}$

$v = 1.81 \times 10^6 \text{ cm/min}$

93. $r = \dfrac{s}{\theta} = \dfrac{2.50}{\frac{0.0008°}{2} \cdot \frac{\pi}{180°}}$

$r = 3.58 \times 10^5 \text{ km}$

Chapter 9

VECTORS AND OBLIQUE TRIANGLES

9.1   Introduction to Vectors

1.

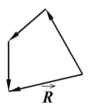

5. (a) scalar, no direction given

    (b) vector, magnitude and direction both given

9.

13.

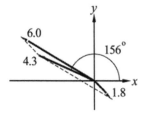

17. 4.3 cm, 156°

21.

25.

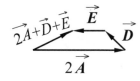

29.

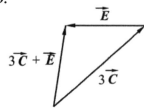

33.

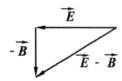

37.

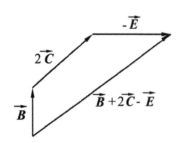

41.

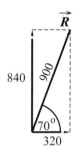

840   900

70°
320

From drawing $\vec{R}$ is approximately 900 lb at 70°

45.

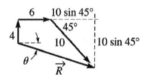

6   10 sin 45°
45°
4   10   10 sin 45°
θ
$\vec{R}$

From the drawing,

$R = 13$ mi

$\theta = 13°$

9.2 Components of Vectors

1.

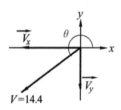

$V_x = V \cos \theta$

$V_x = 14.4 \cos 216° = -11.6$

$V_y = V \sin \theta$

$V_y = 14.4 \sin 216° = -8.46$

5. $V_x = 750 \cos 28° = 662$

   $V_y = 750 \sin 28° = 352$

9. $V_x = -750$

   $V_y = 0$

13. Let $V = 76.8$

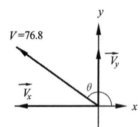

$V_x = V \cos 145.0° = 76.8(-0.819) = -62.9$ m/s

$V_y = V \sin 145.0° = 76.8(0.574) = 44.1$ m/s

17. Let $V = 2.65$

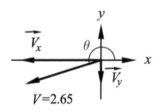

$V_x = V \cos 197.3° = 2.65(-0.955) = -2.53$ mN

$V_y = V \sin 197.3° = 2.65(-0.297) = -0.788$ m/N

9.3   Vector Addition by Components

21.

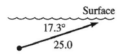

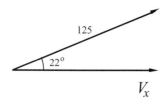

$V_x = 25.0\cos 17.3^\circ = 23.9$ km/h

$V_y = 25.0\sin 17.3^\circ = 7.43$ km/h

25.

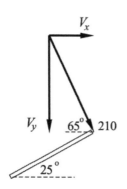

$V_x = 125\cos 22^\circ = 116$ km/h

29.

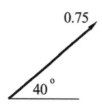

$V_x = 210\cos 65^\circ = 89$ N

$V_y = -210\sin 65^\circ = -190$ N

33.

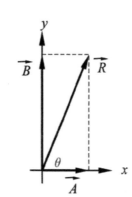

horizontal component $= 0.75\cos 40^\circ$

$= 0.57$ (km/h)/m

vertical component $= 0.75\sin 40^\circ$

$= 0.48$ (km/h)/m

1. $A = 1200 = A_x,\ B = 1750$

$Ay = 0$

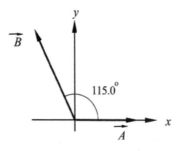

$R_x = A_x + B_x = 1200 + 1750\cos 115^\circ = 460.4$
$R_y = A_y + B_y = 0 + 1750\sin 115^\circ = 1586$
$R = \sqrt{R_x^2 + R_y^2} = \sqrt{460.4^2 + 1586^2} = 1650$

$\theta = \tan^{-1}\dfrac{R_y}{R_x} = \tan^{-1}\dfrac{1586}{460.4} = 73.8^\circ$

5.

$R = \sqrt{3.086^2 + 7.143^2} = \sqrt{60.54} = 7.781$

$\tan\theta = \dfrac{7.143}{3.086} = 2.315$

$\theta = 66.63^\circ \ (\text{with } \vec{A})$

9.

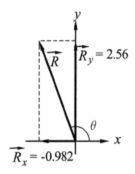

$R_x = -0.982, R_y = 2.56$

$R = \sqrt{R_x^2 + R_y^2} = \sqrt{(-0.982)^2 + 2.56^2} = 2.74$

$\tan\theta_{\text{ref}} = \left|\dfrac{2.56}{-0.982}\right| = 2.61$

$\quad \theta_{\text{ref}} = 69.0°$
$\qquad \theta = 180° - 69.0° = 111.0°$

(θ is in Quad II since R_x is negative and R_y is positive)

13.

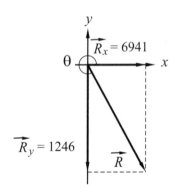

$\quad R_x = 6941, \; R_y = -1246$

$\quad R = \sqrt{6941^2 + (-1246)^2} = 7052$

$\tan\theta_{\text{ref}} = \left|\dfrac{-1246}{6941}\right| = 0.1795$

$\quad \theta_{\text{ref}} = 10.18°$

$\qquad \theta = 360° - 10.18° = 349.82°$

(θ is in QIV since R_x is positive and R_y is negative)

17.

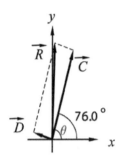

$C = 5650, \theta_C = 76.0°$
$C_x = 5650\cos 76.0° = 1370$
$C_y = 5650\sin 76.0° = 5480$
$D = 1280, \theta_D = 160.0°$
$D_x = 1280\cos 160.0° = -1200$
$D_y = 1280\sin 160.0° = 438$
$R_x = 1370 - 1200 = 170$
$R_y = 5480 + 438 = 5920$
$R = \sqrt{170^2 + 5920^2} = 5920$

$\tan\theta = \dfrac{R_y}{R_x} = \dfrac{5920}{170}, \theta = 88.4°$

21.

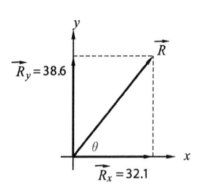

$R_x = A_x + B_x + C_x$
$R_x = 21.9\cos 236.2° + 96.7\cos 11.5°$
$\qquad + 62.9\cos 143.4°$
$R_y = A_y + B_y + C_y$
$R_y = 21.9\sin 236.2° + 96.7\sin 11.5°$
$\qquad + 62.9\sin 143.4°$
$R = \sqrt{R_x^2 + R_y^2} = 50.2$

$\theta = \tan^{-1}\dfrac{R_y}{R_x} = 50.3°$

	Vector	Magnitude	Ref. Angle
25.	A	318	$67.5°$
	B	245	$16.3°$

	x-component	y-component
	$-318\cos 67.5° = -122$	$318\sin 67.5° = 294$
	$245\sin 16.3° = 68.8$	$245\cos 16.3° = 235$
R	-53.2	529

$$\theta_{\text{ref}} = \tan^{-1}\left|\frac{529}{-53.2}\right| = 84.3°$$

$$\theta = 180° - 84.3° = 95.7°$$

$$R = \sqrt{(-53.2)^2 + 529} = 531 \text{ m}$$

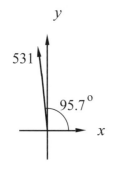

29.

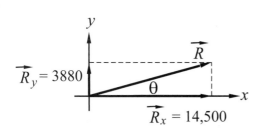

$R_x = 5500 + 6500\cos 15.5° + 3500\cos 37.7°$
$\quad = 14{,}500$
$R_y = 6500\sin 15.5° + 3500\sin 37.7°$
$\quad = 3880$
$R = \sqrt{14{,}500^2 + 3880^2} = 15{,}000 \text{ lb}$

$\theta = \tan^{-1}\dfrac{3880}{14{,}500} = 15.0°$ down from 5500 lb force

33.

$$R = \sqrt{R_x^2 + R_y^2}$$

$$= \sqrt{\left(15\cos 72° + 15\right)^2 + \left(15\sin 72°\right)^2}$$

$$= 24.3 \text{ kg} \cdot \text{m/s}$$

9.4   Applications of Vectors

1.

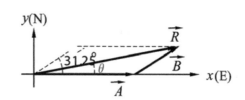

$R_x = A + B_x = 32.50 + 16.18\cos 31.25° = 46.33$
$R_y = 16.18\sin 31.25° = 8.394$
$R = \sqrt{46.33^2 + 8.394^2} = 47.08 \text{ mi}$
$\theta = \tan^{-1}\dfrac{8.394}{46.33} = 10.27°$

The ship is 47.08 mi from start in direction $10.27°$ N of E.

5.

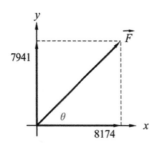

$F_x = 8300\cos 10.0°$
$\quad = 8174 \text{ N}$
$F_y = 8300\sin 10.0° + 6500$
$\quad = 7941 \text{ N}$
$F = \sqrt{8174^2 + 7941^2}$
$\quad = 11{,}000 \text{ N}$

$\tan\theta = \dfrac{7941}{8174} = 0.97, \theta = 44°$ above horizontal

9.

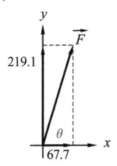

$F_x = 358.2 \cos 37.72° - 215.6 = 67.7$
$F_y = 358.2 \sin 37.72° = 219.1$
$F = \sqrt{67.7^2 + 219.1^2}$
$\quad = 229.4 \, \text{ft}$

$\tan \theta = \dfrac{219.1}{67.74} = 3.234, \ \theta = 72.82°, \ \text{N of E}$

13.

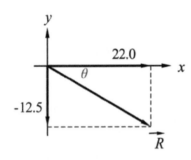

$R = \sqrt{22.0^2 + 12.5^2} = 25.3 \, \text{km/h}$

$\theta = \tan^{-1} \dfrac{12.5}{22.0} = 29.6°$

17.

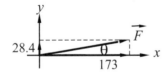

$F_x = 95.0 + 83.0 \cos 20.0° = 173 \, \text{lb}$
$F_y = 83.0 \sin 20.0° = 28.4 \, \text{lb}$
$F = \sqrt{173^2 + 28.4^2} = 175 \, \text{lb}$

$\tan \theta = \dfrac{28.4}{173} = 0.163$

$\theta = 9.3°$, above horizontal and to right.

21. $v_{sx} = 18,250 - 120 \cos 5.20° = 18,130 \, \text{mi/h}$
$v_{sy} = 120 \sin 5.20° = 10 \, \text{mi/h}$
$v = \sqrt{18,130^2 + 10^2} = 18,130 \, \text{mi/h}$

$\tan \theta = \dfrac{10}{18,310} = 0.000055$

$\theta = 0.03°$ from direction of shuttle

25.

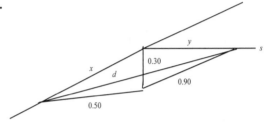

$x = \sqrt{0.50^2 - 0.30^2}, \ y = \sqrt{0.90^2 - 0.30^2}$
$d = \sqrt{x^2 + y^2} = 0.94 \, \text{km}$
$\theta = \tan^{-1} \dfrac{y}{x} = 64.7° \ \text{S of E}$

$w \cos 45° + v \cos 15° = 32$
$w \sin 45° = v \sin 15°$
$v = \dfrac{w \sin 45°}{\sin 15°}$
$w \cos 45° + \dfrac{w \sin 45°}{\sin 15°} \cos 15° = 32$
$w = 9.6 \, \text{km/h}$

29.

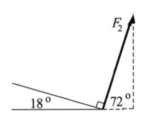

$$F_2 = \frac{1.2F_1}{0.3}$$

$$F_{2v} = F_2 \sin 72°$$

$$= \frac{1.2(240)}{0.3} \sin 72°$$

$$F_{2v} = 910 \text{ N}$$

33.

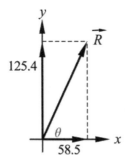

$$x = 75.0 + 75.0 \cos 65.0° - 75.0 \cos 50.0° = 58.5$$
$$y = 75.0 \sin 65.0 + 75.0 \sin 50.0° = 125.4$$
$$R = \sqrt{58.5^2 + 125.4^2} = 138 \text{ mi}$$

$$\theta = \tan^{-1} \frac{125.4}{58.5} = 65.0° \text{ N of E}$$

37.

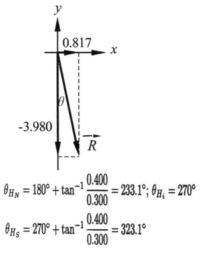

$$\theta_{H_N} = 180° + \tan^{-1} \frac{0.400}{0.300} = 233.1°; \; \theta_{H_i} = 270°$$

$$\theta_{H_S} = 270° + \tan^{-1} \frac{0.400}{0.300} = 323.1°$$

9.5 Oblique Triangles, the Law of Sines

1.

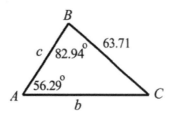

$$C = 180° - (56.29° - 82.94°) = 40.77°$$

$$\frac{b}{\sin 82.94°} = \frac{63.71}{\sin 56.29°} = \frac{c}{\sin 40.77°}$$

$$b = \frac{63.71 \sin 82.94°}{\sin 56.29°} = 76.01$$

$$c = \frac{63.71 \sin 40.77°}{\sin 56.29°} = 50.01$$

5.

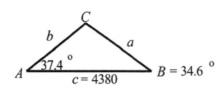

$$c = 4380, \; A = 37.4°, \; B = 34.6°$$
$$C = 180.0° - (37.4° + 34.6°) = 108.0°$$

$$\frac{b}{\sin B} = \frac{c}{\sin C}; \; \frac{b}{\sin 34.6°} = \frac{4380}{\sin 108.0°}$$

$$b = \frac{4380 \sin 34.6°}{\sin 108.0°} = 2620$$

$$\frac{a}{\sin A} = \frac{c}{\sin C}; \; \frac{a}{\sin 37.4°} = \frac{4380}{\sin 108.0°}$$

$$a = \frac{4380 \sin 37.4°}{\sin 108.0°} = 2800$$

9.

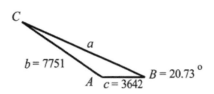

$b = 7751, c = 3642, B = 20.73°$

$$\frac{b}{\sin B} = \frac{c}{\sin C}; \frac{7751}{\sin 20.73°} = \frac{3642}{\sin C}$$

$$\sin C = \frac{3642 \sin 20.73°}{7751} = 0.1663$$

$$C = 9.57°$$

$$A = 180.0° - (20.73° - 9.57°) = 149.70°$$

$$\frac{a}{\sin A} = \frac{b}{\sin B}; \frac{a}{\sin 149.70°} = \frac{7751}{\sin 20.73°}$$

$$a = \frac{7751 \sin 149.70°}{\sin 20.73°} = 11,050$$

13.

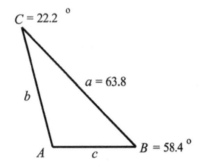

$a = 63.8, B = 58.4°, C = 22.2°$

$$A = 180.0° - 58.4° - 22.2° = 99.4°$$

$$\frac{a}{\sin A} = \frac{b}{\sin B}; \frac{63.8}{\sin 99.4°} = \frac{b}{\sin 58.4°}$$

$$b = \frac{63.8 \sin 58.4°}{\sin 99.4°} = 55.1$$

$$\frac{a}{\sin A} = \frac{c}{\sin C}; \frac{63.8}{\sin 99.4°} = \frac{c}{\sin 22.2°}$$

$$c = \frac{63.8 \sin 22.2°}{\sin 99.4°} = 24.4$$

17.

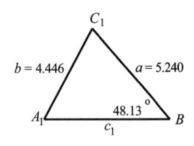

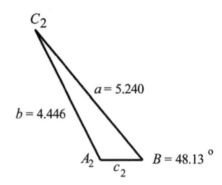

$a = 5.240, b = 4.446, B = 48.13°$

$$\frac{a}{\sin A} = \frac{b}{\sin B}; \frac{5.240}{\sin A} = \frac{4.446}{\sin 48.13°}$$

$$\sin A = \frac{5.240 \sin 48.13°}{4.446} = 0.8776$$

$$A_1 = 61.36°$$

$$C_1 = 180.0° - 48.13° - 61.36° = 70.51°$$

or

$$A_2 = 180.0° - 61.36° = 118.64°$$

and

$$C_2 = 180.0° - 48.13° - 118.64° = 13.23°$$

$$\frac{b}{\sin B} = \frac{c_1}{\sin C_1}; \frac{4.446}{\sin 48.13°} = \frac{c}{\sin 70.51°}$$

$$c_1 = \frac{4.446 \sin 70.51°}{\sin 48.13°} = 5.628$$

or

$$\frac{b}{\sin B} = \frac{c_2}{\sin C_2}; \frac{4.446}{\sin 48.13°} = \frac{c_2}{\sin 13.23°}$$

$$c_2 = \frac{4.446 \sin 13.23°}{\sin 48.13°} = 1.366$$

21.

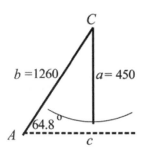

$a = 450$, $b = 1260$, $A = 64.8°$

$$\frac{a}{\sin A} = \frac{b}{\sin B}; \frac{450}{\sin 64.8°} = \frac{1260}{\sin B}$$

$$\sin B = \frac{1260 \sin 64.8°}{450} = 2.53 \text{ (not } \leq 1)$$

Therefore, no solution.

25.

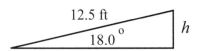

$h = 12.5 \sin 18.0°$

$$\theta = \sin^{-1} \frac{12.5 \sin 18.0°}{22.5}$$

$$\theta = 9.9°$$

29.

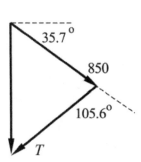

$$\frac{T}{\sin 54.3°} = \frac{850}{\sin 51.3°}; T = \frac{850 \sin 54.3°}{\sin 51.3°} = 880 \text{ N}$$

33.

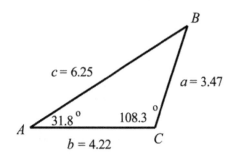

$$\frac{6.25}{\sin 108.3°} = \frac{a}{\sin 31.8°}$$

$$a = \frac{6.25 \sin 31.8°}{\sin 108.3°} = 3.47 \text{ cm}$$

$$B = 180° - 108.3° - 31.8° = 39.9°$$

$$\frac{6.25}{\sin 108.3°} = \frac{b}{\sin 39.9°}$$

$$b = \frac{6.25 \sin 39.9°}{\sin 108.3°} = 4.22 \text{ cm}$$

Perimeter $= 6.25 + 3.47 + 4.22 = 13.94$ cm

37.

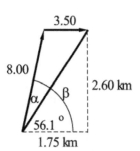

$$\theta = \tan^{-1} \frac{2.60}{1.75} = 56.1°$$

$$\frac{8.00}{\sin 56.1°} = \frac{3.50}{\sin \alpha}$$

$$\alpha = 21.3°,$$

$$\beta = 56.1° + 21.3° = 77.4°$$

with bank downstream

9.6   The Law of Cosines

1.

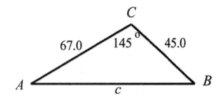

$$c = \sqrt{45.0^2 + 67.0^2 - 2(45.0)(67.0)\cos 145°}$$
$$= 107$$

$$\frac{45.0}{\sin A} = \frac{c}{\sin 145°} = \frac{67.0}{\sin B}$$

$$A = \sin^{-1}\frac{45.0\sin 145°}{c} = 14.0°$$

$$B = \sin^{-1}\frac{67.0\sin 145°}{c} = 21.0°$$

5.

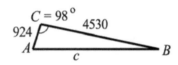

$$a = 4530, b = 924, C = 98.0°$$

$$c = \sqrt{4530^2 + 924^2 + 2(4530)(924)(\cos 98.0°)}$$
$$= 4750$$

$$\frac{c}{\sin C} = \frac{b}{\sin B}; \frac{4750}{\sin 98.0°} = \frac{924}{\sin B}$$

$$\sin B = \frac{924\sin 98.0°}{4750} = -0.193$$

$$B = 11.1°$$
$$A = 180° - 98.0° - 11.1° = 70.9°$$

9.

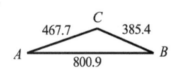

$$a = 385.4, b = 467.7, c = 800.9$$

$$\cos A = \frac{467.7^2 + 800.9^2 - 385.4^2}{2(467.7)(800.9)} = 0.9499$$

$$A = 18.21°$$

$$\cos B = \frac{385.4^2 + 800.9^2 - 467.7^2}{2(385.4)(800.9)} = 0.9253$$

$$B = 22.28°$$
$$C = 180° - 18.21° - 22.28° = 139.51°$$

13.

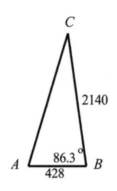

$$a = 2140, c = 428, B = 86.3°$$

$$b = \sqrt{2140^2 + 428^2 - 2(2140)(428)(\cos 86.3°)}$$
$$= 2160$$

$$\frac{b}{\sin B} = \frac{c}{\sin C}; \frac{2160}{\sin 86.3°} = \frac{428}{\sin C}$$

$$\sin C = \frac{428\sin 86.3°}{2160} = 0.198$$

$$C = 11.4°$$
$$A = 180° - 86.3° - 11.4° = 82.3°$$

17.

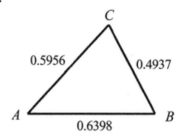

$a = 0.4937, b = 0.5956, c = 0.6398$

$$\cos A = \frac{0.5956^2 + 0.6398^2 - 0.4937^2}{2(0.5956)(0.6398)} = 0.6827$$

$A = 46.94°$

$$\cos B = \frac{0.4937^2 + 0.6398^2 - 0.5956^2}{2(0.4937)(0.6398)} = 0.4723$$

$B = 61.82°$

$C = 180° - 46.94° - 61.82° = 71.24°$

21.

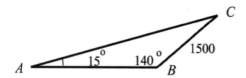

$a = 1500, A = 15°, B = 140°$

$$\frac{a}{\sin A} = \frac{b}{\sin B}; b = \frac{1500 \sin 140°}{\sin 15°} = 3700$$

$C = 180° - 15° - 140° = 25°$

$c = \sqrt{1500^2 + 3700^2 - 2(1500)(3700)(\cos 25°)}$

$= 2400$

25. $c^2 = a^2 + b^2 - 2ab \cos 90°$

$c^2 = a^2 + b^2$

29.

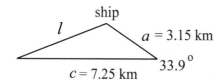

$l = \sqrt{3.15^2 + 7.25^2 - 2(3.15)(7.25)\cos 33.9°}$

$l = 4.96$ km

33.

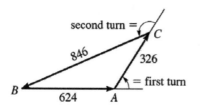

$846^2 = 624^2 + 326^2 - 2(624)(326)\cos A$

$\cos A = -0.5409$

$A = 122.7°$

first turn $= 180° - 122.7° = 57.3°$

$624^2 = 846^2 + 326^2 - 2(846)(326)\cos C$

$\cos C = 0.7843$

$C = 38.3°$

second turn $= 180° - 38.3° = 141.7°$

37.

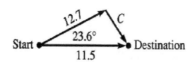

$c^2 = 12.7^2 + 11.5^2 - 2(12.7)(11.5)\cos 23.6°$

$c = 5.09$ km/h

41.

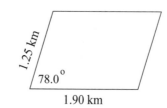

$d = \sqrt{1.25^2 + 1.90^2 - 2(1.25)(1.9)\cos 78.0°}$

$= 2.05$ km

$D = \sqrt{1.25^2 + 1.90^2 - 2(1.25)(1.9)\cos 102.0°}$

$= 2.48$ km

Chapter 9   Review Exercises

1.

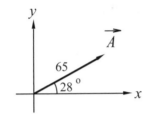

x-component $= 65.0\cos 28.0° = 57.4$

y-component $= 65.0\sin 28.0° = 30.5$

5.

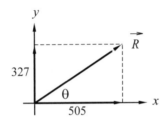

$R = \sqrt{327^2 + 505^2} = 602$

$\theta = \tan^{-1}\dfrac{327}{505} = 32.9°$

9.

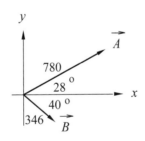

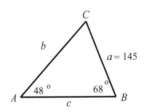

$R_x = 780\cos 28.0° + 346\cos 40.0°$

$R_y = 780\sin 28.0° - 346\sin 40.0°$

$R = \sqrt{R_x^2 + R_y^2} = \sqrt{954^2 + 1442} = 965$

$\theta_R = \tan^{-1}\dfrac{144}{954} = 8.6°$

13.

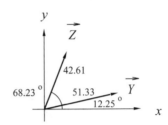

$Y_x = 51.33\cos 12.25° = 5016$

$Y_y = 51.33\sin 12.25° = 10.89$

$Z_x = 42.61\cos 68.23° = 15.80$

$Z_y = -42.61\sin 68.23° = -39.57$

$R_x = 50.16 + 15.80 = 65.98$

$R_y = 10.89 - 39.57 = -28.68$

$R = \sqrt{R_x^2 + R_y^2} = \sqrt{65.98^2 + (-28.68)^2} = 71.94$

$\theta_R = \tan^{-1}\left(\dfrac{-28.68}{65.98}\right) = 336.5°$

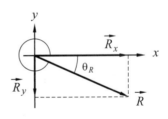

17.

![Triangle with vertices A, B, C; side a = 145 opposite A, side b opposite B, side c along base; angle A = 48°, angle B = 68°]

$$C = 180^\circ - 48.0^\circ - 68.0^\circ = 64.0^\circ$$

$$\frac{145}{\sin 48.0^\circ} = \frac{b}{\sin 68.0^\circ} = \frac{c}{\sin 64.0^\circ}$$

$$b = \frac{145 \sin 68.0^\circ}{\sin 48.0^\circ} = 181,$$

$$c = \frac{145 \sin 64.0^\circ}{\sin 48.0^\circ} = 175$$

21.

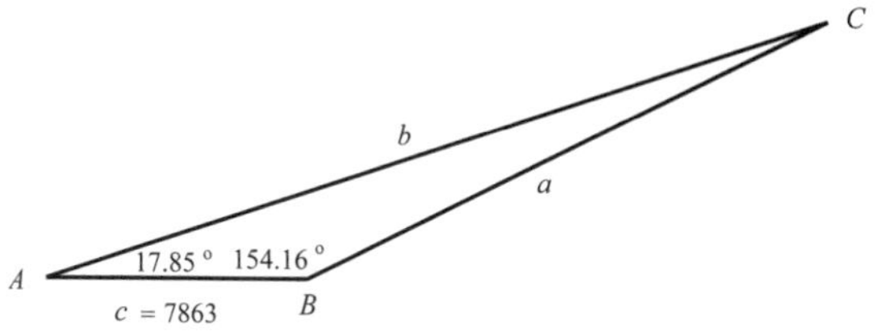

$$C = 180.0^\circ - 17.85^\circ - 154.16^\circ = 7.99^\circ$$

$$\frac{a}{\sin 17.85^\circ} = \frac{b}{\sin 154.16^\circ} = \frac{7863}{\sin 7.99^\circ}$$

$$b = \frac{7863 \sin 154.16^\circ}{\sin 7.99^\circ} = 24{,}660$$

$$a = \frac{7863 \sin 17.85^\circ}{\sin 7.99^\circ} 17{,}340$$

25.

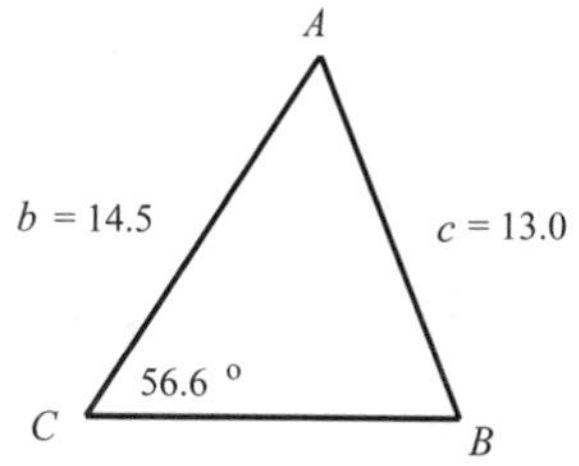

$$\frac{a}{\sin A} = \frac{14.5}{\sin B} = \frac{13.0}{\sin 56.6^\circ}$$

$$\sin B = \frac{14.5 \sin 56.6^\circ}{13.0}$$

$$B = 68.6^\circ \text{ or } 111.4^\circ$$

Case I:

$$B = 68.6^\circ,\ A = 180^\circ - 68.6^\circ - 56.6^\circ = 54.8^\circ$$

$$\frac{a}{\sin 54.8^\circ} = \frac{13.0}{\sin 56.6^\circ} \Rightarrow a = 12.7$$

Case II:

$$B = 111.4^\circ,\ A = 180^\circ - 111.4^\circ - 56.6^\circ = 12.0^\circ$$

$$\frac{a}{\sin 12.0^\circ} = \frac{13.0}{\sin 56.6^\circ} \Rightarrow a = 3.24$$

29.

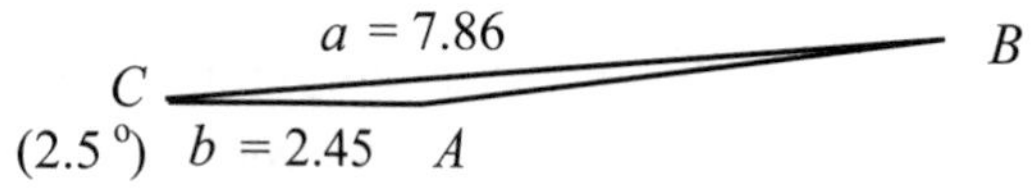

$$c^2 = a^2 + b^2 - 2ab\cos C$$

$$c^2 = 7.86^2 + 2.45^2 - 2(7.86)(2.45)\cos 2.5^\circ$$

$$c = 5.413386814 \Rightarrow c = 5.41$$

$$a^2 = b^2 + c^2 - 2bc\cos A$$

$$7.86^2 = 2.45^2 + c^2 - 2(2.45)(c)\cos A$$

$$\cos A = 0.9979924288$$

$$A = 176.4^\circ$$

$$B = 180^\circ - C - A = 180^\circ - 2.5^\circ - 176.4^\circ$$

$$B = 1.1^\circ$$

33.

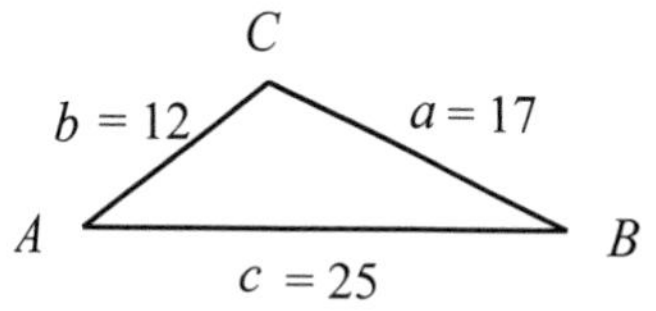

$$a^2 = b^2 + c^2 - 2bc\cos A$$

$$17^2 = 12^2 + 25^2 - 2(12)(25)\cos A$$

$$\cos A = 0.8$$

$$A = 37^\circ$$

$$b^2 = a^2 + c^2 - 2ac\cos B$$

$$12^2 = 17^2 + 25^2 - 2(17)(25)\cos B$$

$$\cos B = 0.9059$$

$$B = 25^\circ$$

$$C = 180^\circ - 37^\circ - 25^\circ = 118^\circ$$

37.

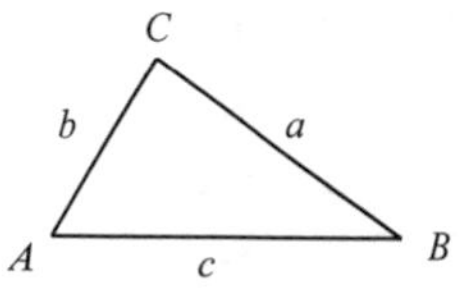

$$a^2 = b^2 + c^2 - 2bc\cos A$$

$$b^2 = a^2 + c^2 - 2ac\cos B$$

$$\underline{c^2 = a^2 + b^2 - 2ab\cos C} \quad \text{add}$$

$$a^2 + b^2 + c^2 = 2a^2 + 2b^2 + 2c^2 - 2bc\cos A$$
$$- 2ac\cos B - 2ab\cos C$$
$$a^2 + b^2 + c^2 = 2bc\cos A + 2ac\cos B + 2ab\cos C$$
$$\frac{a^2 + b^2 + c^2}{2abc} = \frac{\cos A}{a} + \frac{\cos B}{b} + \frac{\cos C}{c}$$

41.

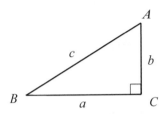

$$\frac{a}{\sin A} = \frac{b}{\sin B} \Rightarrow b = \frac{a\sin B}{\sin A}$$
$$A_t = \frac{1}{2}ab = \frac{1}{2} \cdot a \cdot \frac{a\sin B}{\sin A}$$
$$A_t = \frac{a^2\sin B}{2\sin A}$$

45.

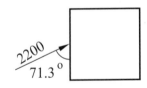

$$v_\perp = 2200\sin 71.3°$$
$$= 2100 \text{ ft/s}$$

49.

vector	x-component	y-component
1300	$-1300\cos 54°$	$1300\sin 54°$
3200	$3200\sin 32°$	$3200\cos 32°$
2100	$-2100\cos 35°$	$-2100\sin 35°$
	-788.6	2561

$$R = \sqrt{(-788.6)^2 + (2561)^2} = 2700 \text{ lb}$$
$$\theta_{ref} = \tan^{-1}\left|\frac{2561}{-788.6}\right|$$
$$\theta = 180° - \theta_{ref} = 107°$$

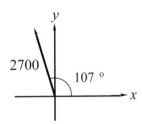

53. upward force $= 0.15\sin 22.5° + 0.20\sin 15.0°$
$$= 0.11 \text{ N}$$

57.

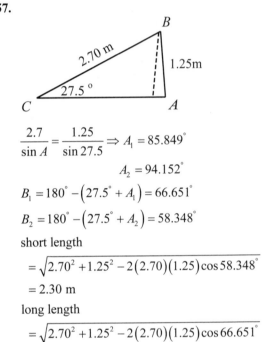

$$\frac{2.7}{\sin A} = \frac{1.25}{\sin 27.5} \Rightarrow A_1 = 85.849°$$
$$A_2 = 94.152°$$
$$B_1 = 180° - (27.5° + A_1) = 66.651°$$
$$B_2 = 180° - (27.5° + A_2) = 58.348°$$
short length
$$= \sqrt{2.70^2 + 1.25^2 - 2(2.70)(1.25)\cos 58.348°}$$
$$= 2.30 \text{ m}$$
long length
$$= \sqrt{2.70^2 + 1.25^2 - 2(2.70)(1.25)\cos 66.651°}$$
$$= 2.49 \text{ m}$$

61.

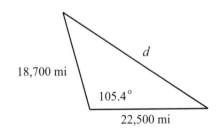

$$d = \sqrt{18,700^2 + 22,500^2 - 2(18,700)(22,500)\cos 105.4°}$$
$$= 32,900 \text{ mi}$$

65.

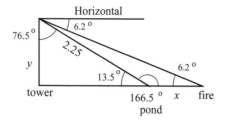

$$\frac{2.25}{\sin 6.2^\circ} = \frac{x}{\sin 7.3^\circ}$$
$$x = 2.65 \text{ km}$$

69.

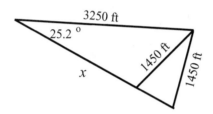

$1450^2 = x^2 + 3250^2 - 2(x)(3250)\cos 25.2^\circ$, using
quadratic formula,

$x = 2510$ ft, 3370 ft, ambiguous case

73.

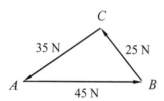

Since the resultant of the three forces is zero the
vectors form a closed triangle with sides of 45, 25
and 35. The angles between the forces may be
found using the law of cosines.

$$25^2 = 35^2 + 45^2 - 2(35)(45)\cos A$$
$$A = 34^\circ$$

$$35^2 = 25^2 + 45^2 - 2(25)(45)\cos B$$
$$B = 51^\circ$$

$$C = 180^\circ - (34^\circ - 51^\circ)$$
$$C = 95^\circ$$

Chapter 10

GRAPHS OF THE TRIGONOMETRIC FUNCTIONS

10.1 Graphs of $y = a \sin x$ and $y = a \cos x$

1. $y = 3 \cos x$

x	0	$\frac{\pi}{6}$	$\frac{\pi}{3}$	$\frac{\pi}{2}$	$\frac{2\pi}{3}$	$\frac{5\pi}{6}$
y	0	2.6	1.5	0	−1.5	−2.6

x	π	$\frac{7\pi}{6}$	$\frac{4\pi}{3}$	$\frac{3\pi}{2}$	$\frac{5\pi}{3}$	$\frac{11\pi}{6}$	2π
y	−3	−2.6	−1.5	0	1.5	2.6	3

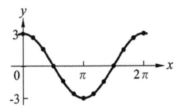

5. $y = -3 \cos x$

x	$-\pi$	$-\frac{3\pi}{4}$	$-\frac{\pi}{2}$	$-\frac{\pi}{4}$	0	$\frac{\pi}{4}$	$\frac{\pi}{2}$	$\frac{3\pi}{4}$	π
y	3	2.1	0	−2.1	−3	−2.1	0	2.1	3

x	$\frac{5\pi}{4}$	$\frac{3\pi}{4}$	$\frac{7\pi}{4}$	2π	$\frac{9\pi}{4}$	$\frac{5\pi}{2}$	$\frac{11\pi}{4}$	3π
y	2.1	0	−2.1	−3	−2.1	0	2.1	3

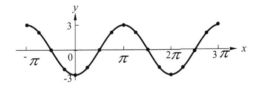

9. $y = \dfrac{5}{2} \sin x$; $\sin x$ has its amplitude value at $x = \dfrac{\pi}{2}$ and $x = \dfrac{3\pi}{2}$ and has intercepts at $x = 0$, $x = \pi$, and $x = 2\pi$. The graph can be sketched with these values.

x	0	$\frac{\pi}{2}$	π	$\frac{3\pi}{2}$	2π
y	0	$\frac{5}{2}$	0	$-\frac{5}{2}$	0

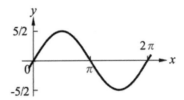

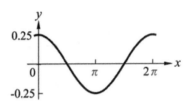

13. $y = 0.8 \cos x$; $\cos x$ has its amplitude value at $x = 0$ and $x = \pi$, and $x = 2\pi$, and has intercepts at $x = \dfrac{\pi}{2}$, and $x = \dfrac{3\pi}{2}$. The graph can be sketched with these values.

x	0	$\frac{\pi}{2}$	π	$\frac{3\pi}{2}$	2π
y	0.8	0	−0.8	0	0.8

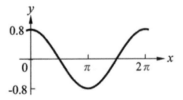

17. $y = -1500 \sin x = -1500(\sin x)$; $\sin x$ has its amplitude value at $x = \frac{\pi}{2}$ and $x = \frac{3\pi}{2}$ and has intercepts at $x = 0$, $x = \pi$, and $x = 2\pi$. (The negative sign will invert the graph values.)

x	0	$\frac{\pi}{2}$	π	$\frac{3\pi}{2}$	2π
y	0	−1500	0	1500	0

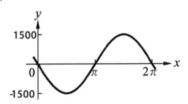

21. $y = -50 \cos x = -50(\cos x)$; $\cos x$ has its
amplitude value at $x = 0$, $x = \pi$, and $x = 2\pi$.
and has intercepts at $x = \frac{\pi}{2}$, $x = \frac{3\pi}{2}$.
(The negative sign will invert the graph values.)

x	0	$\frac{\pi}{2}$	π	$\frac{3\pi}{2}$	2π
y	-50	0	50	0	-50

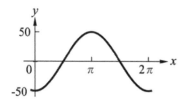

25. Sketch $y = -12 \cos x$ for $x = 0,1,2,3,4,5,6,7$

x	0	1	2	3	4
-12 cos x	-12	-6.5	5.0	11.9	7.8

x	5	6	7
-12 cos x	-3.4	-11.5	-9.0

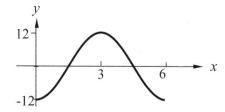

29. $y = a \cos x,\ (\pi, 2)$
$2 = a \cos x \Rightarrow a = -2$
$y = -2 \cos x$

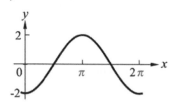

33. The graph passes through $(0,0),(\pi,0),$ and $(2\pi,0)$
with amplitude 4. The graph is $y = 4 \sin x$.

37.

x	$\pm 2.5 \sin x$	$\pm 2.5 \cos x$
0.67	± 1.55	± 1.96

The function is $y = -2.5 \sin x$

10.2 Graphs of $y = a \sin bx$ and $y = a \cos bx$

1. $y = 3 \sin 6x$, amplitude = 3, period = $\frac{2\pi}{6} = \frac{\pi}{3}$

x	0	$\frac{\pi}{12}$	$\frac{\pi}{6}$	$\frac{\pi}{4}$	$\frac{\pi}{3}$
y	0	3	0	-3	0

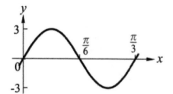

5. Since $\cos bx$ has period $\frac{2\pi}{b}$, $y = 3 \cos 8x$ has a
period of $\frac{2\pi}{8}$, or $\frac{\pi}{4}$.

9. $y = -\cos 16x$ has period of $\frac{2\pi}{16}$, or $\frac{\pi}{8}$.

13. $y = 3 \cos 4\pi x$ has period of $\frac{2\pi}{4\pi}$, or $\frac{2}{4} = \frac{1}{2}$.

17. $y = -\frac{1}{2} \cos \frac{2}{3}x$ has period of $\frac{2\pi}{\frac{2}{3}} = \frac{2\pi}{1} \times \frac{3}{2} = 3\pi$.

21. $y = 3.3 \cos \pi^2 x$ has period of $\frac{2\pi}{\pi^2} = \frac{2}{\pi}$.

25. $y = 3 \cos 8x$ has amplitude of 3 and period $\frac{\pi}{4}$.

x	0	$\frac{\pi}{16}$	$\frac{\pi}{8}$	$\frac{3\pi}{16}$	$\frac{\pi}{4}$
y	3	0	-3	0	3

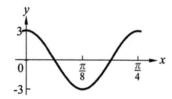

29. $y = -\cos 16x$ has amplitude of $\left|-1\right| = 1$, and

period of $\dfrac{\pi}{8}$.

x	0	$\frac{\pi}{32}$	$\frac{\pi}{16}$	$\frac{3\pi}{32}$	$\frac{\pi}{8}$
y	-1	0	1	0	-1

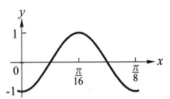

33. $y = 3\cos 4\pi x$ has amplitude of 3 and period of $\dfrac{1}{2}$.

x	0	$\frac{1}{8}$	$\frac{1}{4}$	$\frac{3}{8}$	$\frac{1}{2}$
y	3	0	-3	0	3

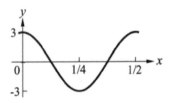

37. $y = -\dfrac{1}{2}\cos\dfrac{2}{3}x$ has amplitude of $\left|-\dfrac{1}{2}\right| = \dfrac{1}{2}$, and

period of 3π.

x	0	$\frac{3\pi}{4}$	$\frac{3\pi}{2}$	$\frac{9\pi}{4}$	3π
y	$-\frac{1}{2}$	0	$\frac{1}{2}$	0	$-\frac{1}{2}$

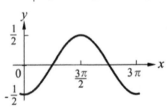

41. $y = 3.3\cos \pi^2 x$ has amplitude of 3.3 and period

of $\dfrac{2}{\pi}$.

x	0	$\frac{1}{2\pi}$	$\frac{1}{\pi}$	$\frac{3}{2\pi}$	$\frac{2}{\pi}$
$\pi^2 x$	0	$\frac{\pi}{2}$	π	$\frac{3\pi}{2}$	2π
$\cos \pi^2 x$	1	0	-1	0	1
$3.3\cos \pi^2 x$	3.3	0	-3.3	0	3.3

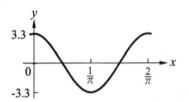

45. $b = \dfrac{2\pi}{1/3} = 6;\ y = \sin 6\pi x$

49. $y = 8\left|\cos\dfrac{\pi}{2}x\right|$

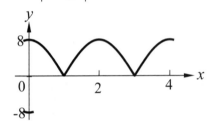

53. $\sin 2x$ has period $\dfrac{2\pi}{2} = \pi$

$\sin 3x$ has period $\dfrac{2\pi}{3}$

The period of $\sin 2x + \sin 3x$ is the least

common multiple of π, $\dfrac{2\pi}{3}$; 2π.

57. $y = 2\cos bx,\ (\pi, 0),\ b > 0$

$0 = 2\cos b\pi \Rightarrow b\pi = \dfrac{\pi}{2} + n\pi$

$b = \dfrac{1}{2} + n$ of which

the smallest is $b = \dfrac{1}{2}$

$y = 2\cos\dfrac{1}{2}x$ is the function.

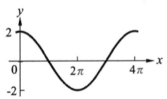

61. $v = 450\cos 3600t$; amplitude is 450; period is

$$\frac{2\pi}{3600} = \frac{\pi}{1800}.$$

$$\left(0.006 \div \frac{1}{1800} = 3.4 \text{ cycles}\right)$$

t	0	$\frac{\pi}{7200}$	$\frac{\pi}{3600}$	$\frac{\pi}{2400}$	$\frac{\pi}{1800}$
$3600t$	0	$\frac{\pi}{2}$	π	$\frac{3\pi}{2}$	2π
$\cos 3600t$	1	0	-1	0	1
$450\cos 3600t$	450	0	-450	0	450

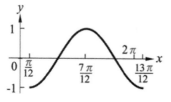

65. $y = -4\sin \pi x$ period of 2, amplitude 4

10.3 Graphs of $y = a \sin\left(bx + c\right)$ and $y = a \cos\left(bx + c\right)$

1. $y = -\cos\left(2x - \frac{\pi}{6}\right)$

  (1) the amplitude is 1

  (2) the period is $\frac{2\pi}{2} = \pi$

  (3) the displacement is $-\frac{-\frac{\pi}{6}}{2} = \frac{\pi}{12}$

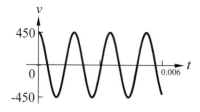

5. $y = \cos\left(x + \frac{\pi}{6}\right)$; $a = 1$, $b = 1$, $c = -\frac{\pi}{6}$

Amplitude is $|a| = 1$; period is $\frac{2\pi}{b} = 2\pi$;

displacement is $-\frac{c}{b} = \frac{\pi}{6}$

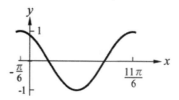

9. $y = -\cos\left(2x - \pi\right)$; $a = -1$, $b = 2$, $c = -\pi$

Amplitude is $|a| = 1$; period is $\frac{2\pi}{b} = \frac{2\pi}{1} = \pi$;

displacement is $-\frac{c}{b} = -\left(\frac{-\pi}{2}\right) = \frac{\pi}{2}$

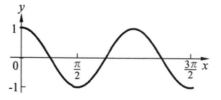

13. $y = 30\cos\left(\frac{1}{3}x + \frac{\pi}{3}\right)$; $a = 30$, $b = \frac{1}{3}$, $c = \frac{\pi}{3}$

Amplitude is $|a| = 30$; period is $\frac{2\pi}{b} = \frac{2\pi}{1/3} = 6\pi$;

displacement is $-\frac{c}{b} = \frac{-\pi/3}{1/3} = -\pi$

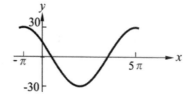

17. $y = 0.08\cos\left(4\pi x - \frac{\pi}{5}\right)$; $a = 0.08$, $b = 4\pi$, $c = -\frac{\pi}{5}$

Amplitude is $|a| = 0.08$; period is $\frac{2\pi}{b} = \frac{2\pi}{4\pi} = \frac{1}{2}$;

displacement is $-\frac{c}{b} = -\left(-\frac{\pi/5}{4\pi}\right) = \frac{1}{20}$

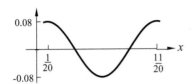

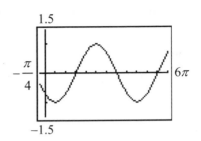

21. $y = 40\cos(3\pi x + 1)$; $a = 40$, $b = 3\pi$, $c = 2$

Amplitude is $|a| = 40$; period is $\dfrac{2\pi}{b} = \dfrac{2\pi}{3\pi} = \dfrac{2}{3}$;

displacement is $-\dfrac{c}{b} = -\dfrac{1}{3\pi}$

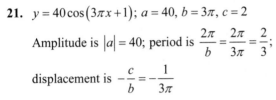

37. $y = 2\sin(2x + c)$, $c > 0$, $\left(-\dfrac{\pi}{8}, 0\right)$

$0 = 2\sin\left(2\left(-\dfrac{\pi}{8}\right) + c\right)$

$-\dfrac{\pi}{4} + c = 0 + n\pi$

$c = \dfrac{\pi}{4} + n\pi$ of which the

smallest is $c = \dfrac{\pi}{4}$

$y = 2\sin\left(2x + \dfrac{\pi}{4}\right)$

$a = 2$

$\text{period} = \dfrac{2\pi}{2} = \pi$

$\text{displacement} = -\dfrac{\frac{\pi}{4}}{2} = -\dfrac{\pi}{8}$

25. $y = -\dfrac{3}{2}\cos\left(\pi x - \dfrac{\pi^2}{6}\right)$; $a = -\dfrac{3}{2}$, $b = \pi$, $c = -\dfrac{\pi^2}{6}$

Amplitude is $|a| = \dfrac{3}{2}$; period is $\dfrac{2\pi}{b} = \dfrac{2\pi}{\pi} = 2$;

displacement is $-\dfrac{c}{b} = -\dfrac{\pi^2/6}{\pi} = -\dfrac{\pi}{6}$

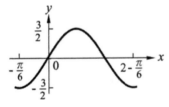

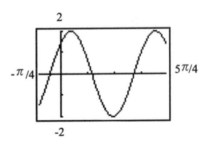

29. cosine, 12, $\dfrac{1}{2}$, $\dfrac{1}{8}$

$a = 12$; $\text{period} = \dfrac{1}{2} = \dfrac{2\pi}{b} \Rightarrow b = 4\pi$

$\text{displacement} = \dfrac{1}{8} = -\dfrac{c}{4\pi} \Rightarrow c = -\dfrac{\pi}{2}$

$y = 12\cos\left(4\pi x - \dfrac{\pi}{2}\right)$

41. $y = 4500\cos(0.025t - 0.25)$; $a = 4500$, $b = 0.25$,

$c = -0.25$

Amplitude $= |a| = 4500$;

$\text{period} = \dfrac{2\pi}{b} = \dfrac{2\pi}{0.025} = 80\pi$;

$\text{displacement} = -\dfrac{c}{b} = -\left(\dfrac{-0.25}{0.025}\right) = 10$

33. Graph $y_1 = \sin\left(\dfrac{x}{2} - \dfrac{3\pi}{4}\right)$ and $y_2 = -\sin\left(\dfrac{3\pi}{4} - \dfrac{x}{2}\right)$.

Graphs are the same.

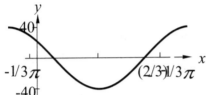

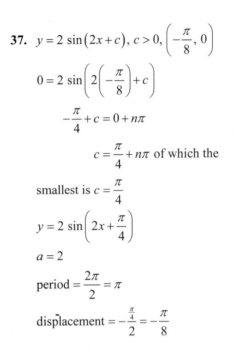

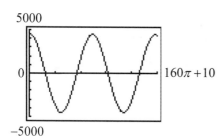

45. $y = a\cos(bx + c)$. Amplitude $= \left|-0.8\right| = 0.8$,

period $= \dfrac{2\pi}{b} = \pi, \; b = 2$

displacement $= -\dfrac{c}{b} = 0, \; c = 0$

$y = -0.8\cos 2x$

10.4   Graphs of $y = \tan x$, $y = \cot x$, $y = \sec x$, $y = \csc x$

1. $y = 5.0\cot 2x$. Graph $y_1 = \dfrac{5.0}{\tan 2x}$.

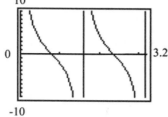

5.

x	$-\frac{\pi}{2}$	$-\frac{\pi}{3}$	$-\frac{\pi}{4}$	$-\frac{\pi}{6}$	0	$\frac{\pi}{6}$	$\frac{\pi}{4}$
$\sec x$	*	2	1.4	1.2	1	1.2	1.4

x	$\frac{\pi}{3}$	$\frac{\pi}{2}$	$\frac{2\pi}{3}$	$\frac{3\pi}{4}$	$\frac{5\pi}{6}$	π
$\sec x$	2	*	-2	-1.4	-1.2	-1

(* = undefined)

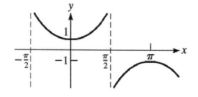

9. For $y = \dfrac{1}{2}\sec x$, first sketch the graph of $y = \sec x$, then multiply the y-values of the secant function by $\dfrac{1}{2}$ and graph.

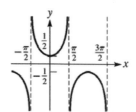

13. For $y = -3\csc x$, sketch the graph of $y = \csc x$, then multiply the y-values by -3, and resketch the graph. It will be inverted.

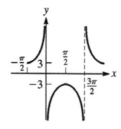

17. Since the period of $\sec x$ is 2π, the period of $y = \dfrac{1}{2}\sec 3x$ is $\dfrac{2\pi}{3}$. Graph $y_1 = 0.5(\cos 3x)^{-1}$.

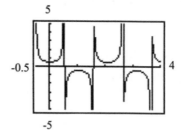

21. Since the period of $\csc x$ is 2π, the period of $y = 18\csc\left(3x - \dfrac{\pi}{3}\right)$ is $\dfrac{2\pi}{3}$. The displacement is $-\left(-\dfrac{\pi/3}{3}\right) = \dfrac{\pi}{9}$. Graph $y_1 = 18\left(\sin\left(3x - \dfrac{\pi}{3}\right)\right)^{-1}$.

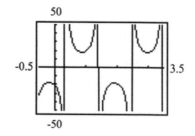

25. $y = \sec(bx) + k$ is a secant function with zero displacement, period 4π and passing through $(0, -3)$ if $\dfrac{2\pi}{b} = 4\pi \Rightarrow b = \dfrac{1}{2}$ and $1 + k = -3 \Rightarrow k = -4$.

$y = \sec\left(\dfrac{1}{2}x\right) - 4$ is the required function

29. $b = (a \sin B) \csc A$

$\quad = \left(4.00 \sin \dfrac{\pi}{4}\right) \csc A$

$\quad = 2.83 \csc A$

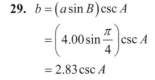

10.5 Applications of the Trigonometric Graphs

1. The displacement of the projection on the y-axis is d and is given by $d = R \cos wt$.

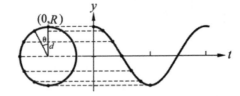

5. $y = R \cos \omega t$

$\quad = 8.30 \cos\left[(3.20)(2\pi)\right]t$

Amplitude is 8.30 cm;

period is $\dfrac{1}{3.20} = 0.3125$ s,

0.625 s for 2 cycles;

displacement is 0 s.

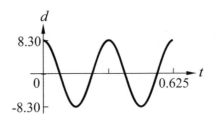

9. $e = E \cos(2\pi ft + \alpha)$

$\quad = 170 \cos\left[2\pi(60.0)t - \dfrac{\pi}{3}\right]$

Amplitude is 170 V

period is $\dfrac{2\pi}{2\pi(60.0)} = 0.016$ s, 0.033 s

for 2 cycles; displacement is $\dfrac{\pi/3}{2\pi(60.0)} = \dfrac{1}{360}$ s

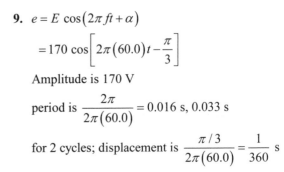

13. $p = p_0 \sin 2\pi ft$

$\quad = 2.80 \sin\left[2\pi(2.30)\right]t$

$\quad = 2.80 \sin 14.45t$

Amplitude is 2.80 lb/in.^{2}, period is $\dfrac{2\pi}{14.45} = 0.435$ s

for 1 cycle, 0.87 s for 2 cycles; displacement is 0 s

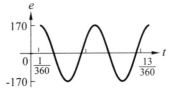

17. $e = 0.014 \cos\left(2\pi ft + \dfrac{\pi}{4}\right)$

$\quad = 0.014 \cos\left[2\pi(0.950)t + \dfrac{\pi}{4}\right]$

Amplitude is 0.014 V

period is $\dfrac{2\pi}{2\pi(0.950)} = 1.05$ s, 2.10 s

for 2 cycles; displacement is $\dfrac{-\pi/4}{2\pi(0.950)} = -0.13$ s

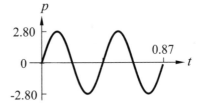

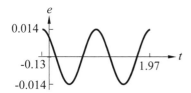

9. Graph $y_1 = x^3 + 10 \sin 2x$.

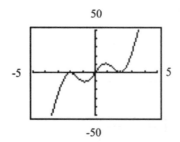

10.6 Composite Trigonometric Curves

13. Graph $y_1 = 20 \cos 2x + 30 \sin x$.

1. $y = 1 + \sin x$

x	-2π	$-\frac{3\pi}{2}$	$-\pi$	$-\frac{\pi}{2}$	0	$\frac{\pi}{2}$	π	$\frac{3\pi}{2}$	2π
y	1	2	1	0	1	2	1	0	1

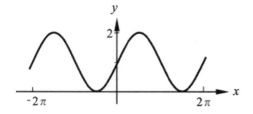

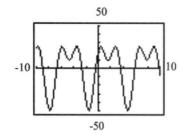

17. Graph $y_1 = \sin \pi x - \cos 2x$.

5. $y = \dfrac{1}{10}x^2 - \sin \pi x$

x	-4	-3.43	-2.55	-1.88	-1.47
y	1.60	0.20	1.64	0	-0.78

x	-1.03	-0.51	0	0.49	0.97	1.53
y	0	1.03	0	-0.98	0	1.23

x	2.15	2.45	2.73	0
y	0	-0.39	4	1.6

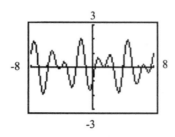

21. In parametric mode graph

$x_{IT} = 3 \sin t$, $y_{IT} = 2 \sin t$

t	$-\frac{\pi}{2}$	$-\frac{\pi}{4}$	0	$\frac{\pi}{4}$	$\frac{\pi}{2}$
x	-3	-2.12	0	2.12	3
y	-2	-1.41	0	1.41	2

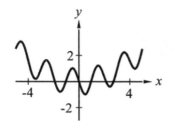

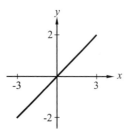

25. In parametric mode graph

$$x_{IT} = \cos \pi\left(t + \frac{1}{6}\right), \ y_{IT} = 2 \sin \pi t$$

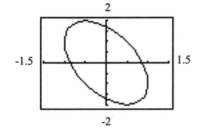

29. In parametric mode graph

$$x_{IT} = t \cos t , \ y_{IT} = t \sin t$$

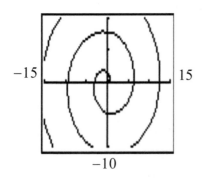

33. $x = \sqrt{t} + 4, \ y = 3\sqrt{t} - 1$

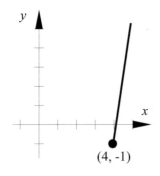

37. $y = 0.4 \sin 4t + 0.3 \cos 4t$

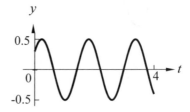

41. 60 beats/min $\times \dfrac{\text{min}}{60 \text{ s}} = 1$ beat/s from which $\omega = 2\pi$

$$p = 100 + 20 \cos(2\pi t)$$

Graph $y_1 = 100 + 20 \cos(2\pi x)$.

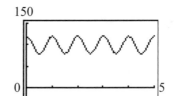

45. $x = 4 \cos \pi t, \ y = 2 \sin 3\pi t$

t	0	$\frac{\pi}{4}$	$\frac{\pi}{2}$	$\frac{3\pi}{4}$	π
x	4	−3.12	0.88	1.75	−3.61
y	0	1.80	1.57	−0.43	−1.94

t	$\frac{5\pi}{4}$	$\frac{3\pi}{2}$	$\frac{7\pi}{4}$	2π
x	3.90	−2.48	−0.028	2.52
y	−1.27	0.84	2.0	0.91

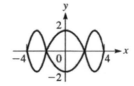

33. Graph $y_1 = 2 \sin x - \cos 2x$.

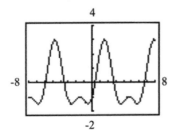

37. Graph $y_1 = \dfrac{\sin x}{x}$.

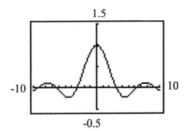

41. From the graph, $a = 2$, period $= \pi = \dfrac{2\pi}{b} \Rightarrow b = 2$,

and displacement $= 0 = -\dfrac{c}{b} = -\dfrac{\pi}{4} \Rightarrow c = \dfrac{\pi}{2}$.

$y = a \sin(bx + c)$ is $y = 2 \sin\left(2x + \dfrac{\pi}{2}\right)$

45. In parametric mode graph

$x_1 = -\cos 2\pi t$, $y_1 = 2 \sin \pi t$

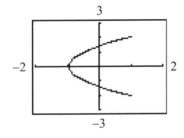

49. Graph $y_1 = 2\left|2 \sin 0.2\pi x\right| - \left|\cos 0.4\pi x\right|$

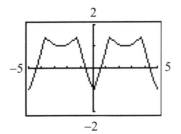

53. The period of $\cos 0.5x$ is $\dfrac{2\pi}{0.5} = 4\pi$.

The period of $\sin 3x$ is $\dfrac{2\pi}{3}$.

The period of $y = 2 \cos 0.5x + \sin 3x$ is the least

common multiple of 4π and $\dfrac{2\pi}{3}$; 4π.

57. $y = 3 \cos bx$, $\left(\dfrac{\pi}{3}, -3\right)$, $b > 0$

$-3 = 3 \cos\left(b \cdot \dfrac{\pi}{3}\right) \Rightarrow b \cdot \dfrac{\pi}{3} = \pi + 2\pi \cdot n$

$b = 3 + 6 \cdot n$

of which the smallest is $b = 3$. Graph

$y = 3 \cos 3x$.

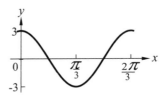

61.

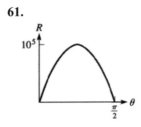

65.

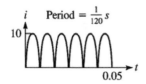

85.

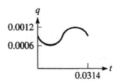

69.

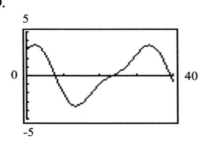

73. $d = a \sec \theta$, $a = 3.00$

$d = 3.00 \sec \theta$

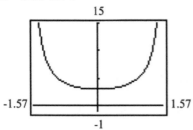

77. $y = 4 \sin 2t - 2 \cos 2t$

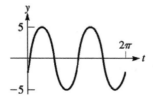

81.

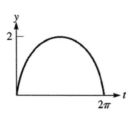

Chapter 11

EXPONENTS AND RADICALS

11.1 Simplifying Expressions with Integral Exponents

1. $\left(x^{-2}y\right)^2\left(\dfrac{2}{x}\right)^2 = \dfrac{x^{-4}y^2}{\left(\dfrac{2}{x}\right)^2} = \dfrac{x^{-4}y^2}{\dfrac{4}{x^2}}$

$\qquad = \dfrac{x^{-4}y^2}{1} \times \dfrac{x^2}{4}$

$\qquad = \dfrac{x^{-2}y^2}{4}$

$\qquad = \dfrac{y^2}{4x^2}$

5. $x^7 \cdot x^{-4} = x^{7+(-4)} = x^3$

9. $5^0 \times 5^{-3} = 5^{0+(-3)} = 5^{-3} = \dfrac{1}{5^3} = \dfrac{1}{125}$

13. $2\left(5an^{-2}\right)^{-1} = 2 \times 5^{-1}a^{-1}n^{(-2)(-1)} = \dfrac{2n^2}{5a}$

17. $-7x^0 = -7 \times 1 = -7$

21. $\left(7a^{-1}x\right)^3 = 7^{-3}a^3x^{-3} = \dfrac{a^3}{7^3x^3} = \dfrac{a^3}{343x^3}$

25. $\left(\dfrac{a}{b^{-2}}\right)^{-3} = \dfrac{3a^{-3}}{\left(b^{-2}\right)^{-3}} = \dfrac{\frac{3}{a^3}}{b^{(-2)(-3)}} = \dfrac{\frac{3}{a^3}}{b^6} = \dfrac{3}{a^3b^6}$

29. $3x^{-2} + 5y^{-2} = \dfrac{3}{x^2} + \dfrac{5}{y^2} = \dfrac{5x^2 + 3y^2}{x^2y^2}$

33. $\left(\dfrac{3a^2}{4b}\right)^{-3}\left(\dfrac{4}{a}\right)^{-5} = \dfrac{3^{-3}a^{-6}}{4^{-3}b^{-3}} \times \dfrac{4^{-5}}{a^{-5}}$

$\qquad = \dfrac{4^3b^3}{3^3a^6} \times \dfrac{a^5}{4^5} = \dfrac{b^3}{432a}$

37. $3a^{-2} + \left(3a^{-2}\right)^4 = \dfrac{3}{a^2} + 3^4a^{-8} = \dfrac{3}{a^2} + \dfrac{18}{a^8}$

$\qquad = \dfrac{3a^6 + 81}{a^8}$

41. $\left(R_1^{-1} + R_2^{-1}\right)^{-1} = \dfrac{1}{\frac{1}{R_1} + \frac{1}{R_2}} = \dfrac{R_1R_2}{R_1 + R_2}$

45. $\dfrac{6^{-1}}{4^{-2} + 2} = \dfrac{\frac{1}{6}}{\frac{1}{4^2} + 2} = \dfrac{\frac{1}{6}}{\frac{1}{16} + 2}\left(\dfrac{48}{48}\right)$

$\qquad = \dfrac{8}{3 + 96} = \dfrac{8}{99}$

49. $2t^{-2} + t^{-1}(t+1) = \dfrac{2}{t^2} + t^0 + t^{-1}$

$\qquad = \dfrac{2}{t^2} + 1 + \dfrac{1}{t}$

$\qquad = \dfrac{2 + t^2 + t}{t^2}$

53. If $x < 0$, then $x^2 > 0$ and $x^{-2} > 0$

If $x < 0$, then $\dfrac{1}{x} < 0$ and $x^{-1} < 0$.

$x^{-2} > 0 > x^{-1}$ or $x^{-2} > x^{-1}$.

Is it ever true that $x^{-2} < x^{-1}$? No.

57. (a) $\left(\dfrac{a}{b}\right)^{-n} = \dfrac{1}{\left(\frac{a}{b}\right)^n} = \dfrac{1}{\frac{a^n}{b^n}} = 1 \times \dfrac{b^n}{a^n} = \left(\dfrac{b}{a}\right)^n$

(b) $\left(\dfrac{3.576}{8.091}\right) = (0.4419725)^{-7} = 303.55182$

$\left(\dfrac{8.091}{3.576}\right) = (2.26258)^7 = 303.55182$

61. Yes, $\left[-2^0 - (-1)^0\right]^0 = [-1 - 1]^0 = [-2]^0 = 1$

65. $\dfrac{1}{\left(2x^2\right)^{-1}} = \dfrac{64}{x^3}$

$\qquad 2x^2 = \dfrac{64}{x^3}$

$\qquad x^5 = 32$

$\qquad x = 2$

69. $\text{kg}\,\text{s}^{-2}\,\text{m}^2 = \dfrac{\text{kg}\,\text{m}^2}{\text{s}^2} = \dfrac{\text{kg}\,\text{m}}{\text{s}^2}\,\text{m} = \text{N}\,\text{m}$

73.
$$v = a^p t^r$$
$$m \times s^{-1} = \left(m \times s^{-2}\right)^p \times s^r$$
$$m^1 \times s^{-1} = m^p \times s^{-2p+r} \Rightarrow p = 1$$
$$\text{and } -2p + r = -1 \Rightarrow -2(1) + r$$
$$= -1 \Rightarrow r = 1$$

11.2   Fractional Exponents

1. $8^{4/3} = \left(8^{1/3}\right)^4 = \left(\sqrt[3]{8}\right)^4 = 2^4 = 16$

5. $25^{1/2} = \sqrt{25} = 5$

9. $100^{25/2} = \left(100^{1/2}\right)^{25} = \left(\sqrt{100}\right)^{25} = 10^{25}$

13. $64^{-2/3} = \dfrac{1}{\left(64^{1/3}\right)^2} = \dfrac{1}{\left(\sqrt[3]{64}\right)^2} = \dfrac{1}{4^2} = \dfrac{1}{16}$

17. $\left(3^6\right)^{2/3} = (3)^{12/3} = 3^4 = 81$

21. $\dfrac{15^{2/3}}{5^2 \times 15^{-1/3}} = \dfrac{15^{2/3+1/3}}{5^2} = \dfrac{15^1}{25} = \dfrac{3}{5}$

25. $125^{-2/3} - 100^{-3/2} = \dfrac{1}{\left(125^{1/3}\right)^2} - \dfrac{1}{\left(100^{1/2}\right)^3}$

$\qquad = \dfrac{1}{\left(\sqrt[3]{125}\right)^2} - \dfrac{1}{\left(\sqrt{100}\right)^3}$

$\qquad = \dfrac{1}{5^2} - \dfrac{1}{10^3} = \dfrac{1}{25} - \dfrac{1}{1000}$

$\qquad = \dfrac{39}{1000}$

29. $17.98^{1/4} = 2.059$

```
17.98^(1/4)
          2.059194748
■
```

33. $B^{2/3} \cdot B^{1/2} = B^{2/3+1/2} = B^{7/6}$

37. $\dfrac{x^{3/10}}{x^{-1/5} x^2} = x^{3/10+1/5-2} = x^{-3/2} = \dfrac{1}{x^{3/2}}$

41. $\left(16a^4 b^3\right)^{-3/4} = 16^{-3/4} a^{4(-3/4)} b^{3(-3/4)}$

$\qquad = \dfrac{1}{16^{3/4} a^3 b^{9/4}} = \dfrac{1}{8a^3 b^{9/4}}$

45. $\dfrac{1}{2}\left(4x^2 + 1\right)^{-1/2}(8x) = \dfrac{4x}{\left(4x^2 + 1\right)^{1/2}}$

49. $\left(T^{-1} + 2T^{-2}\right)^{-1/2} = \dfrac{1}{\left(\dfrac{1}{T} + \dfrac{2}{T^2}\right)^{1/2}}$

$\qquad = \dfrac{1}{\left(\dfrac{T+2}{T^2}\right)} = \dfrac{1}{\dfrac{(T+2)^{1/2}}{\left(T^2\right)^{1/2}}}$

$\qquad = \dfrac{T}{(T+2)^{1/2}}$

53. $\left[\left(a^{1/2}-a^{-1/2}\right)^2+4\right]^{1/2}=\left[\left(a^{1/2}-\dfrac{1}{a^{1/2}}\right)^2+4\right]^{1/2}$

$$=\left[\left(\dfrac{a-1}{a^{1/2}}\right)^2+4\right]^{1/2}$$

$$=\left[\dfrac{(a-1)^2}{a}+4\right]^{1/2}$$

$$=\left[\dfrac{a^2-2a+1+4a}{a}\right]^{1/2}$$

$$=\left[\dfrac{a^2+2a+1}{a}\right]^{1/2}$$

$$=\left[\dfrac{(a+1)^2}{a}\right]^{1/2}=\dfrac{a+1}{a^{1/2}}$$

57. $f(x)=3x^{1/2}$

x	0	1	2	4	9
y	0	3	4.24	6	9

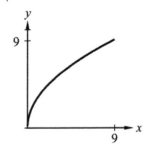

61.   Yes.

$$\left[\left(4^{1/2}\right)^{-3/4}+\left(\dfrac{1}{4^3}\right)^{1/8}\right]^4=\left[\left(2^{-3/4}\right)+4^{-3/8}\right]^4$$

$$=\left[2^{-3/4}+2^{-3/4}\right]^4$$

$$=\left[2\left(2^{-3/4}\right)\right]^4=\left[2^{1/4}\right]^4=2$$

65. $\left(\dfrac{A}{S}\right)^{-1/4}=\left(\dfrac{S}{A}\right)^{1/4}=0.5$

$$\dfrac{S}{A}=0.5^4=\dfrac{1}{16}\Rightarrow\dfrac{A}{S}=16$$

69. $V=\dfrac{1}{3}\pi r^2\cdot h=\dfrac{1}{3}\pi r^2\cdot r$

$$V=\dfrac{1}{3}\pi r^3$$

$$125=\dfrac{1}{3}\pi r^3$$

$$r=4.92\text{ cm}$$

11.3   Simplest Radical Form

1. $\sqrt{a^3b^4}=\sqrt{a^2\left(b^2\right)^2\cdot a}=ab^2\sqrt{a}$

5. $\sqrt{24}=\sqrt{4\cdot6}=\sqrt{4}\cdot\sqrt{6}=2\sqrt{6}$

9. $\sqrt{x^2y^5}=\sqrt{x^2y^4y}=\sqrt{x^2}\sqrt{y^4}\sqrt{y}=xy^2\sqrt{y}$

13. $\sqrt{80R^5TV^4}=\sqrt{16R^4V^4\cdot5RT}$

$$=4R^2V^2\sqrt{5RT}$$

17. $\sqrt[5]{96}=\sqrt[5]{32\cdot3}=\sqrt[5]{32}\sqrt[5]{3}=2\sqrt[5]{3}$

21. $\sqrt[4]{64r^3s^4t^5}=\sqrt[4]{16\cdot4s^4t^4r^3t}$

$$=\sqrt[4]{16}\sqrt[4]{s^4}\sqrt[4]{t^4}\sqrt[4]{4r^3t}$$

$$=2st\sqrt[4]{4r^3t}$$

25. $\sqrt[3]{P}\sqrt[3]{P^2V}=\sqrt[3]{P^3V}=P\sqrt[3]{V}$

29. $\sqrt[3]{\dfrac{9}{12}}=\sqrt[3]{\dfrac{48}{64}}=\dfrac{\sqrt[3]{48}}{\sqrt[3]{64}}=\dfrac{\sqrt[3]{8\cdot6}}{4}=\dfrac{2\sqrt[3]{6}}{4}=\dfrac{\sqrt[3]{6}}{2}$

33. $\sqrt[4]{400}=\sqrt[4]{2^4\cdot5^2}=2\cdot\sqrt[4]{25}=2\sqrt{5}$

37. $\sqrt{4\times10^4}=\sqrt{4}\times=\sqrt{10^4}=2\times10^2=200$

41. $\sqrt[4]{4a^2}=\left(4a^2\right)^{1/4}=\left(2^2\right)^{1/4}\left(a^2\right)^{1/4}=2^{1/2}a^{1/2}=\sqrt{2a}$

45. $\sqrt[4]{\sqrt[3]{16}}=\sqrt[4]{16^{1/3}}=\left(16^{1/3}\right)^{1/4}$

$$=\left(16^{1/4}\right)^{1/3}=2^{1/3}=\sqrt[3]{2}$$

49. $\sqrt{28u^3v^{-5}} = \sqrt{\dfrac{4u^2 \cdot 7u \cdot v}{v^6}} = \dfrac{2u\sqrt{7uv}}{v^3}$

53. $\sqrt{\dfrac{2x}{3c^4}} = \sqrt{\dfrac{2x}{3c^4} \cdot \dfrac{3}{3}} = \dfrac{\sqrt{6x}}{3c^2}$

57. $\sqrt{xy^{-1} + x^{-1}y} = \sqrt{\dfrac{x}{y} + \dfrac{y}{x}} = \sqrt{\dfrac{x^2 + y^2}{xy}}$

$= \sqrt{\dfrac{x^2 + y^2}{xy} \cdot \dfrac{xy}{xy}} = \sqrt{\dfrac{xy(x^2 + y^2)}{x^2y^2}}$

$= \dfrac{\sqrt{xy(x^2 + y^2)}}{xy}$

61. $\sqrt{a^2 + b^2}$ cannot be simplified any further.

65. $\sqrt{a} = a^{1/2} = a^{3/6} = \sqrt[6]{a^3}$

$\sqrt[3]{b} = b^{1/3} = b^{2/6} = \sqrt[6]{b^2}$

$\sqrt[6]{c} = c^{1/6} = \sqrt[6]{c}$

69. $f = \sqrt{\dfrac{T}{4\mu L^2}} = \dfrac{\sqrt{T}}{2L\sqrt{\mu}} \cdot \dfrac{\sqrt{\mu}}{\sqrt{\mu}} = \dfrac{\sqrt{T\mu}}{2L\mu}$

73. $\dfrac{8A}{\pi^2\sqrt{1 + \left(\dfrac{f_0}{f}\right)^2}} = \dfrac{8A}{\pi^2\sqrt{\dfrac{f^2 + f_0^2}{f^2}}}$

$= \dfrac{8A}{\pi^2} \times \dfrac{f^2}{f^2 + f_0^2}$

$= \dfrac{8A}{\pi^2}\sqrt{\dfrac{f^2}{f^2 + f_0^2} \times \dfrac{f^2 + f_0^2}{f^2 + f_0^2}}$

$= \dfrac{8A}{\pi^2}\sqrt{\dfrac{f^2(f^2 + f_0^2)}{(f^2 + f_0^2)^2}}$

$= \dfrac{8Af\sqrt{f^2 + f_0^2}}{\pi^2(f^2 + f_0^2)}$

11.4 Addition and Subtraction of Radicals

1. $3\sqrt{125} - \sqrt{20} + \sqrt{45} = 3\sqrt{25(5)} - \sqrt{4(5)} + \sqrt{9(5)}$

$= 15\sqrt{5} - 2\sqrt{5} + 3\sqrt{5}$

$= 15\sqrt{5} + \sqrt{5}$

$= 16\sqrt{5}$

5. $\sqrt{28} + \sqrt{5} - 3\sqrt{7} = 2\sqrt{7} + \sqrt{5} - 3\sqrt{7}$

$= \sqrt{5} + (2 - 3)\sqrt{7}$

$= \sqrt{5} - \sqrt{7}$

9. $2\sqrt{3t^2} - 3\sqrt{12t^2} = 2t\sqrt{3} - 3t\sqrt{3(4)}$

$= 2t\sqrt{3} - 3(2)t\sqrt{3} = (2 - 6)t\sqrt{3}$

$= -4t\sqrt{3}$

13. $2\sqrt{28} + 3\sqrt{175}$

$= 2\sqrt{4(7)} + 3\sqrt{25(7)} + 2(2)\sqrt{7} + 3(5)\sqrt{7}$

$(4 + 15)\sqrt{7} = 19\sqrt{7}$

17. $3\sqrt{75R} + 2\sqrt{48R} - 2\sqrt{18R}$

$= 3\sqrt{25(3)R} + 2\sqrt{16(3)R} - 2\sqrt{9(2)R}$

$= 3(5)\sqrt{3R} + 2(4)\sqrt{3R} - 2(3)\sqrt{2R}$

$= (15 + 8)\sqrt{3R} - 6\sqrt{2R}$

$= 23\sqrt{3R} - 6\sqrt{2R}$

21. $\sqrt{\dfrac{1}{2}} + \sqrt{\dfrac{25}{2}} - 4\sqrt{18} = \sqrt{\dfrac{1(2)}{2(2)}} + \sqrt{\dfrac{25(2)}{4}} - 4\sqrt{(9)2}$

$= \sqrt{\dfrac{2}{4}} - \sqrt{\dfrac{25(2)}{4}} - 12\sqrt{2}$

$= \left(\dfrac{1}{2} + \dfrac{5}{2} - 12\right)\sqrt{2} = -9\sqrt{2}$

25. $\sqrt[4]{32} - \sqrt[8]{4} = \sqrt[4]{16(2)} - \sqrt[8]{2(2)}$

$= 2\sqrt[4]{2} - 2^{2/8} = 2\sqrt[4]{2} - 2^{1/4}$

$= 2\sqrt[4]{2} - \sqrt[4]{2} = (2 - 1)\sqrt[4]{2}$

$= \sqrt[4]{2}$

29. $\sqrt{6}\sqrt{5}\sqrt{3} - \sqrt{40a^2} = \sqrt{90} - \sqrt{4a^2(10)}$

$$= \sqrt{9(10)} - 2a\sqrt{10}$$

$$= (3 - 2a)\sqrt{10}$$

33. $\sqrt{\dfrac{a}{c^5}} - \sqrt{\dfrac{c}{a^3}} = \sqrt{\dfrac{a(c)}{c^5(c)}} - \sqrt{\dfrac{ca}{a^3(a)}}$

$$= \sqrt{\dfrac{ac}{c^6}} - \sqrt{\dfrac{ac}{a^4}}$$

$$= \left(\dfrac{1}{c^3} - \dfrac{1}{a^2}\right)\sqrt{ac}$$

$$= \dfrac{(a^2 - c^3)\sqrt{ac}}{a^2 c^3}$$

37. $\sqrt{\dfrac{T-V}{T+V}} - \sqrt{\dfrac{T+V}{T-V}}$

$$= \sqrt{\dfrac{(T-V)(T+V)}{(T+V)^2}} - \sqrt{\dfrac{(T+V)(T-V)}{(T-V)^2}}$$

$$= \dfrac{\sqrt{T^2 - V^2}}{T+V} - \dfrac{\sqrt{T^2 - V^2}}{T-V}$$

$$= \dfrac{(T-V)\sqrt{T^2 - V^2} - (T+V)\sqrt{T^2 - V^2}}{(T+V)(T-V)}$$

$$= \dfrac{(T-V-T-V)\sqrt{T^2 - V^2}}{T^2 - V^2}$$

$$= \dfrac{-2V\sqrt{T^2 - V^2}}{T^2 - V^2}$$

$$= \dfrac{-2V\sqrt{T^2 - V^2}}{T^2 - V^2}$$

$$= \dfrac{2V\sqrt{T^2 - V^2}}{V^2 - T^2}$$

41. $3\sqrt{45} + 3\sqrt{75} - 2\sqrt{500}$

$$= 1.384014361\cdots \text{ on calculator}$$

$3\sqrt{45} + 3\sqrt{75} - 2\sqrt{500}$

$$= 3\sqrt{9(5)} + 3\sqrt{25(3)} - 2\sqrt{100(5)}$$

$$= 3(3)\sqrt{5} + 3(5)\sqrt{3} - 2(10)\sqrt{5}$$

$$= 9\sqrt{5} + 15\sqrt{3} - 20\sqrt{5}$$

$$= 15\sqrt{3} - 11\sqrt{5}$$

$$= 1.384014361\cdots \text{ on calculator}$$

45. $2\sqrt[3]{16} - \sqrt[3]{\dfrac{1}{4}} = 2\sqrt[3]{8(2)} - \sqrt[3]{\dfrac{8(2)}{64}}$

$$= 4\sqrt[3]{2} - \dfrac{2}{4}\sqrt[3]{2}$$

$$= \dfrac{7}{2}\sqrt[3]{2} = 4.409723675$$

49. $10\sqrt{11} - \sqrt{1000} = 10\sqrt{11} - \sqrt{100(10)}$

$$= 10\sqrt{11} - 10\sqrt{10}$$

$$= 10\left(\sqrt{11} - \sqrt{10}\right)$$

$$> 0 \text{ since } \sqrt{11} > \sqrt{10}$$

53. Perimeter $= 2(8) + 2(4) - 2(1) - 2(2) + \sqrt{2(1+2)^2}$

$$= 18 + \sqrt{2(3)^2} = 18 + 3\sqrt{2} = 22.2 \text{ ft}$$

11.5   Multiplication and Division of Radicals

1. $\sqrt{2}\left(3\sqrt{5} - 4\sqrt{8}\right) = 3\sqrt{10} - 4\sqrt{16}$

$$= 3\sqrt{10} - 4(4)$$

$$= 3\sqrt{10} - 16$$

5. $\sqrt{3}\sqrt{10} = \sqrt{3(10)} = \sqrt{30}$

9. $\sqrt[3]{4} \cdot \sqrt[3]{2} = \sqrt[3]{4(2)} = \sqrt[3]{8} = 2$

13. $\sqrt{72} \cdot \sqrt{\dfrac{5}{2}} = \sqrt{\dfrac{360}{2}} = \sqrt{180} = \sqrt{36 \cdot 5} = 6\sqrt{5}$

17. $\left(2 - \sqrt{5}\right)\left(2 + \sqrt{5}\right) = 2^2 - \sqrt{5}^2 = 4 - 5 = -1$

21. $\left(3\sqrt{11} - \sqrt{x}\right)\left(2\sqrt{11} + 5\sqrt{x}\right)$

$$= 6\sqrt{11}^2 + 15\sqrt{11x} - 2\sqrt{11x} - 5\sqrt{x}^2$$

$$= 6 \cdot 11 + 13\sqrt{11x} - 5x$$

$$= 66 + 13\sqrt{11x} - 5x$$

25. $\dfrac{\sqrt{6}-3}{\sqrt{6}} = \dfrac{\sqrt{6}-3}{\sqrt{6}}\cdot\dfrac{\sqrt{6}}{\sqrt{6}}\,\dfrac{\sqrt{6}^2-3\sqrt{6}}{\sqrt{6}^2}$

$\qquad = \dfrac{6-3\sqrt{6}}{6} = \dfrac{2-\sqrt{6}}{2}$

29. $\sqrt{2}\sqrt[3]{3} = 2^{1/2}3^{1/3} = 2^{3/6}3^{2/6} = \left(2^3 3^2\right)^{1/6}$

$\qquad = \sqrt[6]{2^3 3^2} = \sqrt[6]{72}$

33. $\dfrac{\sqrt{2}-1}{\sqrt{7}-3\sqrt{2}} = \dfrac{\sqrt{2}-1}{\sqrt{7}-3\sqrt{2}}\cdot\dfrac{\sqrt{7}+3\sqrt{2}}{\sqrt{7}+3\sqrt{2}}$

$\qquad = \dfrac{\sqrt{14}+3\sqrt{2}^2-\sqrt{7}-3\sqrt{2}}{\sqrt{7}^2-3^2\sqrt{2}^2}$

$\qquad = \dfrac{\sqrt{14}+3\cdot2-\sqrt{7}-3\sqrt{2}}{7-9\cdot2}$

$\qquad = \dfrac{\sqrt{14}+6-\sqrt{7}-3\sqrt{2}}{7-18}$

$\qquad = -\dfrac{\sqrt{14}+6-\sqrt{7}-3\sqrt{2}}{11}$

37. $\dfrac{6\sqrt{x}}{\sqrt{x}-\sqrt{5}} = \dfrac{6\sqrt{x}}{\sqrt{x}-\sqrt{5}}\cdot\dfrac{\sqrt{x}+\sqrt{5}}{\sqrt{x}+\sqrt{5}}$

$\qquad = \dfrac{6\sqrt{x}^2+6\sqrt{x}\sqrt{5}}{\sqrt{x}^2-\sqrt{5}^2} = \dfrac{6x+6\sqrt{5x}}{x-5}$

41. $\left(\sqrt{\dfrac{2}{R}}+\sqrt{\dfrac{R}{2}}\right)\left(\sqrt{\dfrac{2}{R}}-2\sqrt{\dfrac{R}{2}}\right)$

$\qquad = \dfrac{2}{R}-2\sqrt{\dfrac{2R}{2R}}+\sqrt{\dfrac{2R}{2R}}-2\left(\dfrac{R}{2}\right)$

$\qquad = \dfrac{2}{R}-2+1-R = \dfrac{2-2R+R-R^2}{R}$

$\qquad = \dfrac{2-R-R^2}{R}$

45. $\dfrac{\sqrt{a}+\sqrt{a-2}}{\sqrt{a}-\sqrt{a-2}}\cdot\dfrac{\sqrt{a}+\sqrt{a-2}}{\sqrt{a}+\sqrt{a-2}}$

$\qquad = \dfrac{a+2\sqrt{a}\sqrt{a-2}+a-2}{a-(a-2)}$

$\qquad = \dfrac{2a+2\sqrt{a}\sqrt{a-2}-2}{2}$

$\qquad = a+\sqrt{a}\sqrt{a-2}-1$

$\qquad = a-1+\sqrt{a(a-2)}$

49. $\dfrac{2\sqrt{6}-\sqrt{5}}{3\sqrt{6}-4\sqrt{5}} = \dfrac{2\sqrt{6}-\sqrt{5}}{3\sqrt{6}-4\sqrt{5}}\cdot\dfrac{3\sqrt{6}+4\sqrt{5}}{3\sqrt{6}+4\sqrt{5}}$

$\qquad = \dfrac{6(6)+8\sqrt{30}-3\sqrt{30}-4(5)}{9(6)-16(5)}$

$\qquad = -\dfrac{16+5\sqrt{30}}{26} = -1.6686972$

53. $\dfrac{x^2}{\sqrt{2x+1}}+2x\sqrt{2x+1}$

$\qquad = \dfrac{x^2}{\sqrt{2x+1}}+\dfrac{2x\sqrt{2x+1}\sqrt{2x+1}}{\sqrt{2x+1}}$

$\qquad = \dfrac{x^2+2x\sqrt{2x+1}^2}{\sqrt{2x+1}} = \dfrac{x^2+4x^2+2x}{\sqrt{2x+1}}$

$\qquad = \dfrac{5x^2+2x}{\sqrt{2x+1}}$

57. $\dfrac{\sqrt{x+h}-\sqrt{x}}{h} = \dfrac{\sqrt{x+h}-\sqrt{x}}{h}\cdot\dfrac{\sqrt{x+h}+\sqrt{x}}{\sqrt{x+h}+\sqrt{x}}$

$\qquad = \dfrac{x+h-x}{h\sqrt{x+h}+h\sqrt{x}}$

$\qquad = \dfrac{h}{h\left(\sqrt{x+h}+\sqrt{x}\right)}$

$\qquad = \dfrac{1}{\sqrt{x+h}+\sqrt{x}}$

61. $x^2-2x-1\big|_{x=1-\sqrt{2}} = \left(1-\sqrt{2}\right)^2-2\left(1-\sqrt{2}\right)-1$

$\qquad = 1-2\sqrt{2}+2-2+2\sqrt{2}-1$

$\qquad = 0$

65. $\dfrac{1}{\sqrt[3]{x^2}+\sqrt[3]{x}+1} = \dfrac{1}{x^{2/3}x^{1/3}+1}\cdot\dfrac{x^{1/3}-1}{x^{1/3}-1}$

$\qquad = \dfrac{x^{1/3}-1}{x-x^{2/3}+x^{2/3}-x^{1/3}+x^{1/3}-1}$

$\qquad = \dfrac{x^{1/3}-1}{x-1} = \dfrac{\sqrt[3]{x}-1}{x-1}$

69.
$$\frac{50}{50+\sqrt{V}} = \frac{50}{50+\sqrt{V}} \cdot \frac{50-\sqrt{V}}{50-\sqrt{V}}$$
$$= \frac{2500-50\sqrt{V}}{2500-V}$$

73.
$$\omega = \frac{1}{\sqrt{LC}}\sqrt{1-\frac{R^2C}{L}}$$
$$= \frac{1}{\sqrt{LC}}\sqrt{\frac{L-R^2C}{L}\cdot\frac{L}{L}}\cdot\frac{\sqrt{LC}}{\sqrt{LC}}$$
$$\omega = \frac{\sqrt{LC}}{LC}\cdot\frac{\sqrt{L^2-LR^2C}}{L}$$
$$= \frac{\sqrt{L^2\left(LC-R^2C^2\right)}}{L^2C}$$
$$\omega = \frac{L\sqrt{LC-R^2C^2}}{L^2C}$$
$$= \frac{\sqrt{LC-R^2C^2}}{LC}$$

Chapter 11 Review Exercises

1. $7a^{-2}b^0 = 7a^{-2}\cdot 1 = 7\cdot\frac{1}{a^2} = \frac{7}{a^2}$

5. $3(25)^{3/2} = 3\left[(25)^{1/2}\right]^3 = 3[25]^3$
$$= 3[5]^3 = 3\cdot 125 = 375$$

9. $\left(\frac{3}{t^2}\right)^{-2} = \frac{1}{\left(\frac{3}{t^2}\right)^2} = \frac{1}{\frac{3^2}{\left(t^2\right)^2}} = \frac{1}{\frac{9}{t^4}} = \frac{t^4}{9}$

13. $\left(2a^{1/3}b^{5/6}\right)^6 = 2^6\cdot\left(a^{1/3}\right)^6\cdot\left(b^{5/6}\right)^6$
$$= 64\cdot a^{6/3}\cdot b^{5/6\cdot 6}$$
$$= 64a^2b^5$$

17. $6L^{-2}-4C^{-1} = \frac{6}{L^2}-\frac{4}{C} = \frac{6C-4L^2}{L^2C}$

21. $\left(a-3b^{-1}\right)^{-1} = \frac{1}{\left(a-3b^{-1}\right)} = \frac{1}{a-\frac{3}{b}}\cdot\frac{b}{b} = \frac{b}{ab-3}$

25. $\left(W^2+2WH+H^2\right)^{-1/2} = \frac{1}{\left(\left(W+H\right)^2\right)^{1/2}} = \frac{1}{W+H}$

29. $\sqrt{68} = \sqrt{4\cdot 17} = \sqrt{4}\cdot\sqrt{17} = 2\sqrt{17}$

33. $\sqrt{9a^3b^4} = \sqrt{9a^2b^4\cdot a} = 3ab^2\sqrt{a}$

37. $\frac{5}{\sqrt{2s}} = \frac{5}{\sqrt{2s}}\cdot\frac{\sqrt{2s}}{\sqrt{2s}} = \frac{5\sqrt{2s}}{2s}$

41. $\sqrt[4]{8m^6n^9} = \sqrt[4]{8m^4\cdot m^2\cdot n^8\cdot n} = mn^2\sqrt[4]{8m^2n}$

45. $\sqrt{36+4} = -2\sqrt{10} = \sqrt{4(10)}-2\sqrt{10}$
$$= 2\sqrt{10}-2\sqrt{10}$$
$$= 0$$

49. $a\sqrt{2x^3}+\sqrt{8a^2x^3} = a\sqrt{2x^2\cdot x}+\sqrt{4\cdot 2\cdot a^2x^2\cdot x}$
$$= ax\sqrt{2x}+2ax\sqrt{2x} = 3ax\sqrt{2x}$$

53. $5\sqrt{5}\left(6\sqrt{5}-35\right) = 30\sqrt{5}^2-5\sqrt{5}\sqrt{35}$
$$= 30\cdot 5-5\sqrt{175} = 150-5\sqrt{25(7)}$$
$$= 150-25\sqrt{7}$$

57. $\left(2-3\sqrt{17B}\right)\left(3+\sqrt{17B}\right)$
$$= 6+2\sqrt{17B}-9\sqrt{17B}-3\sqrt{17B}$$
$$= 6-51B-7\sqrt{17B}$$

61. $\frac{\sqrt{3x}}{2\sqrt{3x}-\sqrt{y}} = \frac{\sqrt{3x}}{\left(2\sqrt{3x}-\sqrt{y}\right)}\cdot\frac{\left(2\sqrt{3x}+\sqrt{y}\right)}{\left(2\sqrt{3x}+\sqrt{y}\right)}$
$$= \frac{2\sqrt{3x}^2+\sqrt{3x}\cdot\sqrt{y}}{4\sqrt{3x}^2-\sqrt{y}^2} = \frac{2\cdot 3x+\sqrt{3xy}}{4\cdot 3x-y}$$
$$= \frac{6x+\sqrt{3xy}}{12x-y}$$

65. $\dfrac{\sqrt{7}-\sqrt{5}}{\sqrt{5}+3\sqrt{7}} = \dfrac{\left(\sqrt{7}-\sqrt{5}\right)}{\left(\sqrt{5}+3\sqrt{7}\right)} \cdot \dfrac{\left(\sqrt{5}-3\sqrt{7}\right)}{\left(\sqrt{5}-3\sqrt{7}\right)}$

$\qquad = \dfrac{\sqrt{35}-3\cdot 7 - 5 + 3\sqrt{35}}{5 - 9\cdot 7}$

$\qquad = \dfrac{4\sqrt{35}-26}{5-63} = \dfrac{4\sqrt{35}-26}{-58} = \dfrac{13-2\sqrt{35}}{29}$

69. $\sqrt{4b^2+1}$ is in simplest form

73. $\left(1+6^{1/2}\right)\left(3^{1/2}+2^{1/2}\right)\left(3^{1/2}-2^{1/2}\right)$

$\qquad = \left(1+6^{1/2}\right)\left(\left(3^{1/2}\right)^2 - \left(2^{1/2}\right)^2\right)$

$\qquad = \left(1+6^{1/2}\right)\left(3-2\right)$

$\qquad = 1+6^{1/2}$

77. $\sqrt{3+n}\left(\sqrt{3+n}-\sqrt{n}\right)^{-1} = \dfrac{\sqrt{3+n}}{\sqrt{3+n}-\sqrt{n}} \cdot \dfrac{\sqrt{3+n}+\sqrt{n}}{\sqrt{3+n}+\sqrt{n}}$

$\qquad = \dfrac{3+n+\sqrt{n}\sqrt{3+n}}{3+n-n}$

$\qquad = \dfrac{3+n+\sqrt{3n+n^2}}{3}$

81. $\left(\sqrt{7}-2\sqrt{15}\right)\left(3\sqrt{7}-\sqrt{15}\right) = 21 - \sqrt{105} - 6\sqrt{105} + 30$

$\qquad = 51 - 7\sqrt{105}$

$\qquad = -20.728655$

85.

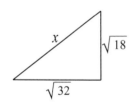

$x^2 = \sqrt{18}^2 + \sqrt{32}^2 = 50$

$x = \sqrt{50}$

perimeter $= \sqrt{18}+\sqrt{32}+\sqrt{50} = 3\sqrt{2}+4\sqrt{2}+5\sqrt{2}$

$\qquad = 12\sqrt{2}$

89. $i = 100\left[\left(\dfrac{C_2}{C_1}\right)^{1/n}-1\right] = 100\left[\left(\dfrac{218.1}{172.2}\right)^{1/10}-1\right]$

$\qquad = 2.391\%$

93. area + increase in area = new area

$x^2 + \text{increase} = \left(x+\sqrt{x}\right)^2 = x^2 + 2x\sqrt{x} + x$

$\text{increase} = x\left(2\sqrt{x}+1\right)$

$\% \text{ increase} = 100\dfrac{\text{increase}}{\text{area}} = \dfrac{100x\left(2\sqrt{x}+1\right)}{x^2}$

$\qquad = \dfrac{100\left(2\sqrt{x}+1\right)}{x}$

97. $\dfrac{\sqrt{A+h}-\sqrt{A}}{h} \cdot \dfrac{\sqrt{A+h}+\sqrt{A}}{\sqrt{A+h}+\sqrt{A}} = \dfrac{A+h-A}{h\left(\sqrt{A+h}+\sqrt{A}\right)}$

$\qquad = \dfrac{1}{\sqrt{A+h}+\sqrt{A}}$

101. $\dfrac{1}{2\pi\sqrt{\dfrac{LC_1C_2}{C_1+C_2}}} \cdot \dfrac{\sqrt{\dfrac{LC_1C_2}{C_1+C_2}}}{\sqrt{\dfrac{LC_1C_2}{C_1+C_2}}}$

$\qquad = \dfrac{\left(C_1C_2\right)}{2\pi LC_1C_2}\sqrt{\dfrac{LC_1C_2}{\left(C_1+C_2\right)} \cdot \dfrac{\left(C_1+C_2\right)}{\left(C_1+C_2\right)}}$

$\qquad = \dfrac{\sqrt{LC_1C_2\left(C_1+C_2\right)}}{2\pi LC_1C_2}$

Chapter 12

COMPLEX NUMBERS

12.1 Basic Definitions

1. $-\sqrt{-3}\sqrt{-12} = -j\sqrt{3}\left(j\sqrt{12}\right) = -j^2\sqrt{36}$
$$= -(-1)6$$
$$= 6$$

5. $\sqrt{-81} = \sqrt{81(-1)} = \sqrt{81}\sqrt{-1} = 9j$

9. $\sqrt{-0.36} = \sqrt{0.36(-1)} = \sqrt{0.36}\sqrt{-1} = 0.6j$

13. $\sqrt{-\dfrac{7}{4}} = \dfrac{\sqrt{-7}}{\sqrt{4}} = \dfrac{\sqrt{7(-1)}}{2} = \dfrac{\sqrt{7}\sqrt{-1}}{2} = j\dfrac{\sqrt{7}}{2}$

17. (a) $\left(\sqrt{-7}\right)^2 = \left(\sqrt{7(-1)}\right)^2 = \left(\sqrt{7}\cdot\sqrt{-1}\right) = \left(\sqrt{7}\cdot j\right)^2$
$$= \sqrt{7}^2\cdot j^2 = 7(-1) = -7$$
  (b) $\left(\sqrt{-7}\right)^2 = \sqrt{49} = 7$

21. $\sqrt{-\dfrac{1}{15}}\sqrt{-\dfrac{27}{5}} = j\sqrt{\dfrac{1}{15}}j\sqrt{\dfrac{27}{5}} = j^2\sqrt{\dfrac{27}{75}}$
$$= -\sqrt{\dfrac{9(3)}{25(3)}} = -\dfrac{3}{5}$$

25. (a) $-j^6 = -\left(j^2\right)^3 = -(-1)^3 = -(-1) = 1$
  (b) $(-j)^6 = \left((-j)^2\right)^3 = (-1)^3 = -1$

29. $j^{15} - j^{13} = j^{12}\cdot j^3 - j^{12}\cdot j = (1)(-j)-(1)(j)$
$$= -j-j = -2j$$

33. $2+\sqrt{-9} = 2+\sqrt{9(-1)} = 2+3j$

37. $\sqrt{-4j^2}+\sqrt{-4} = \sqrt{(-4)(-1)}+2j = 2+2j$

41. $\sqrt{18}-\sqrt{-8} = \sqrt{9\cdot2}-\sqrt{4\cdot2}j$
$$= 3\sqrt{2}-2j\sqrt{2}$$

45. $5j(-3j)\left(j^2\right) = -15j^4 = -15$

49. $\dfrac{\sqrt{-9}-6}{3} = \dfrac{3j-6}{3} = j-2$

53. (a) $2j$ has $-2j$ as conjugate.
  (b) -4 has -4 as conjugate.

57. $\quad 6j-7 = 3-x-yj$
$$6j-7-3 = -x-yj$$
$$6j-10 = -x-yj$$
$$-x = -10 \quad\text{and}\quad -yj = 6j$$
$$x = 10 \qquad\qquad y = \dfrac{6j}{-j} = -6$$

61. $x^2+32 = 0$
$$x^2 = -32$$
$$x = \pm\sqrt{-32}$$
$$x = \pm4j\sqrt{2}$$

65. (a) Since j^4-1, any number multiplied by j^4 is unchanged.
  (b) Since $j^2 = -1$, any number multiplied by j^2 will change sign.

69. $j+j^2+j^3+j^4+j^5+j^6+j^7+j^8$
$$= j+(-1)+(-j)+1+j+(-1)+(-j)+1$$
$$= j-1-j+1+j-1-j+1$$
$$= 0$$

73. $j(x+yj)-jx+yj^2 = -y+xj$, thus the imaginary part of $j(x+yj)$ is x which is the real part of $x+yj$.

12.2 Basic Operations with Complex Numbers

1. $(7-9j)-(6-4j)=7-9j-6+4j=1-5j$

5. $(3-7j)+(2-j)=(3+2)+(-7-1)j=5-8j$

9. $0.23-(0.46-0.9j)+0.67j$
$= 0.23-0.46+0.19j+0.67j$
$= -0.23+0.86j$

13. $(7-j)(j)=49j-7j^2 = 49j-7(-1)=7+49j$

17. $\sqrt{-18}\sqrt{-4}(3j)=(3\sqrt{2}j)(2j)(3j)$
$=18\sqrt{2}j^3 = 18\sqrt{2}j^2 j$
$=18\sqrt{2}(-1)j=-18j\sqrt{2}$

21. $j\sqrt{-7}-j^6\sqrt{112}+3j = j\sqrt{7(-1)}-j^6\sqrt{16(7)}+3j$
$= j^2\sqrt{7}-4j^6\sqrt{7}+3j$
$= (-1)\sqrt{7}-4j^4(j^2)\sqrt{7}+3j$
$= -\sqrt{7}-4(1)(-1)\sqrt{7}+3j$
$= -\sqrt{7}+4\sqrt{7}+3j$
$= 3\sqrt{7}+3j$

25. $(1-j)^3 = (1-j)(1-j)^2 = (1-j)(1-2j+j^2)$
$=1-2j+j^2-j+2j^2-j^3$
$=1-3j+3j^2-j^3$
$=1-3j+3(-1)-(-1)j$
$=-2-2j$

29. $\dfrac{1-j}{3j}\cdot\dfrac{3j}{3j}=\dfrac{3j-3j^2}{9(j^2)}=\dfrac{3j-3(-1)}{9(-1)}=\dfrac{3j+3}{-9}$
$=-\dfrac{1}{3}(1+j)$

33. $\dfrac{j^2-j}{2j-j^8}=\dfrac{-1-j}{2j-1}\cdot\dfrac{-2j-1}{-2j-1}=\dfrac{1+3j+2j^2}{1^2+2^2}$
$=\dfrac{1+3j-2}{5}=\dfrac{-1+3j}{5}$

37. $\left(4j^5-5j^4+2j^3-3j^2\right)^2 = \left(4j-5-2j+3\right)^2$
$=(-2+2j)^2 = 4-8j-4$
$=-8j$

41. $\dfrac{\left(2-j^3\right)^4}{\left(j^8-j^6\right)}+j = \dfrac{(2+j)^4}{(1+1)^3}+j = \dfrac{(2+j)^4}{8}+j$
$=\dfrac{(2+j)(2+j)^2}{8}+j$
$=\dfrac{(3+4j)(3+4j)}{8}+j = \dfrac{-7+24j}{8}+j$
$=-\dfrac{7}{8}+4j$

45. $x^2-4x+13=0$ has
$x_1 = 2+3i,\ x_2 = 2-3i$ from which
$x_1+x_2 = 2+3i+2-3i = 4$

49. $\dfrac{1}{3-j}=\dfrac{1}{3-j}\cdot\dfrac{3+j}{3+j}=\dfrac{3+j}{9+1}=\dfrac{1}{10}(3+j)$
$=\dfrac{3}{10}+\dfrac{1}{10}j$

53. $(x+3j)^2 = 7-24j$
$x^2+6xj-9 = 7-24j$
$x^2+6xj = 16-24j$ requires
$x^2 = 16$ and $6x=-24$
$x=\pm 4$ $x=-4$
The solution is $x=-4$.

57. $E = I \cdot Z = (0.835 - 0.427j)(250 + 170j)$

$\qquad = 208.75 + 141.95j - 106.75j$

$\qquad = -72.59j^2$

$\qquad = 208.75 + 35.2j + 72.59$

$\qquad = 281.34 + 35.2j$ volts

61. (a) $(a + bi) + (a - bi) = a + bi + a = a - bi = 2a,$

$\qquad$ a real number.

$\quad$ (b) $(a + bj) - (a - bj) = 2bj,$ pure imaginary

12.3 Graphical Representation of Complex Numbers

1. Add $5 - 2j$ and $-2 + j$ graphically.

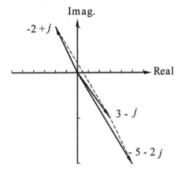

$$5 - 2j + (-2 + j) = 3 - j$$

5. $-4 - 3j$

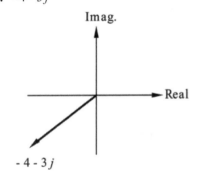

9. $2 + 3 + 4j = 5 + 4j$

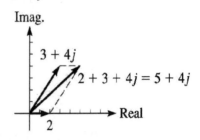

13. $5j - 1(1 - 4j) = 5j - 1 + 4j = -1 + 9j$

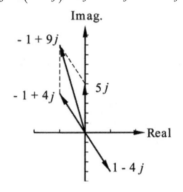

17. $(3 - 2j) - (4 - 6j) = 3 - 2j - 4 + 6j = -1 + 4j$

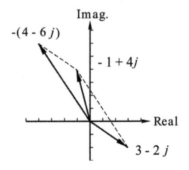

21. $(1.5 - 0.5j) + (3.0 + 2.5j) = 1.5 - 0.5j + 3.0 + 2.5j$

$$= 4.5 + 2.0j$$

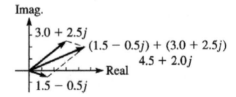

25. $(2j+1)-3j-(j+1)=(2j+1)+(-3j)+(-(j+1))$
$$= 2j+1-3j-j-1$$
$$= -2j$$

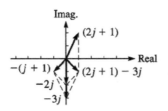

29. $3+2j$

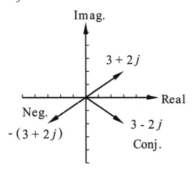

33. $a+bj=3-j$
$$3(a+bj)=9-3j$$
$$-3(a+bj)=-9+3j$$

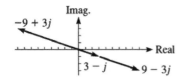

37. $2+4j-(2-4j)=2+4j-2+4j=2(4j)=8j$
The difference of a complex number and its conjugate is twice the imaginary part times the imaginary unit.

12.4   Polar Form of a Complex Number

1. $-3+4j \Rightarrow x=-3,\ y=4 \Rightarrow r=\sqrt{(-3)^2+4^2}=5$
$$\tan\theta_{ref}=\frac{4}{3},\ \theta_{ref}=53.1°,$$
$$\theta=180°-53.1°=126.9°$$
$$5(\cos 126.9° + j\sin 126.9°)$$

5. $30-40j$
$$r=\sqrt{30^2+(-40)^2}$$
$$=50$$
$$\tan\theta_{ref}=\left|\frac{-40}{30}\right|$$
$$=1.33$$
$$\theta_{ref}=53.1°$$
$$\theta=306.9°$$
$$50(\cos 306.9° + j\sin 306.9°)$$

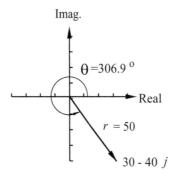

9. $-0.55-0.24j \Rightarrow x=-0.55,\ y=-0.24$
$$r=\sqrt{(-0.55)^2+(-0.24)^2}=0.60$$
$$\tan\theta_{ref}=\frac{0.24}{0.55},\ \theta_{ref}=24°,\theta=180°+24°=204°$$
$$0.60(\cos 204° + j\sin 204°)$$

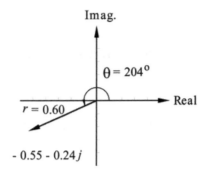

$\theta = 204°$

$r = 0.60$

$-0.55 - 0.24j$

13. $3.514 - 7.256j$

$$r = \sqrt{(3.514)^2 + (-7.256)^2} = 8.062$$

$$\tan\theta_{\text{ref}} = \left|\frac{-7.256}{3.514}\right| = 2.065$$

$$\theta_{\text{ref}} = 64.16°$$

$$\theta = 295.84°$$

$$8.062\left(\cos 295.84° + j \sin 295.84°\right)$$

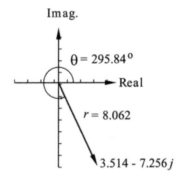

$\theta = 295.84°$

$r = 8.062$

$3.514 - 7.256j$

17. $9j = 0 + 9j$

$$r = \sqrt{0^2 + 9^2} = 9$$

$$\theta = 90° \text{ since } y \text{ is negative}$$

$$9\left(\cos 90° + j \sin 90°\right)$$

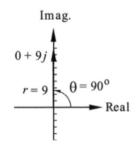

$0 + 9j$

$r = 9$ $\theta = 90°$

21. $160\left(\cos 150.0° + j \sin 150.0°\right)$

$$x = 160 \cos 150.0° = 160(0.866) = -139$$

$$y = 160 \sin 150.0° = 160(0.5) = 80$$

$$-139 + 80j$$

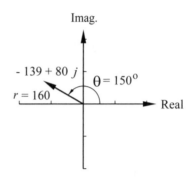

$-139 + 80j$

$\theta = 150°$

$r = 160$

25. $0.08\left(\cos 360° + j \sin 360°\right)$

$$x = 0.08 \cos 360° = 0.088(1) = 0.08$$

$$y = 0.088 \sin 360° = 0.088(0) = 0$$

$$0.08$$

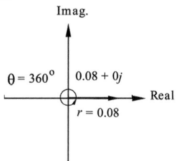

$\theta = 360°$ $0.08 + 0j$

$r = 0.08$

29. $4.75\angle 172.8°$

$$x = 4.75\left(\cos 172.8° + j \sin 172.8°\right)$$

$$y = -4.71 + 0.595j$$

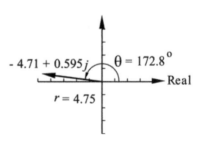

$-4.71 + 0.595j$ $\theta = 172.8°$

$r = 4.75$

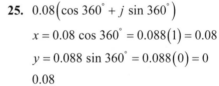

33. $7.32\angle-270° = 7.32\left(\cos\left(-270°\right)+j\sin\left(-270°\right)\right)$

$= 0+7.32j$

Imag.

$0+7.32j$

$r = 7.32$ $\theta = 90°$

Real

37. The argument for any negative real number is $180°$.

41. $r = \sqrt{2.84^2 + (-1.06)^2} = 3.03$

$\tan\theta_{ref} = \dfrac{-1.06}{2.84}, \theta_{ref} = -20.5°, \theta = 339.5°$

$2.84-1.06j = 3.03\left(\cos 339.5° + j\sin 339.5°\right)$

$3.03\angle339.5°$ kV

12.5 Exponential Form of a Complex Number

1. $8.50\angle226.3°$, $r = 8.50$,

$\theta = 226.3°\left(\dfrac{\pi}{180°}\right)$

$\theta = 3.95$

$8.50\angle226.3° = 8.50e^{3.95j}$

5. $3.00\left(\cos 60.0° + j\sin 60.0°\right)$;

$r = 3.00, \theta = 60.0°\cdot\left(\dfrac{\pi}{180°}\right) = 1.05$ rad

$3.00e^{1.05j}$

9. $375.5\left(\cos\left(-95.46°\right)+j\sin\left(-95.46°\right)\right)$; $r = 375.5$

$\theta = -95.46°\cdot\dfrac{\pi}{180°} = -1.666$ rad

$375.5e^{-1.666j} = 375.5e^{4.617j}$

13. $4.06\angle-61.4°$; $r = 4.06$;

$\theta = -61.4° = -1.07$ rad

$4.06e^{-1.07j} = 4.06e^{5.21j}$

17. $3-4j$; $r = \sqrt{3^2 + (-4)^2} = 5$,

$\theta_{ref} = \tan^{-1}\dfrac{-4}{3} = -0.9273$

$\theta = \theta_{ref} + 2\pi = 5.36$ rad

$3-4j = 5e^{5.36j}$

21. $5.90+2.40j$; $r = \sqrt{5.90^2 + 2.40^2} = 6.37$

$\theta = \tan^{-1}\dfrac{2.40}{5.90} = 0.386$

$5.90+2.40j = 6.37e^{0.386j}$

25. $3.00e^{0.500j}$; $r = 3, \theta = 0.500$ rad$=28.6°$

$3.00e^{0.500j} = 3\left(\cos 28.6° + j\sin 28.6°\right)$

$= 2.63+1.44j$

29. $3.20e^{-5.41j}$; $r = 3.20, \theta = -5.41$ rad $= 50°$

$3.20e^{-5.41j} = 3.20\left(\cos 50° + j\sin 50°\right)$

$= 2.06+2.45j$

33. $\left(4.55e^{1.32j}\right)^2 = 4.55^2 e^{2.64j}$ from which

$r = 4.55^2, \theta = 2.64\cdot\dfrac{180°}{\pi}$ and

$\left(4.55e^{1.32j}\right)^2 = r\left(\cos\theta + j\sin\theta\right)$

$= 4.55^2\left(\cos 2.64\cdot\dfrac{180°}{\pi} + j\sin 2.64\cdot\dfrac{180°}{\pi}\right)$

$= -18.2+9.95j$

37.

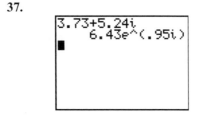

41. $2800 - 1450j; \ r = \sqrt{2800^2 + (-1450)^2} = 3153$

$\tan \theta = \dfrac{-1450}{2800} = -0.5179$

$\theta = -27.38^\circ = \dfrac{-27.38\pi}{180^\circ} = -0.478$

$\left(3153 e^{-0.478j} \right)^{-1} = 3.17 \times 10^{-4} e^{0.478j} \ 1/\Omega$

12.6   Products, Quotients, Powers, and Roots of Complex Numbers

1. $2 + 3j$: $r_1 = \sqrt{2^2 + 3^2} = 3.61$, $\tan \theta_1 = \dfrac{3}{2} \Rightarrow \theta_1 = 56.3^\circ$

$1 + j$: $r_2 = \sqrt{1^1 + 1^2} = 1.41$, $\tan \theta_2 = \dfrac{1}{1} \Rightarrow \theta_2 = 45.0^\circ$

$(2 + 3j)(1 + j) = 3.61(\cos 56.3^\circ + j \sin 56.3^\circ)(1.41)(\cos 45.5^\circ + j \sin 45.0^\circ)$
$\qquad\qquad\qquad = 5.09(\cos 101.3^\circ + j \sin 101.3^\circ)$

5. $[4(\cos 60^\circ + j \sin 60^\circ)][2(\cos 20^\circ + j \sin 20^\circ] = 4 \cdot 2[(\cos(60^\circ + 20^\circ) + j \sin(60^\circ + 20^\circ)]$
$\qquad\qquad\qquad\qquad\qquad\qquad\qquad\qquad = 8(\cos 80^\circ + j \sin 80^\circ)$

9. $\dfrac{8(\cos 100^\circ + j \sin 100^\circ)}{4(\cos 65^\circ + j \sin 65^\circ)} = \dfrac{8}{4}(\cos(100^\circ - 65^\circ) + j \sin(100^\circ - 65^\circ)) = 2(\cos 35^\circ + j \sin 35^\circ)$

13. $\left[0.2 \left(\cos 35^\circ + j \sin 35^\circ \right) \right]^3 = \left[0.2 \left(\cos 3 \cdot 35^\circ + j \sin 3 \cdot 35^\circ \right) \right]^3 = 0.008 \left(\cos 105^\circ + j \sin 105^\circ \right)$

17. $\dfrac{\left(50 \angle 236^\circ \right) \left(2 \angle 84^\circ \right)}{125 \angle 47^\circ} = \dfrac{100 \angle 320^\circ}{125 \angle 47^\circ} = \dfrac{4}{5} \angle 273^\circ$

21. $2.78 \angle 56.8^\circ + 1.37 \angle 207.3^\circ = 2.78(\cos 56.8^\circ + j \sin 56.8^\circ) + 1.37(\cos 207.3^\circ + j \sin 207.3^\circ)$
$\qquad\qquad\qquad = 1.5222 + 2.3262j - 1.2174 - 0.6283j = 0.3048 + 1.6979j.$

$r = \sqrt{0.3048^2 + 1.6979^2} = 1.73, \ \theta = \tan^{-1} \dfrac{1.6979}{0.3048} = 79.8^\circ.$

25. $3 + 4j = \sqrt{3^2 + 4^2} = \left(\cos\left(\tan^{-1}\frac{4}{3}\right) + j\sin\left(\tan^{-1}\frac{4}{3}\right)\right) = 5(\cos 53.1° + j\sin 53.1°)$

$5 - 12j = \sqrt{5^2 + (-12)^2}\left(\cos\left(\tan^{-1}\frac{-12}{5} + 360°\right) + j\sin\left(\tan^{-1}\frac{-12}{5} + 360°\right)\right)$

$\qquad = 13(\cos 292.6° - j\sin 292.6°)$

polar form:

$(3 + 4j)(5 - 12j) = [5(\cos 53.1° + j\sin 53.1°)][13(\cos 292.6° + j\sin 292.6°)]$
$\qquad = 65(\cos(53.1° + 292.6°) + j\sin(53.1° + 292.6°))$
$\qquad = 65(\cos(345.7°) + j\sin(345.7°))$
$\qquad = 63.0 - 16.1j$

rectangular form:

$(3 + 4j)(5 - 12j) = 15 - 36j + 20j - 48j^2 = 15 - 16j + 48 = 63 - 16j$

29. $\dfrac{7}{1 - 3j} = \dfrac{7(\cos 0° + j\sin 0°)}{\sqrt{1^2 + (-3)^2}\left(\cos\left(\tan^{-1}\dfrac{-3}{1} + 360°\right) + j\sin\left(\tan^{-1}\dfrac{-3}{1} + 360°\right)\right)} = \dfrac{7(\cos 0° + j\sin 0°)}{3.16(\cos 288.4° + j\sin 288.4°)}$

$\qquad = \dfrac{7}{3.16}(\cos(0° - 288.4°) + j\sin(0 - 288.4°)) = 2.22(\cos(-288.4°) + j\sin(-288.4°))$

$\qquad = 2.22(\cos 71.6° + j\sin 71.6°) = 0.7 + 2.1j$

Rectangular form:

$\dfrac{7}{1 - 3j} = \dfrac{7}{(1 - 3j)} \cdot \dfrac{(1 + 3j)}{(1 + 3j)} = \dfrac{7 + 21j}{1^2 + 3^2} = \dfrac{7}{10} + \dfrac{21}{10}j$

33. $(3 + 4j) = \sqrt{3^2 + 4^2}\left(\cos\left(\tan^{-1}\frac{4}{3}\right) + j\sin\left(\tan^{-1}\frac{4}{3}\right)\right) = 5(\cos 53.1° + j\sin 53.1°)$

$(3 + 4j)^4 = [5(\cos 53.1° + j\sin 53.1°)]^4 5^4(\cos(4 \cdot 53.1°) + j\sin(4 \cdot 53.1°))$
$\qquad = 625(\cos 212.5° + j\sin 212.5°) = -527 - 336j$

rectangular form:

$(3 + 4j)^4 = [(3 + 4j)^2]^2 = (9 + 24j + (16j^2))^2 = (9 + 24j - 16)^2 = (-7 + 24j)^2$
$\qquad = 49 - 336j + 576j^2 49 - 336j - 576 = -527 - 336j$

37. The two square roots of $4(\cos 60° + j\sin 60°)$ are

$\qquad r_1 = \sqrt{4}\left(\cos\dfrac{60° + 0 \cdot 360°}{2} + j\sin\dfrac{60° + 0 \cdot 360°}{2}\right)$

$\qquad r_1 = 2(\cos 30° + j\sin 30°) = \sqrt{3} + j$

and

$\qquad r_2 = \sqrt{4}\left(\cos\dfrac{60° + 1 \cdot 360°}{2} + j\sin\dfrac{60° + 1 \cdot 360°}{2}\right)$

$\qquad r_2 = 2(\cos 210° + j\sin 210°) = -\sqrt{3} - j$

41. The two square roots of $1 + j = \sqrt{2}(\cos 45° + j \sin 45°)$ are

$$r_1 = \sqrt{\sqrt{2}}\left(\cos \frac{45° + 0 \cdot 360°}{2} + j \sin \frac{4.5° + 0 \cdot 360°}{2}\right)$$

$$r_1 = 2^{1/4}(\cos 22.5° + j \sin 22.5°) = 1.0987 + 0.4551j$$

and

$$r_2 = 2^{1/4}\left(\cos \frac{45° + 1 \cdot 360°}{2} + j \sin \frac{4.5° + 1 \cdot 360°}{2}\right)$$

$$r_2 = 2^{1/4}(\cos 202.5° + j \sin 202.5°) = -1.0987 - 0.4551j$$

45. $-27j;\ x = 0,\ y = -27;\ r = \sqrt{0^2 + 27^2} = 27;\ \theta = 270°$

First root: $(0 - 27j)^{1/3} = [27(\cos 270° + j \sin 270°)]^{1/3} = 3(\cos 90° + j \sin 90°) = 3j$

Second root: $\theta = 360° + 270° = 630°$

$(0 + 27j)^{1/3} = [27(\cos 630° + j \sin 630°)]^{1/3} = 3(\cos 210° + j \sin 210°) = \dfrac{-3\sqrt{3}}{2} - \dfrac{3}{2}j$

Third root: $\theta = 720° + 270° = 990°$

$(0 + 27j)^{1/3} = [27(\cos 990° + j \sin 990°)]^{1/3} = 3(\cos 330° + j \sin 330°) = \dfrac{3\sqrt{3}}{2} - \dfrac{3}{2}j$

49. $-125 = 125\left(\cos 180° + j \sin 180°\right)$

$$(-125)^{1/3} = 125^{1/3}\left[\cos\left(\frac{1}{3} \cdot 180°\right) + j \sin\left(\frac{1}{3} \cdot 180°\right)\right] = \frac{5}{2} + \frac{5\sqrt{3}}{2}j = 2.500 + 4.330j$$

$$(-125)^{1/3} = 125^{1/3}\left[\cos\left(\frac{1}{3} \cdot 540°\right) + j \sin\left(\frac{1}{3} \cdot 540°\right)\right] = -5$$

$$(-125)^{1/3} = 125^{1/3}\left[\cos\left(\frac{1}{3} \cdot 900°\right) + j \sin\left(\frac{1}{3} \cdot 900°\right)\right] = \frac{5}{2} - \frac{5\sqrt{3}}{2}j = 2.500 - 4.330j$$

53.

$$x^3 + 1 = 0$$
$$(x + 1)(x^2 - x + 1) = 0$$
$$x + 1 = 0 \quad \text{or} \quad x^2 - x + 1 = 0$$

$$x = -1 \qquad x = \frac{-(-1) \pm \sqrt{(-1)^2 - 4(1)(1)}}{2(1)} = \frac{1 \pm j\sqrt{3}}{2}$$

The roots are the same as in Example 7.

57. $\dfrac{(8.66\angle 90.0°)(50.0\angle 135.0°)}{10.0\angle 60.0°} = \dfrac{(8.66)(50.0)}{10.0}\angle 90.0° + 135.0° - 60.0°$

$$= 43.3\angle 165.0°, V = 43.3 \text{ V}$$

12.7 An Application to Alternating-Current (ac) Circuits

1. $V_R = IR = 2.00(12.0) = 24.0$ V

$Z = R + j(X_L - X_C) = 12.0 + j(16.0)$

$|Z| = \sqrt{12.0^2 + 16.0^2} = 20.0\,\Omega$

$V_L = IX_L = 2.00(16.0) = 32.0$ V

$V_{RL} = IZ = 2.00(20.0) = 40.0$ V

$\theta = \tan^{-1}\dfrac{X_L}{R} = \tan^{-1}\dfrac{16.0}{12.0} = 53.1°$, voltage

leads current.

5. (a) $|Z| = \sqrt{R^2 + X_L^2} = \sqrt{2250^2 + 1750^2} = 2850\,\Omega$

(b) $\tan\theta = \dfrac{1750}{2250}; \ \theta = 37.9°$

(c) $V_{RLC} = IZ = (0.005\ 75)(2850) = 16.4$ V

9. (a) $X_R = 45.0\,\Omega$

$X_L = 2\pi fL = 2\pi(60)(0.0429) = 16.2\,\Omega$

$Z = 45.0 + 16.2j$

$|Z| = \sqrt{45.0^2 + 16.2^2} = 47.8\,\Omega$

(b) $\tan\theta = \dfrac{16.2}{45.0}; \ \theta = 19.8°$

13. $R = 25.3\,\Omega$

$X_C = 1/(2\pi fC)$

$= 1/\left(2\pi(1.2\times10^6)(2.75\times10^{-9})\right) = 48.2\,\Omega$

$= f = 1200$ kHz $= 1.2\times10^6$ Hz

$Z = R - X_{Cj} = 25.3 - 48.2j$

$|Z| = \sqrt{25.3^2 + (-48.2)^2} = 54.4\,\Omega$

$\tan\theta = \dfrac{-48.2}{25.3}; \ \theta = -62.3°$

17. $L = 12.5\times10^{-6}$ H

$C = 47.0\times10^{-9}$ F

$X_L = X_C$

$2\pi fL = \dfrac{1}{2\pi fC}$

$f = \sqrt{\dfrac{1}{4\pi^2 LC}}$

$= \sqrt{\dfrac{1}{4\pi^2(12.5\times10^{-6})(4.70\times10^{-9})}}$

$= 208$ kHz

21. $P = VI\cos\theta$

$V = 225$ mV

$\theta = -18.0° = 342°$

$Z = 47.3\,\Omega$

$V = IZ$

$I = \dfrac{225\times10^{-3}}{47.3} = 0.00476$ A

$P = (225\times10^{-3})(0.00476)\cos 342°$

$= 0.00102$ W $= 1.02$ mW

Chapter 12 Review Exercises

1. $(6-2j)+(4+j)=6-2j+4+j=10-j$

5. $5j(6-5j)=30j-25j^2=25+30j$

9. $\dfrac{3}{7-6j}=\dfrac{3}{(7-6j)}\cdot\dfrac{(7+6j)}{(7+6j)}=\dfrac{21+18j}{7^2+6^2}=\dfrac{21}{85}+\dfrac{18}{85}j$

13. $\dfrac{5j-(3-j)}{4-2j}=\dfrac{(-3+6j)}{(4-2j)}\cdot\dfrac{(4+2j)}{(4+2j)}$

$=\dfrac{-12-6j+24j+12j^2}{4^2+2^2}$

$=\dfrac{-12+18j-12}{16+4}$

$=\dfrac{-24+18j}{20}$

$=-\dfrac{6}{5}+\dfrac{9}{10}j$

17. $3x-2j=yj-9 \Rightarrow 3x=-9,\ -2=y$

$x=-\dfrac{9}{3}\qquad y=-2$

$x=-3$

$x=-3,\ y=-2$

21.

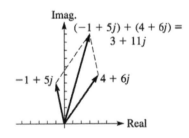

algebraically:

$(-1+5j)+(4+6j)=-1+5j+4+6j$

$=-1+4+5j+6j$

$=3+11j$

25. $1-j \Rightarrow r=\sqrt{1^2+1^2}=\sqrt{2},\ \theta=\tan^{-1}\dfrac{-1}{1}+360°$

$=315°=\dfrac{7\pi}{4}$ rad

polar: $1-j=\sqrt{2}\left(\cos 315°+j\sin 315°\right)$

exponential: $1-j=\sqrt{2}e^{7\pi/4j}$

29. $1.07+4.55j \Rightarrow r=\sqrt{1.07^2+4.55^2}=4.67,$

$\theta=\tan^{-1}\dfrac{4.55}{1.07}=76.8°=1.34$ rad

polar: $1.07+4.55j=4.67\left(\cos 76.8°+j\sin 76.8°\right)$

exponential: $1.07+4.55j=4.67e^{1.34j}$

33. $2\left(\cos 225°+j\sin 225°\right)=-\sqrt{2}-\sqrt{2}j$

37. $0.62\angle-72°=0.62\left(\cos(-72°)+j\sin(-72°)\right)$

$=0.19-0.59j$

41. $200e^{0.25j}=2.00\left(\cos 0.25+j\sin 0.25\right)$

$=1.94+0.495j$

45. $\left[3\left(\cos 32°+j\sin 32°\right)\right]\cdot\left[\left(\cos 52°+j\sin 52°\right)\right]$

$=3\cdot 5\left(\cos\left(32°+52°\right)+j\sin\left(32°+52°\right)\right)$

$=15\left(\cos 84°+j\sin 84°\right)$

49. $\dfrac{24\,(\cos 165°+j\sin 165)°}{\left[3\,(\cos 55°+j\sin 55°)\right]^3}$

$=\dfrac{24\left(\cos 165°+j\sin 165°\right)}{27\left(\cos 165°+j\sin 165°\right)}$

$=\dfrac{24}{27}\cos\left(165°-165°\right)+j\sin\left(165°-165°\right)$

$=\dfrac{8}{9}\left(\cos 0°+j\sin 0°\right)=\dfrac{8}{9}$

53. $0.983\angle 47.2° + 0.366\angle 95.1°$

$\quad = 0.983\left(\cos 47.2° + j \sin 47.2°\right)$

$\quad\quad + 0.366\left(\cos 95.1° + j \sin 95.1°\right)$

$\quad = 0.6679 + 0.7213\,j - 0.03254 + 0.3646\,j$

$\quad = 0.6354 + 1.0859\,j$

in polar form $r = \sqrt{0.6354^2 + 1.0859^2} = 1.26$

$\qquad \theta = \tan^{-1}\dfrac{1.0859}{0.6354} = 59.7°,\ 1.26\angle 59.7°$

57. $\left[2\left(\cos 16° + j \sin 16°\right)\right]^{10}$

$\quad = 2^{10}\left[\cos\left(10\cdot 16°\right) + j \sin\left(10\cdot 16°\right)\right]$

$\quad = 1024\left(\cos 160° + j \sin 160°\right)$

61. $\quad 1 - j = \sqrt{2}\left(\cos 315° + j \sin 315°\right)$ from prob. 25

$\quad \left(1 - j\right)^{10} = \left[\sqrt{2}\left(\cos 315° + j \sin 315°\right)\right]^{10}$

$\quad\quad = \sqrt{2}^{\,10}\left(\cos\left(10\cdot 315°\right) + j \sin\left(10\cdot 315°\right)\right)$

$\quad\quad = 32\left(\cos 3150° + j \sin 3150°\right)$

$\quad\quad = 32\left(\cos 270° + j \sin 270°\right),\ \text{polar form}$

$\quad\quad = 0 - 32\,j,\ \text{rectangular form}$

$\quad \left(1 - j\right)^{10} = \left(\left(1 - j\right)^2\right)^5 = \left(1 - 2j + j^2\right)^5 = \left(1 - 2j - 1\right)^5$

$\quad\quad = \left(-2j\right)^5 = \left(-2\right)^5 \cdot j^5$

$\quad\quad = -32 \cdot j^4 \cdot j = -32j$

65. $-8 = -8 + 0j = 8\left(\cos 180° + j \sin 180°\right)$

$\quad r_1 = \sqrt[3]{8}\left(\cos\dfrac{180° + 360°}{3} + j \sin\dfrac{180° + 0\cdot 360°}{3}\right)$

$\quad = 2\left(\cos 60° + j \sin 60°\right)$

$\quad = 2\left(\dfrac{1}{2} + j\cdot\dfrac{\sqrt{3}}{2}\right) = 1 + j\sqrt{3}$

$\quad r_2 = \sqrt[3]{8}\left(\cos\dfrac{180° + 360°}{3} + j \sin\dfrac{180° + 1\cdot 360°}{3}\right)$

$\quad = 2\left(\cos 180° + j \sin 180°\right) = 2\left(-1 + j\left(0\right)\right) = -2$

$\quad r_3 = \sqrt[3]{8}\left(\cos\dfrac{180° + 2\cdot 360°}{3} + j \sin\dfrac{180° + 2\cdot 360°}{3}\right)$

$\quad = 2\left(\cos 300° + j \sin 300°\right) = 2\left(\dfrac{1}{2} - \dfrac{\sqrt{3}}{2}\,j\right)$

$\quad = 1 - j\sqrt{3}$

69. Rectangular: $40 + 9j$ from the graph

$\quad$ polar: $40 + 9j \Rightarrow r = \sqrt{40^2 + 9^2} = 41,\ \theta = \tan^{-1}\dfrac{9}{40}$

$\qquad\qquad\qquad\qquad\qquad\qquad = 12.7°$

$\quad 40 + 9j = 41\left(\cos 12.7° + j \sin 12.7°\right)$

73. $x^2 - 2x + 4\Big|_{x=5-2j} = \left(5 - 2j\right)^2 - 2\left(5 - 2j\right) + 4$

$\qquad\qquad\qquad = 25 - 20j + 4j^2 - 10 + 4j + 4$

$\qquad\qquad\qquad = 19 - 16j - 4$

$\qquad\qquad\qquad = 15 - 16j$

77. $x^2 + 3jx - 2 = 0$

$\quad x = \dfrac{-3j \pm \sqrt{\left(3j\right)^2 - 4\left(1\right)\left(-2\right)}}{2\left(1\right)}$

$\quad x = \dfrac{\pm\sqrt{-1} - 3j}{2} = \dfrac{\pm j - 3j}{2}$

$\quad x = -j,\ -2j$

81. $x^2 - 2x + 2 = \left(1 - j\right)^2 - 2\left(1 - j\right) + 2$

$\qquad\qquad\qquad = \left(1 - 2j + j^2\right) - 2 + 2j + 2$

$\qquad\qquad\qquad = 1 - 2j - 1 - 2 + 2j + 2$

$\qquad\qquad\qquad = 0 \Rightarrow 1 - j$ is a solution

$\quad x^2 - 2x + 2 = \left(-1 - j\right)^2 - 2\left(-1 - j\right) + 2$

$\qquad\qquad\qquad = \left(1 + 2j + j^2\right) + 2 + 2j + 2$

$\qquad\qquad\qquad = 4j + 4 \Rightarrow -1 - j$ is not a solution

85. $\quad f(x) = 2x - (x-1)^{-1}$

$\quad f(1+2j) = 2(1+2j) - (1+2j-1)^{-1}$

$\qquad\qquad = 2 + 4j - \dfrac{1}{2j} \cdot \dfrac{j}{j}$

$\quad f(1+2j) = 2 + 4.5j$

89. $\ V_L = 60j,\ V_C = -60j$

$\quad V = V_R + V_L + V_C$

$\quad 60 = V_R + 60j - 60j$

$\quad V_R = 60\ \text{V}$

93. $\ 2\pi fL = \dfrac{1}{2\pi fC} \Rightarrow f = \sqrt{\dfrac{1}{4\pi^2 LC}}$

$\qquad\qquad = \sqrt{\dfrac{1}{4\pi^2 (2.65)(18.3 \times 10^{-6})}}$

$\qquad\quad f = 22.9\ \text{Hz}$

97. $\ \dfrac{1}{\mu + j\omega n} = \dfrac{1}{(\mu + j\omega n)} \cdot \dfrac{(\mu - j\omega n)}{(\mu - j\omega n)} = \dfrac{\mu - j\omega n}{\mu^2 + \omega^2 n^2}$

101. Answers may vary.

Chapter 13

EXPONENTIAL AND LOGARITHMIC FUNCTIONS

13.1   Exponential Functions

1. For $x = -\dfrac{3}{2}$, $y = -2\left(4^x\right) = -2\left(4^{-3/2}\right) = -\dfrac{1}{4}$

5. $\left(3\pi\right)^e = 445$

9. (a) $y = -7\left(-5\right)^{-x}$, $-5 < 0$, not an exponential function.

   (b) $y = -7\left(5^{-x}\right)$ is a real number multiple of an exponential function and therefore an exponential function.

13. $y = 9^x$; $x = -2$, $= 9^{-2} = \dfrac{1}{9^2} = \dfrac{1}{81}$

17. $y = 4^x$

x	-3	-2	-1	0	1	2	3
y	$\frac{1}{64}$	$\frac{1}{16}$	$\frac{1}{4}$	1	4	16	64

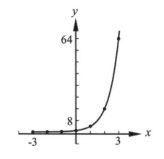

21. $y = -0.5\pi^x$

x	y
-3	27.000
-2	9.000
-1	3.000
0	1.000
1	0.333
2	0.111
3	0.037

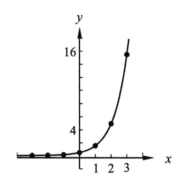

25. $y_1 = 0.1\left(0.25\right)^{2x}$

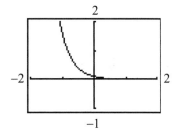

29. a) $y1 = 10^x$
   b) $y2 = 2^x$
   c) $y3 = 1.1^x$

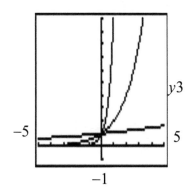

33. $f(x) = b^x$

$f(a+b) = b^{a+b} = b^a \cdot b^b = f(a) \cdot f(b)$

37. Since $x^2 = 2^x \Leftrightarrow x^2 - 2^x = 0$, graph $y_1 = x^2 - 2^x$.

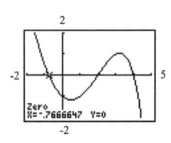

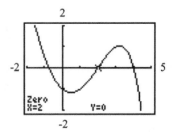

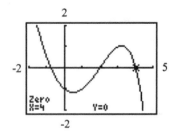

From the graph, $x^2 = 2^2$ for $x = -0.7667, 2, 4$.

41. $V = 250(1.0500)^t = 250(1.0500)^4 = \$303.88$

45. $q = 100e^{-10t}$, graph $y_1 = 100e^{-10x}$.

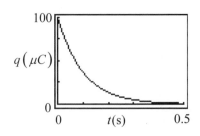

13.2   Logarithmic Functions

1. $32^{4/5} = 16$ in logarithmic form is

$\dfrac{4}{5} = \log_{32} 16.$

5. $3^3 = 27$ has base 3, exponent 3, and number 27.

$\log_3 27 = 3$

9. $7^{-2} = \dfrac{1}{49}$ has base 7, exponent -2, and number $\dfrac{1}{49}$.

$\log_7 \dfrac{1}{49} = -2$

13. $8^{1/3} = 2$ has base 8, exponent $\dfrac{1}{3}$, and number 2.

$\log_8 2 = \dfrac{1}{3}$

17. $\log_3 81 = 4$ has base 3, exponent 2, and number 81.

$3^4 = 81$

21. $\log_{25} 5 = \dfrac{1}{2}$ has base 25, exponent $\dfrac{1}{2}$, and number 5.

$25^{1/2} = 5$

25. $\log_{10} 0.1 = -1$ has base 10, exponent -1, and number 0.1.

$0.1 = 10^{-1}$

29. $\log_4 16 = x$ has base 4, exponent x, and number 16.

$4^x = 16$

$4^x = 4^2$

$x = 2$

33. $\log_7 y = 3$ has base 7, exponent 3, and number 3.

$7^3 = y,\ y = 343$

37. $\log_b 5 = 2$ has base b, exponent 2, and number 5.

$b^2 = 5$, $b = \sqrt{5}$

41. $\log_{10} 10^{0.2} = x$ has base 10, exponent x, and number $10^{0.2}$.

$10^x = 10^{0.2}$, $x = 0.2$

45. Write $y = \log_3 x$ as $3^y = x$ to find values in table.

x	y
0.19	−1.5
0.58	−0.5
1.00	0.0
1.93	0.6
3.00	1.0
5.20	1.5
9.00	2.0

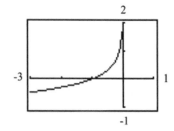

49. $N = 0.2\log_4 v$; $\dfrac{N}{0.2} = \log_4 v$; $4^{N/0.2} = v$

v	N
0.2	−0.232
0.6	−0.0737
0.8	−0.0322
1	0
2	0.1
3	0.159
4	0.2

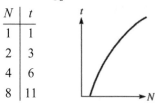

53. Graph $y_1 = -\log_{10}(-x)$.

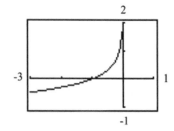

57. (a) $f(x) = \log_5 x$

$f(\sqrt{5}) = \log_5 \sqrt{5} = \log_5 5^{1/2}$

$= \dfrac{1}{2}\log_5 5 = \dfrac{1}{2}$

(b) $f(0)$ does not exist.

61.

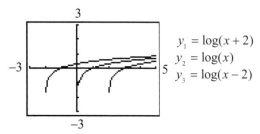

$y_1 = \log(x+2)$
$y_2 = \log(x)$
$y_3 = \log(x-2)$

65. $2 - x > 0$

$x < 2$ is domain

of $f(x) = \log_e(2-x)$

69. $\log_e\left(\dfrac{N}{N_0}\right) = -kt \Rightarrow e^{-kt} = \dfrac{N}{N_0}$, $N = N_0 e^{-kt}$

73. $t = N + \log_2 N$ where $n > 0$ and $t > 0$.

N	t
1	1
2	3
4	6
8	11

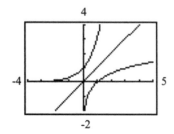

77. Solve $y = 10^{x/2}$ for $x = 2\log_{10} y$. Interchange x and y; $y = 2\log_{10} x$, which is the inverse function. Graph $y_1 = 10^{x/2}$, $y = 2\log_{10} x$.

For each graph to be the mirror image of the other across $y = x$ the calculator window must be "square." One way to do this is ZOOM 5: Square

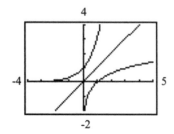

13.3   Properties of Logarithms

1. $\log_4 21 = \log_4\big(3(7)\big) = \log_4 3 + \log_4 7$

5. $\log_3 27 = \log_3 3^3 = 3\log_3 3 = 3(1) = 3$

9. $\log_5 33 = \log_5\big(3 \cdot 11\big) = \log_5 3 + \log_5 11$

13. $\log_2\big(a^3\big) = 3\log_2 a$

17. $8\log_5 \sqrt[4]{y} = 8\log_5 y^{1/4} = 2\log_5 y$

21. $\log_b a + \log_b c = \log_b\big(ac\big)$

25. $-\log_b \sqrt{x} + \log_b x^2 = \log_b \dfrac{x^2}{x^{1/2}} = \log_b x^{3/2}$

29. $\log_2\left(\dfrac{1}{32}\right) = \log_2\left(\dfrac{1}{2^5}\right) = \log_2 2^{-5} = -5\log_2 2 = -5$

33. $6\log_7 \sqrt{7} = 6\log_7 7^{1/2} = 3\log_7 7 = 3$

37. $\log_3 18 = \log_3\big(9 \cdot 2\big) = \log_3 9 + \log_3 2$
$\qquad\quad = \log_3 3^2 + \log_3 2 = 2\log_3 3 + \log_3 2$
$\qquad\quad = 2 + \log_3 2$

41. $\log_3 \sqrt{6} = \log_3\big(3 \cdot 2\big)^{1/2} = \dfrac{1}{2}\log_3\big(3 \cdot 2\big)$
$\qquad\quad = \dfrac{1}{2}\cdot\big[\log_3 3 + \log_3 2\big]$
$\qquad\quad = \dfrac{1}{2}\cdot\big[1 + \log_3 2\big]$

45. $\log_b y = \log_b 2 + \log_b x$
$\quad\;\; \log_b y = \log_b\big(2x\big)$
$\qquad\quad y = 2x$

49. $\log_{10} y = 2\log_{10} 7 - 3\log_{10} x$
$\qquad\quad = \log_{10} 7^2 - \log_{10} x^3$
$\qquad\quad = \log_{10} 49 - \log_{10} x^3$
$\quad \log_{10} y = \log_{10} \dfrac{49}{x^3}$
$\qquad\quad y = \dfrac{49}{x^3}$

53. $\log_2 x + \log_2 y = 1$
$\quad \log_2\big(xy\big) = 1 \Leftrightarrow 2^1 = xy$
$\qquad\qquad\quad y = \dfrac{2}{x}$

57. $\log_{10}\big(x+3\big) = \log_{10} x + \log_{10} 3 = \log_{10}\big(3x\big)$
$\qquad\quad x + 3 = 3x$
$\qquad\qquad 2x = 3$
$\qquad\qquad\; x = \dfrac{3}{2}$ is the only value for which

$\log_{10}\big(x+3\big) = \log_{10} x + \log_{10} 3$ is true

For any other x-value $\log_{10}\big(x+3\big) = \log_{10} x + \log_{10} 3$

is false and thus not true in general. This can be

also be seen from the following graph.

61. $\log_b \sqrt{x^2 y^4} = \log_b\big(x^2 y^4\big)^{1/2}$
$\qquad\qquad = \log_b\big(xy^2\big)$
$\qquad\qquad = \log_b x + 2\log_b y$
$\qquad\qquad = 2 + 2(3) = 8$

65. Graph $y_1 = \log_{10} x - \log_{10}\big(x^2 - 1\big)$, $y_2 = \log_{10} \dfrac{x}{x^2 + 1}$.

The graphs are the same.

69. $\log_e T = \log_e 90.0 - 0.23t$

$0.23t = \log_e 90.0 - \log_e T$

$0.23t = \log_e \left(\dfrac{90.0}{T} \right)$

$e^{0.23t} = \dfrac{90.0}{T}$; $T = \dfrac{90.0}{e^{0.23t}} = 90.0^{-0.23t}$

13.4 Logarithms to the Base 10

1. $\log 0.3654 = -0.4372$

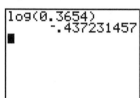

5. $\log 9.24 \times 10^6 = 6.966$

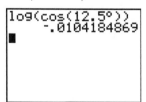

9. $\log \left(\cos 12.5° \right) = -0.0104$

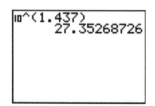

13. $10^{1.437} = 27.35$

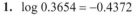

17. $10^{3.30112} = 2000.4$

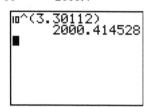

21. $(5.98)(14.3) = 85.5$

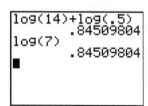

25. $\log 14 + \log 0.5 = \log 7$

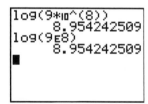

29. $\log 9 \times 10^8 = 8.9542$

33. $\log T = 8$

$T = 10^8 \text{ K}$

37. $\dfrac{\log_b x^2}{\log 100} = \dfrac{2 \log_b x}{\log 10^2} = \dfrac{2 \log_b x}{2 \log 10} = \log_b x$

41. $R = \log \left(\dfrac{I}{I_0} \right)$; $I = 1,600,000,000 I_0$

$R = \log \dfrac{1,600,000,000 I_0}{I_0}$

$= \log 1,600,000,000 = 9.2$

13.5 Natural Logarithms

1. $\ln 200 = \dfrac{\log 200}{\log e} = 5.298$

5. $\ln 1.562 = \dfrac{\log 1.562}{\log e} = \dfrac{0.1937}{0.4343} = 0.4460$

9. $\log_7 42 = \dfrac{\log 42}{\log 7} = \dfrac{1.6232}{0.8451} = 1.92$

13. $\log 7.5 \times 10^2 = \log_{40} 750 = \dfrac{\log 750}{\log 40} = 1.795$

17. $\ln 1.394 = 0.3322$

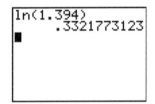

21. $\ln 0.012937^4 = -17.39066$

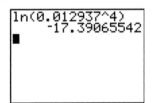

25. $\log 0.68528 = \dfrac{\ln 0.68528}{\ln 10} = -0.16413$

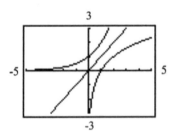

29. $e^{0.0084210} = 1.0085$

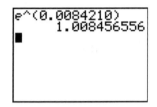

33. $e^{-23.504} = 6.20 \times 10^{-11}$

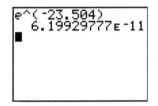

37. $y = 2^x \Leftrightarrow x = \log_2 y \Rightarrow y = \log_2 x$ is the inverse function.

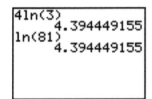

41. $4\ln 3 = \ln 81$

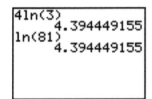

45. $\ln x \left(\log x \right) = 0$

$e^{\ln(\log x)} e^0 = 1$

$\log x = 1$

$10^{\log x} = 10^1$

$x = 10$

49. $\ln\left(e^2\sqrt{1-x}\right) = \ln e^2 + \ln(1-x)^{1/2}$

$$= 2\ln e + \frac{1}{2}\ln(1-x)$$

$$= 2 + \frac{1}{2}\ln(1-x)$$

53. $i = \dfrac{(\ln 2)}{8.5} = 0.081\,55 = 8.155\%$

57. $x = \dfrac{1}{k}\ln(kv_0 t + 1);\ x = 150$ m,

$k = 6.80 \times 10^{-3}$ m

$v_0 = 12.0$ m/s

$kx = \ln(kv_0 t + 1)$

$e^{kx} = kv_0 t + 1$

$kv_0 t = e^{kx} - 1$

$t = \dfrac{e^{kx} - 1}{kv_0} = \dfrac{e^{\left(6.80\times10^{-3}\right)(150)} - 1}{\left(6.80\times10^{-3}\right)(12.0)}$

$t = \dfrac{e^{1.02} - 1}{0.0816} = \dfrac{1.773}{0.0816} = 21.7$ s

13.6 Exponential and Logarithmic Equations

1. $3^{x+2} = 5$

$\log 3^{x+2} = \log 5$

$(x+2)\log 3 = \log 5$

$x + 2 = \dfrac{\log 5}{\log 3}$

$x = \dfrac{\log 5}{\log 3} - 2$

$x = -0.535$

5. $5.50^x = 0.324$

$\ln 5.50^x = \ln 0.324$

$x = \dfrac{\ln 0.324}{\ln 5.50}$

$x = -0.661$

9. $6^{x+1} = 78$

$\ln 6^{x+1} = \ln 78$

$(x+1)\cdot\ln 6 = \ln 78$

$x + 1 = \dfrac{\ln 78}{\ln 6}$

$x = \dfrac{\ln 78}{\ln 6} - 1$

$x = 1.432$

13. $0.6^x = 2^{x^2}$

$\ln\left(0.6^x\right) = \ln 2^{x^2}$

$x \cdot \ln 0.6 = x^2 \cdot \ln 2$

$x^2 \cdot \ln 2 - x\ln 0.6 = 0$

$x\left(x\cdot\ln 2 - \ln 0.6\right) = 0$

$x = 0\ \ \text{or}\ \ x\cdot\ln 2 - \ln 0.6 = 0$

$$x = \frac{\ln 0.6}{\ln 2} = -0.737$$

17. $3\log_8 x = -2$

$\log_8 x = \dfrac{-2}{3}$

$8^{-2/3} = x$

$x = \dfrac{1}{4}$

21. $\log_2 x + \log_2 7 = \log_2 21$

$\log_2 7x = \log_2 21$

$7x = 21$

$x = 3$

25. $\log 12x^2 - \log 3x = 3$

$\log\dfrac{12x^2}{3x} = 3$

$\log 4x = 3$

$4x = 10^3 = 1000$

$x = 250$

$x = 2$ (Since logs are not defined on negatives.)

29. $\dfrac{1}{2}\log(x+2)+\log 5=1$

$\log(x+2)^{1/2}+\log 5=1$

$5\left[(x+2)^{1/2}\right]=10$

$\sqrt{(x+2)}=2$

$x+2=4$

$x=2$

33. $15^{-x}=1.326$. Graph $y_1=15^{-x}-1.326$ and use the zero feature to solve.

$x=-0.1042$

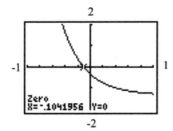

37. $3\ln 2x=2$. Graph $y_1=3\ln 2x-2$ and use the zero feature to solve.

$x=0.9739$

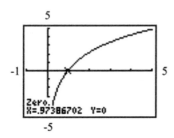

41. Graph $y1=2^{2x}-2^x-6$

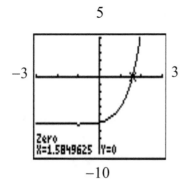

45. $e^x+e^{-x}=3$

$e^{2x}+e^0-3e^x=0$

$\left(e^x\right)^2-3e^x+1=0$, quadratic in e^x

$e^x=\dfrac{-(-3)\pm\sqrt{(-3)^2-4(1)(1)}}{2(1)}=\dfrac{3\pm\sqrt{5}}{2}$

$x=\ln\dfrac{3\pm\sqrt{5}}{2}$

$x=\pm 0.9624$

49. $y=3-4^{x+2}=0$

$4^{x+2}=3$

$\ln 4^{x+2}=\ln 3$

$(x+2)\ln 4=\ln 3$

$x+2=\dfrac{\ln 3}{\ln 4}$

$x=\dfrac{\ln 3}{\ln 4}-2=-1.21$ is the x-intercept

$1975+62.1=2034.1$

In 2035 the amount of DDT will be 25% of original amount.

53. $T=T_0+(37-T_0)0.97^t$

$27=22+(37-22)0.97^t$

$5=15\cdot 0.97^t$

$0.97^t=\dfrac{1}{3}$

$\ln 0.97^t=\ln\dfrac{1}{3}$

$t\ln 0.97=\ln\dfrac{1}{3}$

$t=\dfrac{\ln\frac{1}{3}}{\ln 0.97}=36$

Death occurred 36 hours earlier at noon.

57. $R=\log\left(\dfrac{I}{I_0}\right)$; $R=8.3$

$8.3=\log\left(\dfrac{I}{I_0}\right)$; $10^{8.3}=\dfrac{I}{I_0}$

$\dfrac{I}{I_0}=10^{8.3}=2.0\times 10^8$

61.

$$\ln P = t \ln 0.999 + \ln P_0$$

$$\ln P - \ln P_0 = t \ln 0.999$$

$$\ln \frac{P}{P_0} = \ln 0.999^t;$$

$$\frac{P}{P_0} = 0.999^t$$

$$P = P_0 (0.999)^t$$

65. $y = 2\left(e^{x/4} + e^{-x/4}\right); \ y = 5.8$

Graph $y_1 = 2\left(e^{x/4} + e^{-x/4}\right) - 5.8$ and use the zero

feature to solve.

$x = \pm 3.7 \text{ m}$

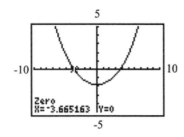

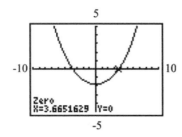

13.7 Graphs on Logarithmic and Semilogarithmic Paper

1. $y = 2\left(3^x\right)$

x	−1	0	2	3	4	5
y	0.67	2	18	54	162	486

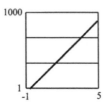

5. $y = 5\left(4^{-x}\right)$

x	0	1	2	3	4
y	5	1.25	3.1×10^{-1}	7.8×10^{-2}	2.0×10^{-2}

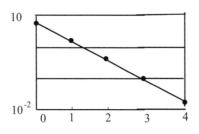

9. $y = 2x^3 + 6x$

x	0	1	2	4	6	8
y	0	8	28	152	468	1072

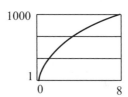

13. $y = x^{2/3}$

x	1	5	10	50	100	500	1000
y	1	2.9	4.6	13.6	21.5	63.0	100

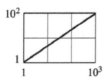

17. $x^2 y^3 = 25$

x	0.1	0.5	1	10	50
y	50	10	5	0.5	0.1

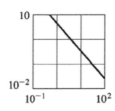

21. $y = 3x^6$, log–log paper

x	1	2	3	4
y	3	192	2187	12,288

25. $x\sqrt{y} = 4$, $y = \dfrac{16}{x^2}$, log–log paper

x	1	25	50	75	100
y	16	0.0256	0.0064	0.00284‾	0.0016

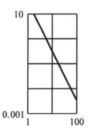

29. $N = N_0 e^{-0.028t}$, $N_0 = 1000$

t	0	25	50	75	100
N	1000	496.6	246.6	122.5	60.81

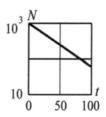

33.

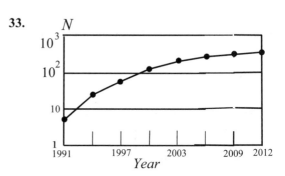

37.

d	0.63	1.3	1.9	2.5	3.8
R	600	190	100	72	46

d	5.0	7.5	10	15
R	29	17	10	6.0

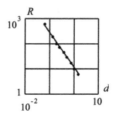

Chapter 13 Review Exercises

1. $\log_{10} x = 4 \Rightarrow x = 10^4 = 10,000$

5. $2\log_{1/2} 8 = x \Rightarrow \log_{1/2} 64 = x \Rightarrow$
$$\left(\frac{1}{2}\right)^x = 64$$
$$x = -6$$

9. $\log_x 36 = 2 \Rightarrow x^2 = 36 = 6^2 = 6$

13. $\log_3 6x = \log_3 6 + \log_3 x = \log_3 (3 \cdot 2) + \log_3 x$
$$= \log_3 3 + \log_3 2 + \log_3 x$$
$$= 1 + \log_3 2 + \log_3 x$$

17. $\log_2 56 = \log_2 \left(2^3 \cdot 7\right)$
$$= \log_2 2^3 + \log_2 7$$
$$= 3\log_2 2 + \log_2 7$$
$$= 3 \cdot 1 + \log_2 7$$
$$= 3 + \log_2 7$$

21. $\log_3 \left(\dfrac{9}{x}\right) = \log_3 9 - \log_3 x$
$$= \log_3 3^2 - \log_3 x$$
$$= 2\log_3 3 - \log_3 x$$
$$= 2 - \log_3 x$$

25. $\log_6 y = \log_6 4 - \log_{6x}$
$$\log_6 y = \log_6 \frac{4}{x}$$
$$y = \frac{4}{x}$$

29. $\log_3 y = \dfrac{1}{2}\log_3 7 + \dfrac{1}{2}\log_3 x$
$$\log_3 y = \log_3 \sqrt{7} + \log_3 \sqrt{x}$$
$$\log_3 y = \log_3 \sqrt{7x}$$
$$y = \sqrt{7x}$$

33. $2\left(\log_4 y - 3\log_4 x\right) = 3$
$$\log_4 y - \log_4 x^3 = \frac{3}{2}$$
$$\log_4 \frac{y}{x^3} = \frac{3}{2}$$
$$\frac{y}{x^3} = 4^{3/2} = 8$$
$$y = 8x^3$$

37. $y = 0.5\left(5^x\right)$

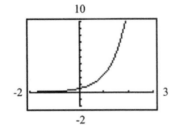

41. $y = \log_{3.15} x$

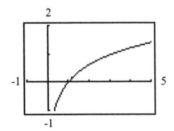

45. $\ln 8.86 = \dfrac{\log_{10} 8.86}{\log_{10} e} = 2.18$

49. $\log_{10} 65.89 = \dfrac{\ln 65.89}{\ln 10} = 1.819$

53. $\quad e^{2x} = 5$
$$\ln e^{2x} = \ln 5$$
$$2x \cdot \ln e = \ln 5$$
$$2x \cdot 1 = \ln 5$$
$$x = \frac{\ln 5}{2} = 0.805$$

57. $\log_4 z + \log_4 6 = \log_4 12$

$$\log(z \cdot 6) = \log_4 12$$

$$6z = 12$$

$$z = 2$$

61. $y = 8^x$

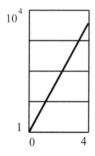

65. $10^{\log_4} = 4$

69. $2\log 3 - \log 6 = \log 1.5$

$$0.1760912591 = 0.1760912591$$

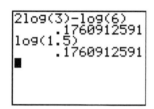

73. $f(x) = 2\log_b x$

$$f(8) = 2\log_b 8 = \log_b 8^2 = \log_b 64 = 3$$

$$b^3 = 64 = 4^3 \Rightarrow b = 4$$

$$f(x) = 2\log_4 x$$

$$f(4) = 2\log_4 4 = 2$$

77. Graph $y_1 = \dfrac{\ln x}{\ln 5} - 2x + 7$ and use the zero feature

to solve.

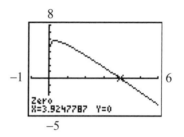

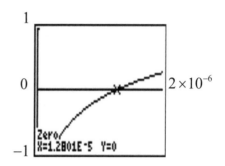

81. a) $A = A_0(1+r)^n$

$$\ln \frac{A}{A_0} = n\ln(1+r)$$

$$n = \frac{\ln \frac{A}{A_0}}{\ln(1+r)}$$

b) $n = \dfrac{\ln \frac{2A_0}{A_0}}{\ln(1+0.04)} = 17.7$ years

85.　$A = A_0 e^{kt}$　$\ln = 0.24e^{kt}$

$$0.16 = 0.24e^{k(2.0)} \Rightarrow k = \ln\left(\frac{0.16}{0.24}\right)^{1/2.0}$$

$$0.08 = 0.24e^{kt}$$

$$t = 5.4 \text{ h}$$

89. $t = 2350(\ln 100 - \ln N)$

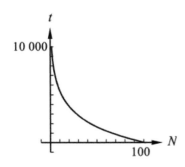

93. $C = B \log_2 (1 + R)$

$$\log_2 (1 + R) = \frac{C}{B}$$

$$1 + R = 2^{C/B}$$

$$R = 2^{C/B} - 1$$

97. $\qquad R = 4520(0.750)^{2.50t}$

$$1950 = 4520(0.750)^{2.50t}$$

$$\frac{1950}{4520} = 0.750^{2.50t}$$

$$\ln \frac{1950}{4520} = \ln (0.750)^{2.50t} = 2.50t \ln (0.750)$$

101. $T = T_1 + (T_0 - T_1) e^{-kt}$

$$e^{-kt} = \frac{T - T_1}{T_0 - T_1}$$

$$-kt = \ln \frac{T - T_1}{T_0 - T_1}$$

$$t = -\frac{1}{k} \ln \frac{T - T_1}{T_0 - T_1}$$

105. $A = 480F^{2.2}$

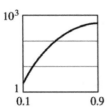

Chapter 14

ADDITIONAL TYPES OF EQUATIONS AND SYSTEMS OF EQUATIONS

14.1 Graphical Solution of Systems of Equations

1. Graph $y_1 = 3x^2 + 6x$.

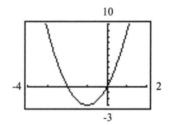

5. $y = 2x$

$x^2 + y^2 = 16 \Rightarrow y = \pm\sqrt{16 - x^2}$.

Graph $y_1 = 2x$, $y_2 = \sqrt{16 - x^2}$, and $y_3 = -\sqrt{16 - x^2}$.

Use the intersect feature to solve.

$x = 1.8,\ y = 3.6;\ x = -1.8,\ y = -3.6$

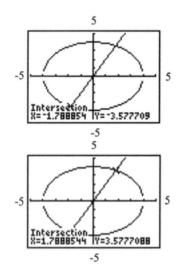

9. $y = x^2 - 2$

$4y = 12x - 7 \Rightarrow y = \dfrac{12x - 7}{4}$

Graph $y_1 = x^2 - 2$ and $y_2 = (12x - 7)/4$.

Use the intersect feature to solve.

$x = 1.5,\ y = 0.2$

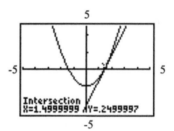

13. $x^2 + y^2 = 25 \Rightarrow y = \pm\sqrt{25 - x^2}$

$y = x^2 - 3$

Graph $y_1 = \sqrt{25 - x^2}$, $y_2 = -\sqrt{25 - x^2}$, $y_3 = x^2 - 3$.

Use the intersect feature to solve.

$x = -2.7,\ y = 4.2;\ x = 2.7,\ y = 4.2$

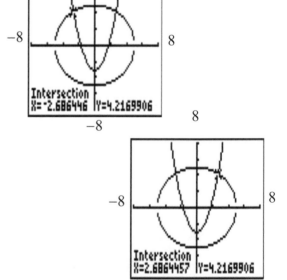

17. $2x^2 + 3y^2 = 19 \Rightarrow y = \pm\sqrt{(19 - 2x^2)/3}$ and

$x^2 + y^2 = 9 \Rightarrow y = \pm\sqrt{9 - x^2}$. Graph

$y_1 = \sqrt{(19 - 2x^2)/3}$, $y_2 = -\sqrt{(19 - 2x^2)/3}$,

$y_3 = \sqrt{9 - x^2}$; $y_4 = -\sqrt{9 - x^2}$ and use the

intersect feature to solve. $x = -2.8$, $y = 1.0$;

$x = 2.8$, $y = 1.0$; $x = -2.8$, $y = -1.0$;

$x = 2.8$, $y = -1.0$

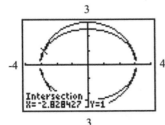

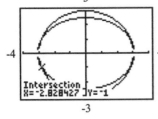

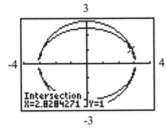

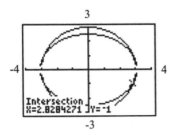

21. $y = x^2$

$y = \sin x$

Graph $y_1 = x^2$ and $y_2 = \sin x$ and use the

intersect feature to solve.

$x = 0.0$, $y = 0.0$ and $x = 0.9$, $y = 0.8$

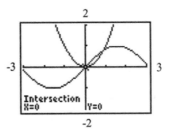

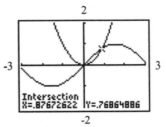

25. $x^2 - y^2 = 7 \Rightarrow y = \pm\sqrt{x^2 - 7}$, $y = 4\log_2 x = \dfrac{4\ln x}{\ln 2}$.

Graph $y_1 = \sqrt{x^2 - 7}$, $y_2 = -\sqrt{x^2 - 7}$, and $y_3 = \dfrac{4\ln x}{\ln 2}$.

Use the intersect feature to solve.

$x = 16.3$, $y = 16.1$

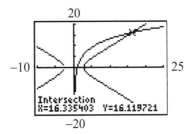

29. $10^{x+y} = 150$

$(x + y)\log 10 = \log 150$

$y = \log 150 - x$

$y = x^2$

Graph $y_1 = \log 150 - x$, $y_2 = x^2$ and use the

intersect feature to solve.

$x = -2.06$, $y = 4.23$; $x = 1.06$, $y = 1.12$

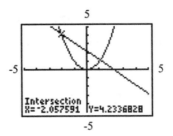

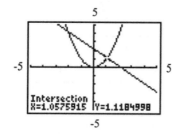

33.

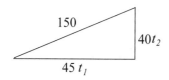

$t_1 + t_2 = 5$

$\left(45t_1\right)^2 + \left(40t_2\right)^2 = 150^2$

Graph $t_2 = 5 - t_1$ and

$$t_2 = \frac{\sqrt{150^2 - \left(45t_1\right)^2}}{40}$$

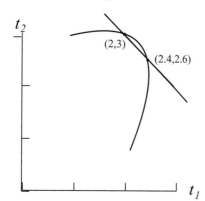

The intersection points are

$t_1 = 2$ h, $t_2 = 3$ h and

$t_1 = 2.4$ h, $t_2 = 2.6$ h

37. $x^2 + y^2 = 41 \Rightarrow y = \pm\sqrt{41 - x^2}$

$\qquad y^2 = 20x + 140 \Rightarrow y = \pm\sqrt{20x + 140}$

Graph $y_1 = \sqrt{41 - x^2}$, $y_2 = -\sqrt{20x + 140}$.

From the graph, there is no intersection. No, the meteorite will not strike the earth.

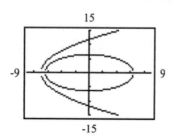

14.2 Algebraic Solution of Systems of Equations

1. $2x + y = 4 \Rightarrow y = 4 - 2x$, substitute into second equation $x^2 - y^2 = 4$

$$x^2 - \left(4 - 2x\right)^2 = 4$$

$$x^2 - \left(16 - 16x + 4x^2\right) = 4$$

$$x^2 - 16 + 16x - 4x^2 = 4$$

$$3x^2 - 16x + 20 = 0$$

$$\left(x - 2\right)\left(3x - 10\right) = 0$$

$$x - 2 = 0 \quad \text{or} \quad 3x - 10 = 0$$

$$x = 2 \qquad\qquad x = \frac{10}{3}$$

$$y = 4 - 2\left(2\right) = 0 \quad y = 4 - 2\left(\frac{10}{3}\right) = -\frac{8}{3}$$

The solutions are $x = 2$, $y = 0$; $x = \dfrac{10}{3}$, $y = -\dfrac{8}{3}$.

5. (1) $y = x + 1$

(2) $y = x^2 + 1$

$\qquad y = y$

$\qquad x^2 + 1 = x + 1$ with y from (1) and (2)

$\qquad x^2 - x = 0$

$\quad x\left(x - 1\right) = 0$

$\qquad\quad x = 0 \quad \text{or} \quad x - 1 = 0$

$\quad y = x + 1 = 1 \qquad\qquad x = 1$

$\qquad\qquad\qquad y = x + 1 = 2$

The solutions are $x = 0$, $y = 1$; $x = 1$, $y = 2$.

9. (1) $\quad x + y = 1 \Rightarrow y = 1 - x$

(2) $x^2 - y^2 = 1$

(2) $x^2 - \left(1 - x\right)^2 = 1$ with $y = 1 - x$ from (1)

$\quad x^2 - 1 + 2x - x^2 = 1$

$\qquad\qquad -1 + 2x = 1$

$\qquad\qquad\qquad 2x = 2$

$\qquad\qquad\qquad x = 1 \quad y = 1 - \left(1\right) = 0$

The solution is $x = 1$, $y = 0$.

13. (1) $wh = 1$

(2) $w + h = 2 \Rightarrow = 2 - w$

(1) $w(2 - w) = 1$ with h from (2)

$2w - w^2 = 1$

$w^2 - 2w + 1 = 0$

$(w - 1)^2 = 0$

$w - 1 = 0$

$w = 1 \Rightarrow (2)\, 1 + h = 2,\ h = 1$

The solution is $w = 1,\ h = 1$.

$\dfrac{-8}{y} + y = 2;\ \dfrac{-8 + y^2}{y} = -2$

$-8 + y^2 = -2y;\ y^2 + 2y - 8 = 0$

$(y + 4)(y - 2) = 0;\ y = -4,\ y = 2$

$x(-4) = -4;\ x = 1$

$x(2) = -4;\ x = -2$

The solutions are $x = 1,\ y = -4;\ x = -2,\ y = 2$.

17. $y = x^2$

$y = 3x^2 - 50$

$y = y \Rightarrow x^2 = 3x^2 - 50$

$2x^2 = 50$

$x^2 = 25$

$x = \pm 5$

$y = 25$

The solutions are $x = 5,\ y = 25,\ x = -5,\ y = 25$.

21. (1) $D^2 - 1 = R \Rightarrow D^2 = 1 + R$

(2) $D^2 - 2R^2 = 1$

$1 + R - 2R^2 = 1$

$R - 2R^2 = 0$

$R(1 - 2R) = 0$

(2) with D^2 from (1)

$R = 0$ or $1 - 2R = 0$

$2R = 1$

$R = \dfrac{1}{2}$

(1) with $R = 0$, $D^2 = 1 + 0$

$D = \pm 1$

(1) with $R = \dfrac{1}{2}$, $D^2 = 1 + \dfrac{1}{2} = \dfrac{3}{2} = \dfrac{6}{4}$

$D = \dfrac{\pm\sqrt{6}}{2}$

Solutions:

$R = 0,\ D = 1$

$R = 0,\ D = -1$

$R = \dfrac{1}{2},\ D = \dfrac{\sqrt{6}}{2}$

$R = \dfrac{1}{2},\ D = \dfrac{-\sqrt{6}}{2}$

25. (1) $x^2 + 3y^2 = 37$

(2) $2x^2 - 9y^2 = 14$

$3 \cdot (1) \Rightarrow\ 3x^2 + 9y^2 = 111$

(2) $\dfrac{2x^2 - 9y^2 = 14}{}$

$5x^2 = 125$

$x^2 = 25$

$x = \pm 5$

(1) $(\pm 5)^2 + 3y^2 = 37$

$25 + 3y^2 = 37$

$3y^2 = 12$

$y^2 = 4$

$y = \pm 2$

Solution: $(5,\, 2),\, (5,\, -2),\, (-5,\, 2),\, (-5,\, -2)$

29. $x - y = a - b \Rightarrow y = x - a + b = x - (a - b)$

$x^2 - y^2 = a^2 - b^2 \Rightarrow x^2 - \big(x - (a - b)\big)^2 = a^2 - b^2$

$x^2 - \big(x^2 - 2(a - b)x + (a - b)^2\big) = a^2 - b^2$

$x^2 - x^2 + 2(a - b)x = a^2 + 2ab - b^2 = a^2 - b^2$

$$2(a-b)x + 2ab - 2a^2 = 0$$
$$(a-b)x - a(a-b) = 0$$
$$x = a$$
$$y = x - a + b = a - a + b$$
$$y = b$$

The solution is $x = a$, $y = b$.

33.

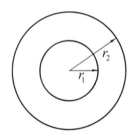

$$r_1 = r_2 - 2.00$$
$$\pi r_2^2 - \pi r_1^2 = 37.7$$
$$\pi r_2^2 - \pi (r_2 - 2.00)^2 = 37.7$$
$$\pi r_2^2 - \pi r_2^2 + 4.00\pi r_2 - 4.00\pi = 37.7$$
$$r_2 = \frac{37.7 + 4.00\pi}{4.00\pi}$$
$$r_2 = 4.00$$
$$r_1 = r_2 - 2.00 = 2.00$$

The radii are 2.00 cm and 4.00 cm.

37.

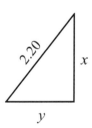

(1) $x + y + 2.20 = 4.60$
$$x + y = 2.40 \Rightarrow y = 2.40 - x$$
(2) $\quad x^2 + y^2 = 2.20^2$
(2) $x^2 + (2.40 - x)^2 = 2.20^2$ from which
$$x = 0.210, \ y = 2.19$$
$$x = 2.19, \ y = 0.210$$

The lengths of the sides of the truss are 2.19 m and 0.210 m.

41.

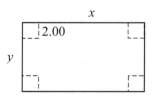

$$xy(2.00) = 224 \Rightarrow y = \frac{112}{x}$$
$$(x + 4.00)(y + 4.00) = 216$$
$$(x + 4.00)\left(\frac{112}{x} + 4.00\right) = 216$$
$$x^2 - 22.0x + 112 = 0$$
$$(x - 8.00)(x - 14.0) = 0$$
$$x = 8.00 \quad \text{or} \quad x = 14.0$$
$$y = 14.0 \qquad\qquad y = 8.0$$

The dimensions of the sheet are 18.00 in. by 12.0 in.

45. Let V_w = wind velocity
$$V_{vv} = \text{velocity with wind} = 610 + V_w$$
$$V_{aw} = \text{velocity against wind} = 610 - V_w$$
$$\frac{3660}{V_{ww}} + 1.6 = \frac{3660}{V_{aw}}$$
$$\frac{3660}{610 + V_w} + 1.6 = \frac{3660}{610 - V_w}$$
$$-1.6_w^2 - 7320v_w + 595,360 = 0$$

Solve using quadratic formula,
$$V_w = 80 \text{ mi/h}$$

14.3 Equations in Quadratic Form

1. $\quad 2x^4 - 7x^2 = 4$

$2x^4 - 7x^2 - 4 = 0$, let $y = x^2$

$2y^2 - 7y - 4 = 0$

$(y-4)(2y+1) = 0$

$y - 4 = 0 \quad$ or $\quad 2y + 1 = 0$

$\quad y = 4 \qquad\qquad y = -\dfrac{1}{2}$

$\quad x^2 = 4 \qquad\qquad x^2 = -\dfrac{1}{2}$

$\quad x = \pm 2 \qquad\qquad x = \pm\dfrac{\sqrt{2}}{2}j$

Check:

$2(\pm 2)^4 - 7(\pm 2)^2 = 32 - 28 = 4$

$2\left(\pm\dfrac{\sqrt{2}}{2}j\right)^4 - 7\left(\pm\dfrac{\sqrt{2}}{2}j\right)^2 = \dfrac{1}{2} + \dfrac{7}{2} = 4$

5. $\quad 3x^{-2} - 7x^{-1} - 6 = 0$

Let $y = x^{-1}$, $y^2 = x^{-2}$, then

$3y^2 - 7y - 6 = 0$

$(y-3)(y+2) = 0$

$y - 3 = 0 \quad$ or $\quad 3y + 2 = 0$

$\quad y = 3 \qquad\qquad 3y = -2, \ y = -\dfrac{2}{3}$

$\quad x^{-1} = 3 \qquad\qquad x^{-1} = -2/3$

$\quad x = \dfrac{1}{3} \qquad\qquad x = \dfrac{-3}{2}$

Check:

$\left(\dfrac{1}{3}\right)^{-2} - 7\left(\dfrac{1}{3}\right)^{-1} - 6 = 0$

$3\left(-\dfrac{3}{2}\right)^{-2} - 7\left(-\dfrac{3}{2}\right)^{-1} - 6 = 0$

9. $\quad 2x - 7\sqrt{x} + 5 = 0$

$\qquad$ Let $y = \sqrt{x}$, $y^2 = x$, then

$2y^2 - 7y + 5 = 0$

$(2y-5)(y-1) = 0$

$2y - 5 = 0 \quad$ or $\quad y - 1 = 0$

$\quad 2y = 5 \qquad\qquad y = 1$

$\quad y = \dfrac{5}{2} \qquad\qquad \sqrt{x} = 1$

$\quad \sqrt{x} = \dfrac{2}{5} \qquad\qquad x = 1$

$\quad x = \dfrac{25}{4}$

Check:

$2\left(\dfrac{25}{4}\right) - 7\sqrt{\dfrac{25}{4}} + 5 = \dfrac{25}{2} - \dfrac{35}{2} + \dfrac{35}{2} + \dfrac{10}{2} = 0$

$2(1) - 7\sqrt{1} + 5 = 2 - 7 + 5 = 0$

13. $\quad x^{2/3} - 2x^{1/3} - 15 = 0$

$\qquad$ Let $y = x^{1/3}$, $y^2 = x^{2/3}$, then

$y^2 - 2y - 15 = 0$

$(y-5)(y+3) = 0$

$y - 5 = 0 \quad$ or $\quad y + 3 = 0$

$\quad y = 5 \qquad\qquad y = -3$

$\quad x = 125 \qquad\qquad x = -27$

Check:

$125^{2/3} - 2(125)^{1/3} - 15 = 25 - 10 - 15 = 0$

$(-27)^{2/3} - 2(-27)^{1/3} - 15 = 9 + 6 - 15 = 0$

17. $\quad (x-1) - \sqrt{x-1} = 20$

$\qquad$ Let $y = \sqrt{x-1}$, $y^2 = x - 1$, then

$y^2 - y - 20 = 0$

$(y-5)(y+4) = 0$

$y - 5 = 0$

$\quad y = 25$

$\sqrt{x-1} = 5$

$\quad x - 1 = 25$

$\quad x = 26$

or

$y + 4 = 0$

$\quad y = -4$ reject since $y = \sqrt{x-1}$

requires $y \geq 0$

Check:

$$(26-1)-\sqrt{26-1} = 25-\sqrt{25}$$
$$= 25-5$$
$$= 20$$

21. $x-3\sqrt{x-2} = 6$

Let $y = \sqrt{x-2}$, $y^2 = x-2 \Rightarrow y^2+2 = x$, then

$$y^2+2-3y = 6$$
$$y^2-3y-4 = 0$$
$$(y-4)(y+1) = 0$$
$$y-4 = 0$$
$$y = 4$$
$$\sqrt{x-2} = 4$$
$$x-2 = 16$$
$$x = 18$$

or

$$y+1 = 0$$
$$y = -1$$

$\sqrt{x-2} = -1$ has no solution since $\sqrt{x-2} \geq 0$

Check:

$$18-3\sqrt{18-2} = 18-3\sqrt{16}$$
$$= 18-3(4)$$
$$= 18-12$$
$$= 6$$

25. $\dfrac{1}{s^2+1}+\dfrac{2}{s^2+3} = 1$

$$s^2+3+2(s^2+1) = (s^2+1)(s^2+3)$$
$$= s^4+4s^2+3$$
$$s^4+4s^2 = s^2+2s^2+2$$
$$s^4+s^2-2 = 0;\ \text{let } s^2 = y,\ s^4 = y^2$$
$$y^2+y-2 = 0$$
$$(y+2)(y-1) = 0$$
$$y+2 = 0 \qquad\qquad y-1 = 0$$
$$y = -2 \qquad\qquad y = 1$$
$$s^2 = -2 \qquad\qquad s^2 = 1$$
$$s = \pm\sqrt{2}j \qquad\qquad s = \pm 1$$

Check:

$$\frac{2}{\left(\pm\sqrt{2}j\right)^2+1}+\frac{2}{\left(\pm\sqrt{2}j\right)^2+3} = \frac{1}{-2+1}+\frac{2}{-2+3}$$
$$= -1+2 = 1$$

$\pm\sqrt{2}j$ are solutions

Check:

$$\frac{1}{\left(\pm 1\right)^2+1}+\frac{2}{\left(\pm 1\right)^2+3} = \frac{1}{1+1}+\frac{2}{1+3} = \frac{1}{2}+\frac{2}{4} = 1$$

± 1 are solutions

29. $x^4-20x^2+64 = 0$, let $y = x^2$

$$y^2-20y+64 = 0$$
$$(y-16)(y-4) = 0$$
$$y-16 = 0 \quad \text{or} \quad y-4 = 0$$
$$y = 16 \qquad\qquad y = 4$$
$$x^2 = 16 \qquad\qquad x^2 = 4$$
$$x = \pm 4 \qquad\qquad x = \pm 2$$

Check:

$$(\pm 4)^4-20(\pm 4)^2+64 = 256-20(16)+64$$
$$= -64+64 = 0$$
$$(\pm 2)^4-20(\pm 2)^2+64 = 16-80+64$$
$$= -64+64 = 0$$

33. $\left(\log x\right)^2-3\log x+2 = 0$

Let $y = \log x$, then $y^2-3y+2 = 0$

$$(y-2)(y-1) = 0$$
$$y = 2 \quad \text{or} \quad y = 1$$
$$\log x = 2 \qquad \log x = 1$$
$$x = 100 \qquad x = 10$$

Both solutions check.

37. $R_T^{-1} = R_1^{-1}+R_2^{-1}$

$$1^{-1} = R_1^{-1}+\sqrt{R_1}^{\,-1}$$
$$1 = R_1^{-1}+R_1^{-1/2}.$$

Let $y = R_1^{-1/2}$, $y^2 = R_1^{-1}$, then

$y^2+y-1 = 0$. The quadratic formula given

$$y = \frac{-1 \pm \sqrt{1^2 - 4(1)(-1)}}{2(1)}$$

$$= \frac{-1 \pm \sqrt{5}}{2}.$$

The negative is rejected since $y > 0$

$$y = \frac{-1 + \sqrt{5}}{2}$$

$$R_1^{-1/2} = \frac{-1 + \sqrt{5}}{2}$$

$$\sqrt{R_1} = \frac{2}{-1 + \sqrt{5}}$$

$$R_1 = \left[\frac{2}{-1 + \sqrt{5}}\right]^2$$

$$\approx 2.62 \ \Omega$$

$$R_2 = R_1^{1/2}$$

$$= \frac{2}{-1 + \sqrt{5}}$$

$$R_2 \approx 1.62 \ \Omega$$

41. $A = lw = 1540$; let $w = \dfrac{1540}{l}$ substitute into second

equation $60.0^2 = l^2 + w^2 \Rightarrow 60.0^2 = l^2 + \dfrac{1540^2}{l^2}$

from which $l = 29.463879$, $l = 52.267388$ and

$$w = \frac{1540}{l} = \frac{1540}{52.267388}$$

$w = 29.46387923$

The dimensions are $l = 52.3$ in., $w = 29.5$ in.

14.4   Equations with Radicals

1. $2\sqrt{3x - 1} = 3$

$4(3x - 1) = 9$

$12x - 4 = 9$

$12x = 13$

$$x = \frac{13}{12}$$

Check: $2\sqrt{3\left(\dfrac{13}{12}\right) - 1} \overset{?}{=} 3$

$$3 = 3$$

5. $\sqrt{x - 8} = 2$; square both sides

$x - 8 = 4$

$x = 12$

Check:

$\sqrt{12 - 8} = \sqrt{4} = 2$

9. $2\sqrt{3x + 2} = 6x$

$\sqrt{3x + 2} = 3x$

$3x + 2 = 9x^2$

$9x^2 - 3x - 2 = 0$

$(3x + 1)(3x - 2) = 0$

$3x + 1 = 0$ or $3x - 2 = 0$

$3x = -1$ $3x = 2$

$$x = \frac{-1}{3} \qquad\qquad x = \frac{2}{3}$$

Check:

$$\sqrt{3 \cdot \frac{1}{3} + 2} \ \bigg|\ 3 \cdot \frac{-1}{3}$$

$\sqrt{-1 + 2}$ -1

$\sqrt{1}$

1

$x = \dfrac{-1}{3}$ is not a solution.

Check:

$$\sqrt{3 \cdot \frac{2}{3} + 2} \ \bigg| \ 3 \cdot \frac{2}{3}$$

$$\sqrt{2+2} \qquad\qquad 2$$

$$\sqrt{4}$$

$$2$$

$x = \dfrac{2}{3}$ is the solution.

13. $\sqrt[3]{y-5} = 3$

$$y - 5 = 3^3$$

$$= 27$$

$$y = 32$$

The solution is $y = 32$, which checks.

17. $\sqrt{x^2 - 9} = 4$

$$x^2 - 9 = 16$$

$$x^2 = 25$$

$$x = \pm 5$$

Check:

$$\sqrt{(\pm 5)^2 - 9} \stackrel{?}{=} 4$$

$$\sqrt{25 - 9} \stackrel{?}{=} 4$$

$$\sqrt{16} \stackrel{?}{=} 4$$

$$4 = 4$$

$x = \pm 5$ is the solution.

21. $\sqrt{5 + \sqrt{x}} = \sqrt{x} - 1$

$$5 + \sqrt{x} = \left(\sqrt{x} - 1\right)^2$$

$$5 + \sqrt{x} = x - 2\sqrt{x} + 1$$

$$x - 3\sqrt{x} - 4 = 0$$

$$\left(\sqrt{x} - 4\right)\left(\sqrt{x} + 1\right) = 0$$

$$x = 16$$

$$\sqrt{x} = -1, \text{ not possible}$$

The solution is $x = 16$ since it checks.

25. $2\sqrt{x+2} - \sqrt{3x+4} = 1$

$$2\sqrt{x+2} = 1 + \sqrt{3x+4}$$

$$4(x+2) = \left(1 + \sqrt{3x+4}\right)^2$$

$$4x + 8 = 1 + 2\sqrt{3x+4} + 3x + 4$$

$$x + 3 = 2\sqrt{3x+4}$$

$$(x+3)^2 = 4(3x+4)$$

$$x^2 + 6x + 9 = 12x + 16$$

$$x^2 - 6x - 7 = 0$$

$$(x-7)(x+1) = 0$$

The solutions are $x = -1$ and $x = 7$, which check.

29. $\sqrt{6x-5} - \sqrt{x+4} = 2 \Rightarrow \sqrt{6x-5} = \sqrt{x+4} + 2$

$$6x - 5 = x + 4 + 4\sqrt{x+4} + 4$$

$$5x - 13 = 4\sqrt{x+4}$$

$$25x^2 - 130x + 169 = 16x + 64$$

$$25x^2 - 146x - 105 = 0 \Rightarrow x = 5, \text{ checks}$$

$$x = \frac{21}{25}, \text{ does not check}$$

The solution is $x = 5$.

33. $\sqrt{x-2} = \sqrt[4]{x-2} + 12$

$$\left(\sqrt{x-2} - 12\right)^2 = \left(\sqrt[4]{x-2}\right)^2$$

$$\sqrt{x-2} = x - 2 - 24\sqrt{x-2} + 144$$

$$\left(25\sqrt{x-2}\right)^2 = (x+142)^2$$

$$625(x-2) = x^2 + 284x + 20,164$$

$$x^2 - 341x + 21,414 = 0$$

$$(x-258)(x-83) = 0$$

The solution is $x = 258$ and $x = 83$ does not check.

37. $\sqrt{2x+1} + 3\sqrt{x} = 9$

$$\sqrt{2x+1} = 9 - 3\sqrt{x}$$

$$2x + 1 = 81 - 54\sqrt{x} + 9x$$

$$54\sqrt{x} = 7x + 80$$

$$2916x = 49x^2 + 1120x + 6400$$

$$49x^2 - 1796x + 6400 = 0$$

$$(x-4)(49x - 1600) = 0$$

$$x - 4 = 0 \quad \text{or} \quad 49x = 1600$$

$$x = 4 \qquad\qquad x = \frac{1600}{49}$$

Check:

$$\sqrt{2(4)+1} + 3\sqrt{4} \overset{?}{=} 9$$

$$3 + 6 \overset{?}{=} 9$$

$$9 = 9$$

$$\sqrt{2\left(\frac{1600}{49}\right)+1} \overset{?}{=} 9$$

$$\frac{57}{7} + \frac{120}{7} \overset{?}{=} 9$$

$$\frac{177}{9} \neq 9$$

$x = 4$ is the solution

41. $\sqrt{x-1} + x = 3 \Rightarrow \sqrt{x-1} = 3 - x$

$$x - 1 = 9 - 6x + x^2$$

$$x^2 - 7x + 10 = 0$$

$$(x-5)(x-2) = 0$$

$$x = 5 \quad \text{or} \quad x = 2$$

$x = 5$ does not check.

The solution is $x = 2$.

The equations, $\sqrt{x-1} = x-3$ and $\sqrt{x-1} = 3-x$,

are different and have different solutions.

45. $kC = \sqrt{R_1^2 - R_2^2} + \sqrt{r_1^2 - r_2^2} - A$

$$kC + A - \sqrt{R_1^2 - R_2^2} = \sqrt{r_1^2 - r_2^2}$$

$$\left(kC + A - \sqrt{R_1^2 - R_2^2}\right) = r_1^2 - r_2^2$$

$$r_1^2 = \left(kC + A - \sqrt{R_1^2 - R_2^2}\right)^2 + r_2^2$$

49.

$$(x + 5.2)^2 = x^2 + 8.3^2$$

$$x^2 + 10.4x + 27.04 = x^2 + 68.89$$

$$10.4x = 41.85$$

$$x = 4.0$$

$$x + 5.2 = 4.0 + 5.2 = 9.2$$

The freighter is 9.2 km from the station.

Chapter 14   Review Exercises

1. Graph $y = \dfrac{6-x}{2}$, $y_2 = 4x^2$, and use intersect.

$x = -0.9$, $y = 3.5$; $x = 0.8$, $y = 2.6$

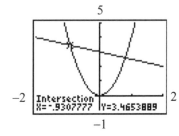

5. Graph $y_1 = x^2 + 1$, $y_2 = \sqrt{\dfrac{29 - 4x^2}{16}}$, $y_3 = -\sqrt{\dfrac{29 - 4x^2}{16}}$

and intersect. $x = -0.6$, $y = 1.3$; $x = 0.6$,

$y = 1.3$

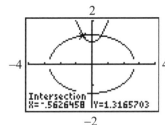

9. Graph $y_1 = x^2 - 2x$, $y_2 = 1 - e^{-x}$, and use intersect.

$x = 0$, $y = 0$; $x = 2.4$, $y = 0.9$

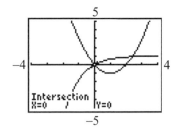

13. (1) $2R = L^2$

(2) $R^2 + L^2 = 3$

$R^2 + 2R = 3$

$R^2 + 2R - 3 = 0$

$(R+3)(R-1) = 0$

$R + 3 = 0 \qquad\qquad R - 1 = 0$

$R = -3 \qquad\qquad R = 1$

$2(-3) = L^2 \qquad\qquad L^2 = 2$

no solution $\qquad\qquad L = \pm\sqrt{2}$

$L = \pm\sqrt{2}$, $R = 1$

17. (1) $4x^2 - 7y^2 = 21$

(2) $x^2 + 2y^2 = 99$

(1) $8x^2 - 14y^2 = 42$

(2) $\underline{7x^2 + 14y^2 = 693}$

$15x^2 \qquad\quad = 735$

$x^2 = 49$

$x = \pm 7$

(2) $(\pm 7)^2 + 2y^2 = 99$

$49 + 2y^2 = 99$

$2y^2 = 50$

$y^2 = 25$

$y = \pm 5$

$x = 7$, $y = \pm 5$; $x = -7$, $y = \pm 5$

21. $x^4 - 20x^2 + 64 = 0$

Let $y = x^2$, $y^2 = x^4$

$y^2 - 20y + 64 = 0$

$(y - 16)(y - 4) = 0$

$y - 16 = 0 \quad$ or $\quad y - 4 = 0$

$y = 16 \qquad\qquad y = 4$

$x^2 = 16 \qquad\qquad x^2 = 4$

$x = \pm 4 \qquad\qquad x = \pm 2$

25. $D^{-2} + 4D^{-1} - 21 = 0$

Let $x = D^{-1}$, $x^2 = D^{-2}$

$x^2 + 4x - 21 = 0$

$(x+7)(x-3) = 0$

$x+7 = 0$ or $x-3 = 0$

$x = -7$ $\qquad x = 3$

$D^{-1} = -7$ $\qquad D^{-1} = 3$

$D = \dfrac{-1}{7}$ $\qquad D = \dfrac{1}{3}$

29. $2(x+1)^2 - 5(x+1) = 3$

$2(x+1)^2 - 5(x+1) - 3 = 0$

$\big(2(x+1)+1\big)\big((x+1)-3\big) = 0$

$2(x+1)+1 = 0$, $x+1-3 = 0$

$2x+3 = 0$ $\qquad x-2 = 0$

$x = \dfrac{-3}{2}$ $\qquad x = 2$

33. $\dfrac{4}{r^2+1} + \dfrac{7}{2r^2+1} = 2$

$4(2r^2+1) + 7(r^2+1) = 2(r^2+1)(2r^2+1)$

$8r^2 + 4 + 7r^2 + 7 = 2(2r^4+3r^2+1)$

$15r^2 + 11 = 4r^4 + 6r^2 + 2$

$4r^4 - 9r^2 - 9 = 0$. Let $x = r^2$, $x^2 = r^4$

$4x^2 - 9x - 9 = 0$

$(4x+3)(x-3) = 0$

$4x+3 = 0$ or $x-3 = 0$

$4x = -3$ $\qquad x = 3$

$x = \dfrac{-3}{4}$ $\qquad r^2 = 3$

$r^2 = \dfrac{-3}{4}$ $\qquad r = \pm\sqrt{3}$

$r = \dfrac{\pm\sqrt{3}}{2} j$

37. $\sqrt{5x+9} + 1 = x$

$\sqrt{5x+9} = x - 1$

$5x+9 = x^2 - 2x + 1$

$x^2 - 7x - 8 = 0$

$(x-8)(x+1) = 0$

$x-8 = 0$ or $x+1 = 0$

$x = 8$ $\qquad x = -1$

Check:

$\begin{array}{l|l} \sqrt{5 \cdot 8 + 9} + 1 & 8 \\ \qquad \sqrt{49} & \\ \quad 7+1 & \\ \qquad 8 & \end{array}$ $\qquad \begin{array}{l|l} \sqrt{5(-1)+9} + 1 & -1 \\ \quad \sqrt{-5+9} + 1 & \\ \qquad \sqrt{4} + 1 & \\ \qquad 2+1 & \\ \qquad 3 & \end{array}$

$x = 8$ is a solution. $\qquad x = -1$ is not a solution.

41. $\sqrt{n+4} + 2\sqrt{n+2} = 3$

$\sqrt{n+4} = 3 - 2\sqrt{n+2}$

$n+4 = 9 - 12\sqrt{n+2} + 4(n+2)$

$n+4 = 9 - 12\sqrt{n+2} + 4n + 8$

$12\sqrt{n+2} = 13 + 3n$

$144(n+2) = 169 + 78n + 9n^2$

$144n + 288 = 169 + 78n + 9n^2$

$9n^2 - 66n - 119 = 0$

From the quadratic formula,

$$n = \frac{-(-66) \pm \sqrt{(-66)^2 - 4(9)(-119)}}{2(9)} = \frac{66 \pm \sqrt{8640}}{18}$$

Check:

$$\sqrt{\frac{66+\sqrt{8640}}{18} + 4} + 2\sqrt{\frac{66+\sqrt{8640}}{18} + 2} = 10.1639\cdots$$

$\dfrac{66+\sqrt{8640}}{18}$ does not check

$$\sqrt{\frac{66-\sqrt{8640}}{18} + 4} + 2\sqrt{\frac{66-\sqrt{8640}}{18} + 2} = 3$$

$\dfrac{66-\sqrt{8640}}{18} = \dfrac{11-4\sqrt{15}}{3}$ is the solution.

45. $x^3 - 2x^{3/2} - 48 = 0$

$x^3 - 2\left(x^3\right)^{1/2} - 48 = 0$, let $x^3 = y$

$y - 2y^{1/2} - 48 = 0$, let $z = y^{1/2}$

$z^2 - 2z - 48 = 0$

$(z-8)(z+6) = 0$

$z - 8 = 0$ or $z + 6 = 0$

$z = 8$ $z = -6$

$y^{1/2} = 8$ $y^{1/2} = -6$, reject

$y = 64$

$x^3 = 64$

$x = 4$

49. $\sqrt[3]{x^3 - 7} = x - 1$

$x^3 - 7 = x^3 - 3x^2 + 3x - 1$

$3x^2 - 3x - 6 = 0$

$x^2 - x - 2 = 0$

$x - 2 = 0$ $x + 1 = 0$

$x = 2$ $x = -1$

53. $\sqrt{\sqrt{x} - 1} = 2$

$\sqrt{x} - 1 = 4$

$\sqrt{x} = 5$

$x = 25$

```
25→X
               25
√(√(X)-1)
                2
■
```

57. $L = \dfrac{h}{2\pi}\sqrt{l(l+1)}$

$L^2 = \dfrac{h^2}{4\pi^2} \cdot l(l+1)$

$\dfrac{4\pi^2 L^2}{h^2} = l^2 + l$

$l^2 + l - \dfrac{4\pi^2 L^2}{h^2} = 0$; from the quadratic formula,

$l = \dfrac{-1 \pm \sqrt{1^2 - 4\cdot 1\left(-\frac{4\pi^2 L^2}{h^2}\right)}}{2(1)} = \dfrac{-1 \pm \sqrt{1 + \frac{16\pi^2 L^2}{h^2}}}{2}$

$l = \dfrac{-1 \pm \sqrt{1 + \frac{16\pi^2 L^2}{h^2}}}{2}$. The + was chosen because $l > 0$

61. (1) $16t_1^2 + 16t_2^2 = 45$ where $t_1, t_2 > 0$

(2) $t_2 = 2t_1$

(1) with t_2 from (2) $16t_1^2 + 16(2t_1)^2 = 45$

$16t_1^2 + 64t_1^2 = 45$

$80t_1^2 = 45$

$t_1^2 = \dfrac{45}{80}$

$t_1 = 0.75$ s

$t_2 = 2 \cdot t_1 = 2(0.75)$

$t_2 = 1.5$ s

65.

perimeter $= 2x + y = 72$

area $= \dfrac{1}{2}y \cdot h = 240 \Rightarrow h = \dfrac{480}{y}$

$x^2 = \dfrac{y^2}{4} + h^2 \Rightarrow x^2 = \dfrac{y^2}{4} + \dfrac{480^2}{y^2}$

$y^4 - 4x^2 y^2 + 4 \cdot 480^2 = 0$

$y = \sqrt{\dfrac{4x^2 \pm \sqrt{\left(4x^2\right)^2 - 4\left(4 - 480^2\right)}}{2}}$

Graph $y_1 = 72 - 2x$

$y_2 = \sqrt{\dfrac{4x^2 + \sqrt{16x^2 - 4\left(4 \cdot 480^2\right)}}{2}}$

$y_3 = \sqrt{\dfrac{4x^2 - \sqrt{16x^2 - 4\left(4 \cdot 480^2\right)}}{2}}$

and use the intersect feature to solve.

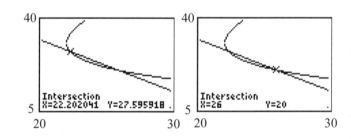

The lengths of the sides the banner are 27.6 in., 22.2 in. and 22.2 in. or 20 in., 26 in. and 26 in.

69. Area $= lw = 1770 \Rightarrow w = \dfrac{1770}{l}$

$l^2 + w^2 = 62^2$

$l^2 + \dfrac{1770^2}{l^2} = 62^2$

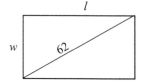

$l^4 - 62^2 l^2 + 1770^2 = 0$

$l = \sqrt{\dfrac{62^2 \pm \sqrt{\left(-62^2\right)^2 - 4\left(1770^2\right)}}{2}} = \begin{cases} 52 \text{ for } +,\ w = 34 \\ 34 \text{ for } -,\ w = 52 \end{cases}$

The dimensions of the rectangle are $l = 52$ mm and $w = 34$ mm.

73. Houston to Mobile: $510 = v \cdot t_1 \Rightarrow t_1 = \dfrac{510}{v}$

Mobile to Houston: $510 = \left(v + 6.0\right) \cdot t_2 \Rightarrow t_2$

$\qquad\qquad = \dfrac{510}{v + 6.0}$

$t_1 + t_2 = 35$

$\dfrac{510}{v} + \dfrac{510}{v + 6.0} = 35 \Rightarrow 510\left(v + 6.0\right) + 510v$

$\qquad\qquad = 35v\left(v + 6.0\right)$

$35v^2 - 810v - 3100 = 0$ from which, using the quadratic formula, $v = 26$, $v = -3.3$ (reject, $v > 0$)

$v = 26$ m/h, Houston to Mobile

$v + 6.0 = 32$ m/h, Mobile to Houston

Chapter 15

EQUATIONS OF HIGHER DEGREE

15.1 The Remainder and Factor Theorems; Synthetic Division

1. Using the remainder theorem find the remainder, for $(3x^3 - x^2 - 20x + 5) \div (x + 3)$.

$R = f(-3) = 3(-3)^3 - (-3)^2 - 20(-3) + 5 = -25$

5.

$$
\begin{array}{r}
x^2 + 2x + 6 \\
x - 2 \overline{\smash{\big)}\, x^3 + 0x^2 + 2x - 8} \\
\underline{x^3 - 2x^2} \\
2x^2 + 2x \\
\underline{2x^2 - 4x} \\
6x - 8 \\
\underline{6x - 12} \\
4
\end{array}
$$

$f(r) = R; \ r = 2$

$f(2) = (2)^3 + 2(2) - 8$

$\quad = 4$

Therefore, $R = 4$

9.

$$
\begin{array}{r}
2x^3 - 2x - 18 \\
x - \frac{3}{2} \overline{\smash{\big)}\, 2x^4 - 3x^3 - 2x^2 - 15x - 16} \\
\underline{2x^4 - 3x^3} \\
-2x^2 - 15x \\
\underline{-2x^2 + 3x} \\
-18x - 16 \\
\underline{-18x + 27} \\
-43
\end{array}
$$

$f(r) = R; \ r = \dfrac{3}{2}$

$f\left(\dfrac{3}{2}\right) = 2\left(\dfrac{3}{2}\right)^4 - 3\left(\dfrac{3}{2}\right)^3 - 2\left(\dfrac{3}{2}\right)^2 - 15\left(\dfrac{3}{2}\right) - 16 = -43$

Therefore, $R = -43$

13. $f(3) = 2 \cdot 3^4 - 7 \cdot 3^3 - 3^2 + 8 = -28$, the remainder

17. $8x^3 + 2x^2 - 32x - 8, \ x - 2; \ r = 2$

$f(2) = 8(2)^3 + 2(2)^2 + 32(2) - 8$

$\quad = 0$

$x - 2$ is a factor since $f(r) = R = 0$.

21. $x^{61} - 1, \ x + 1; \ r = -1$

$f(-1) = (-1)^{61} - 1 = -2 \neq 0$

$x + 1$ is a not factor since $f(r) = R \neq 0$.

25. $(x^3 + 2x^2 - 3x + 4) \div (x + 4) = x^2 + 2x + 5 + \dfrac{-16}{x + 4}$

$R = -16$

$$
\begin{array}{rrrr|r}
1 & 2 & -3 & 4 & \underline{-4} \\
 & -4 & 8 & -20 & \\
\hline
1 & -2 & 5 & -16 &
\end{array}
$$

29. $(x^7 - 128) \div (x - 2)$

$= x^6 + 2x^5 + 4x^4 + 8x^3 + 16x^2 + 32x + 64$

$R = 0$

$$
\begin{array}{rrrrrrrr|r}
1 & 0 & 0 & 0 & 0 & 0 & 0 & -128 & \underline{2} \\
 & 2 & 4 & 8 & 16 & 32 & 64 & 128 & \\
\hline
1 & 2 & 4 & 8 & 16 & 32 & 64 & 0 &
\end{array}
$$

33. $2x^5 - x^3 + 3x^2 - 4; \ x + 1$

$$
\begin{array}{rrrrrr|r}
2 & 0 & -1 & 3 & 0 & -4 & \underline{-1} \\
 & -2 & 2 & -1 & -2 & 2 & \\
\hline
2 & -2 & 1 & 2 & -2 & -2 &
\end{array}
$$

$R = -2,$

$x + 1$ is not a factor.

37. $2Z^4 - Z^3 - 4Z^2 + 1;\ 2Z - 1$

$$\begin{array}{rrrrr|r} 2 & -1 & -4 & 0 & 1 & \underline{1/2} \\ & 1 & 0 & -2 & -1 & \\ \hline 2 & 0 & -4 & -2 & 0 & \end{array}$$

$R = 0,$

$Z - \dfrac{1}{2}$ is a factor $\Rightarrow 2x - 1$ is a factor.

41. $x^4 - 5x^3 - 15x^2 + 5x + 14;\ 7$

$$\begin{array}{rrrrr|r} 1 & -5 & -15 & 5 & 14 & \underline{7} \\ & 7 & 14 & -7 & -14 & \\ \hline 1 & 2 & -1 & -2 & 0 & \end{array}$$

$R = 0$, 7 is a zero.

45. $f(x) = 2x^3 + 3x^2 - 19x - 4 = (x + 4)g(x)$

$\Rightarrow g(x) = \dfrac{2x^3 + 3x^2 - 19x - 4}{x + 4}$

$$\begin{array}{rrrr|r} 2 & 3 & -19 & -4 & \underline{-4} \\ & -8 & 20 & -4 & \\ \hline 2 & -5 & 1 & -8 & \end{array}$$

$g(x) = 2x^2 - 5x + 1 - \dfrac{8}{x + 4}$

49. $f(x) = 2x^3 + kx^2 - x + 14;\ x - 2$

we want $f(r) = R = 0$

$f(2) = 2(2)^3 + k(2)^2 - 2 + 14$

$\quad = 16 + 4k - 2 + 14$

$\quad = 28 + 4k = 0$

$\quad 4k = -28$

$k = -7$, then $x - 2$ will be a factor.

53. Suppose r is a zero of $f(x)$, then $f(r) = 0$. But $f(r) = -g(r) = 0 \Rightarrow g(r) = 0$, so r is also a zero of $g(x)$. Yes, they have the same zeros.

57. (a) $\left(s^3 + 5s^2 + 4s + 20\right) \div (s - 2)$

$$\begin{array}{rrrr} 1 & 5 & 4 & 20 \\ & 2 & 14 & 36 \\ \hline 1 & 7 & 18 & 56 \end{array}$$

$R = 56.\ s - 2$ is not a factor.

(b) $\left(s^3 + 5s^2 + 4s + 20\right) \div (s + 5)$

$$\begin{array}{rrrr} 1 & 5 & 4 & 20 \\ & -5 & 0 & -20 \\ \hline 1 & 0 & 4 & 0 \end{array}$$

$R = 0,\ s + 5$ is a factor.

15.2 The Roots of an Equation

1. $f(x) = (x - 1)^3 (x^2 + 2x + 1) = 0$

$(x - 1)^3 = 0 \quad$ or $\quad x^2 + 2x + 1 = 0$

$\qquad x = 1 \qquad\qquad\quad (x + 1)^2 = 0$

A triple root $\qquad\qquad\ x + 1 = 0$

$\qquad\qquad\qquad\qquad\qquad\ x = -1$, a double root

the five roots are $1,\ 1,\ 1,\ -1,\ -1$

5. $\left(x^2 + 6x + 9\right)\left(x^2 + 4\right) = 0$

$\qquad (x + 3)^2 \left(x^2 + 4\right) = 0$, by inspection

$x = -3$ double root, $x = \pm 2j$

9.

$$\begin{array}{rrrr} 2 & 11 & 20 & 12 \\ & -3 & -12 & -12 \\ \hline 2 & 8 & 8 & 0 \end{array}$$

$2x^3 + 11x^2 + 20x + 12 = \left(x + \dfrac{3}{2}\right)\left(2x^2 + 8x + 8\right)$

$\qquad\qquad\qquad\qquad\qquad = 2\left(x + \dfrac{3}{2}\right)\left(x^2 + 4x + 4\right)$

$\qquad\qquad\qquad\qquad\qquad = 2\left(x + \dfrac{3}{2}\right)(x + 2)(x + 2)$

$r_1 = -\dfrac{3}{2},\ r_2 = -2,\ r_3 = -2$

13. $t^2 - 7t^2 + 17t - 15 = 0$, $r_1 = 2 + j$

1	−7	17	−15
	$2 + j$	$-11 - 3j$	15
1	$-5 + j$	$6 - 3j$	0
	$2 - j$	$-6 + 3j$	
1	−3	0	

$t^3 - 7t^2 + 17t - 15$

$= (x - 3)(x - (2 + j))(x - (2 - j)) = 0$

$r_1 = 3$, $r_2 = 2 + j$, $r_3 = 2 - j$

17.

6	5	−15	0	4	$-\frac{1}{2}$
	−3	−1	8	−4	
6	2	−16	8	0	

$6x^4 + 5x^3 - 15x^2 + 4$

$= \left(x + \dfrac{1}{2}\right)\left(6x^3 + 2x^2 - 16x + 8\right)$

$= 2\left(x + \dfrac{1}{2}\right)\left(3x^3 + x^2 - 8x + 4\right)$

3	1	−8	4	$\frac{2}{3}$
	2	2	−4	
3	3	−6	0	

$6x^4 + 5x^3 - 15x^2 + 4$

$= 2\left(x + \dfrac{1}{2}\right)\left(x - \dfrac{2}{3}\right)\left(x^2 + 3x - 6\right)$

$= 6\left(x + \dfrac{1}{2}\right)\left(x - \dfrac{2}{3}\right)\left(x^2 + x - 2\right)$

$= 6\left(x + \dfrac{1}{2}\right)\left(x - \dfrac{2}{3}\right)(x + 2)(x - 1)$

$r_1 = -\dfrac{1}{2}$, $r_2 = \dfrac{2}{3}$, $r_3 = -2$, $r_4 = 1$

21. $x^5 - 3x^4 + 4x^3 - 4x^2 + 3x - 1 = 0$ (1 is a triple root)

1	−3	4	−4	3	−1	1
	1	−2	2	−2	1	
1	−2	2	−2	1	0	1
	1	−1	1	−1		
1	−1	1	1	0		1
	1	0	1			
1	0	1	0			

$(x - 1)(x^2 + 1)$

The roots are 1, 1, 1, $-j$, j.

25. $x^6 + 2x^5 - 4x^4 - 10x^3 - 41x^2 - 72x - 36 = 0$

$(-1 \text{ is a double root}; \ 2j \text{ is a root})$

1	2	−4	−10	−41	−72	−36	−1
	−1	−1	5	5	36	36	
1	1	−5	−5	−36	−36	0	−1
	−1	0	5	0	36		
1	0	−5	0	−36	0		2j
	2j	−4	−18j	36	0		
1	2j	−9	−18j	0	0	0	−2j
	−2j	0	18j	0	0		
1	0	−9	0	0	0	0	

$(x + 1)^2 (x - 2j)(x + 2j)(x^2 - 9)$

The roots are $-1, -1, 2j, -2j, -3, 3$.

29. $2x^3 + kx^2 - kx - 2$

2	k	$-k$	−2	2
	4	$2k + 8$	$2k + 16$	
2	$k + 4$	$k + 8$	$2k + 14$	

$\Rightarrow 2k + 14 = 0$, $k = -7$

15.3 Rational and Irrational Roots

1. $f(x) = 4x^5 + x^4 + 4x^3 - x^2 + 5x + 6 = 0$ has two sign changes and thus no more than two positive roots.

$f(-x) = -4x^5 + x^4 - 4x^3 - x^2 - 5x + 6 = 0$ has three sign changes and thus no more than three negative roots.

5. $x^3 + 2x^2 - 5x - 6 = 0$; there are 3 roots.

$f(x) = x^3 + x^2 - 5x + 3$; there are at most 2 positive roots.

$f(-x) = -x^3 + x^2 + 5x + 3$; there is one negative root.

Possible rational roots are $\pm 1, \pm 2, \pm, \pm 6$.

9. $3x^3 + 11x^2 + 5x - 3 = 0$; there are three roots.

$f(x) = 3x^3 + 11x^2 + 5x - 3$; there is 1 positive root.

$f(-x) = -3x^3 + 11x^2 - 5x - 3$; there are at most two negative roots.

Possible rational roots are $\pm\dfrac{1}{3}, \pm 1, \pm 3$. We try to find the one positive root first. $\dfrac{1}{3}$ is the first root with 0 remainder.

$$
\begin{array}{rrrr}
3 & 1 & 5 & -3 \\
  & 11 & 4 & 3 \\
\hline
3 & 12 & 9 & 0
\end{array}
$$

Thus, $\dfrac{1}{3}$ is the positive root. The remaining factor $3x^2 + 12x + 9 = 3(x^2 + 4x + 3) = 3(x+1)(x+3)$.

The remaining roots are $-1, -3$.

13. $5n^4 - 2n^3 + 40n - 17 = 0$; there are four roots.

$f(n)$ has 3 sign changes, at most 3 positive roots.

$f(-n) = 5n^4 + 2n^3 - 40n - 16$ has one sign change, at most one negative root.

Possible rational roots: $\pm 16, \pm\dfrac{16}{5}, \pm 2, \pm\dfrac{2}{5}, \pm 4,$

$\pm\dfrac{4}{5}, \pm 8, \pm\dfrac{8}{5}$

$$
\begin{array}{rrrrr}
5 & -2 & 0 & 40 & -16 \underline{\lfloor -2} \\
  & -10 & 24 & -48 & 16 \\
\hline
5 & -12 & 24 & -8 & 0
\end{array}
$$

-2 is a root

$5n^4 - 2n^3 + 40n - 16 = (n+2)(5n^3 - 12n^2 + 24n - 8)$

$$
\begin{array}{rrrr}
5 & -12 & 24 & -8 \; \lfloor\frac{2}{5} \\
  & 2 & -4 & 8 \\
\hline
5 & -10 & 20 & 0
\end{array}
$$

$\dfrac{2}{5}$ is a root

$5n^4 - 2n^3 + 40n - 16$

$= (n+2)\left(n - \dfrac{2}{5}\right)(5n^2 - 10n + 20)$

$n = \dfrac{-(-10) \pm \sqrt{(-10)^2 - 4(5)(20)}}{2(5)} = 1 \pm j\sqrt{3}$

roots: $-2, \dfrac{2}{5}, 1 \pm j\sqrt{3}$

17. $f(D) = D^5 + D^4 - 9D^3 - 5D^2 + 16D + 12 = 0$ has $n = 5$ and therefore five roots. $f(D)$ has two sign changes and therefore at most two positive roots.

$f(-D) = -D^5 + D^4 + 9D^3 - 5D^2 - 16D + 12$

has three sign changes and therefore at most three negative roots.

Possible rational roots $= \dfrac{\text{factors of 12}}{\text{factors of 1}}$

$= \dfrac{\pm 1, \pm 2, \pm 3, \pm 4, \pm 6, \pm 12}{\pm 1}$

$$
\begin{array}{rrrrrr}
1 & 1 & -9 & -5 & 16 & 12 \underline{\lfloor 2} \\
  & 2 & 6 & -6 & -22 & -12 \\
\hline
1 & 3 & -3 & -11 & -6 & 0
\end{array}
$$

2 is a root

$D^5 + D^4 - 9D^3 - 5D^2 + 16D + 12$

$= (D-2)(D^4 + 3D^3 - 3D^2 - 11D - 6)$

$$
\begin{array}{rrrrr}
1 & 3 & -3 & -11 & -6 \underline{\lfloor 2} \\
  & 2 & 10 & 14 & 6 \\
\hline
1 & 5 & 7 & 3 & 0
\end{array}
$$

2 is a root of $D^4 + 3D^3 - 3D^2 - 11D - 6$

$D^5 + D^4 - 9D^3 - 5D^2 + 16D + 12$

$= (D-2)(D-2)(D^3 + 5D^2 + 7D + 3)$

$$
\begin{array}{rrrr}
1 & 5 & 7 & 3 \underline{\lfloor -1} \\
  & -1 & -4 & -3 \\
\hline
1 & 4 & 3 & 0
\end{array}
$$

-1 is a root of $D^3 + 5D^2 + 7D + 3$

$D^5 + D^4 - 9D^3 - 5D^2 + 16D + 12$

$= (D-2)(D-2)(D+1)(D^2 + 4D + 3)$

$= (D-2)(D-2)(D+1)(D+1)(D+3)$

roots: $2, 2, -1, -1, -3$

21. $x^3 - 2x^2 - 5x + 4 = 0$

Graph $y_1 = x^3 - 2x^2 - 5x + 4$ and use the zero
feature to solve. The zeros are -1.86, 0.68, 3.18.

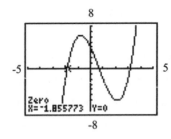

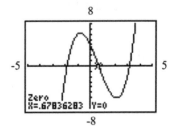

25. $x^3 - 6x^2 + 10x - 4 = 0 \left(0 \text{ and } 1\right)$

Graph $y_1 = x^3 - 6x^2 + 10x - 4$ and use the zero
feature to solve. $x = 0.59$

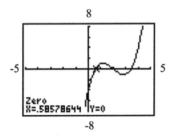

29. $y = x^4 - 11x^2$, $y = 12x - 4$

$x^4 - 11x^2 = 12x - 4$

$x^4 - 11x^2 - 12x + 4 = 0$, use rational root theorem
and synthetic division to solve.

```
  1   0   -11   -12   -4  |-2
         -2     4    14   -4
  ───────────────────────────
  1  -2   -7    2    0
```

```
  1  -2   -7    2  |-2
        -2    8   -2
  ──────────────────────
  1  -4    1    0
```

Solve $x^2 - 4x + 1 = 0$ with quadratic formula.

$$x = \frac{-(-4) \pm \sqrt{(-4)^2 - 4(1)(1)}}{2(1)} = 2 \pm \sqrt{3}$$

The solutions are $x = -2$, $y = -28$; $x = 2 + \sqrt{3}$,
$y = 20 + 12\sqrt{3}$; $x = 2 - \sqrt{3}$, $y = 20 - 12\sqrt{3}$.

33. $f(x) = 2x^4 + x^2 - 22x + 8$ has two sign changes $\Rightarrow$
no more than two positive roots.

$f(-x) = 2x^4 + x^2 + 22x + 8$ has no sign changes $\Rightarrow$
no negative roots. The possible rational roots are

± 1, $\pm \dfrac{1}{2}$, ± 2, ± 4, ± 8. The smallest possible

rational root is $\dfrac{1}{2}$. The largest possible rational

root is 8.

37.

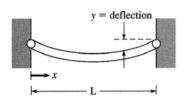

$y = k\left(x^4 - 2Lx^3 + L^3x\right) = 0$ for a deflection of zero.

$$x^4 - 2Lx^3 + L^3x = 0$$
$$x\left(x^3 - 2Lx^2 + L^3\right) = 0, \; x = 0 \text{ is a root}$$
$$x^3 - 2Lx^2 + L^3 = 0$$

```
  1   -2L    0     L³   |L
         L    -L²   -L³
  ──────────────────────────
  1   -L-L²   0,
```

L is a root of $x^3 - 2Lx^2 + L^3 = 0$

$x^3 - 2Lx^2 + L^3 = (x - L)\left(x^2 - Lx - L^2\right)$

$x^2 - Lx - L^2 = 0$ has roots

$$= \frac{-(-L) \pm \sqrt{(-L)^2 - 4(1)(-L^2)}}{2}$$

$$= \frac{L \pm L\sqrt{5}}{2} = \frac{L(1 \pm \sqrt{5})}{2}$$

$\frac{L}{2}(1 - \sqrt{5}) < 0$ and $\frac{L}{2}(1 + \sqrt{5}) > L$; reject both

Beam has a deflection of 0 for $x = 0$ and $x = L$.

41.

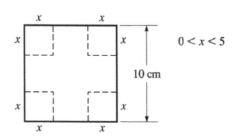

$0 < x < 5$

$V = (10 - 2x)^2 \cdot x = 70 \Rightarrow (100 - 40x + 4x^2) \cdot x = 70$

$4x^3 - 40x^2 + 100x - 70 = 0$

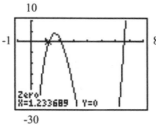

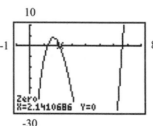

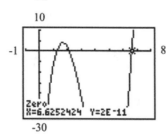

$x = 1.23, 2.14$ cm

$6.62 > 5$ and must be rejected.

45. Let x, $x + 1400$, $x + 5600$ be the values of the resistors, then

$$\frac{1}{1400} = \frac{1}{x} + \frac{1}{x + 1400} + \frac{1}{x + 5600}$$

$$x(x + 1400)(x + 5600) = 1400(x + 1400)(x + 5600)$$
$$+ 1400x(x + 5600) + 1400x(x + 1400)$$

$$x^3 + 2800x^2 - 11,760,000x - 10,976,000,000 = 0$$

2800	$-11,760,000$	$-10,976,000,000$
2800	15,680,000	10,976,000,000
5600	3,920,000	0

Since both roots of $x^2 + 5600x + 3,920,000 = 0$ are negative, the values of the resistors are $2800\,\Omega$, $4200\,\Omega$, $8400\,\Omega$,

---

Chapter 15 Review Exercises

1. $2(1)^3 - 4(1)^2 - (1) + 4 = 1$

$(2x^3 - 4x^2 - x + 4) \div (x - 1)$ has remainder 1

5.
$$
\begin{array}{ccccc}
1 & 1 & 1 & -2 & -3 \,\underline{|-1} \\
 & -1 & 0 & -1 & 3 \\
\hline
1 & 0 & 1 & -3 & 0
\end{array}
$$

remainder $= 0$, theefore $x + 1$ is a factor of $x^4 + x^3 + x^2 - 2x - 3$

9.
$$
\begin{array}{cccc}
1 & 3 & 6 & 1 \,\underline{|1} \\
 & 1 & 4 & 10 \\
\hline
1 & 4 & 10 & 11
\end{array}
$$

$(x^3 + 3x^2 + 6x + 1) \div (x - 1)$

$= (x^2 + 4x + 10) + \dfrac{11}{x - 1}$

13.
$$
\begin{array}{ccccc}
1 & 3 & -20 & -2 & 56 \,\underline{|-6} \\
 & -6 & 18 & 12 & -60 \\
\hline
1 & -3 & -2 & 10 & -4
\end{array}
$$

$\dfrac{x^4 + 3x^3 - 20x^2 - 2x + 56}{x + 6}$

$= x^3 - 3x^2 - 2x + 10 + \dfrac{-4}{x + 6}$

17.
$$\begin{array}{r} 1 \quad 5 \quad 0 \quad -6\underline{|-3} \\ \underline{-3 \quad -6 \quad 18} \\ 1 \quad 2 \quad -6 \quad 12 \end{array}$$

remainder $= 12$, therefore -3 is not a root of

$y^3 + 5y^2 - 6 = 0$

21.
$$\begin{array}{r} 1 \quad -4 \quad -7 \quad 10\underline{|5} \\ \underline{5 \quad 5 \quad -10} \\ 1 \quad 1 \quad -2 \quad 0 \end{array}$$

$x^2 + x - 2 = 0 \Rightarrow (x+2)(x-1) = 0 \Rightarrow x = -2, \; x = 1$

$r_1 = 5, \; r_2 = -2, \; r_3 = 1$

25.
$$\begin{array}{r} 4 \quad 0 \quad -1 \quad -18 \quad 9\underline{|\frac{1}{2}} \\ \underline{2 \quad 1 \quad 0 \quad -9} \\ 4 \quad 2 \quad 0 \quad -18 \quad 0 \end{array}$$

$4p^4 - p^2 - 18p + 9 = \left(p - \dfrac{1}{2}\right)\left(4p^3 + 2p^2 - 18\right)$

$$\begin{array}{r} 4 \quad 2 \quad 0 \quad -18\underline{|\frac{3}{2}} \\ \underline{6 \quad 12 \quad 18} \\ 4 \quad 8 \quad 12 \quad 0 \end{array}$$

$4p^4 - p^2 - 18p + 9$

$= \left(p - \dfrac{1}{2}\right)\left(p - \dfrac{3}{2}\right)\left(4p^2 + 8p + 12\right)$

$= 4\left(p - \dfrac{1}{2}\right)\left(p - \dfrac{3}{2}\right)\left(p^2 + 2p + 3\right)$

$p^2 + 2p + 3$ has roots $= -1 \pm \sqrt{2}\,j$

roots $\dfrac{1}{2}, \dfrac{3}{2}, -1 \pm \sqrt{2}\,j$

29.
$$\begin{array}{r} 3 \quad -1 \quad -11 \quad -12 \quad -4\underline{|-1} \\ \underline{-1 \quad -2 \quad 3 \quad 8 \quad 4} \\ 1 \quad 2 \quad -3 \quad -8 \quad -4 \quad 0 \end{array}$$

$s^5 + 3s^4 - s^3 - 11s^2 - 12s - 4$

$= (s+1)\left(s^4 + 2s^3 - 3s^2 - 8s - 4\right)$

$$\begin{array}{r} 1 \quad 2 \quad -3 \quad -8 \quad -4\underline{|-1} \\ \underline{-1 \quad -1 \quad 4 \quad 4} \\ 1 \quad 1 \quad -4 \quad -4 \quad 0 \end{array}$$

$s^5 + 3s^4 - s^3 - 11s^2 - 12s - 4$

$= (s+1)(s+1)\left(s^3 + s^2 - 4s - 4\right)$

$$\begin{array}{r} 1 \quad 1 \quad -4 \quad -4\underline{|-1} \\ \underline{-1 \quad 0 \quad 4} \\ 1 \quad 0 \quad -4 \quad 0 \end{array}$$

$s^5 + 3s^4 - 11s^3 - 11s^2 - 4$

$= (s+1)(s+1)(s+1)\left(s^2 - 4\right)$

roots: $-1, -1, -1, 2, -2$

33. $x^3 + x^2 - 10x + 8 = 0$ has three roots.

Possible rational roots $= \dfrac{\pm 1, \pm 2, \pm 4, \pm 8}{\pm 1}$

$$\begin{array}{r} 1 \quad 1 \quad -10 \quad 8\underline{|1} \\ \underline{1 \quad 2 \quad -8} \\ 1 \quad 2 \quad -8 \quad 0 \end{array}$$

1 is a root

$x^3 + x^2 - 10x + 8 = (x-1)\left(x^2 + 2x - 8\right)$

$\qquad\qquad\qquad = (x-1)(x+4)(x-2)$

roots: $1, -4, 2$

37. $6x^3 - x^2 - 12x - 5 = 0$ has three roots.

Possible rational roots $= \dfrac{\pm 1, \pm 5}{\pm 1, \pm 2, \pm 3, \pm 6}$

$$\begin{array}{r} 6 \quad -1 \quad -12 \quad -5\underline{|\frac{5}{3}} \\ \underline{10 \quad 15 \quad 5} \\ 6 \quad 9 \quad 3 \quad 0 \end{array}$$

$\dfrac{5}{3}$ is a root

$6x^3 - x^2 - 12x - 5 = \left(x - \dfrac{5}{3}\right)\left(6x^2 + 9x + 3\right)$

$\qquad\qquad\qquad = 3\left(x - \dfrac{5}{3}\right)\left(2x^2 + 3x + 1\right)$

$\qquad\qquad\qquad = 3\left(x - \dfrac{5}{3}\right)(2x+1)(x+1)$

roots: $\dfrac{5}{3}, \dfrac{-1}{2}, -1$

41. From graph, the zeros are $-2, -1, \dfrac{4}{3}, 4 \Rightarrow$ factors

are $f(x) = 3x^4 - 7x^3 - 26x^2 + 16x + 32$

$\qquad = (x+2)(x+1)(3x-4)(x-4)$

45. Since there are six roots, one of which is real, and the complex roots occur in conjugate pairs, the number of complex roots is two or four.

49. $(x-j)(x+j)(x-5) = 0$

$\qquad (x^2+1)(x-5) = 0$

$\qquad x^3 - 5x^2 + x - 5 = 0$

53. $3x^3 - x^2 - 8x - 2 = 0$

Graph $y1 = 3x^3 - x^2 - 8x - 2$ and use zero to solve.

$\Rightarrow$ zeros are

$x_1 = -1.311945234$

$x_2 = -0.2658857202$

$x_3 = 1.911164287$

57. $f(d) = 64d^3 - 144d^2 + 108d - 27$ has $n = 3$ and therefore three roots at most. $f(d)$ has three sign changes and therefore at most three positive roots. $f(-d) = -64d^3 - 144d^2 + 108d - 27$ has two sign changes and therefore no negative roots.

Possible rational roots

$= \dfrac{\pm 1, \pm 3, \pm 9, \pm 27}{\pm 1, \pm 2, \pm 4, \pm 8, \pm 16, \pm 32, \pm 64}$

From the graph, the root is between 0 and 1.

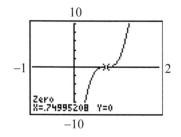

64	−144	108	−27
	48	−72	27
64	−96	36	0

$\dfrac{3}{4}$ is a repeated (multiplicity = 3) root.

$d = 0.75$ cm

61. $h = r + 3.2$, h and $r > 0$.

$V = \pi r^2 h = \pi r^2 (r+3.2) = \pi r^3 + 3.2\pi r^2 = 680$

$\pi r^3 + 3.2\pi r^2 - 680 = 0$ has one sign change and therefore one positive root. From the graph $r = 5.1$ m and $h = 5.1 + 3.2 = 8.3$ m.

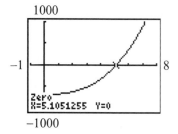

65. $A = 2x \cdot y = 2x \cdot (4 - x^2)$

$A = 8x - 2x^3$. Graph $y1 = 8x - 2x^3$ and use the maximum to solve.

$A_{max} = 6.16$ m^2

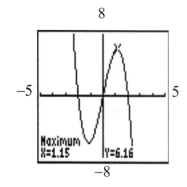

Chapter 16

MATRICES

16.1 Definitions and Basic Operations

1. $\begin{bmatrix} 8 & 1 & -5 & 9 \\ 0 & -2 & 3 & 7 \end{bmatrix} + \begin{bmatrix} -3 & 6 & 4 & 0 \\ 6 & 6 & -2 & 5 \end{bmatrix}$

$= \begin{bmatrix} 8+(-3) & 1+6 & -5+4 & 9+0 \\ 0+6 & -2+6 & 3+(-2) & 7+5 \end{bmatrix}$

$= \begin{bmatrix} 5 & 7 & -1 & 9 \\ 6 & 4 & 1 & 12 \end{bmatrix}$

5. $\begin{bmatrix} x & 2y & z \\ \frac{r}{4} & -s & -5t \end{bmatrix} = \begin{bmatrix} -2 & 10 & -9 \\ 12 & -4 & 5 \end{bmatrix}$

$x = -2, \quad 2y = 10, \quad z = -9$

$\qquad\qquad y = 5,$

$\dfrac{r}{4} = 12, \quad -s = -4, \ -5t = 5$

$r = 48 \qquad s = 4 \qquad t = -1$

9. $\begin{bmatrix} x-3 & x+y \\ x-z & y+z \\ x+t & y-t \end{bmatrix} = \begin{bmatrix} 5 & 3 \\ 4 & -1 \end{bmatrix}$, cannot solve, matrices

have different dimensions.

13. $\begin{bmatrix} 50+(-55) & -82+82 \\ -34+45 & 57+14 \\ -15+26 & 62+(-67) \end{bmatrix} = \begin{bmatrix} -5 & 0 \\ 11 & 71 \\ 11 & -5 \end{bmatrix}$

17. Since A and C do not have the same number of columns, they cannot be added.

21. $-3D+C = -3\begin{bmatrix} 7 & 9 & -6 \\ 2 & -6 & 11 \end{bmatrix} - \begin{bmatrix} -1 & 4 & -7 \\ 2 & -6 & 11 \end{bmatrix}$

$= \begin{bmatrix} -22 & -23 & 11 \\ 14 & -6 & 30 \end{bmatrix}$

25. $B - 3A = \begin{bmatrix} 3 & 12 \\ -9 & -6 \end{bmatrix} - 3\begin{bmatrix} 6 & -3 \\ 4 & -5 \end{bmatrix}$

$= \begin{bmatrix} -15 & 21 \\ -21 & 9 \end{bmatrix}$

29.

```
[C]+3[D]
  [[20  31  -25]
   [-10 -6  35 ]]
```

33.

```
-6[B]-4[A]
  [[-42 -60]
   [38  56 ]]
■
```

37. $-(A-B) = -\begin{bmatrix} -1-4 & 2+1 & 3+3 & 7-0 \\ 0-5 & -3-0 & -1+1 & 4-1 \\ 9-1 & -1-11 & 0-8 & -2-2 \end{bmatrix}$

$= \begin{bmatrix} 5 & -3 & -6 & -7 \\ 5 & 3 & 0 & -3 \\ -8 & 12 & 8 & 4 \end{bmatrix}$

$B - A = \begin{bmatrix} 4+1 & -1-2 & -3-3 & 0-7 \\ 5-0 & 0+3 & -1+1 & 1-4 \\ 1-9 & 11+1 & 8-0 & 2+2 \end{bmatrix}$

$= \begin{bmatrix} 5 & -3 & -6 & -7 \\ 5 & 3 & 0 & -3 \\ -8 & 12 & 8 & 4 \end{bmatrix}$

41. $\begin{bmatrix} 96 & 75 & 0 & 0 \\ 62 & 44 & 24 & 0 \\ 0 & 35 & 68 & 78 \end{bmatrix} + 2\begin{bmatrix} 96 & 75 & 0 & 0 \\ 62 & 44 & 24 & 0 \\ 0 & 35 & 68 & 78 \end{bmatrix}$

$= \begin{bmatrix} 288 & 225 & 0 & 0 \\ 186 & 132 & 72 & 0 \\ 0 & 105 & 204 & 234 \end{bmatrix}$

16.2 Multiplication of Matrices

1. $A = \begin{bmatrix} 1 & 2 \\ 0 & -3 \\ 2 & 1 \end{bmatrix}$, $B = \begin{bmatrix} -1 & 6 & 5 & -2 \\ 3 & 0 & 1 & -4 \end{bmatrix}$

$AB = \begin{bmatrix} 1 & 2 \\ 0 & -3 \\ 2 & 1 \end{bmatrix}\begin{bmatrix} -1 & 6 & 5 & -2 \\ 3 & 0 & 1 & -4 \end{bmatrix}$

$AB = \begin{bmatrix} -1+6 & 6+0 & 5+2 & -2-8 \\ 0-9 & 0+0 & 0-3 & 0+12 \\ -2+3 & 12+0 & 10+1 & -4-4 \end{bmatrix}$

$AB = \begin{bmatrix} 5 & 6 & 7 & -10 \\ -9 & 0 & -3 & 12 \\ 1 & 12 & 11 & -8 \end{bmatrix}$

5. $\begin{bmatrix} 2 & -3 & 1 \\ 0 & 7 & -3 \end{bmatrix}\begin{bmatrix} 90 \\ -25 \\ 50 \end{bmatrix} = \begin{bmatrix} 2(80)+(-3)(-25)+1(50) \\ 0(90)+7(-25)+(-3)(50) \end{bmatrix}$

$= \begin{bmatrix} 305 \\ -325 \end{bmatrix}$

9. $\begin{bmatrix} -1 & 7 \\ 3 & 5 \\ 10 & -1 \\ -5 & 12 \end{bmatrix}\begin{bmatrix} 2 & 1 \\ 5 & -3 \end{bmatrix} = \begin{bmatrix} -1(2) & -1(1)+7(-3) \\ 3(2) & 3(1)+5(-3) \\ 10(2) & 10(1)+(-1)(-3) \\ -5(2) & -5(1)+12(-3) \end{bmatrix}$

$= \begin{bmatrix} 33 & -22 \\ 31 & -12 \\ 15 & 13 \\ 50 & -41 \end{bmatrix}$

13.

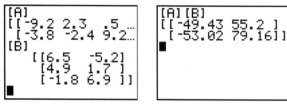

17. $AB = \begin{bmatrix} -10 & 25 & 40 \\ 42 & -5 & 0 \end{bmatrix}\begin{bmatrix} 6 \\ -15 \\ 12 \end{bmatrix} = \begin{bmatrix} 45 \\ 327 \end{bmatrix}$

BA is not possible because the number of columns in B is not equal to the number of rows in A.

21. $AI = \begin{bmatrix} 3 & 9 & -15 \\ 8 & 0 & 4 \\ 6 & -12 & 24 \end{bmatrix}\begin{bmatrix} 1 & 0 & 0 \\ 0 & 1 & 0 \\ 0 & 0 & 1 \end{bmatrix} = \begin{bmatrix} 3 & 9 & -15 \\ 8 & 0 & 4 \\ 6 & -12 & 24 \end{bmatrix}$

$IA = \begin{bmatrix} 1 & 0 & 0 \\ 0 & 1 & 0 \\ 0 & 0 & 1 \end{bmatrix}\begin{bmatrix} 3 & 9 & -15 \\ 8 & 0 & 4 \\ 6 & -12 & 24 \end{bmatrix} = \begin{bmatrix} 3 & 9 & -15 \\ 8 & 0 & 4 \\ 6 & -12 & 24 \end{bmatrix}$

Therefore, $AI = IA = A$

25. $AB = \begin{bmatrix} 1 & -2 & 3 \\ 2 & -5 & 7 \\ -1 & 3 & -5 \end{bmatrix}\begin{bmatrix} 4 & -1 & 1 \\ 3 & -2 & -1 \\ 1 & -1 & -1 \end{bmatrix}$

$= \begin{bmatrix} 1(4)+(-2)(3)+3(1) & 1(-1)+(-2)(-2)+3(-1) \\ 2(4)+(-5)(3)+7(1) & 2(-1)+(-5)(-2)+7(-1) \\ -1(4)+3(3)+(-5)(1) & -1(-1)+3(-2)+(-5)(-1) \end{bmatrix}$

$\begin{matrix} 1(1)+(-2)(-1)+3(-1) \\ 2(1)+(-5)(-1)+7(-1) \\ -1(1)+3(-1)+(-5)(-1) \end{matrix}$

$= \begin{bmatrix} 1 & 0 & 0 \\ 0 & 1 & 0 \\ 0 & 0 & 1 \end{bmatrix}$

Therefore, $B = A^{-1}$ since $AB = I$.

29. $\begin{bmatrix} 3 & 1 & 2 \\ 1 & -3 & 4 \\ 2 & 2 & 1 \end{bmatrix}\begin{bmatrix} -1 \\ 2 \\ 1 \end{bmatrix} = \begin{bmatrix} 3(-1)+1(2)+2(1) \\ 1(-1)+(-3)(2)+4(1) \\ 2(-1)+2(2)+1(1) \end{bmatrix}$

$\neq \begin{bmatrix} 1 \\ -3 \\ 1 \end{bmatrix};$

A is not the proper matrix of solution values.

33.

```
[B]³
   [[1  -2 -6]
    [-3  2  9 ]
    [2   0 -3]]
=[B]■
```

37. $I = \begin{bmatrix} 1 & 0 \\ 0 & 1 \end{bmatrix}, -I = \begin{bmatrix} -1 & 0 \\ 0 & -1 \end{bmatrix}$

$(-I)^2 = \begin{bmatrix} -1 & 0 \\ 0 & -1 \end{bmatrix}\begin{bmatrix} -1 & 0 \\ 0 & -1 \end{bmatrix}$

$= \begin{bmatrix} (-1)(-1)+(0)(0) & (-1)(0)+0(-1) \\ 0(-1)+(-1)(0) & 0(0)+(-1)(-1) \end{bmatrix}$

$= \begin{bmatrix} 1 & 0 \\ 0 & 1 \end{bmatrix} = I$

41. $\begin{bmatrix} 1 & -1 \\ 2 & 1 \end{bmatrix}\begin{bmatrix} x & y \\ z & t \end{bmatrix} = \begin{bmatrix} 2 & -3 \\ 7 & 0 \end{bmatrix}$

$\begin{bmatrix} x-z & y-t \\ 2x+z & 2y+t \end{bmatrix} = \begin{bmatrix} 2 & -3 \\ 7 & 0 \end{bmatrix}$

$\begin{aligned} x - z &= 2 \\ \underline{2x + z} &= \underline{7} \\ 3x &= 9 \Rightarrow y = -1 \\ 2(3) + z &= 7 \Rightarrow z = 1 \\ y - t &= -3 \\ \underline{2y + t} &= \underline{0} \\ 3y &= -3 \Rightarrow y = -1 \\ 2(-1) + t &= 0 \Rightarrow t = 2 \end{aligned}$

$x = 3, \ y = -1, \ z = 1, \ t = 2$

45. $\begin{bmatrix} v_2 \\ i_2 \end{bmatrix} = \begin{bmatrix} 1 & 0 \\ -\frac{1}{R} & 1 \end{bmatrix}\begin{bmatrix} v_1 \\ i_1 \end{bmatrix}$

$\begin{bmatrix} v_2 \\ i_2 \end{bmatrix} = \begin{bmatrix} 1(v_1)+0(i_1) \\ -\frac{1}{R}(v_1)+1(i_1) \end{bmatrix}$

$v_2 = v_1$

$i_2 = -v_1 / R + i_1$

16.3   Finding the Inverse of a Matrix

1. $A = \begin{bmatrix} 2 & -3 \\ 4 & -5 \end{bmatrix}$

$\det A = 2(-5) - (-3)(4) = -10 + 12 = 2$

$A^{-1} = \frac{1}{2}\begin{bmatrix} -5 & 3 \\ -4 & 2 \end{bmatrix} = \begin{bmatrix} -\frac{5}{2} & \frac{3}{2} \\ -2 & 1 \end{bmatrix}$

Check: $AA^{-1} = \begin{bmatrix} 2 & -3 \\ 4 & -5 \end{bmatrix}\begin{bmatrix} -\frac{5}{2} & \frac{3}{2} \\ -2 & 1 \end{bmatrix} = \begin{bmatrix} 1 & 0 \\ 0 & 1 \end{bmatrix}$

5. $\begin{bmatrix} -1 & 5 \\ 4 & 10 \end{bmatrix}$

Interchange the elements of the principal diagonal and change the signs of the off-diagonal elements.

$\begin{bmatrix} 10 & -5 \\ -4 & -1 \end{bmatrix}$

Find the determinant of the original matrix.

$\begin{vmatrix} -1 & 5 \\ 4 & 10 \end{vmatrix} = -30$

Divide each element of the second matrix by -30.

$-\frac{1}{30}\begin{bmatrix} 10 & -5 \\ -4 & -1 \end{bmatrix} = \begin{bmatrix} -\frac{1}{3} & \frac{1}{6} \\ \frac{2}{15} & \frac{1}{30} \end{bmatrix}$

9. $\begin{bmatrix} -50 & -45 \\ 26 & 80 \end{bmatrix}$

Interchange the elements of the principal diagonal and change the signs of the off-diagonal elements.

$\begin{bmatrix} 80 & 45 \\ -26 & -50 \end{bmatrix}$

Find the determinant of the original matrix.

$\begin{vmatrix} -50 & -45 \\ 26 & 80 \end{vmatrix} = -2830$

Divide each element of the second matrix by -2830.

$-\dfrac{1}{2830}\begin{bmatrix} 80 & 45 \\ -26 & -50 \end{bmatrix} = \begin{bmatrix} -\frac{8}{283} & -\frac{9}{566} \\ \frac{13}{1415} & \frac{5}{283} \end{bmatrix}$

13. $\begin{bmatrix} 2 & 4 & | & 1 & 0 \\ -1 & -1 & | & 0 & 1 \end{bmatrix} R1 \rightarrow R1 + R2 \begin{bmatrix} 1 & 3 & | & 1 & 1 \\ -1 & -1 & | & 0 & 1 \end{bmatrix}$

$R2 \rightarrow R2 + R1 \begin{bmatrix} 1 & 3 & | & 1 & 1 \\ 0 & 2 & | & 1 & 2 \end{bmatrix}$

$R2 \rightarrow \frac{1}{2} R2 \begin{bmatrix} 1 & 3 & | & 1 & 1 \\ 0 & 1 & | & \frac{1}{2} & 1 \end{bmatrix}$

$R1 \rightarrow R1 - 3R2 \begin{bmatrix} 1 & 0 & | & -\frac{1}{2} & -2 \\ 0 & 1 & | & \frac{1}{2} & 1 \end{bmatrix};$

$A^{-1} = \begin{bmatrix} -\frac{1}{2} & -2 \\ \frac{1}{2} & 1 \end{bmatrix}$

17. $\begin{bmatrix} 1 & -3 & -2 & | & 1 & 0 & 0 \\ -2 & 7 & 3 & | & 0 & 1 & 0 \\ 1 & -1 & -3 & | & 0 & 0 & 1 \end{bmatrix} \begin{matrix} R2 \rightarrow 2R1 + R2 \\ R3 \rightarrow -R1 + R3 \end{matrix}$

$\begin{bmatrix} 1 & -3 & -2 & | & 1 & 0 & 0 \\ 0 & 1 & -1 & | & 2 & 1 & 0 \\ 0 & 2 & -1 & | & -1 & 0 & 1 \end{bmatrix} \begin{matrix} R1 \rightarrow 3R2 + R1 \\ R3 \rightarrow -2R2 + R3 \end{matrix}$

$\begin{bmatrix} 1 & 0 & -5 & | & 7 & 3 & 0 \\ 0 & 1 & -1 & | & 2 & 1 & 0 \\ 0 & 0 & 1 & | & -5 & -2 & 1 \end{bmatrix} \begin{matrix} R2 \rightarrow R3 + R2 \\ R1 \rightarrow 5R3 + R1 \end{matrix}$

$\begin{bmatrix} 1 & 0 & 0 & | & -18 & -7 & 5 \\ 0 & 1 & 0 & | & -3 & -1 & 1 \\ 0 & 0 & 1 & | & -5 & -2 & 1 \end{bmatrix}; A^{-1} = \begin{bmatrix} -18 & -7 & 5 \\ -3 & -1 & 1 \\ -5 & -2 & 1 \end{bmatrix}$

21.
```
[A]
            [[2   8]
             [-1  6]]
[A]-1▶Frac
      [[3/10 -2/5]
       [1/20 1/10]]
■
```

25.
```
            [[2   4   0 ]
             [3   4  -2]
             [-1  1   2 ]]
[A]-1
[[2.5  -2    -2]
 [-1    1     1 ]
 [1.75 -1.5  -1]]
■
```

29. $BA^{-1} = \begin{bmatrix} 8 & -2 \\ 3 & 4 \end{bmatrix} \begin{bmatrix} 2 & -4 \\ -1 & 3 \end{bmatrix}^{-1} = \begin{bmatrix} 8 & -2 \\ 3 & 4 \end{bmatrix} \cdot \begin{bmatrix} 1.5 & 2 \\ 0.5 & 1 \end{bmatrix}$

$= \begin{bmatrix} 11 & 14 \\ 6.5 & 10 \end{bmatrix}$

33. $CA^{-1} = \begin{bmatrix} 5 & -1 & 0 \\ 2 & -2 & 1 \\ 3 & 0 & 4 \end{bmatrix} \cdot \begin{bmatrix} 1 & -1 & 1 \\ 0 & -2 & 1 \\ -2 & -3 & 0 \end{bmatrix}^{-1}$

$= \begin{bmatrix} 5 & -1 & 0 \\ 2 & -2 & 1 \\ 3 & 0 & 4 \end{bmatrix} \begin{bmatrix} 3 & -3 & 1 \\ -2 & 2 & -1 \\ -4 & 5 & -2 \end{bmatrix}$

$= \begin{bmatrix} 17 & -17 & 6 \\ 6 & -5 & 2 \\ -25 & 29 & -11 \end{bmatrix}$

37. $\dfrac{1}{ad-bc} \cdot \begin{bmatrix} a & b \\ c & d \end{bmatrix} \begin{bmatrix} d & -b \\ -c & a \end{bmatrix}$

$= \dfrac{1}{ad-bc} \cdot \begin{bmatrix} ad-bc & -ab+ab \\ cd-cd & -bc+ad \end{bmatrix}$

$\dfrac{1}{ad-bc} \cdot \begin{bmatrix} ad-bc & 0 \\ 0 & ad-bc \end{bmatrix} = \begin{bmatrix} \frac{ad-bc}{ad-bc} & \frac{0}{ad-bc} \\ \frac{0}{ad-bc} & \frac{ad-bc}{ad-bc} \end{bmatrix}$

$= \begin{bmatrix} 1 & 0 \\ 0 & 1 \end{bmatrix}$

41. $\begin{bmatrix} a_{11} & a_{12} \\ a_{21} & a_{22} \end{bmatrix}$

Interchange the elements of the principal diagonal and change the signs of the off-diagonal elements.

$\begin{bmatrix} a_{22} & -a_{12} \\ -a_{21} & a_{11} \end{bmatrix}$

Find the determinant of the original matrix.

$\begin{bmatrix} a_{11} & a_{12} \\ a_{21} & a_{22} \end{bmatrix} = a_{11}a_{22} - a_{12}a_{21}$

$A^{-1} = \dfrac{1}{a_{11}a_{22} - a_{12}a_{21}} \begin{bmatrix} a_{22} & -a_{12} \\ -a_{21} & a_{11} \end{bmatrix}$

$V = A^{-1}I = \dfrac{1}{a_{11}a_{22} - a_{12}a_{21}} \begin{bmatrix} a_{22} & -a_{12} \\ -a_{21} & a_{11} \end{bmatrix} \begin{bmatrix} i_1 \\ i_2 \end{bmatrix}$

$= \dfrac{1}{a_{11}a_{22} - a_{12}a_{21}} \begin{bmatrix} a_{22}i_1 & -a_{12}i_2 \\ -a_{21}i_1 & a_{11}i_2 \end{bmatrix}$

$v_1 = \dfrac{a_{22}i_1 - a_{12}i_2}{a_{11}a_{22} - a_{12}a_{21}}, \quad v_2 = \dfrac{-a_{21}i_1 + a_{11}i_2}{a_{11}a_{22} - a_{12}a_{21}}$

16.4 Matrices and Linear Equations

1. $2x - y = 7$
$5x - 3y = 19$

$A = \begin{bmatrix} 2 & -1 \\ 5 & -3 \end{bmatrix}, \quad C = \begin{bmatrix} 7 \\ 19 \end{bmatrix}, \quad A^{-1} = \begin{bmatrix} 3 & -1 \\ 5 & -2 \end{bmatrix}$

$A^{-1}C = \begin{bmatrix} 3 & -1 \\ 5 & -2 \end{bmatrix} \begin{bmatrix} 7 \\ 19 \end{bmatrix} = \begin{bmatrix} 2 \\ -3 \end{bmatrix} = \begin{bmatrix} x \\ y \end{bmatrix}$

$x = 2, y = -3$ is the solution.

5. $x + 2y = 7$
$2x + 3y = 11$

$A = \begin{bmatrix} 1 & 2 \\ 2 & 3 \end{bmatrix}, \quad C = \begin{bmatrix} 7 \\ 11 \end{bmatrix}, \quad A^{-1} = \begin{bmatrix} -3 & 2 \\ 2 & -1 \end{bmatrix}$

$A^{-1}C = \begin{bmatrix} -3 & 2 \\ 2 & -1 \end{bmatrix} \begin{bmatrix} 7 \\ 11 \end{bmatrix} = \begin{bmatrix} 1 \\ 3 \end{bmatrix} = \begin{bmatrix} x \\ y \end{bmatrix}$

$x = 1, y = 3$

9. $A = \begin{bmatrix} 2 & -3 \\ 4 & -5 \end{bmatrix}; \quad A^{-1} = \begin{bmatrix} -\frac{5}{2} & \frac{3}{2} \\ -2 & 1 \end{bmatrix}$

$A^{-1}C = \begin{bmatrix} -\frac{5}{2} & \frac{3}{2} \\ -2 & 1 \end{bmatrix} \begin{bmatrix} 3 \\ 4 \end{bmatrix} = \begin{bmatrix} -1.5 \\ -2 \end{bmatrix} = \begin{bmatrix} x \\ y \end{bmatrix}$

$x = -1.5, y = -2$

13. $A = \begin{bmatrix} 1 & -2 & 2 \\ 4 & 9 & 10 \\ -1 & 3 & 7 \end{bmatrix}; \quad A^{-1} = \begin{bmatrix} -33 & 8 & -2 \\ 38 & 9 & 2 \\ -21 & 5 & 1 \end{bmatrix}$

$A^{-1}C = \begin{bmatrix} -33 & -8 & -2 \\ 38 & 9 & 2 \\ -21 & 5 & -1 \end{bmatrix} \begin{bmatrix} -4 \\ -18 \\ -7 \end{bmatrix} = \begin{bmatrix} 2 \\ -4 \\ 1 \end{bmatrix} = \begin{bmatrix} x \\ y \\ z \end{bmatrix}$

$x = 2, y = -4, z = 1$

17. $AB = C \Rightarrow B = A^{-1}C = \begin{bmatrix} 3 & -1 \\ 7 & 2 \end{bmatrix}^{-1} \begin{bmatrix} 4 \\ 18 \end{bmatrix} = \begin{bmatrix} 2 \\ 2 \end{bmatrix}$

$AB = \begin{bmatrix} 3 & -1 \\ 7 & 2 \end{bmatrix} \begin{bmatrix} 2 \\ 2 \end{bmatrix} = \begin{bmatrix} 4 \\ 18 \end{bmatrix}$

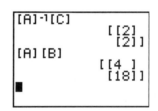

21. $A = \begin{bmatrix} 2 & -1 & -1 \\ 4 & -3 & 2 \\ 3 & 5 & 1 \end{bmatrix}; \quad A^{-1} = \begin{bmatrix} \frac{13}{57} & \frac{4}{57} & \frac{5}{57} \\ \frac{-2}{57} & \frac{-5}{57} & \frac{8}{57} \\ \frac{-29}{57} & \frac{13}{57} & \frac{2}{57} \end{bmatrix}.$

(Inverse from calculator)

$A^{-1}C = \begin{bmatrix} \frac{13}{57} & \frac{4}{57} & \frac{5}{57} \\ \frac{-2}{57} & \frac{-5}{57} & \frac{8}{57} \\ \frac{-29}{57} & \frac{13}{57} & \frac{2}{57} \end{bmatrix} \begin{bmatrix} 7 \\ 4 \\ -10 \end{bmatrix} = \begin{bmatrix} 1 \\ -2 \\ -3 \end{bmatrix} = \begin{bmatrix} x \\ y \\ z \end{bmatrix}$

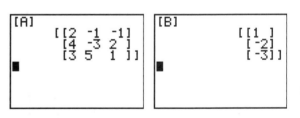

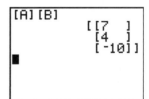

25. $A^{-1}C = \begin{bmatrix} 1 & -5 & 2 & -1 \\ 3 & 1 & -3 & 2 \\ 4 & -2 & 1 & -1 \\ -2 & 3 & -1 & 4 \end{bmatrix}^{-1} \begin{bmatrix} -18 \\ 17 \\ -1 \\ 11 \end{bmatrix}$

$\begin{bmatrix} \frac{-7}{85} & \frac{2}{85} & \frac{23}{85} & \frac{3}{85} \\ \frac{-47}{170} & \frac{-23}{170} & \frac{33}{170} & \frac{4}{85} \\ \frac{-13}{170} & \frac{-57}{170} & \frac{67}{170} & \frac{21}{85} \\ \frac{5}{34} & \frac{1}{34} & \frac{3}{34} & \frac{5}{17} \end{bmatrix}$

$\begin{bmatrix} -18 \\ 17 \\ -1 \\ 11 \end{bmatrix}$

$= \begin{bmatrix} 2 \\ 3 \\ -2 \\ 1 \end{bmatrix} = \begin{bmatrix} x \\ y \\ z \\ t \end{bmatrix}$ (Inverse from calculator)

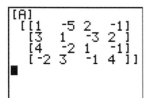

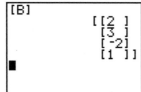

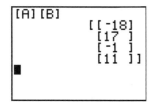

29. $x^2 + y = 2$

$2x^2 - y = 10$

$A = \begin{bmatrix} 1 & 1 \\ 2 & -1 \end{bmatrix}, C = \begin{bmatrix} 2 \\ 10 \end{bmatrix}, A^{-1} = \begin{bmatrix} \frac{1}{3} & \frac{1}{3} \\ \frac{2}{3} & -\frac{1}{3} \end{bmatrix}$

$A^{-1}C = \begin{bmatrix} \frac{1}{3} & \frac{1}{3} \\ \frac{2}{3} & -\frac{1}{3} \end{bmatrix}\begin{bmatrix} 2 \\ 10 \end{bmatrix} = \begin{bmatrix} 4 \\ -2 \end{bmatrix} = \begin{bmatrix} x^2 \\ y \end{bmatrix}$

$x^2 = 4, \quad y = -2$

$x = \pm 2$

$x = 2, y = -2;$ or $x = -2, y = -2$

33. $\begin{bmatrix} A \\ B \end{bmatrix} = \begin{bmatrix} \sin 47.2° & \sin 64.4° \\ \cos 47.2° & -\cos 64.4° \end{bmatrix}^{-1} \begin{bmatrix} 2540 \\ 0 \end{bmatrix} = \begin{bmatrix} 1180 \\ 1160 \end{bmatrix};$

$A = 1180$ N, $B = 1860$ N

37. $\quad x + y = 48$

$0.2x + 0.5y = 0.25(48) = 12$

$A = \begin{bmatrix} 1 & 1 \\ 0.2 & 0.5 \end{bmatrix}, C = \begin{bmatrix} 48 \\ 12 \end{bmatrix}, A^{-1} = \begin{bmatrix} \frac{5}{3} & -\frac{10}{3} \\ -\frac{2}{3} & \frac{10}{3} \end{bmatrix}$

$A^{-1}C = \begin{bmatrix} \frac{5}{3} & -\frac{10}{3} \\ -\frac{2}{3} & \frac{10}{3} \end{bmatrix}\begin{bmatrix} 48 \\ 12 \end{bmatrix} = \begin{bmatrix} 40 \\ 8 \end{bmatrix}$

40 mL of 20% acid and 8 mL of 50% acid will produce 48 mL of 25% acid.

16.5 Gaussian Elimination

1. $3x - 2y = 3 \quad R1 \to \frac{1}{3}R1$

$\underline{2x + y = 4}$

$x - \frac{2}{3}y = 1 \quad R2 \to -2R1 + R2$

$\underline{2x + y = 4}$

$x - \frac{2}{3}y = 1$

$\underline{\frac{7}{3}y = 2} \quad R \to \frac{3}{7}R2$

$x - \frac{2}{3}y = 1$

$\underline{y = \frac{6}{7}}$

$x - \frac{2}{3}\left(\frac{6}{7}\right) = 1$

$x \qquad = \frac{11}{7}$

The solution is $x = \frac{11}{7}$, $y = \frac{6}{7}$.

5. $\quad 5x - 3y = 2 \quad R1 \to \frac{1}{5}R1$

$\underline{-2x + 4y = 3}$

$x - \frac{3}{5}y = \frac{2}{5} \quad R2 \to 2R1 + R2$

$\underline{-2x + 4y = 3}$

$x - \frac{3}{5}y = \frac{2}{5}$

$\underline{\frac{14}{5}y = \frac{19}{5}} \quad R2 \to \frac{5}{14}R2$

$x - \frac{3}{5}y = \frac{2}{5}$

$\underline{y = \frac{19}{14}}$

$x - \frac{3}{5}\left(\frac{19}{14}\right) = \frac{2}{5} \Rightarrow x = \frac{17}{14}$

The solution is $x = \frac{17}{14}$, $y = \frac{19}{14}$.

9.
$$x+3y+3z=-3 \quad R2 \to -2R1+R2$$
$$2x+2y+z=-5 \quad R3 \to 2R1+R3$$
$$\underline{-2x-y+4z=6}$$
$$x+3y+3z=-3$$
$$-4y-5z=1 \quad R2 \to -\tfrac{1}{4}R2$$
$$\underline{5y+10z=0}$$
$$x+3y+3z=-3$$
$$y+\tfrac{5}{4}z=-\tfrac{1}{4} \quad R3 \to -5R2+R3$$
$$\underline{5y+10z=0}$$
$$x+3y+3z=-3$$
$$y+\tfrac{5}{4}z=-\tfrac{1}{4} \quad R3 \to -5R2+R3$$
$$\underline{5y+10z=0}$$
$$x+3y+3z=-3$$
$$y+\tfrac{5}{4}z=-\tfrac{1}{4}$$
$$\underline{\tfrac{15}{4}z=\tfrac{5}{4} \Rightarrow z=\tfrac{1}{3}}$$
$$\underline{y+\tfrac{5}{4}\left(\tfrac{1}{3}\right)=-\tfrac{1}{4} \Rightarrow y=-\tfrac{2}{3}}$$
$$x+3\left(-\tfrac{2}{3}\right)+3\left(\tfrac{1}{3}\right)=-3 \Rightarrow x=-2$$

The solution is $x=-2$, $y=-\tfrac{2}{3}$, $z=\tfrac{1}{3}$.

13.
$$x-4y+z=2 \quad R2 \to -3R1+R2$$
$$\underline{3x-y+4z=-4}$$
$$x-4y+z=2$$
$$\underline{11y+z=-10 \Rightarrow z=-10-11y}$$

There are an unlimited number of solutions.

$x=-3$, $y=-1$, $z=1$; $x=12$, $y=0$, $z=-10$

17.
$$x+3y+z=4$$
$$2x-6y-3z=10 \quad R2 \to -2R1+R2$$
$$\underline{4x-9y+3z=4} \quad R3 \to -4R1+R3$$
$$x+3y+z=4$$
$$-12y-5z=2 \quad R3 \to -\tfrac{21}{12}R2+R3$$
$$\underline{-21y-z=-12}$$
$$x+3y+z=4$$
$$-12y-5z=2$$
$$\underline{\tfrac{31}{4}z=-\tfrac{31}{2} \Rightarrow z=-2}$$
$$-12y-5(-2)=2 \Rightarrow y=\tfrac{2}{3}$$
$$x+3\left(\tfrac{2}{3}\right)+(-2)=4 \Rightarrow x=4$$

The solution is $x=4$, $y=\tfrac{2}{3}$, $z=-2$.

21.
$$6x+10y=-4 \Rightarrow 3x+5y=-2$$
$$3x+5y=-2$$
$$24x-18y=13 \quad R2 \to -8R1+R2$$
$$15x-33y=19 \quad R3 \to -5R1+R3$$
$$\underline{6x+68y=-33 \quad R4 \to -2R1+R4}$$
$$3x+5y=-2$$
$$-58y=29 \Rightarrow y=-\tfrac{1}{2}$$
$$-58y=29 \Rightarrow y=-\tfrac{1}{2}$$
$$\underline{58y=-29 \Rightarrow y=-\tfrac{1}{2}}$$
$$3x+5\left(-\tfrac{1}{2}\right)=-2 \Rightarrow x=\tfrac{1}{6}$$

The solution is $x=\tfrac{1}{6}$, $y=-\tfrac{1}{2}$.

25.
$$s+2t-3u=2$$
$$3s+6t-9u=6 \quad R2 \to -3R1+R2$$
$$\underline{7s+14t-21u=13 \quad R3 \to -7R1+R3}$$
$$s+2t-3u=2$$
$$0=0$$
$$\underline{0=-1}$$

The system is inconsistent.

29.
$$a_1x+b_1y=c_1$$
$$\underline{a_2x+b_2y=c_2 \quad R2 \to -\frac{a_2}{a_1}R1+R2}$$
$$a_1x+b_1y=c_1$$
$$\underline{\left(\frac{-a_2b_1}{a_1}+b_2\right)y=\frac{-a_2c_1}{a_1}+c_2}$$
$$y=\frac{-\frac{a_2c_1}{a_1}+c_2}{-\frac{a_2b_1}{a_1}+b_2}=\frac{a_1c_2-a_2c_1}{a_1b_2-a_2b_1}=\frac{\begin{vmatrix}a_1 & c_1\\ a_2 & c_2\end{vmatrix}}{\begin{vmatrix}a_1 & b_1\\ a_2 & b_2\end{vmatrix}}$$
$$a_1x+b_1\cdot\frac{a_1c_2-a_2c_1}{a_1b_2-a_2b_1}=c_1$$
$$x=\frac{-b_1\left(a_1c_2-a_2c_1\right)}{a_1b_2-a_2b_1}+\frac{c_1\left(a_1b_2-a_2b_1\right)}{a_1\left(a_1b_2-a_2b_1\right)}$$
$$x=\frac{-a_1b_1c_2+a_2b_1c_1+a_1b_2c_1-a_2b_1c_1}{a_1\left(a_1b_2-a_2b_1\right)}$$
$$x=\frac{a_1\left(b_2c_1-b_1c_2\right)}{a_1\left(a_1b_2-a_2b_1\right)}=\frac{\begin{vmatrix}c_1 & b_1\\ c_2 & b_2\end{vmatrix}}{\begin{vmatrix}a_1 & b_1\\ a_2 & b_2\end{vmatrix}}$$

33. $x + y + z = 650$

$-x + 2y - z = 10$ $\qquad R2 \to R1 + R2$

$\underline{3.00x + 2.00y + 2.00z = 1550}$ $\quad R3 \to -3R1 + R3$

$x + y + z = 650$

$3.00y = 660 \Rightarrow y = 220$

$\underline{-y - z = -400}$

$-220 - z = -400 \Rightarrow z = 180$

$x + 220 + 180 = 650 \Rightarrow x = 250$

The production rates are 250 parts/h, 220 parts/h, and 180 parts/h.

16.6 Higher Order Determinants

1. $\begin{vmatrix} 3 & 0 & 0 \\ 1 & 1 & 0 \\ 2 & 1 & 3 \end{vmatrix} = 9$, switch first and third column

$\begin{vmatrix} 0 & 0 & 3 \\ 0 & 1 & 1 \\ 3 & 1 & 2 \end{vmatrix} = -9$

5. $\begin{vmatrix} 3 & -2 & 4 & 2 \\ 5 & -1 & 2 & -1 \\ 3 & -2 & 4 & 2 \\ 0 & 3 & -6 & 0 \end{vmatrix} = 0$, Row 1 and R3 are identical.

9. $\begin{vmatrix} 2 & -3 & -1 \\ -4 & 1 & -3 \\ 1 & -3 & 2 \end{vmatrix} = -40,$

Column 3 of given determinant was multiplied by -1.

13. Expand by first row,

$\begin{vmatrix} 3 & 1 & 0 \\ -2 & 3 & -1 \\ 4 & 2 & 5 \end{vmatrix} = 3\begin{vmatrix} 3 & -1 \\ 2 & 5 \end{vmatrix} - \begin{vmatrix} -2 & -1 \\ 4 & 5 \end{vmatrix} + 0\begin{vmatrix} -2 & 3 \\ 4 & 2 \end{vmatrix}$

$= 3(3(5) - 2(-1)) - (-2(5) - 4(-1))$

$= 57$

17. Expand by first column,

$\begin{vmatrix} 1 & 3 & -3 & 5 \\ 4 & 2 & 1 & 2 \\ 3 & 2 & -2 & 2 \\ 0 & 1 & 2 & -1 \end{vmatrix} = 1\begin{vmatrix} 2 & 1 & 2 \\ 2 & -2 & 2 \\ 1 & 2 & -1 \end{vmatrix} - 4\begin{vmatrix} 3 & -3 & 5 \\ 2 & -2 & 2 \\ 1 & 2 & -1 \end{vmatrix}$

$+ 3\begin{vmatrix} 3 & -3 & 5 \\ 2 & 1 & 2 \\ 1 & 2 & -1 \end{vmatrix} - 0\begin{vmatrix} 3 & -3 & 5 \\ 2 & 1 & 2 \\ 2 & -2 & 2 \end{vmatrix}$

$= 1(12) - 4(12) + 3(-12) = -72$

21.

$\begin{vmatrix} 3 & 0 & 0 \\ -2 & 1 & 4 \\ 4 & -2 & 5 \end{vmatrix}$ $\quad R2 \to -\dfrac{4}{5}R3 + R2$

$\begin{vmatrix} 3 & 0 & 0 \\ -2 & 1 & 4 \\ 4 & -2 & 5 \end{vmatrix} = \begin{vmatrix} 3 & 0 & 0 \\ -\frac{26}{5} & \frac{13}{5} & 0 \\ 5 & -2 & 5 \end{vmatrix} = 3\left(\dfrac{13}{5}\right)(5) = 39$

25.

$\begin{vmatrix} 4 & 3 & 6 & 0 \\ 3 & 0 & 0 & 4 \\ 5 & 0 & 1 & 2 \\ 2 & 1 & 1 & 7 \end{vmatrix}$ $\begin{array}{l} R2 \to -\frac{3}{4}R1 + R2 \\ R3 \to -\frac{5}{4}R1 + R3 \\ R4 \to -\frac{1}{2}R1 + R4 \end{array}$

$\begin{vmatrix} 4 & 3 & 6 & 0 \\ 3 & 0 & 0 & 4 \\ 5 & 0 & 1 & 2 \\ 2 & 1 & 1 & 7 \end{vmatrix} = \begin{vmatrix} 4 & 3 & 6 & 0 \\ 0 & -\frac{9}{4} & -\frac{9}{2} & 4 \\ 0 & -\frac{15}{4} & -\frac{13}{2} & 2 \\ 0 & -\frac{1}{2} & -2 & 7 \end{vmatrix}$ $\begin{array}{l} R3 \to -\frac{15}{9}R2 + R3 \\ R4 \to -\frac{2}{9}R2 + R4 \end{array}$

$= \begin{vmatrix} 4 & 3 & 6 & 0 \\ 0 & -\frac{9}{4} & -\frac{9}{2} & 4 \\ 0 & 0 & 1 & -\frac{14}{3} \\ 0 & 0 & -1 & \frac{55}{9} \end{vmatrix}$ $\quad R4 \to R3 + R4$

$= \begin{vmatrix} 4 & 3 & 6 & 0 \\ 0 & -\frac{9}{4} & -\frac{9}{2} & 4 \\ 0 & 0 & 1 & -\frac{14}{3} \\ 0 & 0 & 0 & \frac{13}{9} \end{vmatrix} = 4\left(-\frac{9}{4}\right)(1)\left(\frac{13}{9}\right) = -13$

29.

$\begin{vmatrix} 1 & 2 & 0 & 1 & 0 \\ 0 & 2 & 1 & 0 & 1 \\ 1 & 0 & -1 & 1 & -1 \\ -2 & 0 & -1 & 2 & 1 \\ 1 & 0 & 2 & -1 & -2 \end{vmatrix}$ $\begin{array}{l} R3 \to -R1 + R3 \\ R4 \to 2R1 + R4 \\ R5 \to -R1 + R5 \end{array}$

$$=\begin{vmatrix}1&2&0&1&0\\0&2&1&0&1\\0&-2&-1&0&-1\\0&4&-1&4&1\\0&-2&2&2&-2\end{vmatrix}\begin{matrix}\\\\R3\to R2+R3\\R4\to-2R2+R4\\R5\to R2+R5\end{matrix}$$

$$=\begin{vmatrix}1&2&0&1&0\\0&2&1&0&1\\0&0&0&0&0\\0&4&-1&4&1\\0&-2&2&-2&-2\end{vmatrix}R3\Leftrightarrow R5$$

$$=-\begin{vmatrix}1&2&0&1&0\\0&2&1&0&1\\0&-2&2&-2&-2\\0&4&-1&4&1\\0&0&0&0&0\end{vmatrix}\begin{matrix}\\\\R3\to R2+R3\\R4\to-2R2+R4\\\end{matrix}$$

$$=-\begin{vmatrix}1&2&0&1&0\\0&2&1&0&1\\0&0&3&-2&-1\\0&0&-3&4&-1\\0&0&0&0&0\end{vmatrix}R4\to R3+R4$$

$$=-\begin{vmatrix}1&2&0&1&0\\0&2&1&0&1\\0&0&3&-2&-1\\0&0&0&2&-2\\0&0&0&0&0\end{vmatrix}=-(1)(2)(3)(2)(0)=0$$

33.
$$\begin{aligned}x+2y-z&=6\\y-2z-3t&=-5\\3x-2y+t&=2\\2x+y+z-t&=0\end{aligned}$$

$$\begin{vmatrix}1&2&-1&0\\0&1&-2&-3\\3&-2&0&1\\2&1&1&-1\end{vmatrix}=61$$

$$x=\dfrac{\begin{vmatrix}6&2&-1&0\\-5&1&-2&-3\\2&-2&0&1\\0&1&1&-1\end{vmatrix}}{61}=\dfrac{61}{61}=1$$

$$y=\dfrac{\begin{vmatrix}1&6&-1&0\\0&-5&-2&-3\\3&2&0&1\\2&0&1&-1\end{vmatrix}}{61}=\dfrac{122}{61}=2$$

$$z=\dfrac{\begin{vmatrix}1&2&6&0\\0&1&-5&-3\\3&-2&2&1\\2&1&0&-1\end{vmatrix}}{61}=\dfrac{-61}{61}=-1$$

$$t=\dfrac{\begin{vmatrix}1&2&-1&6\\0&1&-2&-5\\3&-2&0&2\\2&1&1&0\end{vmatrix}}{61}=\dfrac{183}{61}=3$$

The solution is $x=1$, $y=2$, $z=-1$, $t=3$.

37.
$$\begin{aligned}D+E+2F\phantom{{}+2F}&=1\\2D-E+G&=-2\\D-E-F-2G&=4\\2D-E+2F-G&=0\end{aligned}$$

$$\begin{vmatrix}1&1&2&0\\2&-1&0&1\\1&-1&-1&-2\\2&-1&2&-1\end{vmatrix}=-18$$

$$D=\dfrac{\begin{vmatrix}1&1&2&0\\-2&-1&0&1\\4&-1&-1&-2\\0&-1&2&-1\end{vmatrix}}{-18}=\dfrac{-18}{-18}=1$$

$$E=\dfrac{\begin{vmatrix}1&1&2&0\\2&-2&0&1\\1&4&-1&-2\\2&0&2&-1\end{vmatrix}}{-18}=\dfrac{-36}{-18}=2$$

41.
$$\begin{vmatrix} 2a & 2b & 2c \\ 2d & 2e & 2f \\ 2g & 2h & 2i \end{vmatrix} = 2\begin{vmatrix} a & b & c \\ 2d & 2e & 2f \\ 2g & 2h & 2i \end{vmatrix}$$

$$= 2(2)\begin{vmatrix} a & b & c \\ d & e & f \\ 2g & 2h & 2i \end{vmatrix}$$

$$= 2(2)(2)\begin{vmatrix} a & b & c \\ d & e & f \\ g & h & i \end{vmatrix}$$

The value of the determinant is changed by a factor of 8.

45.
$$\begin{vmatrix} C & -1 & 0 & 0 \\ -1 & C & -1 & 0 \\ 0 & -1 & C & -1 \\ 0 & 0 & -1 & C \end{vmatrix} = C^4 - 3C^2 + 1 = 0,$$

solve graphically,

$C = 0.618, 1.618$

Chapter 16 Review Exercises

1. $\begin{pmatrix} 2a \\ a-b \end{pmatrix} = \begin{pmatrix} 8 \\ 5 \end{pmatrix}$; $2a = 8$;

$a = 4;\ a - b = 5;\ 4 - b = 5;\ b = -1$

5. $\begin{bmatrix} \cos\pi & \sin\frac{\pi}{6} \\ x+y & x-y \end{bmatrix} = \begin{bmatrix} x & y \\ a & b \end{bmatrix}$

$x = \cos\pi = -1,\ y = \sin\frac{\pi}{6} = \frac{1}{2}$

$a = x + y = -1 + \frac{1}{2} = -\frac{1}{2}$

$b = x - y = -1 - \frac{1}{2} = -\frac{3}{2}$

9. $B - A = \begin{bmatrix} -1 & 0 \\ 4 & -6 \\ -3 & -2 \\ 1 & -7 \end{bmatrix} - \begin{bmatrix} 2 & -3 \\ 4 & 1 \\ -5 & 0 \\ 2 & -3 \end{bmatrix} = \begin{bmatrix} -3 & 3 \\ 0 & -7 \\ 2 & -2 \\ -1 & -4 \end{bmatrix}$

13.

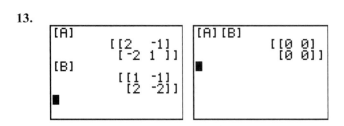

17. $\begin{bmatrix} 2 & -5 \\ 2 & -4 \end{bmatrix}$

Interchange elements of principal diagonal and change signs of off-diagonal elements.

$\begin{bmatrix} -4 & 5 \\ -2 & 2 \end{bmatrix}$

Find the determinant of original matrix.

$\begin{vmatrix} 2 & -5 \\ 2 & -4 \end{vmatrix} = 2$

Divide each element of second matrix by 2.

$\frac{1}{2}\begin{bmatrix} -4 & 5 \\ -2 & 2 \end{bmatrix} = \begin{bmatrix} -2 & \frac{5}{2} \\ -1 & 1 \end{bmatrix}$

21. $\begin{bmatrix} 1 & 1 & -2 & | & 1 & 0 & 0 \\ -1 & -2 & 1 & | & 0 & 1 & 0 \\ 0 & 3 & 4 & | & 0 & 0 & 1 \end{bmatrix}$

$R1+R2 \to R2 \begin{bmatrix} 1 & 1 & -2 & | & 1 & 0 & 0 \\ 0 & -1 & -1 & | & 1 & 1 & 0 \\ 0 & 3 & 4 & | & 0 & 0 & 1 \end{bmatrix}$

$3R2+R3 \to R3 \begin{bmatrix} 1 & 1 & -2 & | & 1 & 0 & 0 \\ 0 & -1 & -1 & | & 1 & 1 & 0 \\ 0 & 0 & 1 & | & 3 & 3 & 1 \end{bmatrix}$

$R2+R1 \to R1 \begin{bmatrix} 1 & 0 & -3 & | & 2 & 1 & 0 \\ 0 & -1 & -1 & | & 1 & 1 & 0 \\ 0 & 0 & 1 & | & 3 & 3 & 1 \end{bmatrix}$

25. $A = \begin{bmatrix} 2 & -3 \\ 4 & -1 \end{bmatrix}$; $C = \begin{bmatrix} -9 \\ -13 \end{bmatrix}$; $\begin{vmatrix} 2 & -3 \\ 4 & -1 \end{vmatrix} = 10$;

$A^{-1} = \dfrac{1}{10}\begin{bmatrix} -1 & 3 \\ -4 & 2 \end{bmatrix} = \begin{bmatrix} -\frac{1}{10} & \frac{3}{10} \\ -\frac{4}{10} & \frac{2}{10} \end{bmatrix}$

$A^{-1}C = \begin{bmatrix} -\frac{1}{10} & \frac{3}{10} \\ -\frac{4}{10} & \frac{2}{10} \end{bmatrix}\begin{bmatrix} -9 \\ -13 \end{bmatrix} = \begin{bmatrix} \frac{9}{10} & -\frac{39}{10} \\ \frac{36}{10} & -\frac{26}{10} \end{bmatrix}\begin{bmatrix} -3 \\ 1 \end{bmatrix}$

$x = -3,\ y = 1$

29. $A = \begin{bmatrix} 2 & -3 & 2 \\ 3 & 1 & -3 \\ 1 & 4 & 1 \end{bmatrix}$; $C = \begin{bmatrix} 7 \\ -6 \\ -13 \end{bmatrix}$

$\begin{bmatrix} 2 & -3 & 2 & | & 1 & 0 & 0 \\ 3 & 1 & -3 & | & 0 & 1 & 0 \\ 1 & 4 & 1 & | & 0 & 0 & 1 \end{bmatrix}$

$-3R3 + R2 \rightarrow R2 \begin{bmatrix} 2 & -3 & -2 & | & 1 & 0 & 0 \\ 3 & 1 & -3 & | & 0 & 1 & 0 \\ 0 & -11 & -6 & | & 0 & 1 & -3 \end{bmatrix}$

$-3R1 + 2R2 \rightarrow R2 \begin{bmatrix} 2 & -3 & 2 & | & 1 & 0 & 0 \\ 0 & 11 & -12 & | & -3 & 2 & 0 \\ 0 & -11 & -6 & | & 1 & 0 & -3 \end{bmatrix}$

$\frac{1}{2}R1 \rightarrow R1 \begin{bmatrix} 1 & -\frac{3}{2} & 1 & | & \frac{1}{2} & 0 & 0 \\ 0 & -11 & -12 & | & -3 & 2 & 0 \\ 0 & -11 & -6 & | & 0 & 1 & -3 \end{bmatrix}$

$R2 + R3 \rightarrow R3 \begin{bmatrix} 1 & -\frac{3}{2} & 1 & | & \frac{1}{2} & 0 & 0 \\ 0 & 11 & -12 & | & -3 & 2 & 0 \\ 0 & 0 & -18 & | & -3 & 3 & -3 \end{bmatrix}$

$\frac{1}{11}R2 \rightarrow R2 \begin{bmatrix} 1 & -\frac{3}{2} & 1 & | & \frac{1}{2} & 0 & 0 \\ 0 & 1 & -\frac{12}{11} & | & -\frac{3}{11} & \frac{2}{11} & 0 \\ 0 & 0 & -18 & | & -3 & 3 & -3 \end{bmatrix}$

$\frac{3}{2}R2 + R1 \rightarrow R1 \begin{bmatrix} 1 & 0 & -\frac{7}{11} & | & \frac{1}{11} & \frac{3}{11} & 0 \\ 0 & 1 & -\frac{12}{11} & | & -\frac{3}{11} & \frac{2}{11} & 0 \\ 0 & 0 & -18 & | & -3 & 3 & -3 \end{bmatrix}$

$-\frac{1}{18}R3 \rightarrow R3 \begin{bmatrix} 1 & 0 & -\frac{7}{11} & | & \frac{1}{11} & \frac{3}{11} & 0 \\ 0 & 1 & -\frac{12}{11} & | & -\frac{3}{11} & \frac{2}{11} & 0 \\ 0 & 0 & 1 & | & \frac{1}{6} & -\frac{1}{6} & \frac{1}{6} \end{bmatrix}$

$\frac{12}{11}R3 + R2 \rightarrow R2 \begin{bmatrix} 1 & 0 & -\frac{7}{11} & | & \frac{1}{11} & \frac{3}{11} & 0 \\ 0 & 1 & 0 & | & -\frac{1}{11} & 0 & \frac{2}{11} \\ 0 & 0 & 1 & | & \frac{1}{6} & -\frac{1}{6} & \frac{1}{6} \end{bmatrix}$

$\frac{7}{11}R3 + R1 \rightarrow R1 \begin{bmatrix} 1 & 0 & 0 & | & \frac{13}{66} & \frac{11}{66} & \frac{7}{66} \\ 0 & 1 & 0 & | & -\frac{1}{11} & 0 & \frac{2}{11} \\ 0 & 0 & 1 & | & \frac{1}{6} & -\frac{1}{6} & \frac{1}{6} \end{bmatrix}$

$A^{-1}C = \begin{bmatrix} \frac{13}{66} & \frac{11}{66} & \frac{7}{66} \\ -\frac{1}{11} & 0 & \frac{2}{11} \\ \frac{1}{6} & -\frac{1}{6} & \frac{1}{6} \end{bmatrix}\begin{bmatrix} 7 \\ -6 \\ -13 \end{bmatrix}$

$= \begin{bmatrix} \frac{91}{66} & \frac{66}{66} & -\frac{91}{66} \\ -\frac{7}{11} & 0 & -\frac{26}{11} \\ \frac{7}{6} & 1 & -\frac{13}{6} \end{bmatrix} = \begin{bmatrix} -1 \\ -3 \\ 0 \end{bmatrix}$

33. $2x - 3y = -9$

$\underline{4x - y = -13}\quad R2 \rightarrow -2R1 + R2$

$2x - 3y = -9$

$\underline{\qquad 5y = \ \ 5} \Rightarrow y = 1$

$2x - 3(1) = -9 \Rightarrow x = -3$

The solution is $x = -3,\ y = 1$.

$\begin{bmatrix} 1 & 0 & \frac{-7}{11} & | & -1 \\ 0 & 1 & \frac{-12}{11} & | & -3 \\ 0 & 0 & 6 & | & 0 \end{bmatrix}\ \frac{1}{6}R3 \rightarrow R3$

$\begin{bmatrix} 1 & 0 & \frac{-7}{11} & | & -1 \\ 0 & 1 & \frac{-12}{11} & | & -3 \\ 0 & 0 & 1 & | & 0 \end{bmatrix}\ \begin{array}{l}\frac{12}{11}R3 + R2 \rightarrow 2 \\ \frac{7}{11}R3 + R1 \rightarrow R1\end{array}$

$\begin{bmatrix} 1 & 0 & 0 & | & -1 \\ 0 & 1 & 0 & | & -3 \\ 0 & 0 & 1 & | & 0 \end{bmatrix}$

37. $x + 2y + 3z = 1$

$3x - 4y - 3z = 2\quad R2 \rightarrow -3R1 + R2$

$\underline{7x - 6y + 6z = 2}\quad R3 \rightarrow -7R1 + R3$

$x + 2y + \ 3z = \ 1$

$-10y - 12z = -1$

$\underline{-20y - 15z = -5}\quad R3 \rightarrow -2R2 + R3$

$x + \ 2y + \ 3z = \ 1$

$-10y - 12z = -1$

$\underline{\qquad\qquad 9z = -3} \Rightarrow \ z = -\frac{1}{3}$

$-10y - 12\left(-\frac{1}{3}\right) = -1 \Rightarrow y = \frac{1}{2}$

$x + 2\left(\frac{1}{2}\right) + 3\left(-\frac{1}{3}\right) = 1 \Rightarrow x = 1$

The solution is $x = 1,\ y = \frac{1}{2},\ z = -\frac{1}{3}$

41. $2u - 3v + 2w = 7$

$3u + v - 3w = -6$

$u + 4v + w = -13$

$\begin{vmatrix} 2 & -3 & 2 \\ 3 & 1 & -3 \\ 1 & 4 & 1 \end{vmatrix} = 66$

$u = \dfrac{\begin{vmatrix} 7 & -3 & 2 \\ -6 & 1 & -3 \\ -13 & 4 & 1 \end{vmatrix}}{66} = \dfrac{-66}{66} = -1$

$v = \dfrac{\begin{vmatrix} 2 & 7 & 2 \\ 3 & -6 & -3 \\ 1 & -13 & 1 \end{vmatrix}}{66} = \dfrac{-198}{66} = -3$

$w = \dfrac{\begin{vmatrix} 2 & -3 & 7 \\ 3 & 1 & -6 \\ 1 & 4 & -13 \end{vmatrix}}{66} = \dfrac{0}{66} = 0$

The solution is $u = -1$, $v = -3$, $w = 0$.

45. $3x - 2y + z = 6$

$2x + 0y + 3z = 3$

$4x - y + 5z = 6$

$A = \begin{bmatrix} 3 & -2 & 1 \\ 2 & 0 & 3 \\ 4 & -1 & 5 \end{bmatrix}$, $X = \begin{bmatrix} x \\ y \\ z \end{bmatrix}$, $C = \begin{bmatrix} 6 \\ 3 \\ 6 \end{bmatrix}$

$X = A^{-1}C = \begin{bmatrix} 3 \\ 1 \\ -1 \end{bmatrix}$ check: $AX = \begin{bmatrix} 6 \\ 3 \\ 6 \end{bmatrix}$

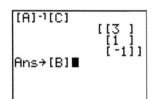

49. $3x - y + 6z - 2t = 8$

$2x + 5y + z + 2t = 7$

$4x - 3y + 8z + 3t = -17$

$3x + 5y - 3z + t = 8$

$A = \begin{bmatrix} 3 & -1 & 6 & -2 \\ 2 & 5 & 1 & 2 \\ 4 & -3 & 8 & 3 \\ 3 & 5 & -3 & 1 \end{bmatrix}$, $X = \begin{bmatrix} x \\ y \\ z \\ t \end{bmatrix}$, $C = \begin{bmatrix} 8 \\ 7 \\ -17 \\ 8 \end{bmatrix}$

$X = A^{-1}C = \begin{bmatrix} -\frac{1}{3} \\ 3 \\ \frac{2}{3} \\ -4 \end{bmatrix}$ check: $AX = \begin{bmatrix} 8 \\ 7 \\ -17 \\ 8 \end{bmatrix}$

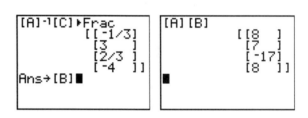

53. $A^2 = \begin{bmatrix} 1 & 0 \\ 3 & 4 \end{bmatrix}\begin{bmatrix} 1 & 0 \\ 3 & 4 \end{bmatrix} = \begin{bmatrix} 1 & 0 \\ 15 & 16 \end{bmatrix}$ from calculator

$A^3 = A^2 A = \begin{bmatrix} 1 & 0 \\ 15 & 16 \end{bmatrix}\begin{bmatrix} 1 & 0 \\ 3 & 4 \end{bmatrix} = \begin{bmatrix} 1 & 0 \\ 63 & 64 \end{bmatrix}$

from calculator

$A^4 = A^3 A = \begin{bmatrix} 1 & 0 \\ 63 & 64 \end{bmatrix}\begin{bmatrix} 1 & 0 \\ 3 & 4 \end{bmatrix} = \begin{bmatrix} 1 & 0 \\ 255 & 256 \end{bmatrix}$

from calculator

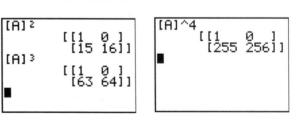

57. $\begin{vmatrix} 4 & 2 & 3 \\ 1 & -5 & -2 \\ 0 & 4 & -3 \end{vmatrix}$

$$= 4\left(-5(-3)-4(-2)\right)-2\left(1(-3)-(0)(-2)\right)$$
$$+3\left(1(4)-(0)(-5)\right) = 110$$

61. $\begin{vmatrix} 4 & 2 & 3 \\ 1 & -5 & -2 \\ -3 & 4 & -1 \end{vmatrix} \begin{array}{l} -\frac{1}{4}R1 + r2 \to R2 \\ \frac{3}{4}R1 + R3 \to R3 \end{array}$

$\begin{vmatrix} 4 & 2 & 3 \\ 0 & -\frac{11}{2} & -\frac{11}{4} \\ 0 & \frac{11}{2} & -\frac{3}{4} \end{vmatrix} R2 + R3 \to R3$

$\begin{vmatrix} 4 & 2 & 3 \\ 0 & -\frac{11}{2} & -\frac{11}{4} \\ 0 & 0 & -\frac{7}{2} \end{vmatrix} = 4\left(-\frac{11}{2}\right)\left(-\frac{7}{2}\right) = 77$

65. $N = \begin{bmatrix} 0 & -1 \\ 1 & 0 \end{bmatrix}$; N^{-1}

$$= \frac{1}{0-(-1)}\begin{bmatrix} 0 & 1 \\ -1 & 0 \end{bmatrix} = 1\begin{bmatrix} 0 & -1 \\ -1 & 0 \end{bmatrix} = -N$$

69. $AB \cdot B^{-1} = \begin{bmatrix} -1 & 3 \\ 0 & 2 \end{bmatrix} \cdot \begin{bmatrix} 1 & 2 \\ 3 & 4 \end{bmatrix}$

$$A = \begin{bmatrix} 8 & 10 \\ 6 & 8 \end{bmatrix}$$

73. Using the calculator to perform matrix operations,

$$(2A)^{-1} = \begin{bmatrix} \frac{1}{2} & \frac{1}{3} \\ 0 & \frac{1}{6} \end{bmatrix} \text{ and } \frac{A^{-1}}{2} = \begin{bmatrix} \frac{1}{2} & \frac{1}{3} \\ 0 & \frac{1}{6} \end{bmatrix}$$

which shows $(2A)^{-1} = \dfrac{A^{-1}}{2}$.

77. $\begin{bmatrix} F \\ T \end{bmatrix} = \begin{bmatrix} 0.500 & -0.866 \\ 0.866 & 0.500 \end{bmatrix}^{-1}\begin{bmatrix} 0 \\ 350 \end{bmatrix} = \begin{bmatrix} 303.113337 \\ 175.0077003 \end{bmatrix}$

$F = 303$ lb, $T = 175$ lb

81. $0.500F - 0.866T = \quad 0 \quad \frac{-0.866}{0.500}R1 + R2 \to R2$
$0.866F + 0.500T = 350$

$0.500F - 0.866T = \quad 0$
$1.999912T = 350 \Rightarrow T = 175$ lb

$0.500F - 0.866(175) = 0 \Rightarrow F = 303$ lb

85. $0.6A + 0.4B + 0.3C = 0.44(100)$
$0.3A + 0.3B + 0.7C = 0.38(100)$
$0.1A + 0.3B \qquad = 0.18(100)$

$$\begin{bmatrix} A \\ B \\ C \end{bmatrix} = \begin{bmatrix} 0.6 & 0.4 & 0.3 \\ 0.3 & 0.3 & 0.7 \\ 0.1 & 0.3 & 0 \end{bmatrix}^{-1}\begin{bmatrix} 44 \\ 38 \\ 18 \end{bmatrix} = \begin{bmatrix} 30 \\ 50 \\ 20 \end{bmatrix}$$

Use 30 gm of A, 50 gm of B, and 20 gm of C.

89. $\left[\begin{bmatrix} R_1 & -R_2 \\ -R_2 & R_1 \end{bmatrix} + R_2\begin{bmatrix} 1 & 0 \\ 0 & 1 \end{bmatrix}\right]\begin{bmatrix} i_1 \\ i_2 \end{bmatrix} = \begin{bmatrix} 6 \\ 0 \end{bmatrix}$

$\left[\begin{bmatrix} R_1 & -R_2 \\ -R_2 & R_1 \end{bmatrix} + \begin{bmatrix} R_2 & 0 \\ 0 & R_2 \end{bmatrix}\right]\begin{bmatrix} i_1 \\ i_2 \end{bmatrix} = \begin{bmatrix} 6 \\ 0 \end{bmatrix}$

$\begin{bmatrix} R_1 + R_2 & -R_2 \\ -R_2 & R_1 + R_2 \end{bmatrix}\begin{bmatrix} i_1 \\ i_2 \end{bmatrix} = \begin{bmatrix} 6 \\ 0 \end{bmatrix}$

$\begin{bmatrix} i_1(R_1 + R_2) - i_2 R_2 \\ -i_1 R_2 + i_2(R_1 + R_2) \end{bmatrix} = \begin{bmatrix} 6 \\ 0 \end{bmatrix}$

$i_1(R_1 + R_2) - i_2 R_2 = 6$
$-i_1 R_2 + i_2(R_1 + R_2) = 0$

INEQUALITIES

17.1 Properties of Inequalities

1. $x+1<0$ is true for all values of x less than -1. Therefore, the values of x that satisfy this inequality are written as $x<-1$.

5. $4+3<9+3$; $7<12$; property 1

9. $\dfrac{4}{-1}>\dfrac{9}{-1}$; $-4>-9$; property 3

13. $x>-2$ **17.** $1<x<7$

21. $x<1$ or $3<x\le5$

25. x is greater than 0 and less than or equal to 2.

29. $x<3$

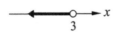

33. $0\le x<5$

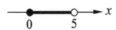

37. $x<-1$ or $1\le x<4$

41. $t\le-5$ and $t\ge-5$

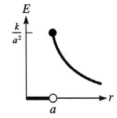

45. $0<a<b$ given
$0<a^2<ab$ since $a>0$
$0<ab<b^2$ since $b>0$
$a^2<ab<b^2$
$a^2<b^2$ is an absolute inequality

49. Note: $x^2=|x|^2$.
For $x>0$, $y<0 \Rightarrow xy<0<|x||y|$
$2|x||y|>2xy$
$x^2+2|x||y|+y^2>x^2+2xy+y^2$
$|x|^2+2|x||y|+|y|^2>(x+y)^2$
$(|x|+|y|)^2>|x+y|^2$
$|x|+|y|>|x+y|$

53. $110^2\le L^2\le120^2$
$70^2\le w^2\le80^2$
$110^2+70^2\le L^2+w^2\le120^2+80^2$
$110^2+70^2\le d^2\le120^2+80^2$
$130\le d\le144$ yd

57. $18{,}000<v<25{,}000$ mi/h

61. $E=0$ for $0\le r<a$
$E=\dfrac{k}{r^2}$ for $r\ge a$

17.2 Solving Linear Inequalities

1. $21-2x\geq 15$

$-2x\geq -6$

$x\leq 3$

5. $x-3>-4$

$x>-4+3$

$x>-1$

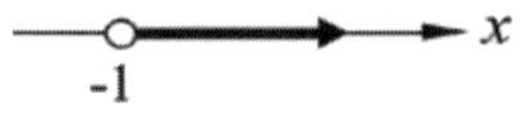

9. $3x-5\leq -11$

$3x\leq -11+5$

$3x\leq -6$

$x\leq -2$

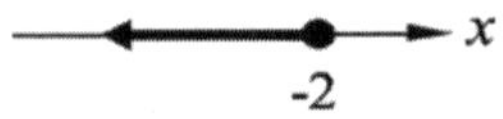

13. $\frac{4x-5}{2}\leq x$

$4x-5\leq 2x$

$4x-2x\leq 5$

$2x\leq 5$

$x\leq \frac{5}{2}$

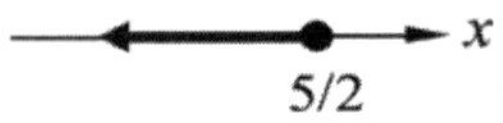

17. $2.50(1.50-3.40x)<3.84-8.45x$

$3.75-8.50x<3.84-8.45x$

$-0.09<0.05x$

$x>-1.80$

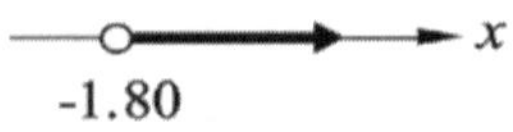

21. $-1\leq 2x+1\leq 3$

$-2\leq 2x\leq 2$

$-1\leq x\leq 1$

25. $2x<x-1\leq 3x+5$

$0<-x-1\leq x+5$

$0<-x-1$ and $-x-1\leq x+5$

$x<-1$ and $-6\leq 2x$

$x<-1$ and $x\geq -3$

$-3\leq x<-1$

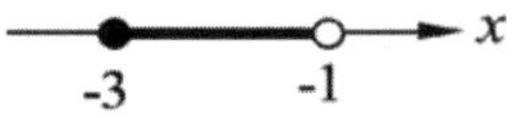

29. $3x-2<8-x$

$4x<10$

$x<\frac{5}{2}$

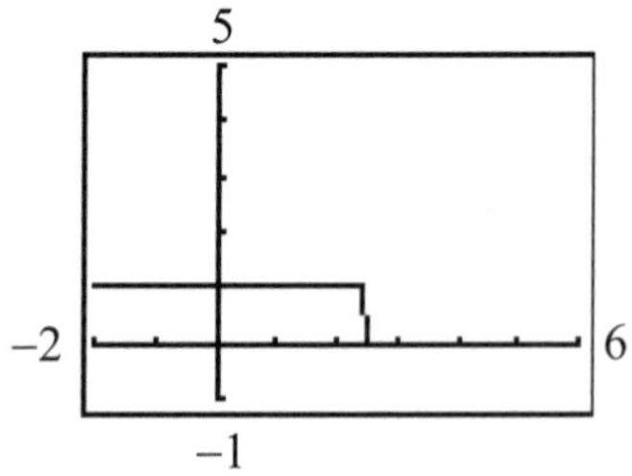

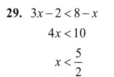

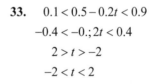

33. $0.1<0.5-0.2t<0.9$

$-0.4<-0.;2t<0.4$

$2>t>-2$

$-2<t<2$

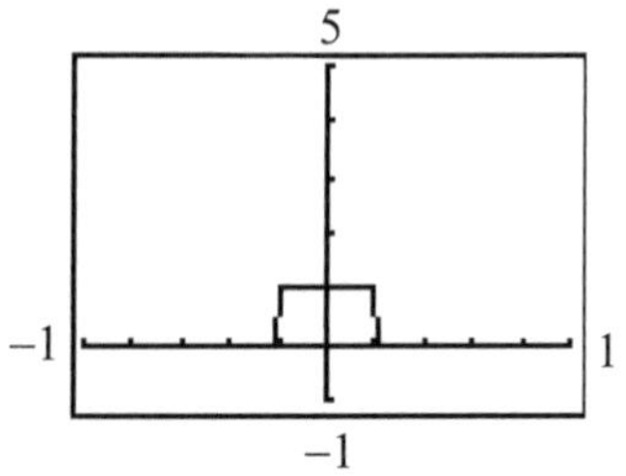

37. Graph $y1 = -2(2.5^x) + 5 \geq 3, \; x \leq 0$

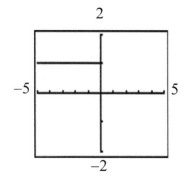

41. $x^2 - kx + 9 = 0$ has roots $\dfrac{k \pm \sqrt{k^2 - 36}}{2}$ which will

be imaginary for $k^2 - 36 < 0 \Rightarrow (k+6)(k-6) < 0$

which requires:

$(k+6 > 0 \text{ and } k-6 < 0) \text{ or } (k+6 < 0 \text{ and } k-6 > 0)$

$\quad (k > -6 \text{ and } k < 6) \qquad\qquad (k < -6 \text{ and } k > 6)$

$\quad (-6 < k < 6) \qquad\qquad\quad (\text{no values in common})$

The roots are imaginary for $-6 < k < 6$.

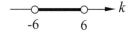

45. $|5 - (-2)| = |5 + 2| = |7| = 7$

$\big||-5| - |-2|\big| = |-5 - 2| = |-7| = 7$

$\Rightarrow |5 - (-2)| = \big||-5| - |-2|\big|$

49. $25n > 350 + 15n$

$10n > 350$

for $n > 35$ h, the second position pays more.

53. $100 < 130(1.42)w < 150$

$0.54 \text{ m} < w < 0.81 \text{ m}$

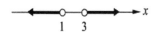

57. $0 \leq x \leq 800 - 300$

$0 \leq x \leq 500$

$y = x + 400 - 200 = x + 200$

$x = y - 200$

$0 \leq y - 200 \leq 500$

$200 \leq y \leq 700$

17.3 Solving Nonlinear Inequalities

1. $\qquad x^2 + 3 > 4x$

$x^2 - 4x + 3 > 0$

$(x - 3)(x - 1) > 0$

The critical values are 1, 3.

	$(x-3)(x-1)$		Sign
$x < 1$	−	−	+
$1 < x < 3$	−	+	−
$x > 3$	+	+	+

$x^2 - 4x + 3 > 0$ when $x < 1$ or $x > 3$

5. $\qquad x^2 - 16 < 0$

$(x + 4)(x - 4) < 0$

The critical values are $x = -4$ and $x = 4$.

	$(x+4)(x-4)$		Sign
$x < -4$	−	−	+
$x - 4 < x < 4$	+	−	−
$x > 4x$	+	+	+

$x^2 - 16 < 0$ for $-4 < x < 4$

9.
$$2x^2 - 12 \le -5x$$
$$2x^2 + 5x - 12 < 0$$
$$(2x-3)(x+4) \le 0$$

The critical values are $x = \dfrac{3}{2}$, $x = -4$.

	$(2x-3)(x+4)$		Sign
$x < -4$	$-$	$-$	$+$
$-4 < x < 3/2$	$-$	$+$	$-$
$0 < x < 3/2$	$+$	$+$	$+$

$(2x-3)(x+4) \le 0$ for $-4 \le x \le \dfrac{3}{2}$

$$\underset{\substack{-4 \qquad\quad 3/2}}{\xrightarrow{\hspace{2cm}\bullet\!\!-\!\!-\!\!\bullet\hspace{1cm}}}\; x$$

13. $R^2 + 4 > 0$

$R^2 + 4$ is never less than 4,

so all values of R are solutions.

$$\underset{0}{\xleftrightarrow{\hspace{3cm}}}\; R$$

17.
$$s^3 + 2s^2 - s \ge 2$$
$$s^2(s+2) - 1(s+2) \ge 0$$
$$(s^2 - 1)(s+2) \ge 0$$
$$(s+1)(s-1)(s+2) \ge 0$$

The critical values are $s = -1$, $s = 1$, $s = -2$.

	$(s-1)(s+1)(s+2)$			Sign
$s < -2$	$-$	$-$	$-$	$-$
$-2 < s < -1$	$-$	$-$	$+$	$+$
$-1 < s < 1$	$-$	$+$	$+$	$-$
$s > 1$	$+$	$+$	$+$	$+$

$(s+1)(s-1)(s+2) \ge 0$ for

$-2 \le s \le -1$ or $s \ge 1$

$$\underset{\substack{-2 \qquad -1 \qquad 1}}{\xrightarrow{\hspace{1cm}\bullet\!-\!-\!\bullet\quad\bullet\hspace{1cm}}}\; s$$

21.
$$\frac{x^2 - 6x - 7}{x+5} > 0$$
$$\frac{(x-7)(x+1)}{x+5} > 0$$

The critical values are $x = 7$, $x = -1$, $x = -5$.

	$(x-7)(x+1)(x+5)$			Sign
$x < -5$	$-$	$-$	$-$	$-$
$-5 < x < -1$	$-$	$-$	$+$	$+$
$-1 < x < 7$	$-$	$+$	$+$	$-$
$x > 7$	$+$	$+$	$+$	$+$

$\dfrac{(x-7)(x+1)}{x+5} > 0$ for $-5 < x < -1$ or $x > 7$

$$\underset{\substack{-5 \qquad -1 \quad 7}}{\xrightarrow{\hspace{1cm}\circ\!-\!-\!\circ\;\circ\hspace{1cm}}}\; x$$

25.
$$3x^2 + 5x \ge 2$$
$$3x^2 + 5x - 2 \ge 0$$
$$(3x-1)(x+2) \ge 0$$

The critical values are $x = \dfrac{1}{3}$, $x = -2$.

	$(3x-1)(x+2)$		Sign
$x < -2$	$-$	$-$	$+$
$-2 < x < 1/3$	$-$	$+$	$-$
$x > 1/3$	$+$	$+$	$+$

$(3x-1)(x+2) \ge 0$ for $x \le -2$ or $x \ge \dfrac{1}{3}$

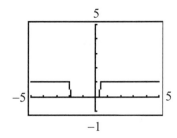

29.
$$\frac{6-x}{3-x-4x^2} \ge 0$$
$$\frac{6-x}{(1+x)(3-4x)} \ge 0; \left(x \ne -1, \, x \ne \frac{3}{4}\right)$$

The critical values are $x = 6$, $x = -1$ and $x = \dfrac{3}{4}$.

	$(6-x)/(1+x)(3-4x)$			Sign
$x < -1$	$+$	$-$	$+$	$-$
$-1 < x < 3/4$	$+$	$+$	$+$	$+$
$3/4 < x < 6$	$+$	$+$	$-$	$-$
$x > 6$	$-$	$+$	$-$	$+$

$\dfrac{6-x}{(1+x)(3-4x)} \geq 0$ for $-1 < x < \dfrac{3}{4}$ or $x \geq 6$

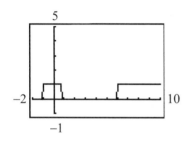

33. $\sqrt{(x-1)(x+2)}$ is real if $(x-1)(x+2) \geq 0$

The critical values are $x = 1$ and $x = -2$.

	$(x-1)(x+2)$		Sign
$x < -2$	$-$	$-$	$+$
$-2 < x < 1$	$-$	$+$	$-$
$x > 1$	$+$	$+$	$+$

$(x-1)(x+2) > 0$ for $x \leq -2$ or $x \geq 1$

37. To solve $x^3 - x > 2$ using a graphing calculator,

let $y_1 = x^3 - x - 2$.

$y > 0$ for $x > 1.52$

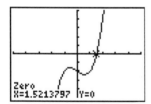

41. To solve $2^x > x + 2$ using a graphing calculator,

let $y_1 = 2^x - x - 2$.

$y > 0$ for $x < -1.69$, $x > 2.00$

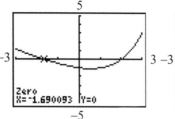 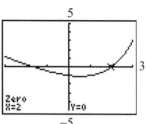

45. $\qquad x^2 > x$

$x^2 - x > 0$

$x(x-1) > 0$

The critical values are $x = 0$, $x = 1$.

	$x(x-1)$		Sign
$x < 0$	$-$	$-$	$+$
$0 < x < 1$	$+$	$-$	$-$
$x > 1$	$+$	$+$	$+$

$x(x-1) > 0$ for $x < 0$ or $x > 1$

Is $x^2 > x$ for all x? No.

$x^2 > x$ for $x < 0$ or $x > 1$.

$x^2 > x$ is not true for $0 \leq x \leq 1$.

49. $\qquad\qquad 2^{x+2} > 3^{2x-3}$

$\log 2^{x+2} > \log 3^{2x-3}$

$(x+2)\log 2 > (2x-3)\log 3$

$x\log 2 + 2\log 2 > 2x\log 3 - 3\log 3$

$x\log 2 + \log 4 > 2x\log 3 - \log 27$

$\log 4 + \log 27 > x(2\log 3 - \log 2)$

$\log 108 > x(\log 9 - \log 2)$

$\log 108 > x\log\dfrac{9}{2}$

$x < \dfrac{\log 108}{\log\frac{9}{2}}$

53. $p = 9 + 5t - t^2 > 15 \Rightarrow$

$t^2 - 5t + 6 < 0$

$(t-2)(t+3) < 0$

	$(t-2)(t+3)$		Sign
$0 < t < 2$	−	−	+
$2 < t < 3$	−	+	−
$t > 3$	+	+	+

$p > 15$ for $2 < t < 3$ min

57. $C > 1.00 \Rightarrow 1 > 1.00 C^{-1} \Leftrightarrow 1.00 C^{-1} < 1$

$C^{-1} = C_1^{-1} + C_2^{-1} < 1$

$C_1^{-1} + 4.00^{-1} < 1 \Rightarrow C_1^{-1} < 0.750$

$C_1 > 1.33 \ \mu F$

61.

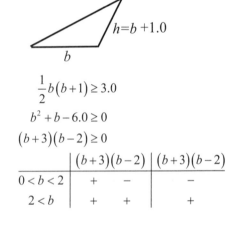

$\frac{1}{2} b(b+1) \geq 3.0$

$b^2 + b - 6.0 \geq 0$

$(b+3)(b-2) \geq 0$

	$(b+3)(b-2)$		$(b+3)(b-2)$
$0 < b < 2$	+	−	−
$2 < b$	+	+	+

$b \geq 2.$

$h - 1 \geq 2$

$h \geq 3.0$ cm

17.4 Inequalities Involving Absolute Values

1. $|2x-1| < 5$

$-5 < 2x - 1 < 5$

$-4 < 2x < 6$

$-2 < x < 3$

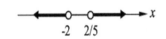

5. $|5x+4| > 6$

$5x + 4 < -6$ or $5x + 4 > 6$

$5x < -10$ or $5x > 2$

$x < -2$ or $x > \frac{2}{5}$

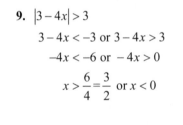

9. $|3-4x| > 3$

$3 - 4x < -3$ or $3 - 4x > 3$

$-4x < -6$ or $-4x > 0$

$x > \frac{6}{4} = \frac{3}{2}$ or $x < 0$

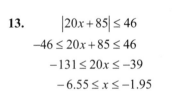

13. $|20x+85| \leq 46$

$-46 \leq 20x + 85 \leq 46$

$-131 \leq 20x \leq -39$

$-6.55 \leq x \leq -1.95$

17. $8 + 3|3 - 2x| < 11$

$3|3 - 2x| < 3$

$|3 - 2x| < 1$

$-1 < 3 - 2x < 1$

$-4 < -2x < -2$

$2 > x > 1$

$1 < x < 2$

21. $\left|\dfrac{3R}{5} + 1\right| < 8$

$-8 < \dfrac{3R}{5} + 1 < 8$

$-15 < R < \dfrac{35}{3}$

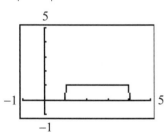

25. $|2x - 5| < 3 \Leftrightarrow 1 < x < 4$

29. $|x^2 + x - 4| > 2$

$x^2 + x - 4 > 2$ or $x^2 + x - 4 < -2$

$x^2 + x - 6 > 0$ or $x^2 + x - 2 < 0$

(A) $(x + 3)(x - 2) > 0$

(B) $(x - 1)(x + 2) < 0$

(A) Critical values are $x = -3$, $x = 2$.

	$(x + 3)(x - 2)$		Sign
$x < -3$	$-$	$-$	$+$
$-3 < x < 2$	$+$	$-$	$-$
$x > 2$	$+$	$+$	$+$

$(x + 3)(x - 2) > 0$ for $x < -3$ or $x > 2$

(B) Critical values are $x = 1$, $x = -2$.

	$(x - 1)(x + 2)$		Sign
$x < -2$	$-$	$-$	$+$
$-2 < x < 1$	$-$	$+$	$-$
$x > 1$	$+$	$+$	$+$

$(x - 1)(x + 2) < 0$ for $-2 < x < 1$

The solution consists of values of x that are in

(A) or (B): $x < -3$, $-2 < x < 1$, $x > 2$

33. Solve for x if $|x| < a$ and $a \le 0$.

$|x| < a \le 0$

$|x| < 0$, no values since $|x| \ge 0$.

37. $|x - 1| < 4 \Leftrightarrow$

$-4 < x - 1 < 4$

$1 < x + 4 < 9 \Rightarrow$

$a = 1$, $b = 9$

41. $|p - 2,000,000| \le 200,000$

$-200,000 \le p - 2,000,000 \le 200,000$

$1,800,000 \le p \le 2,200,000$ barrels

The production will be at least 1,800,000 barrels but not greater than 2,200,000 barrels.

45. $3.675 - 0.002 \le d \le 3.675 + 0.002$

$-0.002 \le d - 3.675 \le 0.002$

$|d - 3.675| \le 0.002$ cm

17.5   Graphical Solution of Inequalities with Two Variables

1. $y < 3 - x$

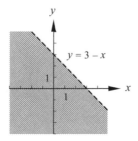

5. $y \geq 2x + 5$; graph $y = 2x + 5$. Use a solid line to indicate that points on it satisfy the inequality. Shade the region above the line.

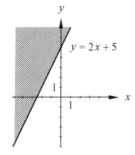

9. $4y < x^2$; graph $y = 1/4x^2$. Use a dashed curve to indicate that points on it do not satisfy the inequality. Shade the region below the curve.

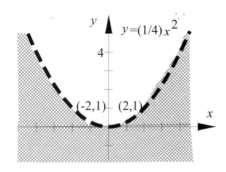

13. $y < 32x - x^4$; graph $y = 32 - x^4$. Use a dashed curve to indicate that points on it do not satisfy the inequality. Shade the region below the curve.

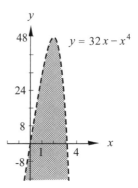

17. $y > 1 + \sin 2x$; graph $y = 1 + \sin 2x$. Use a dashed curve to indicate that the points on it do not satisfy the inequality. Shade the region above the curve.

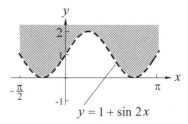

21. $|y| > |x|$. For $y > 0$, $|y| > |x|$ becomes $y = |x|$.
Graph $y = |x|$ with dashed line and shade region
above graph.
For $y < 0$, $|y| > |x|$ becomes $-y > |x| \Rightarrow$
$y < -|x|$. Graph $y = -|x|$ with dashed line and
shade region below graph. The solution is both
regions.

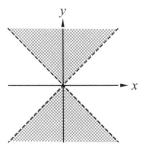

25. $y \le 2x^2$ and $y > x - 2$. Graph $y = 2x^2$ using a solid
curve. Shade the region below the curve. Graph
$y = x - 2$ using a dashed line. Shade the region
above the line. The region where the shadings
overlap satisfies both inequalities.

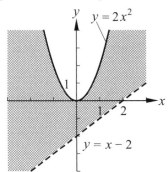

29. $y \ge 0$ and $y \le \sin x$; $0 \le x \le 3\pi$. Graph $y = \sin x$
using a solid curve. Shade the region below the
curve and above the x-axis for $0 \le x \le 3\pi$.

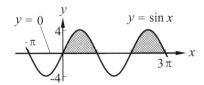

33. For $y > 0$, $x < 0$ shade points in QIII left of x-axis
and above y-axis. Graph $y = x$ with solid line and
shade region below line. There are no points in
both regions.

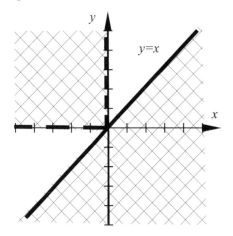

37. $y \ge 1 - x^2$. Graph $y_1 = 1 - x^2$.
The boundary line is solid.

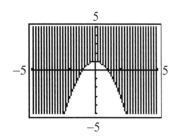

41. $y > x^2 + 2x - 8$. Graph $y_1 = x^2 + 2x - 8$.
$y < \dfrac{1}{x} - 2$. Graph $y_2 = \dfrac{1}{x} - 2$.
The boundary lines are dashed.

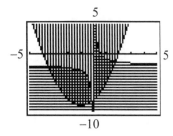

45. $4x - 2y + 5 = 0$

$$2y = 4x + 5$$

$$y = 2x + \frac{5}{2}$$

The region below the line $4x - 2y = 0$ is defined by the inequality $y < 2x + \frac{5}{2}$.

49. $y \geq 2x^2 - 6$ defines the region on and above the parabola $y = 2x^2 - 6$. $y = x - 3$ defines all points on the graph of the line $y = x - 3$. The solution to the system $y \geq 2x^2 - 6$ and $y = x - 3$ is the intersection of the graph of the parabola and the line together with the points on the line above the parabola.

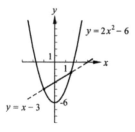

$y = 2x^2 - 6$

$y = x - 3$

53. $p_R = Ri^2$; $R = 0.5\Omega$, $p > p_R$, $p > 0.5i^2$

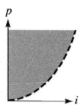

17.6   Linear Programming

1. Maximize $F = 2x + 3y$ subject to $x \geq 0$,
$y \geq 0$, $x + y \leq 6$, $2x + y \leq 8$.

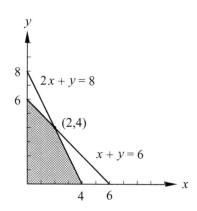

point	value of F
$(0, 0)$	0
$(0, 6)$	18
$(2, 4)$	16
$(4, 0)$	8

5. Maximum P: $P = 3x + 5y$ subject to
$x \geq 0$, $y \geq 0$
$2x + y \leq 6$

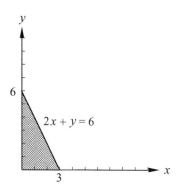

$2x + y = 6$

Vertex	$P = 3x + 5y$
$(0, 0)$	0
$(0, 6)$	30
$(3, 0)$	9

max $P = 30$ at $(0, 6)$

9. Minimum C: $C = 4x + 6y$ subject to
$x \geq 0$, $y \geq 0$
$x + y \geq 5$
$x + 2y \geq 7$

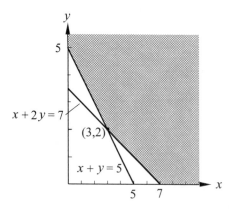

$x + 2y = 7$

$(3, 2)$

$x + y = 5$

Vertex	$C = 4x + 6y$
$(0, 5)$	30
$(3, 2)$	24
$(7, 0)$	28

min $C = 24$ at $(3, 2)$

13. Maximum P: $P = 9x + 2y$ subject to

$x \geq 0,\ y \geq 0$

$2x + 5y \leq 10$

$4x + 3y \leq 12$

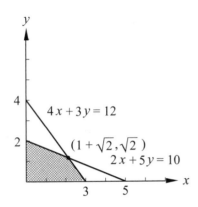

Vertex	$P = 9x + 2y$
$(0, 0)$	0
$(0, 2)$	4
$(1+\sqrt{2},\ \sqrt{2})$	$9 + 11\sqrt{2} = 15.5$
$(3, 0)$	27

max $P = 27$ at $(3, 0)$

17. x = number of newspaper ads

y = number of radio ads

$N = 8000x + 6000y$

$x \geq 0,\ y \geq 0$

$50x + 150y \leq 9000$

$2(50x) \leq 150y \Rightarrow y \geq \dfrac{2x}{3}$

Vertex	$N = 8000x + 6000y$
$(0, 0)$	0
$(0, 60)$	36,000
$(60, 40)$	720,000

max $N = 720,000$ people with 60 news ads and 40 radio ads.

21. x = cereal A, y = cereal B, c = cost

$c = 12x + 18y$

$x + 2y \geq 10$; $y = -\dfrac{1}{2}x + 5$; graph: shade region above the graph.

$5x + 3y \geq 30$; $y = -\dfrac{5}{3}x + 10$; graph: shade regiobn above graph.

Minimum c is the intersection of $y = -\dfrac{1}{2}x + 5$

and $y = -\dfrac{5}{3}x + 10$.

Solve simultaneously by substitution:

$-\dfrac{1}{2}x + 5 = -\dfrac{5}{3}x + 10$; $x = \dfrac{30}{7}$ oz, $y = \dfrac{20}{7}$ oz

OR

Use coordinates of vertices:

$c = 12(10) + 18(0) = 120$ cents

$c = 12(0) + 18(10) = 180$ cents

$c = 12\left(\dfrac{30}{7}\right) + 18\left(\dfrac{20}{7}\right) = \dfrac{720}{7} = 103$ cents

The minimum cost occurs when $x = \dfrac{30}{7}$, $y = \dfrac{20}{7}$;

i.e., $4\dfrac{2}{7}$ oz of A, $2\dfrac{6}{7}$ oz of B.

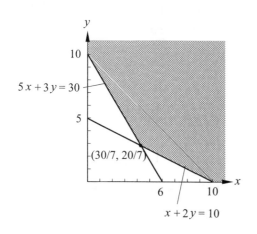

$5x + 3y = 30$

$(30/7, 20/7)$

$x + 2y = 10$

	$2x-1$	$3-x$	$x+4$	$\frac{(2x-1)(3-x)}{(x+4)}$
$x < -4$	$-$	$+$	$-$	$+$
$-4 < x < \frac{1}{2}$	$-$	$+$	$+$	$-$
$\frac{1}{2} < x < 3$	$+$	$+$	$+$	$+$
$x > 3$	$+$	$-$	$+$	$-$

Solution: $x < -4$ or $\frac{1}{2} < x < 3$

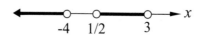

$-4 \quad 1/2 \quad 3$

13. $\quad |3x + 2| \le 4$

$-4 \le 3x + 2 \le 4$

$-6 \le 3x \le 2$

$-2 \le x \le \frac{2}{3}$

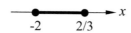

$-2 \qquad 2/3$

Chapter 17   Review Exercises

1. $2x - 12 > 0 \Rightarrow 2x > 12 \Rightarrow x > 6$

6

5. $\qquad 5x^2 + 9x < 2$

$(x+2)(5x-1) < 0$

	$(x+2)$	$(5x-1)$	Sign
$x < -2$	$-$	$-$	$+$
$-2 < x < \frac{1}{5}$	$+$	$-$	$-$
$x > \frac{1}{5}$	$+$	$+$	$+$

$-2 < x < \frac{1}{5}$

$-2 \qquad 1/5$

9. $\dfrac{(2x-1)(3-x)}{(x+4)} > 0$

Critical values are $-4, \dfrac{1}{2}, 3$.

17. $5 - 3x > 0 \Leftrightarrow 3x < 5 \Leftrightarrow x < \dfrac{5}{3}$

Graph $y_1 = 5 - 3x > 0$.

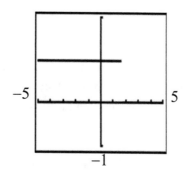

2

-5 $\qquad\qquad$ 5

-1

21. $\dfrac{8-R}{2R+1} \le 0$

Critical values are $-\dfrac{1}{2}, 8$.

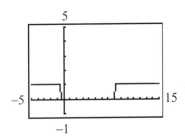

	$8-R$	$2R+1$	$\frac{8-R}{2R+1}$
$R < -\frac{1}{2}$	$+$	$+$	$-$
$-\frac{1}{2} < R < 8$	$-$	$+$	$+$
$R \geq 8$	$-$	$+$	$-$

$R < \dfrac{-1}{2}$ or $R \geq 8$

Graph $y_1 = (8-x)(2x+1) \leq 0$

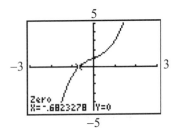

25. $x^3 + x + 1 < 0$; Graph $y_1 = x^3 + x + 1$ and use the zero feature. $(x < -0.68)$

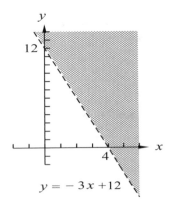

29. $y > 12 - 3x$. Graph $y = 12 - 3x$. Use a dashed line to indicate that points on it do not satisfy the inequality. Shade the region above the line.

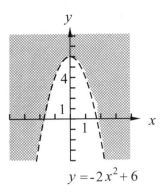

33. $2x^2 > 6 - y$

$y > 6 - 2x^2$. Graph $y = 6 - 2x^2$. Use a dashed line to indicate points on boundary line are not part of solution. Shade region above line.

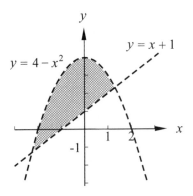

37. $y > x+1$, $y < 4 - x^2$. Graph $y = x+1$ and $y = 4 - x^2$. Use a dashed line to indicate boundary line is not part of solution. Shade region below parabola and above line.

41. $y < 3x + 5$. Graph $y_1 = 3x + 5$ and shade below the line. Boundary line is not part of solution.

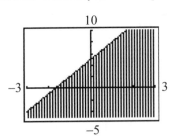

45. $y < 32x - x^4$. Graph $y_1 = 32x - x^4$ and shade below curve. Boundary line is not part of solution.

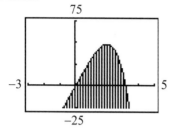

49. $\sqrt{3-x}$ is a real number for $3 - x \geq 0$
$\Rightarrow 3 \geq x \Leftrightarrow x \leq 3$.

53. Maximize P: $P = 2x + 9y$ subject to
$x \geq 0,\ y \geq 0$
$x + 4y \leq 13$
$3y - x \leq 8$

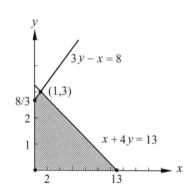

Vertex	$P = 2x + 9y$
$(0, 0)$	0
$\left(0, \frac{8}{3}\right)$	24
$(1, 3)$	29
$(13, 0)$	26

max $P = 29$ at $(1, 3)$

57. When is $|a + b| < |a| + |b|$. There are 4 cases.
(1) $a \geq 0,\ b \geq 0 \Rightarrow a + b \geq 0$, the given inequality is
$$a + b < a + b$$
$$0 < 0,\ F$$

(2) $a < 0,\ b < 0 \Rightarrow a + b < 0$, the given inequality is
$$-(a+b) < -a + (-b)$$
$$-(a+b) < -(a+b)$$
$$0 < 0,\ F$$

(3) $a < 0,\ b > 0,\ |a| > b \Rightarrow a + b < 0$, the given inequality is
$$-(a+b) < -a + b$$
$$-a - b < -a + b$$
$$-b < b,\ T$$

(4) $a < 0,\ b > 0,\ |a| < b \Rightarrow a + b > 0$, the given inequality is
$$a + b < -a + b$$
$$a < -a,\ T$$

Note: in cases (3) and (4) a and b can be reversed without loss of generality.
$|a + b| < |a| + |b|$ when a and b have opposite signs.

61. $(x - 5)(x + 2) < 0 \Rightarrow x^2 - 3x - 10 < 0$

65.

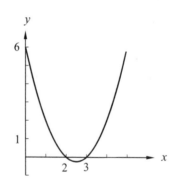

69. $2(x + 6.0) + 2(x + 10.0) \geq 2(4x)$
$$2x + 12 + 2x + 20 \geq 8x$$
$$32 \geq 4x$$
$$8.0 \geq x$$
$$0 < x \leq 8.0 \text{ cm}$$

73. $p = 101 + 10.1d > 500$
$$10.1d > 399$$
$$d > 39.5 \text{ m}$$

77. $p = Ri^2 = 12.0i^2$

$\quad\quad 2.50 < 12.0i^2 < 8.00$

$\quad\quad 0.456 < i < 0.816$ A

81.

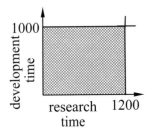

85.

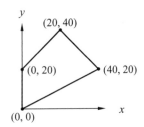

Write the equation of each line:

(1) $y = x + 20$

(2) $y = -x + 60$

(3) $y = \dfrac{1}{2}x$

The park region is described by the inequalities.

$$x \geq 0$$
$$y \geq 0$$
$$y \leq x + 20$$
$$y \leq -x + 60$$

Chapter 18

VARIATION

18.1　Ratio and Proportion

1. $\dfrac{24\text{ ft}}{15\text{ ft}} = \dfrac{8}{5}$

5. $\dfrac{96\text{ h}}{3\text{ days}} = \dfrac{96\text{ h}}{72\text{ h}} = \dfrac{4}{3}$

9. $\dfrac{0.14\text{ kg}}{3500\text{ mg}} = \dfrac{0.14\text{ kg}}{0.0035\text{ kg}} = 40$

13. $\dfrac{2.6\text{ W}}{9.6\text{ W}} = 0.27 = 27\%$

17. $C = \dfrac{q}{V} = \dfrac{5.00\ \mu C}{200\text{ V}} \cdot \dfrac{C}{10^{6}\ \mu C} = 2.5 \times 10^{-8}\text{ C}$

21. $\dfrac{487\text{ lb/ft}^{3}}{62.4\text{ lb/ft}^{3}} = 7.80$

25. $\dfrac{8460\text{ N}}{9.80\text{ m/s}^{2}} = \dfrac{8460\text{ m}\cdot\text{kg/s}^{2}}{9.80\text{ m/s}^{2}} = 863\text{ kg}$

29. $\dfrac{6.00}{R_{2}} = \dfrac{62.5}{15.0}$

$62.5 R_{2} = 90.0;\ R_{2} = 1.44\ \Omega$

33. $\dfrac{1.00\text{ hp}}{746\text{ W}} = \dfrac{x}{250\text{ W}}$; $746x = 250;\ x = 0.335\text{ hp}$

37. $45.0\text{ km/h} = 45.0\dfrac{\text{km}}{\text{h}} \times \dfrac{1000\text{ m/km}}{3600\text{ s/h}} = 12.5\text{ m/s}$

41. exact speed $= \dfrac{5.0\text{ km}}{(4\text{ min }54\text{ sec})\left(\dfrac{\text{h}}{60\text{ min}}\right)}$

$\dfrac{\text{exact} - \text{measured}}{\text{exact}} = \dfrac{\text{exact} - 58}{\text{exact}} = 5.3\%$ too low

45.

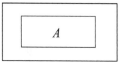

$A = \dfrac{1}{3}(1160 - A)$

$3A = 1160 - A$

$4A = 1160$

$A = 290\text{ in.}^{2}$

remaining area $= 870\text{ in.}^{2}$

49. $\dfrac{17}{595} = \dfrac{500}{x};\ 17x = 297,500;\ x = 17,500$ chips

---

18.2　Variation

1. $C = kd = \pi d$

　　$k = \pi$

5. $v = kr$

9. $P = \dfrac{k}{\sqrt{A}}$

13. The area varies directly as the square of the radius.

17. $V = kH^{2};\ 2 = k \cdot 64^{2}$

$k = \dfrac{2}{64^{2}};\ V = \dfrac{2H^{2}}{64^{2}} = \dfrac{H^{2}}{2048}$

21. $y = kx;\ 200 = k(80);\ k = 2.5;\ y = 2.5x$

$y = 2.5(10) = 25$

25. $y = \dfrac{kx}{z}$; $60 = \dfrac{k(4)}{10}$; $k = 150$; $y = \dfrac{150x}{z}$

$y = \dfrac{150(6)}{0.5} = 1800$

29. $A = k_1 x$, $B = k_2 x$

$A + B = k_1 x + k_2 x = (k_1 + k_2)x$ which shows $A + B$

varies directly as x.

33. $H = km$; $2.93 \times 10^5 = k(875)$; $k = 335$ J/g

$H = 335\,m$; $H = 335(625) = 2.09 \times 10^5$ J

37. $F = ks$; $10.0 = k(4.00)$; $k = \dfrac{2.50}{4.00} = 2.50$ N/cm

$6 = 2.50s$; $s = 2.40$ in.

41. $t = \dfrac{k}{A}$; $2.0 = \dfrac{k}{48}$; $k = 48(2.0) = 96$ h·in.²

$t = \dfrac{96}{68} = 1.4$ h

45. $P = ks^3$; $5200 = k(12.0)^3$; $k = 3.01$ hp·h³/mi³

$P = 3.01(15)^3 = 10{,}200$ hp

49.
$$F = \dfrac{k}{d}$$

$$0.750 \text{ N} = \dfrac{k}{1.25 \text{ cm}}$$

$$k = \dfrac{15}{16} \text{ N·cm}$$

$$F = \dfrac{\frac{15}{16}}{d}$$

$$F = \dfrac{\frac{15}{16}}{1.75}$$

$$F = 0.536 \text{ N}$$

53. $R = \dfrac{kl}{A}$; $0.200 = \dfrac{k(225)}{0.0500}$

$k = 4.44 \times 10^{-5}$ Ω·in.²/ft

$R = \dfrac{4.44 \times 10^{-5}\, l}{A}$

57. $P = kRi^2$; $10.0 = k(40.0)(0.500)^2$; $k = \dfrac{10.0}{10.0} = 1.00$

$P = 20.0(2.00^2) = 80.0$ W

$$\dfrac{kP}{r_1^2} = \dfrac{2kP}{(d - r_1)^2}$$

$$\dfrac{1}{r_1^2} = \dfrac{2}{d^2 - 2r_1 d + r_1^2}$$

$$2r_1^2 = d^2 - 2r_1 d + r_1^2$$

$r_1^2 + 2dr_1 - d^2 = 0$

Solve using the quadratic formula:

$$r_1 = \dfrac{-2d \pm \sqrt{4d^2 + 4d^2}}{2}$$

$$= \dfrac{-2d \pm 2\sqrt{2}\,d}{2}$$

$$= -d \pm \sqrt{2}\,d$$

$$= -2.4d,\ 0.41d \text{ (ignore negative value)}$$

61. Note: Make sure calculator is in rad mode.

$x = k\omega^2 (\cos \omega t)$

$-11.4 = k(0.524^2)\cos\left[(1.00)(0.524)\right]$

$k = \dfrac{-11.4}{0.238} = -47.95261498$ ft

$x = k(0.524)^2 \cos\left[(2.00)(0.524)\right]$

$= -6.57$ ft/s²

Chapter 18   Review Exercises

1. $\dfrac{4 \text{ Mg}}{20 \text{ kg}} = \dfrac{4000 \text{ kg}}{20 \text{ kg}} = 200$

5. $\pi = \dfrac{c}{d}$

$\dfrac{4.2736}{1.3603} = 3.1417$

9. $p = \dfrac{F}{A} = \dfrac{37.4}{2.25^2}$

$p = 7.39 \text{ lb/in.}^2$

13. Commission rate $= \dfrac{\text{Commission}}{\text{Selling price}} = \dfrac{20,900}{380,000}$

$= 0.055$

17. $\dfrac{1.3 \text{ in.}}{20.0 \text{ mi}} = \dfrac{6.0 \text{ in.}}{x}$

$x = 92 \text{ mi}$

21. $10 \text{ min} = 600 \text{ s}$

$\dfrac{60 \text{ pages}}{45 \text{ s}} = \dfrac{p}{600 \text{ s}} \Rightarrow p = 800 \text{ pages}$

25. $\dfrac{25.0 \text{ ft}}{2.00 \text{ in.}} = \dfrac{x}{5.75 \text{ in.}}; \; x = 71.9 \text{ ft}$

29. $\dfrac{80.0}{98.0} = \dfrac{x}{37.0} \Rightarrow x = 30.2 \text{ kg}$

33. $y = kx^2; \; 27 = k\left(3^2\right); \; k = 3; \; y = 3x^2$

37. $\dfrac{F_1}{F_2} = \dfrac{L_2}{L_1}; \; F_1 = 4.50 \text{ lb}, \; F_2 = 6.75 \text{ lb}, \; L_1 = 17.5 \text{ in.}$

$\dfrac{4.50}{6.75} = \dfrac{L_2}{17.5}$

$(6.75)L_2 = (4.50)(17.5)$

$L_2 = 11.7 \text{ in.}$

41. $R = kA$

$850 = k(900)$

$k = 0.94$

$R = 0.94 \text{ A}$

45. $F = \dfrac{k}{L}$

$250 = \dfrac{k}{22} \Rightarrow k = 5500$

$F = \dfrac{5500}{L}$

49. $v = k\sqrt{p}$

$120 = k\sqrt{100} \Rightarrow k = 12$

$v = 12\sqrt{p} = 12\sqrt{64}$

$v = 96 \text{ ft/s}$

53. $p = kA; \; 30.0 = k(8.00); \; k = 3.75\dfrac{\text{hp}}{\text{in.}^2}$

$p = 3.75 \text{ A}; \; p = 3.75(6.00) = 22.5 \text{ hp}$

57. $\dfrac{\left(\log 8000\right)^2}{\left(\log 2000\right)^2} = 1.4$ times longer to sort 8000 numbers.

61. $f = \dfrac{v}{\lambda};$

$v = f\lambda = 90.0 \times 10^6 (3.29)$

$= 299 \times 10^6$

$= 2.99 \times 10^8 \text{ m/s}$

65. $r = k\sqrt{\lambda}$

$3.56 = k\sqrt{575} \Rightarrow k = \dfrac{3.56}{\sqrt{575}}$

$r = k\sqrt{\lambda} = \dfrac{3.56}{\sqrt{575}}\left(\sqrt{483}\right)$

$r = 3.26 \text{ cm}$

69. $d = kv^2$

$52 = k(32)^2 \Rightarrow k = \dfrac{52}{32^2}$

$d = \dfrac{52}{32^2} v^2 = \dfrac{52}{32^2}(55)^2$

$d = 150$ ft

73. $A = k(1+r)^n = k(1+0.04)^n = k(1.04)^n$

$185.03 = k(1.04)^{10} \Rightarrow k = 125$

$A = 125(1.04)^n$

$A = 125(1.04)^0 = \$125,$ value of original

investment

$A = 125(1.04)^{40} = \$600.13,$ current value of

investment

77. $I = k\cos^2\theta$

$0.025 = k\cos^2 12.0° \Rightarrow k = \dfrac{0.025}{\cos^2 12.0°}$

$I = \dfrac{0.025}{\cos^2 12.0°}\cos^2\theta = \dfrac{0.025}{\cos^2 12.0°}\cos^2 20.0°$

$I = 0.023$ W/m^2

Chapter 19

SEQUENCES AND THE BINOMIAL THEOREM

19.1   Arithmetic Sequences

1. $a_1 = 5$, $a_{32} = -88$, $n = 32$

$-88 = 5 + (32-1)d$

$31d = -93$

$d = -3$

5. 4, 6, 8, 10, 12   **9.** $a_8 = 1 + (8-1) = 22$

13. $a_{80} = -0.7 + (80-1)0.4 = 30.9$

17. $64 = 4 + (n-1)4 \Rightarrow n = 16$

$S_{16} = \dfrac{16}{2}(4 + 64) = 544$

21. $d = 13 - 5 = 8$

$45 = 8n - 3$

$n = 6$

$S_6 = \dfrac{6}{2}(5 + 45) = 150$

25. $a_{30} = a_1 + (29)(3) = a_1 + 87$

$1875 = \dfrac{30}{2}(a_1 + a_1 + 87)$

$125 = 2a_1 + 87$

$a_1 = 19;\ a_{30} = 106$

29. $a_3 = a_1 + 2d \Rightarrow 6k = a_1 + 2(0.5k)$

$a_1 = 5k$

$a_n = 5k + (n-1)(0.5k)$

$S_n = \dfrac{n}{2}\Big[5k + \big(5k + (n-1)(0.5k)\big)\Big]$

$104k = \dfrac{n}{2}(5k + 5k + 0.5kn - 0.5k)$

$208k = n(9.5k + 0.5kn)$

33. $d = \dfrac{720 - 560}{10 - 6} = 40$, $a_6 = 560 = a_1 + (5)(40)$

$a_1 = 360$

$S_{10} = 5(360 + 720) = 5400$

37. $a_{n+1} = a_n + 2 \Rightarrow d = 2$

$a_n = a_1 + (n-1)d = 3 + (n-1)2 = 2n + 1$

41. a_1, b, c, a_4, a_5

$b + d = c \Rightarrow d = c - b$

$a_1 = b - d = b - (c - b) = 2b - c$

$a_4 = c + d = c + (c - b) = 2c - b$

$a_5 = c + 2d = c + 2(c - b) = 3c - 2b$

45. $3 - x$, $-x$, $\sqrt{9 - 2x}$; $9 - 2x \geq 0 \Rightarrow x \leq \dfrac{9}{2}$

$3 - x + d = -x$

$d = -3$

$-x + (-3) = \sqrt{9 - 2x}$

$x^2 + 6x + 9 = 9 - 2x$

$x^2 + 8x + 9 = 0$

$x(x + 8) = 0 \Rightarrow x = -8;\ x = 0$, reject

Check: $x = -8$

$3 - (-8)$, $-(-8)$, $\sqrt{9 - 2(-8)}$

11, 8, 5, A.S. with $d = -3$

Check: $x = 0$

$3 - 0$, -0, $\sqrt{9 - 2(0)}$

3, 0, 3 is not an A.S.

49. Area losses for next 8 years are: 500 m^2; 600 m^2, $700 \text{ m}^2, \ldots, a_8 = 500 + 7(100) = 1200 \text{ m}^2$

$$S_8 = \frac{8}{2}(500 + 1200) = 6800 \text{ m}^2$$

Area in 8 years $= 9500 - 6800 = 2700 \text{ m}^2$

53. $a_1 = 12$, $d = 4$, $a_n = 300$

$$a_n = 12 + (n-1)4$$

$$300 = \frac{n}{2}(12 + a_n)$$

$$600 = n(12 + 8 + 4n)$$

$$600 = 12n + 8n + 4n^2$$

$$4n^2 + 20n - 600 = 0$$

Use the quadratic formula to solve for n:

$$n = \frac{-20 \pm \sqrt{400 - 4(4)(600)}}{2(4)} = \frac{-20 \pm 100}{8}$$

$$= 10 \text{ rows (ignore negative values)}$$

57. $1.85, 5.55, 9.25, 12.95, \cdots$ is an AS with

$$a_1 = 1.85, \ d = 3.7$$

$$S_{10} = \frac{10}{2}(1.85 + 1.85 + 9(3.7)) = 185 \text{ m}$$

61. $S_n = \frac{n}{2}(a_1 + a_n) = \frac{n}{2}\left[a_1 + (a_1 + (n-1))d\right]$

$$= \frac{n}{2}\left[2a_1 + (n-1)d\right]$$

19.2   Geometric Sequences

1. Find a_{10} for $a_1 = \sqrt{3}$, $r = \sqrt{3}$

$$a_n = a_1 r^{n-1}$$

$$a_{10} = a_1 r^{10-1} = \sqrt{3}\left(\sqrt{3}\right)^9$$

$$a_{10} = 243$$

5. $\dfrac{1}{6}, \dfrac{1}{6} \cdot 3, \dfrac{1}{6} \cdot 3^2, \dfrac{1}{6} \cdot 3^3, \dfrac{1}{6} \cdot 3^4$

$$\frac{1}{6}, \frac{1}{2}, \frac{3}{2}, \frac{9}{2}, \frac{27}{2}$$

9. $r = -25 \div 125 = -0.2$, $a_1 = 125$, $n = 7$

$$a_7 = 125(-0.2)^{7-1} = \frac{1}{125}$$

13. $10^{100}, -10^{98}, 10^{96}, \ldots n = 51$

$$r = \frac{-10^{98}}{10^{100}} = -10^{-2}$$

$$a_{51} = 10^{100} \cdot \left(-10^{-2}\right)^{50} = 1$$

17. $384, 192, 96, \cdots$

$$384 \cdot r = 192 \Rightarrow r = \frac{1}{2}$$

$$S_7 = \frac{384\left(1 - \left(\frac{1}{2}\right)^7\right)}{1 - \frac{1}{2}} = 762$$

21. $a_2 = \dfrac{1}{4} = \dfrac{1}{16}r \Rightarrow r = 4$

$$a_n = \left(\frac{1}{16}\right)(4)^{n-1} = \left(\frac{1}{16}\right)(4)^{n-1} = 64 \Rightarrow n = 6$$

$$S_6 = \frac{\frac{1}{16}(1 - 4^6)}{1 - 4} = \frac{\frac{1}{16}(1 - 4096)}{-3} = \frac{4095}{48} = \frac{1365}{16}$$

25. $27 = a_1 r^{4-1}$; $a_1 = \dfrac{27}{r^3}$

$$40 = a_1 \frac{\left(1 - r^4\right)}{1 - r}$$

$$= a_1 \frac{\left(1 + r^2\right)\left(1 + r\right)\left(1 - r\right)}{1 - r}$$

$$= a_1\left(1 + r^2\right)\left(1 + r\right)$$

Substitute a from first equation in second equation:

$40 = \dfrac{27}{r^3}\left(1+r^2\right)\left(1+r\right);\ 40r^3 = 27 + 27r + 27r^2 +$

$27r^3;$

$13r^3 - 27r^2 - 27r - 27 = 0$

Using synthetic division, 3 gives a remainder of

zero. Therefore, $r = 3,\ a_1\left(3^{4-1}\right);\ 27a_1 = 27;\ a_1 = 1$

29. $3,\ 3^{x+1},\ 3^{2x+1},\cdots$ is a G.S. since

$\dfrac{3^{x+1}}{3} = 3^x$

$\dfrac{3^{2x+1}}{3^{x+1}} = 3^x.$

$a_1 = 3,\ r = 3^x$

$a_{20} = 3\cdot\left(3^x\right)^{20-1} = 3^{19x+1}$

33. G.S: $2,\ 6,\ 2x+8,\cdots$

$6 = 2r \Rightarrow r = 3$

$6\cdot r = 2x + 8 \Rightarrow 6\cdot 3 = 2x + 8 \Rightarrow x = 5$

37. $S_n = \dfrac{a_1\left(1-r^n\right)}{1-r},\ r \neq 1$

$S_3 = 7a_1 = \dfrac{a_1\left(1-r^3\right)}{1-r} \Rightarrow 7 - 7r = 1 - r^3$

$r^3 - 7r + 6 = 0$

$\left(r+3\right)\left(r-2\right)\left(r-1\right) = 0$

$r = -3,\ r = 2,\ r = 1$ reject since $r \neq 1$

41. $r = 1 - 0.082 = 0.918,\ a_1 = 100\%,\ n = 51$

$a_{51} = 100\left(0.918\right)^{50} = 1.4\%$

45. $a_1 = 250;\ n = 8\times 12 = 96$ months;

$r = 1 + 0.004 = 1.004,\ n = 97$

$a_{96} = 250\left(1.004\right)^{96} = \$366.76$

49. $\dfrac{100}{15} = \dfrac{20}{3}$

After 100 km, $0.88^{20/3} = 0.43$, 43% of signal

remains.

53. $a_1 = 0.01$ (dollars); $r = 2;\ n = 27\left(a_1 + 26\text{ more}\right)$

$a_{27} = 0.01\left(2^{27-1}\right) = 0.01\left(2^{26}\right) = \$671,088.64$

57. $a_n = a_1 r^{n-1} = a_1 r^n r^{-1} = a_1 r^{n/r}$

$a_1 r^n = a_n r$

$S_n = \dfrac{a_1\left(1-r^n\right)}{1-r} = \dfrac{a_1 - a_1 r^n}{1-r} = \dfrac{a_1 - r a_n}{1-r}$

61. A.S.: $8, x, y,\cdots \Rightarrow \left.\begin{array}{r}8+d=x\\x+d=y\end{array}\right\} \Rightarrow y = 2x - 8$

G.S: $x, y, 36,\cdots \Rightarrow \left.\begin{array}{r}rx=y\\ry=36\end{array}\right\} \Rightarrow y^2 = 36x$

$\left.\begin{array}{l}y^2 = 36x\\y = 2x-8\end{array}\right\} \Rightarrow x = 16,\ y = 24;\ x = 1,\ y = -6$

A.S: $8, 16, 24,\cdots$

G.S: $16, 24, 36,\cdots$

A.S: $8, 1, -6,\cdots$

G.S: $1, -6, 36,\cdots$

19.3 Infinite Geometric Series

1. Given the G.S. $4 + \dfrac{1}{2} + \dfrac{1}{16} + \dfrac{1}{128} + \cdots$ find the sum.

$a_1 = 4,\ r = \dfrac{1}{8}$

$S = \dfrac{a}{1-r} = \dfrac{4}{1-\frac{1}{8}}$

$S = \dfrac{32}{7}$

5. $S = \dfrac{a_1}{1-r}$

$\dfrac{25}{0.6} = \dfrac{0.5}{1-r}$

$r = \dfrac{1}{5}$

9. $a_1 = 4,\ r = \dfrac{7}{8},\ S = \dfrac{4}{1-\frac{7}{8}} = 32$

13. $a_1 = 2 + \sqrt{3}$, $r = \dfrac{1}{2+\sqrt{3}}$

$$S = \frac{2+\sqrt{3}}{1 - \frac{1}{2+\sqrt{3}}} = \frac{2+\sqrt{3}}{\frac{2+\sqrt{3}-1}{2+\sqrt{3}}}$$

$$= \frac{\left(2+\sqrt{3}\right)\left(2+\sqrt{3}\right)}{1+\sqrt{3}} = \frac{7+4\sqrt{3}}{1+\sqrt{3}} \times \frac{1-\sqrt{3}}{1-\sqrt{3}}$$

$$= \frac{7 - 7\sqrt{3} + 4\sqrt{3} - 12}{1-3}$$

$$= \frac{-5 - 3\sqrt{3}}{-2} = \frac{1}{2}\left(5 + 3\sqrt{3}\right)$$

17. $0.49999\ldots = 0.4 + 0.09 + 0.009 + 0.0009 + \cdots$

$$= 0.4 + \frac{0.09}{1 - \frac{1}{10}} = 0.5$$

21. $0.0273273\ldots = 0.0273 + 0.0000273 + \cdots$

$a_1 = 0.0273$, $r = 0.001$

$$S = \frac{0.0273}{1 - 0.001} = \frac{91}{3330}$$

25. $0.36666\ldots = 0.3 + 0.06666\ldots$

For the G.S. $0.066\ldots$, $a = 0.06$, $r = 0.1$

$$S = \frac{0.06}{1 - 0.1} = \frac{0.06}{0.9} = \frac{1}{15}$$

Therefore,

$$0.36666\ldots = \frac{3}{10} + \frac{1}{15} = \frac{11}{30}$$

29. $50,\, a_2,\, 2, \cdots \Rightarrow 50 \cdot r = a_2 r = 2 \Rightarrow$

$$\frac{a_2}{50} = \frac{2}{a_2} \Rightarrow a_2 = \pm 10$$

$50,\, 10,\, 2, \cdots$ has $r = \dfrac{1}{5} \Rightarrow S = \dfrac{50}{1 + \frac{1}{5}} = \dfrac{125}{2}$

$50,\, -10,\, 2, \cdots$ has $r = -\dfrac{1}{5} \Rightarrow S = \dfrac{50}{1 + \frac{1}{5}} = \dfrac{125}{3}$

33. $a_1 = 5.882$ g, $r = \dfrac{5.782}{5.882} = 0.9830$

$$S = \frac{5.882}{1 - 0.9830} = \frac{5.882}{0.0170} = 346 \text{ g}$$

37. $0.9^2 + 0.9^2 (0.1)^2 + 0.9^2 (0.1)^4 + \cdots$

$$= 0.9^2 \left(1 + 0.1^2 + 0.1^4 + \cdots\right)$$

$$= 0.9^2 \left(\frac{1}{1 - 0.1^2}\right) = 0.818181\cdots$$

81.8% of light is transmitted through the second pane.

19.4 The Binomial Theorem

1. $(2x+3)^5 = (2x)^5 + 5(2)^4(3) + \dfrac{5(4)}{2!}(2x)^3(3)^2$

$$+ \frac{5(4)(3)}{3!}(2x)^2(3)^3$$

$$+ \frac{5(4)(3)(2)}{4!}(2x)^1(3)^4 + (3)^5$$

$(2x+3)^5 = 32x^5 + 240x^4 + 720x^3 + 1080x^2$

$$+ 810x + 243$$

5. $(2x-3)^4 = (2x)^4 + 4(2x)^3(-3) + \dfrac{4(3)}{2}(2x)^2(-3)^2$

$$+ \frac{4(3)(2)}{6}(2x)(-3)^3 + \frac{4(3)(2)(1)}{24}(-3)^4$$

$$= 16x^4 - 96x^3 + 216x^2 - 216x + 81$$

9. $\left(n+2\pi^3\right)^5 = n^5 + 5n^4\left(2\pi^3\right) + \dfrac{5(4)}{2!}n^3\left(2\pi^3\right)^2$

$$+ \frac{5(4)(3)}{3!}n^2\left(2\pi^3\right)^3$$

$$+ \frac{5(4)(3)(2)}{4!}n\left(2\pi^3\right)^4 + \left(2\pi^3\right)^5$$

$\left(n+2\pi^3\right)^5 = n^5 + 10\pi^3 n^4 + 40\pi^6 n^3 + 80\pi^9 n^2$

$$+ 80\pi^{12}n + 32\pi^{15}$$

13. From Pascal's triangle, the coefficients for $n = 4$ are 1, 4, 6, 4, 1.

$(5x-3)^4 = \left[5x + (-3)\right]^4$

$$= 1(5x)^4 + 4(5x)^3(-3) + 6(5x)^2(-3)^2$$

$$+ 4(5x)(-3)^3 + (-3)^4$$

$$= 625x^4 - 1500x^3 + 1350x^2 - 540x + 81$$

17. $(x+2)^{10} = x^{10} + 10x^9(2) + \dfrac{(10)(9)}{2}x^8(2)^2$

$+\dfrac{(10)(9)(8)}{6}x^7(2)^3 + \cdots$

$= x^{10} + 20x^9 + 180x^8 + 960x^7 + \cdots$

21. $\left(x^{1/2} - 4y\right)^{12} = \left(x^{1/2}\right)^{12} - 12\left(x^{1/2}\right)^{11}(4y)$

$+\dfrac{12 \cdot 11}{2!}\left(x^{1/2}\right)^{10}(4y)^2$

$-\dfrac{12 \cdot 11 \cdot 10}{3!}\left(x^{1/2}\right)^9(4y)^3$

$= x^6 - 48x^{11/2}y + 1056x^5y^2$

$-14{,}080x^{9/2}y^3 + \cdots$

25. $(1.05)^6 = (1+0.05)^6$

$= 1^6 + 6(1)^5(0.05) + \dfrac{6(5)}{2!}(1)^4(0.05)^2 = 1.3375$

$= 1.338$ to 3 decimal places using three terms

29. $(1+x)^8 = 1 + 8x + \dfrac{8(7)}{2}x^2 + \dfrac{8(7)(6)}{6}x^3 + \cdots$

$= 1 + 8x + 28x^2 + 56x^3 + \cdots$

33. $\sqrt{1+x} = (1+x)^{1/2} = 1 + \dfrac{1}{2}x + \dfrac{\frac{1}{2}\left(-\frac{1}{2}\right)}{2}x^2$

$+\dfrac{\frac{1}{2}\left(-\frac{1}{2}\right)\left(-\frac{3}{2}\right)}{6}x^3 + \cdots$

$= 1 + \dfrac{1}{2}x - \dfrac{1}{8}x^2 + \dfrac{1}{16}x^3 \cdots$

37. (a) $17! + 4! = 3.557 \times 10^{14}$

(b) $21! = 5.109 \times 10^{19}$

(c) $17! \times 4! = 8.536 \times 10^{15}$

(d) $68! = 2.480 \times 10^{96}$

```
17!+4!
    3.556874281E14
21!
    5.109094217E19
17!*4!
    8.536498274E15
■
```

```
68!
    2.480035542E96
■
```

41. The term involving b^5 will be the sixth term.

$r = 5, \ n = 8$

The sixth term is $\dfrac{8(7)(6)(5)(4)}{5(4)(3)(2)}a^3b^5 = 56a^3b^5$.

45. $(2x-1)^3 + 3(2x-1)^2(3-2x) + 3(2x-1)(3-2x)^2$

$+(3-2x)^3$

$= \left((2x-1) + (3-2x)\right)^3 = (2x-1+3-2x) = (2)^3 = 8$

49.

n	coefficients							sum of coefficients = 0
1				1	-1			
2			1	-2	1			
3		1	-3	3	-1			
4	1	-4	6	-4	1			
5	1	-5	10	-10	5	-1		
6	1	-6	15	-20	15	-6	1	
7	1	-7	21	-35	35	-21	7	-1

53. $V = A(1-r)^5$; expand $(1-r)^5$

$1^5 + 5(1)^4(-r) + \dfrac{5(4)}{2}(1)^3(-r)^2$

$+\dfrac{5(4)(3)}{6}(1)^2(-r)^3 \dfrac{5(4)(3)(2)}{24}(1)(-r)^4 + (-r)^5$

$= 1 - 5r + 10r^2 - 10r^3 + 5r^4 - r^5$

$A(1-r)^5 = A\left(1 - 5r + 10r^2 - 10r^3 + 5r^4 - r^5\right)$

57. $\dfrac{k}{(r+h)^2} - \dfrac{k}{r^2} = \dfrac{k}{r^2\left(1+\frac{h}{r}\right)} - \dfrac{k}{r^2}$

$\dfrac{k}{(r+h)^2} - \dfrac{k}{r^2} = \dfrac{k}{r^2}\left[\left(1+\dfrac{h}{r}\right)^{-2} - 1\right]$

$= \dfrac{k}{r^2}\left[1 - 2\left(\dfrac{h}{r}\right)\right.$

$\left. + \dfrac{-2(-2-1)}{2!}\left(\dfrac{h}{r}\right)^2 - 1\right]$

$r = \dfrac{k}{r^2}\left[-\dfrac{2h}{r} + 3\left(\dfrac{h}{r}\right)^2\right]$

$= -\dfrac{2kh}{r^3} + \dfrac{3kh^2}{r^4}$

Chapter 19   Review Exercises

1. $d = 5; \; a_n = 1 + (17-1)5 = 1 + 80 = 81$

5. $d = 8 - 3.5 = 4.5; \; a_{16} = -1 + (16-1)(4.5) = 66.5$

9. $S_n = \dfrac{15}{2}(-4 + 17) = \dfrac{15}{2}(13) = \dfrac{195}{2}$

13. $a_n = 17 + (9-1)(-2) = 17 - 16 = 1$

$S_n = \dfrac{9}{2}(1+17) = 81$

17. $S_n = \dfrac{n}{2}(a_1 + a_n); \; a_1 = 80, \; a_n = -25, \; S_n = 220$

$220 = \dfrac{n}{2}(80 - 25) \qquad\qquad a_n = a_1 + (n-1)d$

$440 = n(80 - 25) = 55n \qquad -25 = 80 + (8-1)d$

$\qquad\qquad\qquad\qquad\qquad = 80 + 7d$

$n = \dfrac{440}{55} = 8 \qquad\qquad\qquad d = \dfrac{-105}{7} = -15$

21. $S_{12} = \dfrac{12}{2}(-1 + 32) = 6(31) = 186$

25. $r = \dfrac{6}{9} = \dfrac{2}{3}; \; S = \dfrac{a_1}{1-r} = \dfrac{0.9}{1-\frac{2}{3}} = 2.7$

29. $0.030303\ldots = 0.03 + 0.0003 + 0.000003\ldots$

$a_a = 0.03; \; r = 0.01$

$S = \dfrac{0.03}{1 - 0.01} = \dfrac{0.03}{0.99} = \dfrac{3}{99} = \dfrac{1}{33}$

33. $(x-5)^4 = [x + (-5)]^4$

$= x^4 + 4x^3(-5) + \dfrac{4(3)x^2(-5)^2}{2}$

$+ \dfrac{4(3)(2)}{3(2)}x(-5)^3 + (-5)^4$

$= x^4 - 20x^3 + 150x^2 - 500x + 625$

37. $(a + 3e)^{10} = a^{10} + 10a^{10-1}(3e)$

$+ \dfrac{10(10-1)}{2!}a^{10-2}(3e)^2$

$+ \dfrac{10(10-1)(10-2)}{3!}a^{10-3}(3e)^3 + \cdots$

$= a^{10} + 10a^9(3e) + 45a^8(9e^2)$

$+ 120a^7(27e^3) + \cdots$

$= a^{10} + 30a^9e + 405a^8e^2 + 3240a^7e^3 + \cdots$

41. $(1+x)^{12} = 1 + 12x + \dfrac{12(12-1)}{2!}x^2$

$+ \dfrac{12(12-2)}{3!}x^3 + \cdots$

$= 1 + 12x + 66x^2 + 220x^3 + \cdots$

45. $\left[1 + (-a^2)\right]^{1/2} = 1 + \dfrac{1}{2}(-a^2) + \dfrac{\frac{1}{2}\left(-\frac{1}{2}\right)(-a^2)^2}{2}$

$+ \dfrac{\frac{1}{2}\left(-\frac{1}{2}\right)\left(-\frac{3}{2}\right)(-a^2)}{3(2)}$

$= 1 - \dfrac{1}{2}a^2 - \dfrac{1}{8}a^4 - \dfrac{1}{16}a^6 - \cdots$

49. $a = 2, \; d = 2, \; n = 1000$

$a_n = 2 + (1000-1)2 = 2 + 999(2) = 2000$

$S = \dfrac{1000}{2}(2 + 2000) = 1{,}001{,}000$

53. Let a_3, a_4, a_5, a_6, a_7 be the GS 6, a_4, 9, a_6, 12

Suppose $r =$ common ratio, the GS is 6, $6r$, $6r^2$,

$6r^3$, $6r^4$ which gives

$9 = 6r^2$ and $6r^4 = 12$

$r^2 = \dfrac{9}{6}$ $\qquad$ $r^4 = 2$

$\qquad\qquad$ $r^2 = \sqrt{2}$

Since $\frac{9}{6} \neq \sqrt{2}$, 6, a_4, 9, a_6, 12 cannot be a GS.

57. $(1.06)^{-6} = (1+0.06)^{-6} = 1^{-6} + \dfrac{-6}{1} 1^{-7} (0.06)^1$

$\qquad + \dfrac{-6(-7)}{1 \cdot 2} 1^{-8} (0.06)^2$

$\qquad = 0.7156$

61. $\big((3x-2)+(5-3x)\big)^4 = (3x-2+5-3x)^4$

$\qquad\qquad\qquad\qquad = 3^4 = 81$

65. 24, 22, 20, $\cdots$ is an AS with $a_1 = 24$, $d = -=2$

$a_n = a_1 + (n-1)d \Rightarrow 24 + (n-1)(n-2) = 4 \Rightarrow$

$n = 11$

69. There are 15 pieces with a distance between

$d = \dfrac{224.0}{14}$.

Total length of pieces $= 10.0(15) + \dfrac{d}{\tan 84.8^\circ}$

$\qquad\qquad + \dfrac{2d}{\tan 84.8^\circ} + \cdots + \dfrac{14d}{\tan 84.8^\circ}$

$= 10.0(15) + \dfrac{d}{\tan 84.8^\circ} \cdot$

$(1+2+3+\cdots+14) = 302.9$ in.

73. 0.0040, 0.0040(2), $0.0040(2)^2$, $\cdots$, $0.0040(2)^{40}$

$0.0040(2)^{40} = 4.4 \times 10^9$ in. $= 69,000$ mi

77. $10 + 10(0.9) + 10(0.9)^2 + \cdots = \dfrac{10}{1-0.9} = 100$ cm

81. $250(0.4)^4 = \$6.40$

85. Let $x = \dfrac{a-1}{2} m^2$ and $y = \dfrac{a}{a-1}$

$(1+x)^y = 1 + yx + \dfrac{y(y-1)}{2} x^2 \ldots$ (3 terms)

$= 1 + \left(\dfrac{a}{a-1}\right)\left(\dfrac{a-1}{2}m^2\right) + \dfrac{\left(\frac{a}{a-1}\right)\left(\frac{a}{a-1}-1\right)}{2}\left(\dfrac{a-1}{2}m^2\right)^2$

$= 1 + \dfrac{a}{2}m^2 + \dfrac{\left(\frac{a}{a-1}\right)\left(\frac{1}{a-1}\right)}{2}\left(\dfrac{(a-1)^2}{2^2}m^4\right)$

$= 1 + \dfrac{a}{2}m^2 + \dfrac{a}{2^3}m^4 = 1 + \dfrac{1}{2}am^2 + \dfrac{1}{8}am^4$

89. Let $a_1 = 1000$ units. If 75% are killed, 25% remain

after the first application.

$r = \dfrac{a_2}{a_1} = \dfrac{250}{1000} = 0.25$

If 99% are destroyed, 0.1% remain.

$0.001 \times 1000 = 1$ insect remains.

$\qquad 1 = 1000(0.25)^n$

$\qquad 0.001 = 0.25^n$

$\log 0.001 = \log 0.25^n$

$\log 0.001 = n \log 0.25$

$\qquad n = \dfrac{\log 0.001}{\log 0.25} = 5$ applications

93. For odd n, the middle term of an arithmetic

sequence is $a + \dfrac{n-1}{2}d$.

$S_n = a + (a+d) + (a+2d) + \cdots + (a+(n-1)d)$,

$\qquad n$ odd

$S_n = \dfrac{a + (a+(n-1)d)}{2} n$

$\dfrac{S_n}{n} = \dfrac{a + a + (n-1)d}{2} = \dfrac{2a + (n-1)d}{2}$

$\qquad = a + \dfrac{(n-1)d}{2}$

$\dfrac{S_n}{n} =$ middle term.

97. Let $A =$ initial deposit,

$t =$ time of a compounding period.

$V = A + Art = A(1+rt)$, after one compounding

period

$V = A(1+rt) + A(1+rt)rt = A(1+rt)^2$, after two

compounding periods

$V = A(1 + rt)^n$, at the end of one year.

For $A = \$1000$ and $r = 0.1 = 10\%$

$V = 1000(1 + 0.1t)^n$.

As n, the number of compounding periods, increases t, the length of a compounding period decreases. The product $nt = 1$. For example, if the compounding is done monthly $n = 12$ and $t = \frac{1}{12}$, so that $nt = 12 \cdot \frac{1}{2} = 1$.

Thus, $V = 1000\left(1 + 0.1 \cdot \frac{1}{n}\right)^n$ which will increase as n increases the more compounding periods the more interest. Write V as

$$V = 1000\left(1 + \frac{0.1}{n}\right)^{\frac{0.1}{0.1}n} = 1000\left[\left(1 + \frac{0.1}{n}\right)^{\frac{1}{0.1}n}\right]^{0.1}$$

As n increases $\left(1 + \frac{0.1}{n}\right)^{\frac{1}{0.1}n}$ approaches e. Hence,

the maximum value is $V_{max} = 1000 \cdot e^{0.1} = \$1105.17$

as compared with $V = 1000\left(1 + 0.1 \cdot \frac{1}{12}\right)^{12}$

$= \$1104.71$ for monthly compounding.

ADDITIONAL TOPICS IN TRIGONOMETRY

20.1 Fundamental Trigonometric Identities

1. $\sin x = \dfrac{\tan x}{\sec x}$

$= \dfrac{\frac{\sin s}{\cos x}}{\frac{1}{\cos x}}$

$= \dfrac{\sin x}{\cos x} = \cdot \dfrac{\cos x}{21}$

$= \sin x$

5. Verify $\sin^2 \theta + \cos^2 \theta = 1$ for $\theta = \dfrac{4\pi}{3}$

$\left(\sin \dfrac{4\pi}{3} \right)^2 = \left(-\dfrac{1}{2}\sqrt{3} \right)^2 = \dfrac{3}{4}$

$\left(\cos \dfrac{4\pi}{3} \right)^2 = \left(-\dfrac{1}{2} \right)^2 = \dfrac{1}{4}$

$\dfrac{3}{4} + \dfrac{1}{4} = 1$

9. $\cos\theta \cot\theta \left(\sec\theta - 2\tan\theta \right)$

$= \cos\theta \dfrac{\cos\theta}{\sin\theta} \left(\dfrac{1}{\cos\theta} - 2\dfrac{\sin\theta}{\cos\theta} \right)$

$= \cot\theta - 2\cos\theta$

13. $\sin x + \sin x \tan^2 x = \sin x \left(1 + \tan^2 x \right) = \sin x \sec^2 x$

$= \sin x \cdot \dfrac{1}{\cos x} \cdot \sec x = \tan x \sec x$

17. $\csc^4 y - 1 = \left(\csc^2 y + 1 \right) \left(\csc^2 y - 1 \right)$

$= \left(\csc^2 y + 1 \right) \left(\cot^2 y \right)$

21. $\sin x \sec x = \sin x \dfrac{1}{\cos x} = \dfrac{\sin x}{\cos x} = \tan x$

25. $\sin x \left(1 + \cot^2 x \right) = \sin x \left(\csc^2 x \right) = \sin x \left(\dfrac{1}{\sin^2 x} \right)$

$= \dfrac{1}{\sin x} = \csc x$

29. $\cot\theta \sec^2\theta - \dfrac{1}{\tan\theta} = \dfrac{\cos\theta}{\sin\theta}\dfrac{1}{\cos^2\theta} - \dfrac{\cos\theta}{\sin\theta}$

$= \dfrac{1}{\sin\theta\cos\theta} - \dfrac{\cos\theta}{\sin\theta}$

$= \dfrac{1-\cos^2\theta}{\sin\theta\cos\theta} = \dfrac{\sin^2\theta}{\sin\theta\cos\theta}$

$= \dfrac{\sin\theta}{\cos\theta} = \tan\theta$

33. $\cos^2 x - \sin^2 x = 1 - \sin^2 x - \sin^2 x = 1 - 2\sin^2 x$

37. $2\sin^4 x - 3\sin^2 x + 1 = \left(2\sin^2 x - 1 \right)\left(\sin^2 x - 1 \right)$

$= \left(2\sin^2 x - 1 \right)\left(-\cos^2 x \right)$

$= \cos^2 x \left(1 - 2\sin^2 x \right)$

41. $1 + \sin^2 x + \sin^4 x \cdots + = $ infinite series

$a_1 = 1, \; r = \sin^2 x$

$S = \dfrac{1}{1 - \sin^2 x} = \dfrac{1}{\cos^2 x} = \sec^2 x$

45. $\cot x \left(\sec x - \cos x \right) = \dfrac{\cos x}{\sin x} \cdot \dfrac{1}{\cos x} - \dfrac{\cos x}{\sin x} \cdot \cos x$

$= \dfrac{1}{\sin x} - \dfrac{\cos^2 x}{\sin x}$

$= \dfrac{1 - \cos^2 x}{\sin x}$

$= \dfrac{\sin^2 x}{\sin x} = \sin x$

49. $\dfrac{\cos x + \sin x}{1 + \tan x} = \dfrac{\cos x + \sin x}{1 + \frac{\sin x}{\cos x}} \cdot \dfrac{\cos x}{\cos x}$

$= \dfrac{(\cos x + \sin x) \cdot \cos x}{\cos x + \sin x} = \cos x$

53. $\tan x = \dfrac{\sin x}{\cos x} = \dfrac{\frac{1}{\csc x}}{\frac{\sqrt{\csc^2 x - 1}}{\csc x}} = \dfrac{1}{\sqrt{\csc^2 x - 1}}$

57.

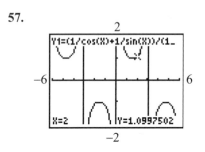

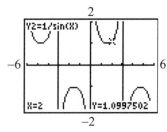

61. No. $\dfrac{2\cos^2 x - 1}{\sin x \cos x} \neq \tan x - \cot x$

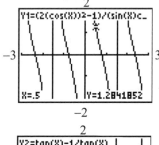

65. $l = a\csc\theta + a\sec\theta$

$= a(\csc\theta + \sec\theta)$

$= a\left(\dfrac{1}{\sin\theta} + \dfrac{1}{\cos\theta}\right)$

$= a\left(\dfrac{1}{\sin\theta} + \dfrac{\sin\theta}{\cos\theta} \cdot \dfrac{1}{\sin\theta}\right)$

$= a\left(\dfrac{1}{\sin\theta} + \dfrac{\tan\theta}{\sin\theta}\right)$

69. $\sin^2 x(1 - \sec^2 s) + \cos^2 x(1 + \sec^4 x)$

$= \sin^2 x - \sin^2 x \sec^2 x + \cos^2 x + \cos^2 x \sec^4 x$

$= \sin^2 x - \dfrac{\sin^2 x}{\cos^2 x} + \cos^2 x + \dfrac{\cos^2 x}{\cos^4 x}$

$= \sin^2 x - \tan^2 x + \cos^2 x + \sec^2 x$

$= 1 - \tan^2 x + \sec^2 x$

$= 1 - (\sec^2 x - 1) + \sec^2 x$

$= 1 - \sec^2 x + 1 + \sec^2 x = 2$

73. $\sec^2\theta + \csc^2\theta = \left(\dfrac{r}{x}\right)^2 + \left(\dfrac{r}{y}\right)^2$

$= \dfrac{r^2}{x^2} + \dfrac{r^2}{y^2}$

$= \dfrac{r^2 y^2 + r^2 x^2}{x^2 y^2}$

$= \dfrac{r^2(y^2 + x^2)}{x^2 y^2}$

$= \dfrac{r^4}{x^2 y^2} = \sec^2\theta \csc^2\theta$

77. $x = 2\tan\theta; \ \sqrt{4 + (2\tan\theta)^2} = \sqrt{4 + 4\tan^2\theta}$

$\sqrt{4(1 + \tan^2\theta)} = \sqrt{4 + 4\tan^2\theta}$

$= \sqrt{4(1 + \tan^2\theta)}$

$= \sqrt{4\sec^2\theta}$

$= 2\sec\theta$

20.2   The Sum and Difference Formulas

1. $\sin\alpha = \dfrac{12}{13}$ (α in first quadrant) and $\sin\beta = -\dfrac{3}{5}$

for β in third quadrant.

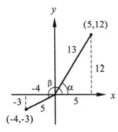

$\cos(\alpha+\beta) = \cos\alpha\cos\beta - \sin\alpha\sin\beta$

$$= \frac{15}{13}\cdot\frac{-4}{5} - \frac{12}{13}\cdot\frac{-3}{5} = \frac{16}{65}$$

5.   Given: $15° = 60° - 45°$

$\cos(\alpha-\beta) = \cos\alpha\cos\beta + \sin\alpha\sin\beta$

$\cos 15° = \cos(60° - 45°)$

$$= \cos 60°\cos 45° + \sin 60°\sin 45°$$

$$= \frac{1}{2}\times\frac{\sqrt{2}}{2} + \frac{\sqrt{3}}{2}\times\frac{\sqrt{2}}{2}$$

$$= \frac{\sqrt{2}}{4} + \frac{\sqrt{6}}{4} = \frac{\sqrt{2}+\sqrt{6}}{4} = 0.9659$$

9. Using the results of exercise 7:

$\cos\alpha = \dfrac{3}{5}$

$\cos\beta = -\dfrac{12}{13}$

$\sin\alpha = \dfrac{4}{5}$

$\sin\beta = \dfrac{5}{13}$

$\cos(\alpha+\beta) = \cos\alpha\cos\beta - \sin\alpha\sin\beta$

$$= \frac{3}{5}\left(-\frac{12}{13}\right) - \frac{4}{5}\left(\frac{5}{13}\right)$$

$$= \frac{-36-20}{65} = -\frac{56}{65}$$

13. $\cos\pi\cos x + \sin\pi\sin x = (-1)\cos x + (0)\sin x$

$$= -\cos x$$

17. $\tan(x-\pi) = \dfrac{\tan x + \tan\pi}{1 - \tan x \times \tan\pi} = \tan x$

21. $\sin 122°\cos 32° - \cos 122°\sin 32°$ is of the form

$\sin\alpha\cos\beta - \cos\alpha\sin\beta$, where $\alpha = 122°$ and

$\beta = 32°$

$\sin\alpha\cos\beta - \cos\alpha\sin\beta = \sin(\alpha-\beta)$ so

$\sin 122°\cos 32° - \cos 122°\sin 32°$

$$= \sin(122° - 32°)$$

$$= \sin 90° = 1$$

25. $\sin(x+y)\sin(x-y)$

$$= (\sin x\cos y + \cos x\sin y)(\sin x\cos y - \cos x\sin y)$$

$$= \sin^2 x\cos^2 y - \cos^2 x\sin^2 y$$

$$= \sin^2 x(1-\sin^2 y) - (1-\sin^2 x)(\sin^2 y)$$

$$= \sin^2 x - \sin^2 x\sin^2 y - \sin^2 y + \sin^2 x\sin^2 y$$

$$= \sin^2 x - \sin^2 y$$

29.

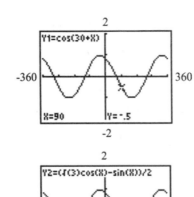

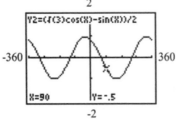

33. $\tan(\alpha \pm \beta)$

$$= \frac{\sin(\alpha \pm \beta)}{\cos(\alpha \pm \beta)} = \frac{\sin\alpha\cos\beta \pm \cos\alpha\sin\beta}{\cos\alpha\cos\beta \mp \sin\alpha\sin\beta}$$

$\left(\text{divide numerator and denominator by}\right.$

$\left.\cos\alpha\cos\beta\right)$

$$= \frac{\dfrac{\sin\alpha\cos\beta}{\cos\alpha\cos\beta} \pm \dfrac{\cos\alpha\sin\beta}{\cos\alpha\cos\beta}}{\dfrac{\cos\alpha os\beta}{\cos\alpha\cos\beta} \mp \dfrac{\sin\alpha\sin\beta}{\cos\alpha\cos\beta}}$$

$$= \frac{\dfrac{\sin\alpha}{\cos\alpha} \pm \dfrac{\sin\beta}{\cos\beta}}{1 \mp \dfrac{\sin\alpha}{\cos\alpha} \times \dfrac{\sin\beta}{\cos\beta}}$$

$$= \frac{\tan\alpha \pm \tan\beta}{1 \mp \tan\alpha\tan\beta}$$

37. $\alpha + \beta = x;\ \alpha - \beta = y;\ \alpha = \dfrac{1}{2}(x+y);\ \beta = \dfrac{1}{2}(x-y)$

$\sin x + \sin y$

$$= \sin(\alpha + \beta) + \sin(\alpha - \beta)$$

$$= 2\left[\frac{1}{2}\sin(\alpha + \beta) + \frac{1}{2}\sin(\alpha - \beta)\right]$$

$$= 2\sin\alpha\cos\beta = 2\sin\frac{1}{2}(x+y)\cos\frac{1}{2}(x-y)$$

41. $\sin(x + 30^\circ)\cos x - \cos(x + 30^\circ)\sin x$

$$= \left(\sin x\cos 30^\circ + \sin 30^\circ\cos x\right)\cos x$$

$$\quad -\left(\cos x\cos 30^\circ - \sin x\,\sin 30^\circ\right)\sin x$$

$$= \frac{\sqrt{3}}{2}\sin x\cos x + \frac{1}{2}\cos^2 x - \frac{\sqrt{3}}{2}\sin x\cos x + \frac{1}{2}\sin^2 x$$

$$= \frac{1}{2}\left(\cos^2 x + \sin^2 x\right) = \frac{1}{2}$$

45. $\sin 75^\circ = \sin\left(30^\circ + 45^\circ\right)$

$$= \sin 30^\circ\cos 45^\circ + \sin 45^\circ\cos 30^\circ$$

$$= \frac{1}{2}\cdot\frac{\sqrt{2}}{2} + \frac{\sqrt{2}}{2}\cdot\frac{\sqrt{3}}{2}$$

$$= \frac{\sqrt{2} + \sqrt{6}}{4}$$

$$\sin 75^\circ = \sin\left(135^\circ - 60^\circ\right)$$

$$= \sin 135^\circ\cos 60^\circ - \sin 60^\circ\cos 135^\circ$$

$$= \frac{\sqrt{2}}{2}\cdot\frac{1}{2} - \frac{\sqrt{3}}{2}\cdot\left(-\frac{\sqrt{2}}{2}\right)$$

$$= \frac{\sqrt{2} + \sqrt{6}}{4}$$

49. $V = 20\sqrt{2}\sin\left(120\pi t + \pi/4\right)$

$$= 20\sqrt{2}\left(\sin 120\pi t\cos\pi/4 + \sin\pi/4\cos 120\pi t\right)$$

$$= 20\sqrt{2}\left(\tfrac{1}{\sqrt{2}}\right)\left(\sin 120\pi t + \cos 120\pi t\right)$$

$$= 20\sin 120\pi t + 20\cos 120\pi t$$

$$= V_1 + V_2$$

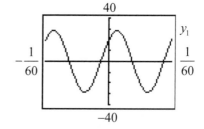

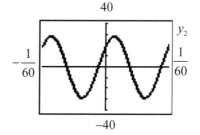

53.

$$\tan\alpha\left(R+\cos\beta\right)=\sin\beta$$

$$R\tan\alpha+\tan\alpha\cos\beta=\sin\beta$$

$$R\tan\alpha=\sin\beta-\tan\alpha\cos\beta$$

$$R=\frac{\sin\beta-\tan\alpha\cos\beta}{\tan\alpha}=\frac{\sin\beta}{\tan\alpha}-\frac{\sin\alpha\cos\beta}{\cos\alpha\tan\alpha}$$

$$=\frac{\sin\beta\cos\alpha-\sin\alpha\cos\beta}{\cos\alpha\tan\alpha}$$

$$=\frac{\sin\left(\beta-\alpha\right)}{\cos\alpha\frac{\sin\alpha}{\cos\alpha}}$$

$$=\frac{\sin\left(\beta-\alpha\right)}{\sin\alpha}$$

20.3   Double-Angle Formulas

1. If $\alpha=\dfrac{\pi}{3}$,

$$\tan\frac{2\pi}{3}=\tan\left(2\cdot\frac{\pi}{3}\right)=\frac{2\tan\frac{\pi}{3}}{1-\tan^2\frac{\pi}{3}}$$

$$=\frac{2\left(\sqrt{3}\right)}{1-\left(\sqrt{3}\right)^2}$$

$$=-\sqrt{3}$$

5. $60°=2\left(30°\right);\ \sin 2\alpha=2\sin\alpha\cos\alpha$

$$\sin 2\left(30°\right)=2\sin 30°\cos 30°$$

$$=2\left(\frac{1}{2}\right)\left(\frac{\sqrt{3}}{2}\right)$$

$$=\frac{\sqrt{3}}{2}$$

9. $\sin 258°=-0.9781476$

$$\sin 258°=\sin 2\left(129°\right)$$

$$=2\sin 129°\cos 129°$$

$$=-0.9781476$$

13. $\tan\dfrac{2\pi}{5}=\dfrac{2\tan\frac{\pi}{5}}{1-\tan^2\frac{\pi}{5}}=3.0777$

17. $\sin x=0.5\ (\text{QII})\Rightarrow x=\dfrac{5\pi}{6}$

$$\tan 2x=\frac{2\tan x}{1-\tan^2 x}=\frac{2\tan\frac{5\pi}{6}}{1-\tan^2\frac{5\pi}{6}}=-\sqrt{3}$$

21. $1-2\sin^2 4x=\cos 2\left(4x\right)=\cos 8x$

25. $8\sin^2 2x-4=4\left(2\sin^2 2x-1\right)$

$$=-4\left(1-2\sin^2 2x\right)$$

$$=-4\cos 2\left(2x\right)$$

$$=-4\cos 4x$$

29. $\dfrac{\sin 3x}{\sin x}-\dfrac{\cos 3x}{\cos x}=\dfrac{\sin 3x\cos x-\sin x\cos 3x}{\sin x\cos x}$

$$=\frac{\sin\left(3x-x\right)}{\sin x\cos x}$$

$$=\frac{\sin 2x}{\sin x\cos x}$$

$$=\frac{2\sin x\cos x}{\sin x\cos x}=2$$

33. $\dfrac{\cos x-\tan x\sin x}{\sec x}=\dfrac{\cos x-\frac{\sin x\sin x}{\cos x}}{\frac{1}{\cos x}}$

$$=\frac{\cos^2 x-\sin^2 x}{\cos x}\times\frac{\cos x}{1}$$

$$=\cos^2 x-\sin^2 x=\cos 2x$$

37. $1-\cos 2\theta=1-\left(1-2\sin^2\theta\right)=2\sin^2\theta$

$$=\frac{2}{\csc^2\theta}$$

$$=\frac{2}{1+\cot^2\theta}$$

41. Both graphs are the same.

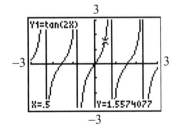

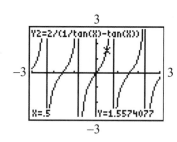

45. $\sin 3x$

$$= \sin(2x + x) = \sin 2x \cos x + \cos 2x \sin x$$

$$= (2 \sin x \cos x)(\cos x) + (\cos^2 x - \sin^2 x)(\sin x)$$

$$= 2 \sin x \cos^2 x + \sin x \cos^2 x - \sin^3 x$$

$$= 3 \sin x \cos^2 x - \sin^3 x$$

$$= 3 \sin x (1 - \sin^2 x) - \sin^3 x$$

$$= 3 \sin x - 4 \sin^3 x$$

49. $\cos 2x + \sin 2x \tan x = \cos^2 x - \sin^2 x + 2 \sin x \cos x$

$$\cdot \frac{\sin x}{\cos x}$$

$$= \cos^2 x - \sin^2 x + 2 \sin^2 x$$

$$= \cos^2 x + \sin^2 x = 1$$

53. $y = 4 \sin x \cos x = 2(2 \sin x \cos x)$

$$y = 2 \sin 2x$$

$$A = 2, \text{ period } = \frac{\pi}{2} = \pi$$

57. Let $\alpha = \beta = \theta$.

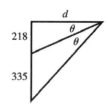

$$\tan \theta = \frac{218}{d}$$

$$\tan 2\theta = \frac{553}{d}$$

$$= \frac{2 \tan \theta}{1 - \tan^2 \theta}$$

$$\frac{553}{d} = \frac{2 \cdot \frac{218}{d}}{1 - \left(\frac{218}{d}\right)^2}$$

$$1 - \left(\frac{218}{d}\right)^2 = \frac{436}{553}$$

$$\left(\frac{218}{d}\right)^2 = 1 - \frac{436}{553}$$

$$\frac{218}{d} = \sqrt{1 - \frac{436}{553}}$$

$$d = \frac{218}{\sqrt{1 - \frac{436}{553}}}$$

$$= 474 \text{ m}$$

61. $p = vi \sin \omega t \sin\left(\omega t - \frac{\pi}{2}\right)$

$$= vi \sin \omega t \left[\sin \omega t \cos \frac{\pi}{2} - \cos \omega t \sin \frac{\pi}{2} \right]$$

$$= vi \sin \omega t \left[\sin \omega t (0) - \cos \omega t (1) \right]$$

$$= -vi \sin \omega t \cos \omega t = -\frac{1}{2} vi (2 \sin \omega t \cos \omega t)$$

$$= -\frac{1}{2} vi \sin 2\omega t$$

20.4   Half-Angle Formulas

1. $\sqrt{\frac{1 + \cos 114°}{2}} = \cos \frac{1}{2}(114°) = \cos 57°$

$$\sqrt{\frac{1 + \cos 114°}{2}} = 0.544639035, \text{ calculator}$$

$$\cos 57° = 0.544639035, \text{ calculator}$$

5. $\sin 105° = \sin \frac{210°}{2} = \sqrt{\frac{1 - \cos 210°}{2}} = \sqrt{\frac{1 - (-\sqrt{3}/2)}{2}}$

$$= 0.9659$$

9. $\sqrt{\frac{1 - \cos 236°}{2}} = \sin \frac{1}{2}(236°) = \sin 118°$

$$= 0.8829476$$

13. $\sin\dfrac{\alpha}{2} = \sqrt{\dfrac{1-\cos\alpha}{2}}$

$\sqrt{\dfrac{1-\cos 6x}{2}} = \sin\dfrac{6x}{2} = \sin 3x$

17. $\sqrt{4-4\cos 10\theta} = \sqrt{4\left(1-\cos\left(2(5\theta)\right)\right)}$

$= 2\sqrt{2}\sqrt{\dfrac{1-\cos 5\theta}{2}}$

$= 2\sqrt{2}\sin 5\theta$

21. $\sin\dfrac{\alpha}{2} = \sqrt{\dfrac{1-\cos\alpha}{2}} = \sqrt{\dfrac{1-\frac{12}{13}}{2}}$

$= \sqrt{\dfrac{1}{13}\cdot\dfrac{1}{2}} = \sqrt{\dfrac{1}{26}}$

$= \sqrt{\dfrac{1}{26}\cdot\dfrac{26}{26}} = \dfrac{1}{26}\sqrt{26}$

25. $\csc\dfrac{\alpha}{2} = \dfrac{1}{\sin\frac{\alpha}{2}} = \dfrac{1}{\pm\sqrt{\frac{1-\cos\alpha}{2}}} = \dfrac{\sqrt{2}}{\pm\sqrt{1-\cos\alpha}}$

$= \pm\sqrt{\dfrac{2}{1-\cos\alpha}} = \pm\sqrt{\dfrac{2}{1-\frac{1}{\sec\alpha}}}$

$= \pm\sqrt{\dfrac{2\sec\alpha}{\sec\alpha-1}}$

29. $\dfrac{1-\cos\alpha}{2\sin\frac{\alpha}{2}} = \dfrac{1-\cos\alpha}{2\sqrt{\frac{1-\cos\alpha}{2}}} \times \dfrac{\sqrt{\frac{1-\cos\alpha}{2}}}{\sqrt{\frac{1-\cos\alpha}{2}}}$

$= \dfrac{(1-\cos\alpha)\sqrt{\frac{1-\cos\alpha}{2}}}{2\left(\frac{1-\cos\alpha}{2}\right)}$

$= \sqrt{\dfrac{1-\cos\alpha}{2}} = \sin\dfrac{\alpha}{2}$

33. $2\sin^2\dfrac{\alpha}{2} - \cos^2\dfrac{\alpha}{2} = \dfrac{1-3\cos\alpha}{2}$

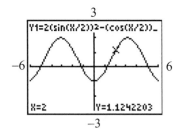

37. Find $\tan\theta$ if $\sin\dfrac{\theta}{2} = \dfrac{3}{5}$.

$\sin\dfrac{\theta}{2} = \dfrac{3}{5} \Rightarrow \dfrac{\theta}{2}$ in QI or $\dfrac{\theta}{2}$ in QII

For $\dfrac{\theta}{2}$ in QI, $\sin\dfrac{\theta}{2} = \dfrac{3}{5} \Rightarrow \cos\dfrac{\theta}{2} = \dfrac{4}{5}$

$\sin 2\cdot\dfrac{\theta}{2} = \sin\theta = 2\sin\dfrac{\theta}{2}\cos\dfrac{\theta}{2} = 2\cdot\dfrac{3}{5}\cdot\dfrac{4}{5} = \dfrac{24}{25}$

$\cos 2\cdot\dfrac{\theta}{2} = \cos\theta = \cos^2\dfrac{\theta}{2} - \sin^2\dfrac{\theta}{2} = \dfrac{16}{25} - \dfrac{9}{25} = \dfrac{7}{25}$

$\tan\theta = \dfrac{\sin\theta}{\cos\theta} = \dfrac{\frac{24}{25}}{\frac{7}{25}} = \dfrac{24}{7}$

For $\dfrac{\theta}{2}$ in QII, $\sin\dfrac{\theta}{2} = \dfrac{3}{5} \Rightarrow \cos\dfrac{\theta}{2} = \dfrac{-4}{5}$

$\sin 2\cdot\dfrac{\theta}{2} = \sin\theta = 2\sin\dfrac{\theta}{2}\cos\dfrac{\theta}{2} = 2\cdot\dfrac{3}{5}\cdot\dfrac{-4}{5} = \dfrac{-24}{25}$

$\cos 2\cdot\dfrac{\theta}{2} = \cos\theta = \cos^2\dfrac{\theta}{2} - \sin^2\dfrac{\theta}{2} = \dfrac{16}{25} - \dfrac{9}{25} = \dfrac{7}{25}$

$\tan\theta = \dfrac{\sin\theta}{\cos\theta} = \dfrac{-\frac{24}{25}}{\frac{7}{25}} = -\dfrac{24}{7}$

if $\sin\dfrac{\theta}{2} = \dfrac{3}{5}$, $\tan\theta = \pm\dfrac{24}{7}$

41. $90° < \theta < 180° \Rightarrow 45° < \dfrac{\theta}{2} < 90°$

$\sin\theta = \dfrac{4}{5} \Rightarrow \cos\theta = -\dfrac{3}{5}$

$\cos\dfrac{\theta}{2} = \sqrt{\dfrac{1+\cos\theta}{2}} = \sqrt{\dfrac{1+\left(-\frac{3}{5}\right)}{2}} = \dfrac{\sqrt{5}}{5}$

45. $\sin^2\omega t = \sin^2\left[\left(\dfrac{1}{2}\right)(2\omega t)\right]$

$= \left(\sqrt{\dfrac{1-\cos 2\omega t}{2}}\right)^2$

$= \dfrac{1-\cos 2\omega t}{2}$

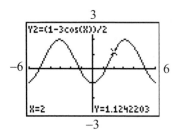

20.5 Solving Trigonometric Equations

1. $\tan\theta - 1 = 0,\ 0 \le \theta < 2\pi$

$\tan\theta = 1$

$\theta = \tan^{-1} 1 = \dfrac{\pi}{4},\ \dfrac{5\pi}{4}$

5. $\sin x - 1 = 0,\ 0 \le x < 2\pi;\ \sin x = 1;\ x = \dfrac{\pi}{2}$

9. $4 - \sec^2 x = 0 \Rightarrow 4 - \dfrac{1}{\cos^2 x} = 0$

$4\cos^2 x - 1 = 0;\ \ 0 \le x < 2\pi$

$4\cos^2 x = 1;\ \cos^2 x = \dfrac{1}{4};\ \cos x = \pm\dfrac{1}{2}$

$x = \dfrac{\pi}{3},\ \dfrac{2\pi}{3},\ \dfrac{4\pi}{3},\ \dfrac{5\pi}{3}$

13. $\sin 2x \sin x + \cos x = 0,\ \ 0 \le x < 2\pi$

$(2\sin x \cos x)(\sin x) + \cos x = 0$

$2\sin^2 x \cos x + \cos x = 0;\ \cos x(2\sin^2 x + 1) = 0$

$\cos x = 0;\ x = \dfrac{\pi}{2},\ \dfrac{3\pi}{2};\ 2\sin x + 1 = 0;\ 2\sin^2 x = -1$

$\sin^2 x = -\dfrac{1}{2}$ which has no real solution; thus,

$x = \dfrac{\pi}{2},\ \dfrac{3\pi}{2}$

17. $4\tan x - \sec^2 x = 0;\ 4\tan x - (1 + \tan^2 x) = 0$

$4\tan x - 1 - \tan^2 x = 0;\ \tan^2 x - 4\tan x + 1 = 0$

$\tan^2 x - 4\tan x = -1;\ \tan^2 x - 4\tan x + 4 = -1 + 4$

(completing the square)

$(\tan x - 2)^2 = 3;\ \tan x - 2 = \pm\sqrt{3}$

$\tan x = 2 \pm \sqrt{3} = 3.732,\ 0.2679$

$x = \tan^{-1} 0.2679 = 0.2618,\ \pi + 0.2618 = 3.403$

$x = \tan^{-1} 3.732 = 1.309,\ \pi + 1.309 = 4.451$

$x = 0.2618,\ 1.309,\ 3.403,\ 4.451$

21. $\tan x + 1 = 0;\ \tan x = -1;\ x_{\text{ref}} = \dfrac{\pi}{4}$

(tan negative QII, QIV)

$x = \pi - \dfrac{\pi}{4} = \dfrac{3\pi}{4} \approx 2.36$ or

$x = 2\pi - \dfrac{\pi}{4} = \dfrac{7\pi}{4} \approx 5.50$

Graph $y_1 = \tan x + 1$. Use zero feature to solve.

$x = 2.36,\ 5.50.$

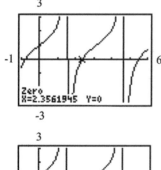

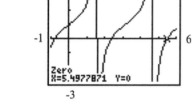

25. $4 - 3\csc^2 x = 4 - \dfrac{3}{\sin^2 x} = 0$

$4\sin^2 x - 3 = 0;\ 4\sin^2 x = 3;\ \sin^2 x = \dfrac{3}{4};$

$\sin x = \pm\sqrt{\dfrac{3}{4}} = \pm\dfrac{\sqrt{3}}{2};\ x_{\text{ref}} = \dfrac{\pi}{3}$

(sin positive or negative-all quadrants)

$x = \dfrac{\pi}{3} = 1.05;\ \pi - \dfrac{\pi}{3} = \dfrac{2\pi}{3} = 2.09$

$x = \pi + \dfrac{\pi}{3} = \dfrac{4\pi}{3} = 4.19;\ x = 2\pi - \dfrac{\pi}{3} = \dfrac{5\pi}{3} = 5.24$

Graph $y_1 = 4\sin^2 xx - 3$. Use zero feature to solve.

$x = 1.05,\ 2.09,\ 4.19,$ and $5.24.$

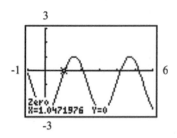

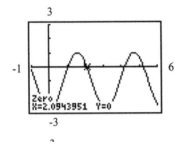

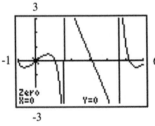

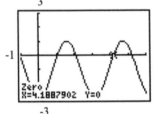

29. $2\sin x - \tan x = 0$; $2\sin x - \dfrac{\sin x}{\cos x} = 0$;

$\sin x\left(2 - \dfrac{1}{\cos x}\right) = 0$; $\sin x = 0$; $x = 0.00$;

$x = \pi = 3.14$

$2 - \dfrac{1}{\cos x} = 0$; $\dfrac{1}{\cos x} = 2$; $\cos x = \dfrac{1}{2}$; $x_{\text{ref}} = \dfrac{\pi}{3}$;

$x = \dfrac{\pi}{3} = 1.05$; $x = 2\pi - \dfrac{\pi}{3} = \dfrac{5\pi}{3} = 5.24$

Graph $y_1 = 2\sin x - \tan x$. Use zero feature

to solve. $x = 0.00, 1.05, 3.14, 5.24$.

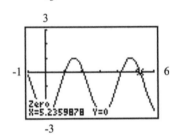

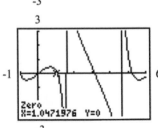

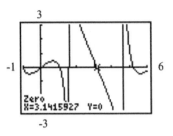

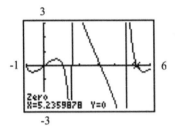

33. $\tan x + 3\cot x = 4$; $\tan x + \dfrac{3}{\tan x} = 4$

$\tan^2 x + 3 = 4\tan x$; $\tan^2 x - 4\tan x + 3 = 0$;

$(\tan x - 1)(\tan x - 3) = 0$

$\tan x = 1$; $x = \dfrac{\pi}{4} = 0.7854$ or

$x = \pi + \dfrac{\pi}{4} = \dfrac{5\pi}{4} = 3.927$;

$\tan x - 3 = 0$; $\tan x = 3$; $x = 1.249$ or

$x = \pi + 1.249 = 4.391$

Graph $y_1 = \tan x + 3/\tan x - 4$. Use zero feature

to solve. $x = 0.79, 1.25, 3.93, 4.39$.

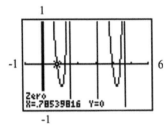

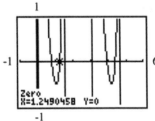

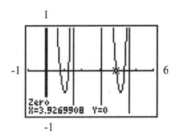

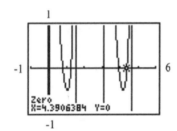

37.
$$2\sin 2x - \cos x \sin^3 x = 0$$
$$2(2\sin x \cos x) = \cos x \sin^3 x = 0$$
$$\sin x \cos x (4 - \sin^2 x) = 0$$

$\sin x \cos x = 0$ $4 - \sin^2 x = 0$

$\sin x = 0$ $\cos x = 0$ $\sin x = \pm 2$

$x = 0, \pi$ $x = \dfrac{\pi}{2}, \dfrac{3\pi}{2}$ no solution

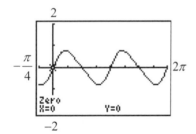

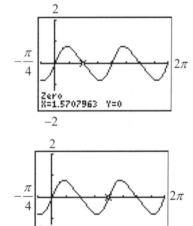

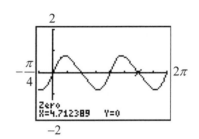

41. $\sin\theta + \cos\theta + \tan\theta + \cot\theta + \sec\theta + \csc\theta = f(\theta) = 1$

θ is a positive acute angle $\Rightarrow 0 < \theta < \dfrac{\pi}{2}$ for which

$0 < \sin\theta < 1$, $0 < \cos\theta < 1$, $\tan\theta > 0$, $\cot\theta > 0$,

$\sec\theta > 1$, $\csc\theta > 1$ which implies $f(\theta) = \sin\theta$

$+ \cos\theta + \tan\theta + \cot\theta + \sec\theta + \csc\theta > 1 \neq 1$.

$f(\theta) = 1$ has no solution.

45.

[Triangle diagram with vertices B, A, C. Side from B to C labeled $2x$, side from C to A labeled x.]

$$\dfrac{2x}{\sin A} = \dfrac{x}{\sin B}$$

$$\dfrac{2}{\sin A} = \dfrac{1}{\sin(A - 60)} \Rightarrow 2\sin(A - 60) = \sin A$$

$$2(\sin A \cos 60 - \sin 60 \cos A) = \sin A$$

$$2\left(\dfrac{1}{2}\sin A - \dfrac{\sqrt{3}}{2}\cos A\right) = \sin A \Rightarrow \cos A = 0, \; A = 90°$$

$B = A - 60° = 90° - 60° = 30°$, $A + B + C = 180° \Rightarrow C = 60°$

The angles are $30°$, $60°$, $90°$.

49. $y = 2.30\cos 0.1t - 1.35\sin 0.2t$

$$2.30\cos 0.1t - 1.35\sin 0.2t = 0$$

$$2.30\cos 0.1t - 1.35(2\sin 0.1t \cos 0.1t) = 0$$

$$\cos 0.1t(2.30 - 2.70\sin 0.1t) = 0$$

$$\cos 0.1t = 0, \; \sin 0.1t = 2.30 / 2.70$$

For $\cos 0.1t = 0$, $0.1t = \dfrac{\pi}{2}, \dfrac{3\pi}{2}$

$t = 15.7$ s, 47.1 s

For $\sin 0.1t = 2.30 / 2.70$

$0.1t = 1.0195$, $\pi - 1.0195 = 2.122$

$t = 10.2$ s, 21.2 s

53.

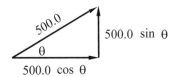

$500.0 \cos\theta + 500.0 \sin\theta = 700.0$

$\sqrt{1 - \sin^2\theta} + \sin\theta = 1.400$

$1 - \sin^2\theta = 1.400^2 - 2.800 \sin\theta + \sin^2\theta$

$2\sin^2\theta - 2.800 \sin\theta + 4.00^2 - 2.800 \sin\theta + \sin^2\theta$

$\sin\theta = 0.6000$	$\sin\theta = 0.8000$
$\theta = 36.87°$	$\theta = 53.13°$

The components are 300 N and 400 N.

57. $2\sin 2x = x^2 + 1$; $-x^2 + 2\sin x - 1 = 0$

Graph $y_1 = -x^2 + 2\sin x - 1$. Use zero feature

to solve. $x = 0.29, 0.95$.

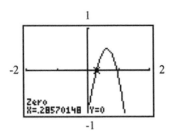

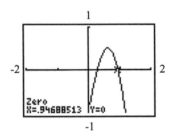

61. $x \tan x = 2.00$, $0 < x < \dfrac{\pi}{2}$; $x \tan x - 2.00 = 0$

Graph $y_1 = x \tan x - 2.00$. Use zero feature to solve.

$x = 1.08$.

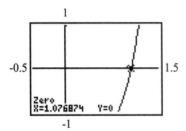

20.6 The Inverse Trigonometric Functions

1. $y = \tan^{-1} 3A$ is read as "y is the angle whose tangent is $3A$." In this case, $3A = \tan y$.

5. y is an angle whose cotangent is $3x$.

9. y is five times the angle whose cosine is $2x - 1$.

13. $\tan^{-1} 1 = \dfrac{\pi}{4}$ since $\tan \dfrac{\pi}{4} = 1$ and $-\dfrac{\pi}{2} < \dfrac{\pi}{4} < \dfrac{\pi}{2}$ \

17. Let $\sec^{-1} 0.5 = x$, then

$\sec x = 0.5$

$\dfrac{1}{\cos x} = 0.5$

$\cos x = 2$

and since $-1 \le \cos x \le 1$ there is no value for x.

21. $\sin\left(\tan^{-1} \sqrt{3}\right) = \sin \dfrac{\pi}{3} = \dfrac{1}{2}\sqrt{3}$

25. $\cos\left[\tan^{-1}(-5)\right] = \dfrac{1}{\sqrt{26}} = 0.1961$

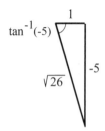

29. $\tan^{-1} x = \sin^{-1}\dfrac{2}{5}$

$x = \tan\left(\sin^{-1}\dfrac{2}{5}\right)$

$x = \dfrac{2}{\sqrt{21}}$

33. $\tan^{-1}(-3.7321) = -1.3090$

37. $\tan\left[\cos^{-1}(-0.6281)\right] = \tan 2.250 = -1.2389$

41. $y = \sin 3x;\ 3x = \sin^{-1} y;\ x = \dfrac{1}{3}\sin^{-1} y$

45. $1 - y = \cos^{-1}(1 - x)$

$\cos(1 - y) = 1 - x$

$x = 1 - \cos(1 - y)$

49. Let $\alpha = \sin^{-1} x,\ \beta = \cos^{-1} y$

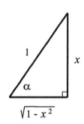

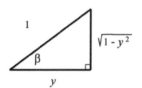

$\sin\left(\sin^{-1} x + \cos^{-1} y\right) = \sin(\alpha + \beta)$

$= \sin\alpha\cos\beta + \sin\beta\cos\alpha$

$= xy + \sqrt{1 - y^2}\,\sqrt{1 - x^2}$

$= xy + \sqrt{1 - y^2 - x^2 + x^2 y^2}$

53. $\sin\left(2\sin^{-1} x\right) = \sin 2\theta = 2\sin\theta\cos\theta$

$= 2\left(\dfrac{x}{1}\right)\left(\dfrac{\sqrt{1 - x^2}}{1}\right) = 2x\sqrt{1 - x^2}$

In triangle, θ is set up such that its sine is $2x$. This gives an opposite side of $2x$, hypotenuse 1, and adjacent side $\sqrt{1 - 4x^2}$.

57.

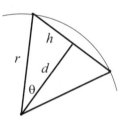

$r^2 = h^2 + d^2 \Rightarrow h = \sqrt{r^2 - d^2}$

$\cos\theta = \dfrac{d}{r} \Rightarrow \theta = \cos^{-1}\dfrac{d}{r}$

$A_{\text{segment}} = A_{\text{sector}} - A_{\text{triangle}}$

$A_{\text{segment}} = \dfrac{1}{2}r^2(2\theta) - 2\left(\dfrac{1}{2}dh\right)$

$A_{\text{segment}} = r^2\theta - d\sqrt{r^2 - d^2} = r^2\cos^{-1}\dfrac{d}{r} - d\sqrt{r^2 - d^2}$

61. Let $\alpha = \sin^{-1}\dfrac{3}{5}$ and $\beta = \sin^{-1}\dfrac{5}{13};\ \sin\alpha = \dfrac{3}{5}$

$\cos\alpha = \sqrt{1 - \tfrac{9}{25}} = \sqrt{\dfrac{16}{25}} = \dfrac{4}{5};\ \sin\beta = \dfrac{5}{13}$

$\cos\beta = \sqrt{1 - \tfrac{25}{169}} = \sqrt{\dfrac{144}{169}} = \dfrac{12}{13}$

$$\sin^{-1}\frac{3}{5}+\sin^{-1}\frac{5}{13}=\alpha+\beta$$

$$\sin(\alpha+\beta)=\sin\alpha\cos\beta+\cos\alpha\sin\beta$$

$$=\frac{3}{5}\left(\frac{12}{13}\right)+\frac{4}{5}\left(\frac{5}{13}\right)$$

$$=\frac{36}{65}+\frac{20}{65}=\frac{56}{65}$$

65. $\sin\left(\sin^{-1}x+\cos^{-1}y\right)=\sin\left(\sin^{-1}x\right)\cos\left(\cos^{-1}y\right)$

$$+\sin\left(\cos^{-1}y\right)\cos\left(\sin^{-1}x\right)$$

$$=xy+\sqrt{1-y^2}\cdot\sqrt{1-x^2}$$

$$=xy+\sqrt{1-x^2-y^2+x^2y^2}$$

69. $y=\sin^{-1}x+\sin^{-1}(-x),\ -1\le x\le 1,$

$$-\frac{\pi}{2}\le y\le\frac{\pi}{2}$$

$$\sin y=\sin\left(\sin^{-1}x\right)+\sin\left(\sin^{-1}(x)\right)$$

$$\sin y=x+(-x)=0$$

$$y=0$$

$$\sin^{-1}x+\sin^{-1}(-x)=0$$

73. $\tan B=\frac{y}{b}\Rightarrow y=b\tan B$

$$\tan A=\frac{y}{a}=\frac{b\tan B}{a}$$

$$A=\tan^{-1}\frac{b\tan B}{a}$$

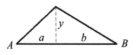

77. $\tan\alpha=\frac{y}{x}\Rightarrow a=\tan^{-1}\left(\frac{y}{x}\right)$

$$\tan(\alpha+\theta)=\frac{y+50}{x}\Rightarrow\alpha+\theta=\tan^{-1}\left(\frac{y+50}{x}\right)$$

$$\tan^{-1}\left(\frac{y}{x}\right)+\theta=\tan^{-1}\left(\frac{y+50}{x}\right)$$

$$\theta=\tan^{-1}\left(\frac{y+50}{x}\right)-\tan^{-1}\left(\frac{y}{x}\right)$$

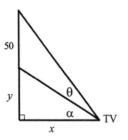

Chapter 20   Review Exercises

1. $\sin 120° = \sin\left(90° + 30°\right)$

$= \sin 90° \cos 30° + \cos 90° \sin 30°$

$= 1\left(\dfrac{\sqrt{3}}{2}\right) + 0\left(\dfrac{1}{2}\right) = \dfrac{\sqrt{3}}{2} = \dfrac{1}{2}\sqrt{3}$

5. $\cos \pi = \cos 2\left(\dfrac{\pi}{2}\right) = \cos^2 \dfrac{\pi}{2} - \sin^2 \dfrac{\pi}{2} = 0 - (1)^2 = -1$

9. $\sin 14° \cos 38° + \cos 14° \sin 38°$

$= \sin\left(14° + 38°\right)$

$= \sin 52° = 0.7880108$

```
sin(14)*cos(38)+
cos(14)*sin(38)
         .7880107536
sin(52)
         .7880107536
■
```

13. $\cos 73° \cos\left(-142°\right) + \sin 73° \sin\left(-142°\right)$

$= \cos 73° \cos 142° - \sin 73° \sin 142°$

$= \cos\left(73° + 142°\right) = \cos 215°$

$= -0.8191520443$

```
cos(73)cos(-142)
+sin(73)sin(-142
)
        -.8191520443
cos(215)
        -.8191520443
■
```

17. $\sin 2x \cos 3x + \cos 2x \sin 3x$

$= \sin \alpha \cos \beta + \cos \alpha \sin \beta$

$= \sin\left(\alpha + \beta\right)$ where $\alpha = 2x,\ \beta = 3x$

$\sin\left(\alpha + \beta\right) = \sin\left(2x + 3x\right) = \sin 5x$

19. $8 \sin 6x \cos 6x = 4\left(2 \sin 6x \cos 6x\right) = 4 \sin 12x$

20. $\dfrac{\tan x + \tan 2x}{1 - \tan x \tan 2x} = \tan\left(x + 2x\right) = \tan 3x$

21. $2 - 4\sin^2 6x = 2\left(1 - 2\sin^2 6x\right)$

$= 2\left(-2\sin^2 \alpha\right) = 2\left(\cos 2\alpha\right)$

$= 2 \cos 12x$ where $\alpha = 6x$

25. $\sin^{-1}\left(-1\right) = -\dfrac{\pi}{2}$ since $\sin\left(-\dfrac{\pi}{2}\right) = -1$ and

$-\dfrac{\pi}{2} \le -\dfrac{\pi}{2} \le \dfrac{\pi}{2}$

29. $\tan\left[\sin^{-1}\left(-0.5\right)\right] = \tan\left(-\dfrac{\pi}{6}\right) = -\dfrac{\sqrt{3}}{3}$

33. $\dfrac{\sec y}{\cos y} - \dfrac{\tan y}{\cot y} = \sec^2 y - \tan^2 y$

$= 1 + \tan^2 y - \tan^2 y$

$= 1$

37. $\dfrac{\sec^4 x - 1}{\tan^2 x} = \dfrac{\left(\sec^2 x - 1\right)\left(\sec^2 x + 1\right)}{\tan^2 x}$

$\dfrac{\left(\sec^2 x + 1\right)\left(\tan^2 x\right)}{\tan^2 x} = \sec^2 x + 1$

$1 + \tan^2 x + 1 = 2 + \tan^2 x$

41. $\dfrac{1 - \sin^2 \theta}{1 - \cos^2 \theta} = \dfrac{\cos^2 \theta}{\sin^2 \theta}$ since $\sin^2 \theta + \cos^2 \theta = 1$

$= \left(\dfrac{\cos \theta}{\sin \theta}\right)^2$

$= \left(\cot \theta\right)^2 = \cot^2 \theta$

45. $\dfrac{\sec x}{\sin x} - \sec x \sin x = \dfrac{1}{\cos x \sin x} - \dfrac{\sin x}{\cos x} \cdot \dfrac{\sin x}{\sin x}$

$= \dfrac{1 - \sin^2 x}{\sin x \cos x} = \dfrac{\cos^2 x}{\sin x \cos x}$

$= \dfrac{\cos x}{\sin x} = \cot x$

49. $\dfrac{\sin x \cot x + \cos x}{2 \cot x} = \dfrac{\sin x \cdot \frac{\cos x}{\sin x} + \cos x}{2 \frac{\cos x}{\sin x}} \cdot \dfrac{\sin x}{\sin x}$

$= \dfrac{\sin x \cos x + \sin x \cos x}{2 \cos x}$

$= \dfrac{2 \sin x \cos x}{2 \cos x} = \sin x$

53.

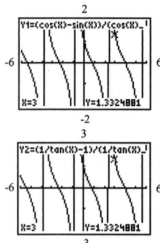

57.

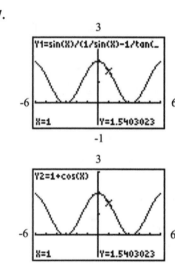

61. $y = 2\cos 2x;\ \dfrac{y}{2} = \cos 2x$

$\cos^{-1}\dfrac{y}{2} = 2x;\ \dfrac{1}{2}\cos^{-1}\dfrac{1}{2}y = x$

65. $3(\tan x - 2) = 1 + \tan x$

$3\tan x - 1 = 1 + \tan x$

$2\tan x = 7$

$\tan x = \dfrac{7}{2}$

$x = \tan^{-1}\dfrac{7}{2} = 1.2925$

Since $\tan x$ is positive in QIII also,

$\pi + 1.2925 = 4.4341$ is also a value for x that is within the specified range of values for x.

69. $2\sin^2\theta + 3\cos\theta - 3 = 0,\ 0 \le \theta < 2\pi$

$2(1 - \cos^2\theta) + 3\cos\theta - 3 = 0$

$2 - 2\cos^2\theta + 3\cos\theta - 3 = 0$

$2\cos^2\theta - 3\cos\theta + 1 = 0$

$(\cos\theta - 1)(2\cos\theta - 1) = 0$

$\cos\theta - 1 = 0$ or $2\cos\theta - 1 = 0$

$\theta = 0$ $\qquad\qquad \cos\theta = \dfrac{1}{2}$

$\qquad\qquad\qquad\qquad \theta = \dfrac{\pi}{3}, \dfrac{5\pi}{3}$

73. $\sin 2x = \cos 3x$

$\sin 2x = \cos(2x + x)$

$2\sin x\cos x = \cos 2x\cos x - \sin 2x\sin x$

$2\sin x\cos x = \cos 2x\cos x - 2\sin x\cos x\sin x$

$2\sin x\cos x - \cos 2x\cos x + 2\sin^2 x\cos x = 0$

$\cos x(2\sin x - \cos 2x + \sin^2 x) = 0$

$\cos x = 0 \Rightarrow x = \dfrac{\pi}{2}, \dfrac{3\pi}{2}$ or

$2\sin x - (1 - 2\sin^2 x) + 2\sin^2 x = 0$

$4\sin^2 x + 2\sin x - 1 = 0$

$\sin x = \dfrac{-2 \pm \sqrt{2^2 - 4(4)(-1)}}{2(4)} = \dfrac{-1 \pm \sqrt{5}}{4}$

$\sin x = \dfrac{-1 + \sqrt{5}}{4}$ or $\sin x = \dfrac{-1 - \sqrt{5}}{4}$

$x = \dfrac{\pi}{10}, \dfrac{9\pi}{10} \qquad x = \dfrac{13\pi}{10}, \dfrac{17\pi}{10}$

$\sin 2x = \cos 3x$ has solutions

$\left\{\dfrac{\pi}{10}, \dfrac{\pi}{2}, \dfrac{9\pi}{10}, \dfrac{13\pi}{10}, \dfrac{3\pi}{2}, \dfrac{17\pi}{10}\right\}$

77. $\tan x + \cot x = \dfrac{\sin x}{\cos x} + \dfrac{\cos x}{\sin x}$

$= \dfrac{\sin^2 x + \cos^2 x}{\sin x\cos x}$

$= \dfrac{1}{\sin x} \dfrac{1}{\cos x}$

$= \csc x\sec x$, identity

81. $x + \ln x - 3\cos^2 x = 2$

$x + \ln x - 3\cos^2 x - 2 = 0$

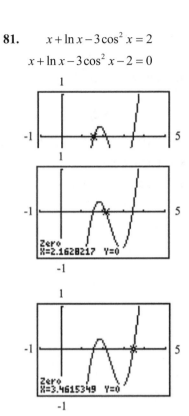

85. $\tan\left(\cot^{-1} x\right) = \tan \theta = \dfrac{1}{x}$

$\theta = \cot^{-1} x$

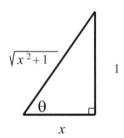

89.

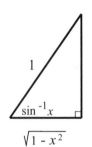

$\cos\left(\sin^{-1} x + \tan^{-1} y\right) = \cos\left(\sin^{-1} x\right)\cos\left(\tan^{-1} y\right)$

$-\sin\left(\sin^{-1} x\right)\sin\left(\tan^{-1} y\right)$

$= \sqrt{1 - x^2}\,\dfrac{1}{\sqrt{1 + y^2}} - x\dfrac{y}{\sqrt{+y^2}}$

$= \dfrac{\sqrt{1 - x^2} - xy}{\sqrt{1 + y^2}}$

93. $\dfrac{x}{\sqrt{1 + x^2}} = \dfrac{\tan \theta}{\sqrt{1 + \tan^2 \theta}} = \dfrac{\tan \theta}{\sqrt{\sec^2 \theta}} = \dfrac{\tan \theta}{\sec \theta}$

$= \dfrac{\sin \theta}{\cos \theta} \cdot \cos \theta = \sin \theta$

97. $\cos\dfrac{\theta}{2} = \sqrt{\dfrac{1 + \cos \theta}{2}} = \sqrt{\dfrac{1 + \frac{4}{5}}{2}} = \dfrac{3\sqrt{10}}{10}$

101. $2\left(\tan^2 x + \sin^2 x - \sec^2 x\right) + \cos 2x$

$= 2\left(\sec^2 x - 1 + \sin^2 x - \sec^2 x\right) + \cos^2 x - \sin^2 x$

$= 2\left(-\cos^2 x\right) + \cos^2 x - \sin^2 x = -\cos^2 x - \sin^2 x$

$= -1$

105. Let $\dfrac{A}{C} = \cos \alpha$ and $\dfrac{B}{C} = \sin \alpha$

$\dfrac{A^2}{C^2} + \dfrac{B^2}{C^2} = \cos^2 \alpha + \sin^2 \alpha = 1$

$A^2 + B^2 = C^2 \Rightarrow C = \sqrt{A^2 + B^2}$

$y = A\sin 2t + B\cos 2t$

$= C\cos \alpha \sin 2t + C\sin \alpha \cos 2t$

$y = C\left(\cos \alpha \sin 2t + \sin \alpha \cos 2t\right)$

$y = C\sin\left(2t + \alpha\right)$

109. $R = \sqrt{Rx^2 + Ry^2}$

$= \sqrt{\left(A\cos \theta - B\sin \theta\right)^2 + \left(A\sin \theta + B\cos \theta\right)^2}$

$= \sqrt{A^2 \cos^2 \theta + A^2 \sin^2 \theta + B^2 \cos^2 \theta + B^2 \sin^2 \theta}$

$= \sqrt{A^2\left(\cos^2 \theta + \sin^2 \theta\right) + B^2\left(\cos^2 \theta + \sin^2 \theta\right)}$

$= \sqrt{\left(\cos^2 \theta + \sin^2 \theta\right)\left(A^2 + B^2\right)} = \sqrt{\left(A^2 + B^2\right)}$

113.
$$\omega t = \sin^{-1}\frac{\theta-\alpha}{R}$$
$$\sin(\omega t) = \frac{\theta-\alpha}{R}$$
$$R\sin(\omega t) = \theta-\alpha$$
$$\theta = R\sin(\omega t)+\alpha$$

117. $p = VI\cos\phi\cos^2\omega t - VI\sin\phi\cos\omega t\sin\omega t$
$$p = VI\cos\omega t\left(\cos\phi\cos\omega t - \sin\phi\sin\omega t\right)$$
$$p = VI\cos\omega t\cos(\omega t+\phi)$$

121.

$$\tan\theta = \frac{6.4}{y} \Rightarrow y = \frac{6.4}{\tan\theta}$$
$$\sin\theta = \frac{6.4}{x} \Rightarrow x = \frac{6.4}{\sin\theta}$$
$$2x+2y = 51.2$$
$$2\cdot\frac{6.4}{\sin\theta}+2\cdot\frac{6.4}{\tan\theta} = 51.2$$
$$\frac{1}{\sin\theta}+\frac{1}{\tan\theta} = 4.00 \Rightarrow 1+\cos\theta = 4\sin\theta \Rightarrow \theta = 28.1°$$
$$x = \frac{6.4}{\sin 28.1°} = 13.6 \text{ ft}$$

125. $\cos\left(2\sin^{-1}0.40\right) = 1-2\sin^2\left(\sin^{-1}0.40\right)$
$$= 1-2(0.40)^2$$
$$= 0.68$$

Chapter 21

PLANE ANALYTIC GEOMETRY

21.1   Basic Definitions

1. The distance between $(3, -1)$ and $(-2, 5)$ is

$$d = \sqrt{(3-(-2))^2 + (-1-5)^2}$$
$$= \sqrt{61}$$

5. Given: $(x_1, y_1) = (3, 8); (x_2, y_2) = (-1, -2)$

$$d = \sqrt{(x_2 - x_1)^2 + (y_2 - y_1)^2}$$
$$= \sqrt{(-1-3)^2 + (-2-8)^2}$$
$$= \sqrt{(-4)^2 + (-10)^2}$$
$$= \sqrt{16 + 100} = \sqrt{116}$$
$$= \sqrt{4 \times 29} = 2\sqrt{29}$$

9. Given: $(x_1, y_1) = (-12, 20); (x_2, y_2) = (32, -13)$

$$d = \sqrt{(x_2 - x_1)^2 + (y_2 - y_1)^2}$$
$$= \sqrt{(32+12)^2 + (-13-20)^2}$$
$$= \sqrt{(44)^2 + (-33)^2}$$
$$= \sqrt{1936 + 1089} = \sqrt{3025}$$
$$= 55$$

13. Given: $(x_1, y_1) = (1.22, -3.45);$
$(x_2, y_2) = (-1.07, -5.16)$

$$d = \sqrt{(x_2 - x_1)^2 + (y_2 - y_1)^2}$$
$$= \sqrt{(-1.07 - 1.22)^2 + (-5.16 - (-3.45))^2}$$
$$= \sqrt{(-2.29)^2 + (-5.16 + 3.45)^2}$$
$$= \sqrt{(-2.29)^2 + (-1.71)^2} = \sqrt{8.1682}$$
$$= 2.86$$

17. Given: $(x_1, y_1) = (4, -5); (x_2, y_2) = (4, -8)$

$$m = \frac{y_2 - y_1}{x_2 - x_1} = \frac{-8 - (-5)}{4 - 4}$$

Since $x_2 - x_1 = 4 - 4 = 0$, the slope is undefined.

21. Given: $(x_1, y_1) = (\sqrt{32}, \sqrt{18});$
$(x_2, y_2) = (-\sqrt{50}, \sqrt{8})$

$$m = \frac{y_2 - y_1}{x_2 - x_1} = \frac{\sqrt{8} - (\sqrt{18})}{-\sqrt{50} - \sqrt{32}} = \frac{1}{9}$$

25. Given: $\alpha = 30°; m = \tan \alpha, 0° < \alpha < 180°$

$$\tan 30° = \frac{\sqrt{3}}{3}$$

29. Given: $m = 0.364; m = \tan \alpha; 0.364 = \tan \alpha;$
$\alpha = 20.0°$

33. Given: $(x, y) = (6, -1); (x_1, y_1) = (4, 3)$
$(x_2, y_2) = (-5, 2); (x_3, y_3) = (-7, 6)$

$$m_1 = \frac{y - y_1}{x - x_1} = \frac{-1 - 3}{6 - 4} = \frac{-4}{2} = -2$$
$$m_2 = \frac{y_2 - y_3}{x_2 - x_3} = \frac{2 - 6}{-5 - (-7)} = \frac{-4}{-5 + 7} = \frac{-4}{2} = -2$$

$m_1 = m_2$ for all parallel lines.

37. Given: distance between $(-1, 3)$ and $(11, k)$ is 13.

$$d = \sqrt{(x_1 - x_2)^2 + (y_1 - y_2)^2}$$
$$13 = \sqrt{(-1 - 11)^2 + (3 - k)^2}$$
$$= \sqrt{(-12)^2 + (3 - k)^2}$$
$$169 = 144 + (3 - k)^2;$$
$$(3 - k)^2 = 25; 3 - k = \pm 5$$
$$-k = -3 \pm 5; k = -2, 8$$

41. $d_1 = \sqrt{(9-7)^2 + [4-(-2)]^2}$

$= \sqrt{2^2 + 6^2} = \sqrt{40} = 2\sqrt{10}$

$d_2 = \sqrt{(9-3)^2 + (4-2)^2} = \sqrt{6^2 + 2^2}$

$= \sqrt{40} = 2\sqrt{10}$

$d_1 = d_2$ so the triangle is isosceles.

45. $d_1 = \sqrt{(3-5)^2 + (-1-3)^2} = \sqrt{(-2)^2 + (-4)^2}$

$= \sqrt{4+16} = \sqrt{20}$

$m_1 = \dfrac{y-y_1}{x-x_1} = \dfrac{5-3}{3-(-1)} = \dfrac{5-3}{3+1} = \dfrac{2}{4} = \dfrac{1}{2}$

$d_2 = \sqrt{(5-1)^2 + (3-5)^2} = \sqrt{(4)^2 + (-2)^2}$

$= \sqrt{16+4} = \sqrt{20}$

$m_2 = \dfrac{y-y_1}{x-x_1} = \dfrac{5-1}{3-5} = \dfrac{4}{-2} = -2$

$m_1 = \dfrac{-1}{m_2},\ m_1 \perp m_2$

$A = \dfrac{1}{2} d_1 d_2 = \dfrac{1}{2}\sqrt{20}\sqrt{20} = \dfrac{1}{2}(20) = 10$

49. $\left(\dfrac{-4+6}{2}, \dfrac{9+1}{2}\right) = \left(\dfrac{2}{2}, \dfrac{10}{2}\right) = (1, 5)$

53. The distance between (x, y) and $(0, 0) = 3$.

$\sqrt{(x-0)^2 + (y-0)^2} = 3$

$x^2 + y^2 = 9$

57. $m = \dfrac{y_2 - y_1}{x_2 - x_1} = \dfrac{5-0}{-2-x} = 3$

$x = -\dfrac{11}{3}$

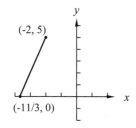

61. $P(-1, 6),\ Q(2, 1),\ R(5, 5)$

$d(P, Q) = \sqrt{(-1-2)^2 + (6-1)^2} = \sqrt{34}$

$d(Q, R) = \sqrt{(2-5)^2 + (1-5)^2} = 5$

P and R are not equidistant from Q

65. $d(D, E) = \sqrt{730^2 + (900+640)^2} = 1700$ km

21.2   The Straight Line

1. $m = -2, (x_1, y_1) = (4, -1)$

$y - y_1 = m(x - x_1)$

$y - (-1) = -2(x - 4)$

$y + 1 = -2x + 8$

$y + 2x - 7 = 0$

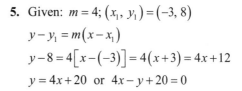

5. Given: $m = 4;\ (x_1, y_1) = (-3, 8)$

$y - y_1 = m(x - x_1)$

$y - 8 = 4[x - (-3)] = 4(x + 3) = 4x + 12$

$y = 4x + 20$ or $4x - y + 20 = 0$

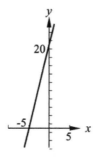

9. Given: $(x_1, y_1) = (-7, 12)$ $\alpha = 45°$

$m = \tan \alpha = \tan 45° = 1$

$y - y_1 = m(x - x_1)$

$y - 12 = 1(x + 7) = x + 7; \ y = x + 19$

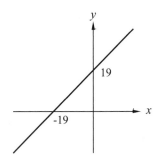

13. Parallel to y-axis and 3 units
left of y-axis.

$x = -3$

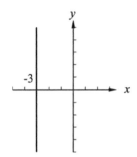

17. $\dfrac{\alpha - 2}{0 - 5} = \dfrac{2 - 0}{5 - \alpha} \Rightarrow \alpha^2 - 7a = 0$

$\alpha(\alpha - 7) = 0 \Rightarrow \alpha = 7$

$y - y_1 = m(x - x_1)$

$y - 0 = \dfrac{7 - 2}{-5}(x - 5)$

$x + y - 7 = 0$

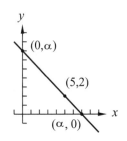

21. Given: $4x - y = 8, \ y = 4x - 8, \ m = 4, \ b = -8$

When $x = 0, \ y = -8$

$y = 0, \ x = 2$

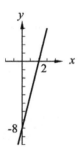

25. Given: $3x - 2y - 1 = 0$

$3x - 2y - 1 = 0; \ -2y = -3x + 1$

$y = \dfrac{-3}{-2}x + \dfrac{1}{-2}; \ y = \dfrac{3}{2}x - \dfrac{1}{2}$

Slope $= \dfrac{3}{2} = m;$

y-intercept $= -\dfrac{1}{2} = b$

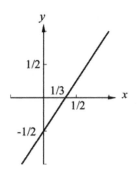

29. $3x - 2y + 5 = 0; \ -2y = -3x - 5;$

$y = \dfrac{-3}{-2}x + \dfrac{-5}{-2}; \ y = \dfrac{3}{2}x + \dfrac{5}{2};$

slope $= \dfrac{3}{2} = m_1$

$4y = 6x - 1; \ y = \dfrac{6}{4}x - \dfrac{1}{4};$

$y = \dfrac{3}{2}x - \dfrac{1}{4}; \ \text{slope} = \dfrac{3}{2} = m_2$

$m_1 = m_2$ for all parallel lines.

33. $5x + 2y = 3 \Rightarrow y = \dfrac{-5}{2} \cdot x + \dfrac{3}{2}$

$10y = 7 - 4x \Rightarrow y = \dfrac{-4}{10}x + \dfrac{7}{10}$

$m_1 \cdot m_2 = \dfrac{-5}{2} \cdot \dfrac{-4}{10} = 1 \neq -1$

$m_1 \neq m_2$

Lines are neither perpendicular nor parallel.

37. Given: $4x - ky = 6 \| 6x + 3y + 2 = 0$

$6x + 3y + 2 = 0;\ 3y = -6x - 2$

$y = \dfrac{-6}{3}x - \dfrac{2}{3};\ y = -2x - \dfrac{2}{3};$ slope is -2

$4x - ky = 6;\ -ky = -4x + 6$

$y = \dfrac{-4}{-k}x + \dfrac{6}{-k};\ y = \dfrac{4}{k}x - \dfrac{6}{k};$ slope is $\dfrac{4}{k}$

Since the lines are parallel, the slopes are equal.

$\dfrac{4}{k} = -2;\ 4 = -2k;\ k = -2$

41.

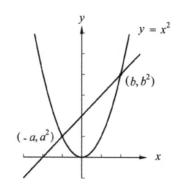

$m = \dfrac{b^2 - a^2}{b - (-a)}$

$m = \dfrac{(b+a)(b-a)}{b+a}$

$m = b - a$

45. $8x + 10y = 3 \Rightarrow y_1 = -\dfrac{4}{5}x + \dfrac{3}{10}$

$2x - 3y = 5 \Rightarrow y_2 = \dfrac{2}{3}x - \dfrac{5}{3}$

$4x - 6y = -3 \Rightarrow y_3 = \dfrac{2}{3}x + \dfrac{1}{2}$

$5y + 4x = 1 \Rightarrow y_4 = -\dfrac{4}{5}x + \dfrac{1}{5}$

$m_1 = m_4 = -\dfrac{4}{5}$ and $m_2 = m_3 = \dfrac{2}{3}$

showing the lines form a parallelogram.

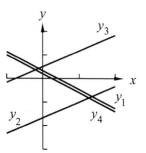

49. $4x - 2y = k \Rightarrow y = 2x - \dfrac{k}{2} = \begin{cases} 2x+2, & k=-4 \\ 2x, & k=0 \\ 2x-2, & k=4 \end{cases}$

All have slope $m = 2$.

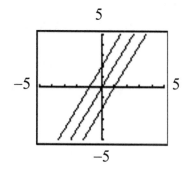

53. $\dfrac{F - 32}{212 - 32} = \dfrac{R - 0}{80 - 0}$

$F = \dfrac{9}{4}R + 32$

$$
\begin{array}{cc}
212\!+ & +\!80 \\
F\!+ & R\!+ \\
32\!+ & +\!0
\end{array}
$$

57. $x = 0;\ T = 3°\text{C}$

$T = kx + T_1$

$3 = 0 + T_1$

$T = kx + 3$

$x = 15$ cm; $T = 23°$ C

$23° = k(15) + 3$

61. $m = \tan\left(180^{\circ} - 0.0032^{\circ}\right)$

$b = 24\,\mu\text{m} = 24 \times 10^{-6}\,\text{m} = 2.4 \times 10^{-5}\,\text{m}$

$m = -5.6 \times 10^{-5}$

$y = mx + b = -5.6 \times 10^{-5}\,x + 2.4 \times 10^{-5}$

$y = \left(-5.6x + 2.4\right)10^{-5}$

65. $n = 1200\sqrt{t} + 0$

$m = 1200$

$b = 0$

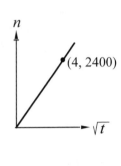

t	$\sqrt{t}$	h
0	0	0
1	1	1200
4	2	2400

69. Slope is found by measuring between points. The vertical displacement and the horizontal displacement between the extreme points is in a 1 to 2 ratio; $m = \frac{1}{2}$.

Since the graph is linear, the log equation is of the form $\log y = m \log x + \log a$, where a is the intercept $\left(1, a\right)$.

$y = ax^n$; $y = 3x^6$

$a = 3$, $n = 4$

x	y
1.0	3.0
1.1	4.4
1.2	6.2
1.3	8.6

$\log y = \log a + n \log x$

$\log y = \log 3 + 3 \log x$

Verify

(1) Slope is $\dfrac{\log y - \log a}{\log x} = 4.$

Vertical and horizontal measures in millimeters between points are shown. Each slope is 4.

(2) The intercept is $a = 3$.

The line crosses the vertical axis at $x = 1.0$,
$y = 3.0.$

21.3 The Circle

1. $\left(x-1\right)^2 + \left(y+1\right)^2 = 16$ has center at $\left(1, -1\right)$ and $r = 4$

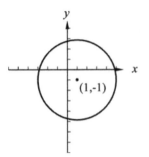

5. $\left(x-2\right)^2 + \left(y-1\right)^2 = 25$

$C\left(2, 1\right)$, radius is 5.

9. $\left(x-h\right)^2 + \left(y-k\right)^2 = r^2$; $C\left(0, 0\right)$, $r = 3$

$\left(x-0\right)^2 + \left(y-0\right)^2 = 3^2$, $x^2 + y^2 = 9$

13. $\left(x-12\right)^2 + \left(y-\left(-15\right)\right)^2 = 18^2$; $C\left(12, -15\right)$, $r = 18$

$\left(x-12\right)^2 + \left(y+15\right)^2 = 324$;

$x^2 + y^2 - 24x + 30y + 45 = 0$

17. Concentric with $\left(x-2\right)^2 + \left(y-1\right)^2 = 4$ gives center at $\left(2, 1\right)$.

$r = \sqrt{\left(2-4\right)^2 + \left(1-\left(-1\right)\right)^2} = 2\sqrt{2}$

The equation is $\left(x-2\right)^2 + \left(y-1\right)^2 = 8$

21. The center is $\left(-2, 2\right)$ and radius is 2.

$\left(x-h\right)^2 + \left(y-k\right)^2 = r^2$

$\left(x+2\right)^2 + \left(y-2\right)^2 = 2^2$

$x^2 + 4x + 4 + y^2 - 4y + 4 = 4$

$x^2 + y^2 + 4x - 4y + 4 = 0$

25. $x^2 + (y-3)^2 = 4$ is the same as

$(x-0)^2 + (y-3)^2 = 2^2$, so

Therefore, $h = 0$, $k = 3$, $r = 2$

$C(0, 3)$

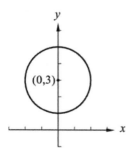

29. $2x^2 + 2y^2 - 16 = 4x$

$x^2 + y^2 - 2x - 8 = 0$

$x^2 - 2x + 1 + y^2 = 9$

$(x-1)^2 + (y-0)^2 = 9$

$h = 1$, $k = 0$, $r = 3$

$C(1, 0)$

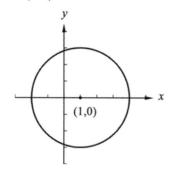

33. $4x^2 + 4y^2 - 9 = 16y$

$4x^2 + 4y^2 - 16y = 9$

$4(x-0)^2 + 4(y^2 - 4y + 4) = 9 + 16$

$(x-0)^2 + (y-2) = \dfrac{25}{4}$

$h = 0$, $k = 2$, $r = \dfrac{5}{2}$

$C(0, 2)$

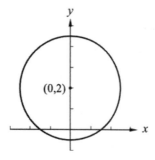

37. $(-x)^2 + y^2 = 100$; $x^2 + y^2 = 100$

Symmetrical to y-axis

$(-x)^2 + (-y)^2 = 100$; $x^2 + y^2 = 100$

Symmetrical to origin

$x^2 + (-y)^2 = 100$; $x^2 + y^3 = 100$

Symmetrical to x-axis

41.

$x^2 + x^2 = 32 \Rightarrow x = 4$

$A = (2x)^2 = 8^2 = 64$

45. $m_{TL} = -\dfrac{1}{\frac{0+3}{0-4}} = \dfrac{4}{3}$

equation of TL: $y + 3 = \dfrac{4}{3}(x-4) \Rightarrow$

$4x - 3y - 25 = 0$

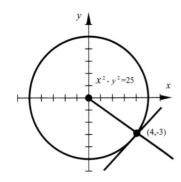

49. $d_2 = 2d_1$

By Pythagoras:

$d_2 = \sqrt{(2-x)^2 + (4-y)^2}$

$d_1 = \sqrt{(x-0)^2 + (y-0)^2}$

$d_2 = 2d_1$

$\sqrt{(2-x)^2 + (4-y)^2} = 2\sqrt{x^2+y^2}$

$(2-x)^2 + (4-y)^2 = 4(x^2+y^2)$

$4-4x+x^2+16-8y+y^2 = 4x^2+4y^2$

$3x^2+4x+3y^2+8y-20 = 0$

This is the equation of a circle.

53. (a) $y = \sqrt{9-(x-2)^2} \Rightarrow y^2 = 9-(x-2)^2 \Rightarrow$

$(x-2)^2 + (y-0)^2 = 3^2$ which is a circle with

center at (2, 0) and radius = 3. $y = \sqrt{9-(x-2)^2}$

represents the top half of this circle, a semicircle.

(b) $y = -\sqrt{9-(x-2)^2}$ represents the bottom

half of the same circle, also a semicircle.

(c) Yes, for each x in the domain there is only one

y.

57. $(x-h)^2 + (y-k)^2 = p > 0$ is a circle

$(x-h)^2 + (y-k)^2 = p = 0$ is the point (h, k).

Since $(x-h)^2 + (y-k)^2 \geq 0 \neq p < 0$ it does not

exist.

61. $x^2 + y^2 = 14.5$ is a circle with center (0, 0),

$r = \sqrt{14.5}$

$x^2 + y^2 - 19.6y + 86 = 0$

$x^2 + y^2 - 19.6y + 96.04 = -86 + 96.04 = 10.04$

$(x-0)^2 + (y-9.8)^2 = 10.04$, circle with center

(0, 9.8),

$r = \sqrt{10.04}$

distance between circles $= 9.8 - \sqrt{10.04} - \sqrt{14.5}$

$\qquad\qquad\qquad\qquad\qquad = 2.82$ in.

65. The center has coordinates $(0, -7)$ the radius $= 1$.

The equation is $x^2 + (y+7)^2 = 1$.

21.4   The Parabola

1. $y^2 = 20x \Rightarrow 4p = 20 \Rightarrow p = 5$. $F(5, 0)$;

directrix $x = -5$

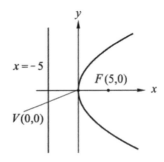

5. $y^2 = 4x$

$y^2 = 4px$

$y^2 = 4x = 4(1)x$; $p = 1$

$F(1, 0)$; directrix $x = -1$

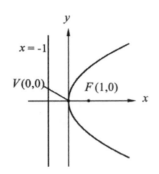

9. $x^2 = 72y$

$x^2 = 4py$

$x^2 = 72y = 4(18)y; \ p = 18$

$F(0, 18)$; directrix $y = -18$

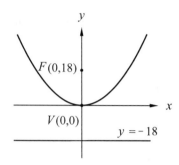

13. $2y^2 - 5x = 0$

$2y^2 = 5x$

$y^2 = \dfrac{5}{2}x = 4px$

$p = \dfrac{5}{8}$

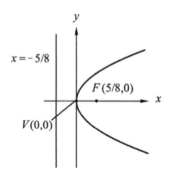

17. $F(3, 0)$; directrix $x = -3, \ p = 3$

$y^2 = 4px$

$y^2 = 4(3)x;$

$y^2 = 12x$

21. $V(0, 0)$, directrix $y = -0.16$

$F(0, 0.16), \ p = 0.16$

$x^2 = 4py = 4(0.16)y$

$x^2 = 0.64y$

25. $V(0, 0)$

Therefore, $x^2 = 4py$

$(-1)^2 = 4p(8); \ 1 = 32p; \ p = \dfrac{1}{32}$

Therefore, $x^2 = \dfrac{1}{8}y.$

29. $y^2 = 4px \Rightarrow 3^2 = 4p(3) \Rightarrow p = \dfrac{3}{4}$

$y^2 = 4\left(\dfrac{3}{4}\right)x$

$y^2 = 3x$

33. $y^2 = 2x \Rightarrow x = \dfrac{y^2}{2}, \ x^2 = -16y$

$$\left(\dfrac{y^2}{2}\right)^2 = -16y$$

$$y^4 + 64y = 0$$

$$y(y^3 + 64) = 0$$

$y = 0$ or $y^3 + 64 = 0 \Rightarrow y^3 = -64$

$2x = 0^2, \ x = 0$ $y = -4$

$(-4)^2 = 2x \Rightarrow x = 8$

The points of intersection are $(0, 0)$ and $(8, -4)$.

37. $y^2 + 2x + 8y + 13 = 0$; solve for y

$y^2 + 8y + (2x + 13) = 0$

$y = \dfrac{-8 \pm \sqrt{8^2 - 4(2x + 13)}}{2}$

$y = \dfrac{-8 \pm \sqrt{12 - 8x}}{2}$

$y_1 = -4 + \sqrt{3 - 2x}, \ y_2 = -4 - \sqrt{3 - 2x}$

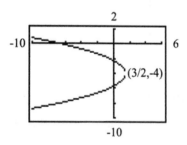

41. $y^2 = 4px$

When $x = p$; $y^2 = 4p^2$; $y = 2p$

Therefore, latus rectum intersects parabola at $(p, 2p)$. Therefore, length of latus rectum is $2(2p) = 4p$.

45.

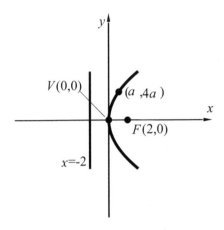

$4(2)(a) = (4a)^2 \Rightarrow 8a - 16a^2 = 0$

$a = 0$, $a = \dfrac{1}{2}$

49.

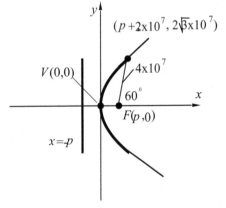

$4p(p + 2 \times 10^7) = (2\sqrt{3} \times 10^7)^2 \Rightarrow p = 1 \times 10^7$ km

10^7 km is the closest distance the comet comes to the sun.

53. $x^2 = 4py$, $10^2 = 4p(3)$, $4p = \dfrac{100}{3}$, $x^2 = \dfrac{100}{3}y$

$\dfrac{100}{3}(340) = 11{,}333 \neq 110^2 = 12{,}100$. No, the path does not lead directly to the moon.

57. The graph is parabolic since it can be transformed into the form $f^2 = 4pA$.

$f = 0.065\sqrt{A}$

$\quad = 0.065\sqrt{200}$

$\quad = 0.92$

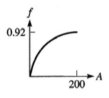

21.5 The Ellipse

1. $\dfrac{x^2}{25} + \dfrac{y^2}{36} = 1,$

$a^2 = 36, a = 6,$

$b^2 = 25, b = 5$

$V(0, \pm 6),$ minor axis: $(\pm 5, 0)$

$a^2 = b^2 + c^2$

$36 = 25 + c^2$

$c = \sqrt{11}$

$F\left(0, \pm \sqrt{11}\right)$

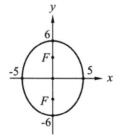

5. $\dfrac{x^2}{25} + \dfrac{y^2}{144} = 1$

$a^2 = 144, a = 12,$

$b^2 = 25, b = 5$

$V(0, \pm 12),$ minor axis: $(\pm 5, 0)$

$a^2 = b^2 + c^2$

$144 = 25 + c^2$

$c = \sqrt{119} \approx 10.9$

$F\left(0, \pm \sqrt{119}\right)$

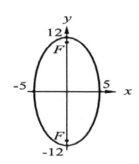

$V(0, \pm 6), F\left(0, \pm \sqrt{11}\right),$ x-intercepts $(\pm 5, 0)$

9. $4x^2 + 9y^2 = 324$

$\dfrac{x^2}{81} + \dfrac{y^2}{36} = 1$

$a^2 = 81, b^2 = 36$

$c^2 = 81 - 36 = 45, c = 3\sqrt{5}$

$V(\pm 9, 0), F\left(\pm 3\sqrt{5}, 0\right),$

y-intercepts $(0, \pm 6)$

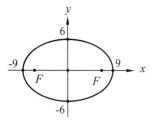

13. $y^2 = 8\left(2 - x^2\right)$

$8x^2 + y^2 = -16$

$\dfrac{8x^2}{16} + \dfrac{y^2}{16} = 1$

$\dfrac{x^2}{2} + \dfrac{y^2}{16} = 1$

$\dfrac{y^2}{16} + \dfrac{x^2}{2} = 1$

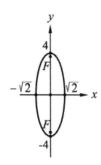

$a^2 = 16, b^2 = 2, c^2 = 16 - 2 = 14$

$V(0, \pm 4), F\left(0, \pm \sqrt{14}\right),$ x-intercepts $\left(\pm \sqrt{2}, 0\right)$

17. $V(15, 0)$; $F(9, 0)$

$a = 15$, $a^2 = 225$;

$c = 9$, $c^2 = 81$; $a^2 - c^2 = b^2$

$b^2 = 144$; $\dfrac{x^2}{a^2} + \dfrac{y^2}{b^2} = 1$;

$\dfrac{x^2}{225} + \dfrac{y^2}{144} = 1$

$144x^2 + 225y^2 = 32,400$

21. $F(8, 0) \Rightarrow c = 8$

end of minor axis: $(0, 12) \Rightarrow b = 12$

$a^2 = b^2 + c^2 \Rightarrow a^2 = 12^2 + 8^2 \Rightarrow a = \sqrt{208}$

$\dfrac{x^2}{\sqrt{208}^2} + \dfrac{y^2}{12^2} = 1$

$\dfrac{x^2}{208} + \dfrac{y^2}{144} = 1$

25. $(x_1, y_1) = (2, 2)$, $(x_2, y_2) = (1, 4)$

$\dfrac{x^2}{b^2} + \dfrac{y^2}{a^2} = 1$

Substitute: $\dfrac{4}{b^2} + \dfrac{4}{a^2} = 1$

Therefore, $4a^2 + 4b^2 = a^2 b^2$

$\dfrac{1}{b^2} + \dfrac{16}{a^2} = 1$

Therefore, $a^2 + 16b^2 = a^2 b^2$

$a^2 + 16b^2 = a^2 b^2$

$16b^2 = a^2 b^2 - a^2 = a^2\left(b^2 - 1\right)$

Therefore, $a^2 = \dfrac{16b^2}{b^2 - 1}$

Substitute:

$4a^2 + 4b^2 = a^2 b^2$; $\dfrac{64b^2}{b^2 - 1} + 4b^2 = \dfrac{16b^4}{b^2 - 1}$

$64b^2 + 4b^4 - 4b^2 = 16b^4$; $-12b^4 + 60b^2 = 0$

$12b^2\left(-b^2 + 5\right) = 0$

$b^2 = 5$

Therefore, $a^2 = \dfrac{16(5)}{4} = 20$

Therefore, $\dfrac{y^2}{20} + \dfrac{x^2}{5} = 1$

or: $5y^2 + 20x^2 = 100$; $4x^2 + y^2 = 20$

29. $\quad 4x^2 + 9y^2 = 40$, $\quad y^2 = 4x$

$\quad 4x^2 + 9(4x) = 40$

$\quad x^2 + 9x - 10 = 0$

$\quad (x + 10)(x - 1) = 0$

33. $4x^2 + 3y^2 + 16x - 18y + 31 = 0$; solve for y

$3y^2 - 18y + \left(4x^2 + 16x + 31\right) = 0$

$y = \dfrac{18 \pm \sqrt{(-18)^2 - 4(3)\left(4x^2 + 16x + 31\right)}}{2(3)}$

$\quad = \dfrac{18 + \sqrt{-48x^2 - 192x - 48}}{2}$

$y_1 = 3 + \dfrac{\sqrt{-12x^2 - 48x - 12}}{3}$

$y_2 = 3 - \dfrac{\sqrt{-12x^2 - 48x - 12}}{3}$

37. $x^2 + y^2 = 1$, $k > 0$

Therefore, $\dfrac{x^2}{1} + \dfrac{y^2}{\frac{1}{k}} = 1$

$\dfrac{1}{k} = a^2 = 1$

Therefore, $\sqrt{\dfrac{1}{k}} = a > 1 \Rightarrow \dfrac{1}{k} > 1$

Therefore, $k < 1$

41. $100x^2 + 49y^2 < 4900$

$100x^2 + 49y^2 = 4900$

$\dfrac{x^2}{49} + \dfrac{y^2}{100} = 1$ is the elliptical dashed boundary

Test point $(0, 0)$

$100(0)^2 + 49(0)^2 < 4900$

$\qquad\qquad 0 < 4900$, T

Shade inside ellipse

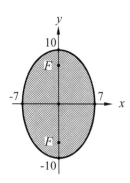

45. $8x + 9y = 25 \Rightarrow y = \dfrac{25 - 8x}{9}$

$4x^2 + 9\left(\dfrac{25 - 8x}{9}\right)^2 = 25 \Rightarrow x^2 - 4x + 4 = 0$

from which $x = 2$, $y = \dfrac{25 - 8(2)}{9} = 1$

$8x + 9y = 25$ is the tangent to $4x^2 + 9y^2 = 25$

at $(2, 1)$.

49. $a = \dfrac{9.0 + 3.5}{2} = 6.25$

$c = 6.25 - 3.5 = 2.75$

$e = \dfrac{c}{a} = \dfrac{2.75}{6.5} = 0.44$

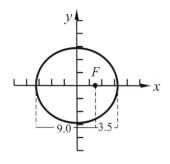

53. $36x^2 + 225y^2 = 8100$

Two people must be separated by a distance $= 2c$.

$\dfrac{36x^2}{8100} + \dfrac{225y^2}{8100} = 1$

$\dfrac{x^2}{225} + \dfrac{y^2}{36} = 1$

$a^2 = 225;\ a = 15$

$b^2 = 36;\ b = 6$

$c^2 = a^2 - b^2 = 225 - 36 = 189;\ c = 13.748$

$2c = 27.5$ m

57. $9x^2 + 20y^2 = 180;\ \dfrac{9x^2}{180} + \dfrac{20y^2}{180} = 1;\ \dfrac{x^2}{20} + \dfrac{y^2}{9} = 1$

$a = \sqrt{20} = 2\sqrt{5} \approx 4.5;\ b = 3;$

$V = $ area of end $\times$ length

$V = \pi ab \times 20.0$

$V = \pi\left(2\sqrt{5}\right)(3)(20.0) = 843$ ft^3

21.6   The Hyperbola

1. $\dfrac{y^2}{16} - \dfrac{x^2}{4} = 1$

$a^2 = 16,\ a = 4$

$b^2 = 4,\ b = 2$

$c^2 = a^2 + b^2 = 20$

$c = 2\sqrt{5}$

$V(0, \pm 4)$

conjugate axis: $(\pm 2, 0)$

$F\left(0, \pm 2\sqrt{5}\right)$

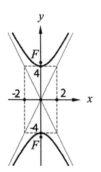

5. $\dfrac{y^2}{9} - \dfrac{x^2}{1} = 1$

$a^2 = 9,\ a = 3$

$b^2 = 1,\ b = 1$

$c^2 = 10;\ c = \sqrt{10}$

$V\left(0,\ \pm 3\right),\ F\left(0,\ \pm\sqrt{10}\right)$

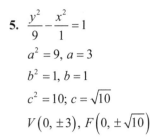

9. $9x^2 - y^2 = 4$

$\dfrac{x^2}{\frac{4}{9}} - \dfrac{y^2}{4} = 1;$

$a = \dfrac{2}{3},\ b = 2$

$c = \sqrt{\dfrac{4}{9} + 4} = \dfrac{2\sqrt{10}}{3}$

$V\left(\pm\dfrac{2}{3},\ 0\right),\ F\left(\pm\dfrac{2\sqrt{10}}{3},\ 0\right)$

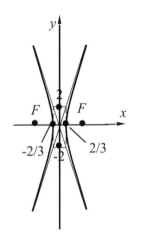

13. $y^2 = 4\left(x^2 + 1\right)$

$4x^2 - y^2 + 4 = 0$

$4x^2 - y^2 = -4$

$\dfrac{4x^2}{-4} - \dfrac{y^2}{-4} = \dfrac{-4}{-4}$

$-x^2 + \dfrac{y^2}{4} = 1$

$\dfrac{y^2}{4} - \dfrac{x^2}{1} = 1$

$a^2 = 4;\ b^2 = 1;\ c^2 = 5$

$V\left(0,\ \pm 2\right),\ F\left(0,\ \pm\sqrt{5}\right)$

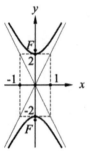

17. $V\left(3,\ 0\right);\ F\left(5,\ 0\right)$

$a = 3;\ c = 5;\ a^2 = 9;\ c^2 = 25$

$b^2 = c^2 - a^2 = 25 - 9 = 16$

$\dfrac{x^2}{a^2} - \dfrac{y^2}{b^2} = 1;\ \dfrac{x^2}{9} - \dfrac{y^2}{16} = 1;$

$16x^2 - 9y^2 = 144$

21. $\left(x,\ y\right)$ is $\left(2,\ 3\right);\ F\left(2,\ 0\right),\ \left(-2,\ 0\right);\ c = \pm 2,\ c^2 = 4$

$d_1 = \sqrt{\left(2 - \left(-2\right)\right)^2 + \left(3 - 0\right)^2}$

$\quad\ = \sqrt{4^2 + 3^2} = \sqrt{16 + 9}$

$\quad\ = \sqrt{25} = 5$

$d_2 = \sqrt{\left(2 - 2\right)^2 + \left(3 - 0\right)^2}$

$\quad\ = \sqrt{0 + 9} = \sqrt{9} = 3$

$d_1 - d_2 = 2a;\ 5 - 3 = 2a;\ 2 = 2a;\ 1 = a;\ a^2 = 1$

$c^2 = 4;\ b^2 = c^2 - a^2 = 3$

$\dfrac{x^2}{a^2} - \dfrac{y^2}{b^2} = 1;\ \dfrac{x^2}{1} - \dfrac{y^2}{3} = 1$

$3x^2 - y^2 = 3$

25. $V(1, 0) \Rightarrow a = 1, a^2 = 1$

Asymptote $y = \dfrac{b}{a}x = \dfrac{b}{1}x = 2x \Rightarrow b = 2, b^2 = 4$

$\dfrac{x^2}{1} - \dfrac{y^2}{4} = 1$

29. $xy = 2$; $y = \dfrac{2}{x}$

x	y
$\pm\frac{1}{2}$	± 4
± 1	± 2
± 2	± 1
± 4	$\pm\frac{1}{2}$
± 8	$\pm\frac{1}{4}$

33. $2x^2 + y^2 = 17$, $y^2 - x^2 = 5 \Rightarrow y^2 = x^2 + 5$

$2x^2 + x^2 + 5 = 17$

$3x^2 = 12$

$x^2 = 4$

$x = \pm 2$

$y^3 = x^2 + 5 = 4 + 5 = 9 \Rightarrow 9 = \pm 3$

points of intersection $(2, \pm 3), (-2, \pm 3)$

37. $x^2 - 4y^2 + 4x + 32y - 64 = 0$; solve for y

$4y^2 - 34y + \left(-x^2 - 4x + 64\right) = 0$

$y = \dfrac{32 \pm \sqrt{(-32)^2 - 4(4)\left(-x^2 - 4x + 64\right)}}{2(4)}$

$= \dfrac{32 \pm \sqrt{16x^2 + 64x}}{8}$

$y_1 = 4 + 0.5\sqrt{x^2 + 4x}$, $y_2 = 4 - 0.5\sqrt{x^2 + 4x}$

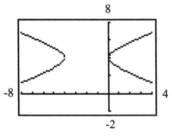

41. $V(0, 1), F\left(0, \sqrt{3}\right); c^2 = a^2 + b^2$ where $c = \sqrt{3}$ and

$a = 1; b^2 = \sqrt{3}^2 - 1^2 = 2$

$\dfrac{y^2}{1^2} - \dfrac{x^2}{\sqrt{2}^2} = 1$

The transverse axis of the first equation is length $2a = 2\sqrt{1}$ along the y-axis. Its conjugate axis is length $2b = 2\sqrt{2}$ along the x-axis.

The transverse axis of the conjugate hyperbola is length $2\sqrt{2}$ along the x-axis, and its conjugate axis is length $2\sqrt{1}$ along the y-axis.

The equation, then, is $\dfrac{x^2}{\sqrt{2}^2} - \dfrac{y^2}{\sqrt{1}^2} = 1$

$\dfrac{x^2}{2} - \dfrac{y^2}{1} = 1$ or $x^2 - 2y^2 = 2$

45. $\dfrac{x^2}{169} + \dfrac{y^2}{144} = 1 \Rightarrow \dfrac{x^2}{13^3} + \dfrac{y^2}{12^2} = 1$

$13^3 = 12^2 + c^2 \Rightarrow c = 5$

$\dfrac{x^2}{a^2} - \dfrac{y^2}{b^2} = 1,$

$c^2 = a^2 + b^2 \Rightarrow 5^2 = a^2 + b^2, 0 < a < 5$

$b^2 = 25 - a^2$

$\dfrac{\left(4\sqrt{2}\right)^2}{a^2} - \dfrac{9}{25 - a^2} = 1$

$a^4 - 66a^2 + 800 = 1$

$\left(a^2 - 16\right)\left(a^2 - 50\right) = 0$

$a^4 - 16 = 0$ or $a^2 - 50 = 0$

$a = 4$ $\qquad a = \sqrt{50} > 5$, reject

$b^2 = 25 - 4^2 = 9 \Rightarrow b = 3$

$\dfrac{x^2}{16} - \dfrac{y^2}{9} = 1$

49. For $a > 0, V(0, -a)$

For $c > 0, c = a + 3$

$c^2 = a^2 + b^2$

$(a + 3)^2 = a^2 + b^2$

$a^2 + 6a + 9 = a^2 + b^2$

$b^2 3(2a + 3)$

$$\frac{y^2}{a^2} - \frac{x^2}{b^2} = 1$$

$$\frac{\left(-(a+3)\right)^2}{a^2} - \frac{9^2}{b^2} = 1$$

$$\frac{(a+3)^2}{a^2} - \frac{81}{3(2a+3)} = 1$$

$$3(2a+3)(a+3)^2 - 81a^2 = 3a^2(2a+3)$$

$$(6a+9)(a^2+6a+9) - 81a^2 = 6a^3 + 9a^2$$

$$6a^3 + 9a^2 + 36a^2 + 54a + 54a + 81 - 81a^2 = 6a^3 + 9a^2$$

$$45a^2 - 108a - 81 = 0$$

$$(a-3)(5a+3) = 0$$

$$a - 3 = 0 \quad \text{or} \quad 5a + 3 = 0$$

$$a = 3 \qquad\qquad a = -\frac{3}{5} \text{ reject, } a > 0$$

$$a^2 = 9$$

$$b^2 = 3(2a+3) = 3(2(3)+3)$$

$$b^2 = 27$$

$$\frac{y^2}{9} - \frac{x^2}{27} = 1$$

53. $V = iR$ (Ohm's law)

$6.00 = iR$

Therefore, $i = \dfrac{6.00}{R}$

R	i
0.5	12
1	6
2	3
3	2
4	1.5
6	1
9	0.7
12	0.5

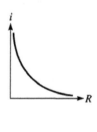

21.7   Translation of Axes

1. $\dfrac{(x-3)^2}{25}-\dfrac{(y-2)^2}{9}=1$, hyperbola: $a=5,\ b=3$

Center: $(3,\,2)$. Transverse axis parallel to x-axis.

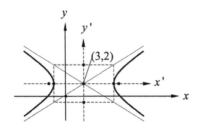

5. $\dfrac{(x-1)^2}{4}-\dfrac{(y-2)^2}{25}=1$; eq. (21-32), hyperbola

Center: $(1,\,2)$; $a=2$; $b=5$

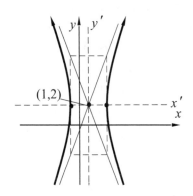

9. $(x+3)^2=-12(y-1)$; eq. (21-29), parabola

$x'=x+3;\ y'=y-1$

$x'^2=-12y'$

Origin O' at $(h,\,k)=(-3,\,1)$

$x'^2=4(3y')$; therefore $p=3$

Vertex $(-3,\,1)$, focus $(-3,\,-2)$, directrix $y=4$

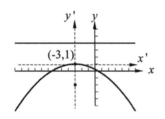

13. $F(12,\,0),\ V(6,\,0),\ p=6$

$(y-k)^2=4p(x-h)^2$

$(y-0)^2=4\cdot6(x-6)$

$y^2=24(x-6)$

17. Ellipse: center $(-2,\,1)$, vertex $(-2,\,5)$, passes through $(0,\,1)$.

$\dfrac{(y-1)^2}{4^2}+\dfrac{(x+2)^2}{b^2}=1$

$\dfrac{(1-1)^2}{4^2}+\dfrac{(0+2)^2}{b^2}=1$

$b^2=4$

$\dfrac{(y-1)^2}{16}+\dfrac{(x+2)^2}{4}=1$

21. Hyperbola: $V(2,1),\ V(-4,\,1),\ F(-6,\,1)$

Center: $(h,\,k)=(-1,\,1)$

$2a=6;\ a=3;\ c=5$

Therefore, $b^2=c^2-a^2=25-9=16$

Transverse axis parallel to x-axis.

Therefore, $\dfrac{(x-h)^2}{a^2}-\dfrac{(y-k)^2}{b^2}=1$

$\dfrac{(x+1)^2}{9}-\dfrac{(y-1)^2}{16}=1$

or $16x^2-9y^2+32x+18y-137=0$

25. $x^2+4y=24$

$x^2=-4y+24$

$x^2=-4(y-6)$ is a parabola with $V(0,\,6)$

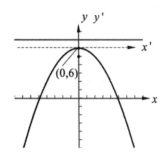

29. $9x^2 - y^2 + 8y = 7$

$9x^2 - y^2 + 8y - 7 = 0$

$9x^2 - \left(y^2 - 8y + 7 + 9\right) = -9$

$9x^2 - \left(y - 4\right)^2 = -9$

$\dfrac{9x^2}{-9} + \dfrac{\left(y - 4\right)^2}{-9} = 1$

$-x^2 + \dfrac{\left(y - 4\right)^2}{9} = 1$

$\dfrac{\left(y - 4\right)^2}{9} - \dfrac{x^2}{1} = 1$

Hyperbola, $\left(h, k\right)$ is $\left(0, 4\right)$; $a = 3$; $b = 1$

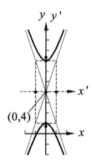

33. $4x^2 - y^2 + 32x + 10y + 35 = 0$

$4\left(x^2 + 8x\right) - \left(y^2 - 10y\right) = -35$

$4\left(x^2 + 8x + 16\right) - \left(y^2 - 10y + 25\right) = -35 + 64 - 25$

$\dfrac{\left(x + 4\right)^2}{1^2} - \dfrac{\left(y - 5\right)^2}{2^2} = 1$, hyperbola

$C\left(-4, 5\right)$

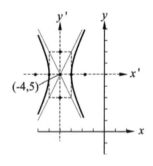

37. $5x^2 - 3y^2 + 95 = 40x$

$5\left(x^2 - 8x + 16\right) - 3y^2 = -95 + 80 = -15$

$\dfrac{y^2}{5} - \dfrac{\left(x - 4\right)^2}{3} = 1$, hyperbola, center $\left(4, 0\right)$

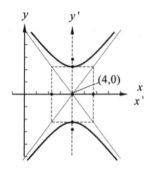

41. Hyperbola: asymptotes: $x - y = -1$ or $x + 1 = y$, and $x + y = -3$ or $y = -x - 3$; vertices $\left(3, -1\right)$ and $\left(-7, -1\right)$. The center is at the point of inter-action of the asymptotes. The equation for the asymptotes are solved simultaneously by adding, $2y = -2$; $y = -1$; $-1 = x + 1$; $x = -2$. Therefore, the coordinates of the center are $\left(-2, -1\right)$. Since the slopes are 1 and -1, $a = b$, where a is the distance from the center $\left(-2, -1\right)$ to the vertex $\left(3, -1\right)$; $a = 5$, $b = 5$.

$\dfrac{\left(x - h\right)^2}{a^2} - \dfrac{\left(y - k\right)^2}{b^2} = 1;$

$\dfrac{\left[x - \left(-2\right)\right]^2}{25} - \dfrac{\left[y - \left(-1\right)\right]^2}{25} = 1$

$\dfrac{\left(x + 2\right)^2}{25} - \dfrac{\left(y + 1\right)^2}{25} = 1$

$x^2 + 4x + 4 - \left(y^2 + 2y + 1\right) = 25;$

$x^2 + 4x + 4 - 2y - 1 = 25$

$x^2 - y^2 + 4x - 2y - 22 = 0$

45. Parabola: vertex and focus on x-axis.

$y^2 = 4p\left(x - h\right)$

49. $i = 2 + \sin\left(2\pi t - \dfrac{\pi}{3}\right)$

$i = i' + 2$

$t = t' + \dfrac{1}{6}$

$i' + 2 = 2 + \sin\left(2\pi\left(t' + \dfrac{1}{6}\right) - \dfrac{\pi}{3}\right)$

$i' = \sin\left(2\pi t' + \dfrac{\pi}{3} - \dfrac{\pi}{3}\right)$

$i' = \sin\left(2\pi t'\right)$

53. First ellipse:

$a = 4$, $b = 3$, therefore $c = \sqrt{7}$

Center $(0, 0)$

$\dfrac{y^2}{a^2} + \dfrac{x^2}{b^2} = 1$; $\dfrac{y^2}{16} + \dfrac{x^2}{9.0} = 1$

Second ellipse:

$a = 4$, $b = 3$, therefore $c = \sqrt{7}$

Center $(7.0, 0.0)$

$\dfrac{(x-h)^2}{a^2} + \dfrac{(y-k)^2}{b^2} = 1$; $\dfrac{(x-7)^2}{16} + \dfrac{y^2}{9.0} = 1$

17. $2xy + x - 3y = 6$

$A = 0$; $B \neq 0$; $C = 0$; hyperbola

21. $(x+1)^2 + (y+1)^2 = 2(x+y+1)$

$x^2 + 2x + 1 + y^2 + 2y + 1 = 2x + 2y + 2$

$x^2 + y^2 = 0$, point $(0, 0)$ is the only solution

25. $x^2 = 8(y - x - 2)$

$x^2 = 8y - 8x - 16$

$x^2 + 8x - 8y + 16 = 0$

$A \neq 0$; $B = 0$;

$C = 0$; parabola

$x^2 + 8x - 8y + 16 = 0$

$x^2 + 8x + 16 = 8y$

$(x+4)^2 = 4(2)y$; $p = 2$

Vertex $(-4, 0)$, focus $(-4, 2)$

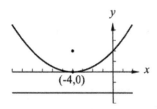

21.8   The Second-Degree Equation

1. $2x^2 = 3 + 2y^2$

$2x^2 - 2y^2 - 3 = 0$, A, C have different signs, $B = 0$,

hyperbola

5. $2x^2 - y^2 - 1 = 0$

A and C have different signs, $B = 0$; hyperbola

9. $2.2x^2 - (x + y) = 1.6$

$2.2x^2 - x - y = 1.6$

$A \neq 0$; $C = 0$; $B = 0$; parabola

13. $36x^2 = 12y(1 - 3y) + 1$

$36x^2 = 12y - 36y^2 + 1$

$36x^2 + 0 \cdot xy + 36y^2 - 12x - 0 \cdot y - 1 = 0$,

$A = C$, $B = 0$, circle

29. $y^2 + 42 = 2x(10 - x)$

$y^2 + 42 = 20x - 2x^2$

$y^2 + 2x^2 - 20x + 42 = 0$; ellipse

$\dfrac{y^2}{2} + x^2 - 10x = -21$

$\dfrac{y^2}{2} + x^2 - 10x + 25 = -21 + 25$

$\dfrac{y^2}{2} + (x - 5)^2 = 4$

$\dfrac{y^2}{8} + \dfrac{(x-5)^2}{4} = 1$

(h, k) at $(5, 0)$, $V\left(5, \pm 2\sqrt{2}\right)$

$a = \sqrt{8} = 2\sqrt{2}$; $b = 2$

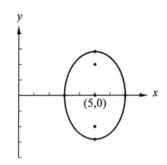

41. In $Ax^2 + Bxy + Cy^2 + Dx + Ey + F = 0$,
$A = B = C = 0$, $D \neq 0$, $E \neq 0$, $F \neq 0$, then
the equation is $Dx + Ey + F = 0$ whose locus is a
straight line.

45. Flashlight parallel to floor, hyperbola.

49. Shape of curve of lake is one branch of a hyperbola.

33. $4\left(y^2 - 4x - 2\right) = 5\left(4y - 5\right)$

$4y^2 - 16x - 8 = 20y - 25$

$4y^2 - 20y - 16x + 17 = 0$

$A = 0;\ C = 4;\ B = 0;$ parabola

$y^2 - 5y - 4x + \dfrac{17}{4} = 0$

$y_1 = \dfrac{5 + \sqrt{25 - 4\left(-4x + \frac{17}{4}\right)}}{2} = \dfrac{5 + \sqrt{16x + 8}}{2}$

$y_2 = \dfrac{5 - \sqrt{16x + 8}}{2}$

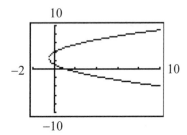

37. (a) If $k = 1$, $x^2 + ky^2 = a^2$; $x^2 + (1)y^2 = a^2$

$x^2 + y^2 = a^2$ (circle)

(b) If $k < 0$, $x^2 + ky^2 = a^2$; $x^2 - |k|y^2 = a^2$

$\dfrac{x^2}{a^2} - \dfrac{y^2}{a^2 / |k|} = 1$ (hyperbola)

(c) If $k > 0\,(k \neq 1)$, $x^2 + ky^2 = a^2$

$\dfrac{x^2}{a^2} + \dfrac{y^2}{a^2 / k} = 1$ (ellipse)

21.9   Rotation of Axes

1. $x^2 - y^2 = 25$, $\theta = 45°$; $x = x'\cos 45° - y'\sin 45°$

$= \dfrac{x'}{\sqrt{2}} - \dfrac{y'}{\sqrt{2}}$; $y = x'\sin 45° + y'\cos 45° = \dfrac{x'}{\sqrt{2}} + \dfrac{y'}{\sqrt{2}}$

$x^2 - y^2 = \left(\dfrac{x'}{\sqrt{2}} - \dfrac{y'}{\sqrt{2}}\right)^2 - \left(\dfrac{x'}{\sqrt{2}} + \dfrac{y'}{\sqrt{2}}\right)^2 = 25$;

$\dfrac{x'^2}{2} - \dfrac{2x'y'}{2} + \dfrac{y'^2}{2} - \dfrac{x'^2}{2} - \dfrac{2x'y'}{2} - \dfrac{y'^2}{2} = 25$

$2x'y' + 25 = 0$, hyperbola

5. $x^2 + 2xy + x - y - 3 = 0$

$B^2 - 4AC = 2^2 - 4(1)(0) = 4 > 0$, hyperbola

9. $13x^2 + 10xy + 13y^2 + 6x - 42y - 27 = 0$

$B^2 - 4AC = 10^2 - 4(13)(13) = -576 < 0$, ellipse

13. $xy = 8$, $A = C = 0$

$\tan 2\theta = \dfrac{B}{A-C} = \dfrac{1}{0-0} \Rightarrow \theta = 45°$

$\left(\dfrac{x'-y'}{\sqrt{2}}\right)\left(\dfrac{x'+y'}{\sqrt{2}}\right) = 8$

$x'^2 - y'^2 = 16$, hyperbola

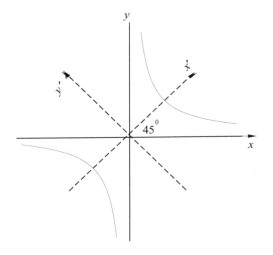

17. $11x^2 - 6xy + 19y^2 = 20$

$\tan 2\theta = \dfrac{B}{A-C} = \dfrac{-6}{11-19} = \dfrac{6}{8} = \dfrac{3}{4}$;

$\cos 2\theta = \dfrac{4}{5}$

$\sin\theta = \sqrt{\dfrac{1-\cos 2\theta}{2}} = \sqrt{\dfrac{1-\frac{4}{5}}{2}} = \dfrac{1}{\sqrt{10}}$;

$\cos\theta = \sqrt{\dfrac{1+\cos 2\theta}{2}} = \sqrt{\dfrac{1+\frac{4}{5}}{2}} = \dfrac{3}{\sqrt{10}}$

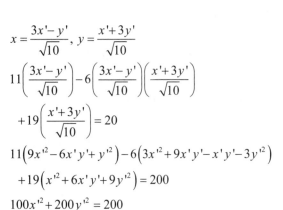

$x = \dfrac{3x'-y'}{\sqrt{10}}$, $y = \dfrac{x'+3y'}{\sqrt{10}}$

$11\left(\dfrac{3x'-y'}{\sqrt{10}}\right) - 6\left(\dfrac{3x'-y'}{\sqrt{10}}\right)\left(\dfrac{x'+3y'}{\sqrt{10}}\right)$

$+ 19\left(\dfrac{x'+3y'}{\sqrt{10}}\right) = 20$

$11\left(9x'^2 - 6x'y' + y'^2\right) - 6\left(3x'^2 + 9x'y' - x'y' - 3y'^2\right)$

$+ 19\left(x'^2 + 6x'y' + 9y'^2\right) = 200$

$100x'^2 + 200y'^2 = 200$

$x'^2 + 2y'^2 = 2$, ellipse

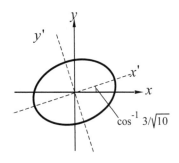

21.
$$2x^2 + xy + y^2 = 0, \quad B^2 - 4AC = 1^2 - 4(2)(1)$$
$$= -7 < 0, \text{ ellipse}$$

$$\tan 2\theta = \frac{1}{2-1} \Rightarrow \sin\theta = \frac{\sqrt{2-\sqrt{2}}}{2}, \quad \cos\theta = \frac{\sqrt{2+\sqrt{2}}}{2}$$

$$2\left(\frac{x'\sqrt{2+\sqrt{2}}}{2} - \frac{y'\sqrt{2-\sqrt{2}}}{2}\right)^2 + \left(\frac{x'\sqrt{2+\sqrt{2}}}{2} - \frac{y'\sqrt{2-\sqrt{2}}}{2}\right)$$

$$\left(\frac{x'\sqrt{2-\sqrt{2}}}{2} + \frac{y'\sqrt{2+\sqrt{2}}}{2}\right) + \left(\frac{x'\sqrt{2-\sqrt{2}}}{2} + \frac{y'\sqrt{2+\sqrt{2}}}{2}\right)^2$$

$$= 0 \Rightarrow$$

$$x'^2\left(4+\sqrt{2}\right) + y'^2\left(4-\sqrt{2}\right) = 0 \Rightarrow \text{graph is single point } (0, 0)$$

25. (a) $Ax^2 + Bxy + Cy^2 + Dx + Ey + F = 0$

$\quad\quad A = -C$

$\quad\quad B^2 - 4AC = B^2 - 4(-C)(C)$

$\quad\quad = B^2 + 4C^2 > 0, \text{ hyperbola}$

$\quad$ (b) $A \neq 0, B \neq 0, C = 0$

$\quad\quad \Rightarrow Ax^2 + Bxy + Dx + Ey + F = 0$

$\quad\quad B^2 - 4AC = B^2 - 4A(0) = B^2 > 0, \text{ hyperbola}$

21.10   Polar Coordinates

1. $\left(3, \dfrac{\pi}{3}\right)$ and $\left(3, -\dfrac{5\pi}{3}\right)$ represent the same point.

$\left(-3, \dfrac{\pi}{3}\right)$ and $\left(3, -\dfrac{2\pi}{3}\right)$ are on the opposite side

of the pole.

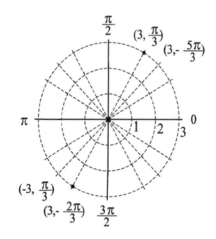

5. $\left(3, \dfrac{\pi}{6}\right); r = 3, \theta = \dfrac{\pi}{6}$

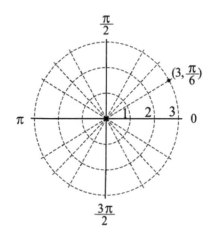

9. $\left(-8, \dfrac{7\pi}{6}\right)$; negative r is reversed in direction

from positive r.

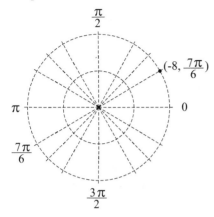

13. $(2, 2)$; $\dfrac{\pi}{180°} = \dfrac{2}{\theta}$; $\theta = \dfrac{360}{\pi} = 114.6°$

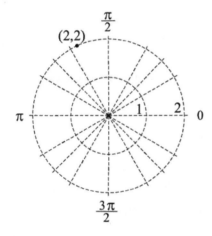

17. $\left(\sqrt{3}, 1\right)$ is (x, y), QI

$\tan \theta = \dfrac{y}{x}$

$\theta = \tan^{-1} \dfrac{y}{x} = \tan^{-1} \dfrac{1}{\sqrt{3}} = \tan^{-1} \dfrac{\sqrt{3}}{3}$;

$\theta = 30° = \dfrac{\pi}{6}$

$r = \sqrt{x^2 + y^2} = \sqrt{\left(\sqrt{3}\right)^2 + 1^2}$

$\quad = \sqrt{3 + 1} = \sqrt{4} = 2$

(r, θ) is $\left(2, \dfrac{\pi}{6}\right)$

21. $(0, 4) \Rightarrow \theta = \dfrac{\pi}{2}$, $r = 4$

$(0, 4) = \left(4, \frac{\pi}{2}\right)$

25. $(-3.0, -0.40)$, QIII

$x = -3.0 \cos -0.4 = -2.76$

$y = -3.0 \sin -0.4 = +1.17$

Therefore, $(3.0, -0.40)$ is $(-2.76, +1.17)$

29. $x = 3$

$r \cos \theta = x = 3$; $r = \dfrac{3}{\cos \theta} = 3 \sec \theta$

33. $x^2 + (y - 2)^2 = 4$

$x^2 + y^2 - 4y + 4 = 4$

$r^2 - 4 \cdot r \sin \theta = 0$

$r = 4 \sin \theta$

37. $x^2 + y^2 = 6y$

$r^2 = 6 \cdot r \sin \theta$

$r = 6 \sin \theta$

41. $r = \sin \theta$; $r^2 = r \sin \theta$; $r^2 = x^2 + y^2$

$x^2 + y^2 = r^2 = r \sin \theta = y$; $x^2 + y^2 - y = 0$,

circle

45. $r = \dfrac{2}{\cos \theta - 3 \sin \theta}$

$r \cos \theta - 3r \sin \theta = 2$

$x - 3y = 2$, line

49. $r = 2(1 + \cos \theta)$; $x = r \cos \theta$; $\dfrac{x}{r} = \cos \theta$

$r^2 = x^2 + y^2$; $r = \sqrt{x^2 + y^2}$

$r = 2(1 + \cos \theta) = 2\left(1 + \dfrac{x}{r}\right) = 2 + \dfrac{2x}{r}$; $r^2 = 2r + 2x$

Multiply through by r.

$x^2 + y^2 = 2\sqrt{x^2 + y^2} + 2x;\ x^2 + y^2 - 2x = 2\sqrt{x^2 + y^2}$

$\left(x^2 + y^2 - 2x\right)^2 = 4\left(x^2 + y^2\right)$

$x^4 + y^4 - 4x^3 + 2x^2 y^2 - 4xy^2 + 4x^2 = 4x^2 + 4y^2$

$x^4 + y^4 - 4x^3 + 2x^2 y^2 - 4xy^2 + 4x^2 - 4x^2 - 4y^2 = 0$

$x^4 + y^4 - 4x^3 + 2x^2 y^2 - 4xy^2 - 4y^2 = 0$

53. As the graph shows the point $\left(2, 3\pi/4\right)$ is on the curve $r = 2\sin 2\theta$ even though $\left(2, 3\pi/4\right)$ is not a solution to $r = 2\sin 2\theta$. $\left(2, 3\pi/4\right)$ and $\left(-2, 7\pi/4\right)$ are the same point and $\left(-2, 7\pi/4\right)$ is a solution to $r = 2\sin 2\theta$.

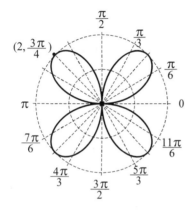

57. Each pair of vertices subtends a central angle of $\dfrac{\pi}{3}$ at the pole. The coordinatges of the other vertices are $(2, 0), \left(2, \dfrac{\pi}{3}\right), \left(2, \dfrac{2\pi}{3}\right), \left(2, \dfrac{4\pi}{3}\right), \left(, \dfrac{5\pi}{3}\right)$

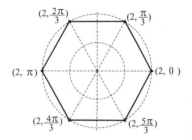

61. $r = 3 - \sin\theta;\ y = r\sin\theta$

$\sqrt{x^2 + y^2} = 3 - \dfrac{y}{r} = 3 - \dfrac{y}{\sqrt{x^2 + y^2}}$

$\sqrt{x^2 + y^2} = \dfrac{3\sqrt{x^2 + y^2} - y}{\sqrt{x^2 + y^2}}$

$x^2 + y^2 = 3\sqrt{x^2 + y^2} - y;\ x^2 + y^2 + y = 3\sqrt{x^2 + y^2}$

Square both sides.

$x^4 + 2x^2 y^2 + 2x^2 y + y^4 + 2y^3 + y^2 = 9\left(x^2 + y^2\right)$

$\qquad\qquad\qquad\qquad\qquad\qquad = 9x^2 + 9y^2$

Therefore,

$x^4 + 2x^2 y^2 + 2x^2 y + y^4 + 2y^3 - 9x^2 - 8y^2 = 0$

21.11   Curves in Polar Coordinates

1. The graph of $\theta = \dfrac{5\pi}{6}$ is a straight line through the pole. $\theta = \dfrac{5\pi}{6}$ for all possible values of r.

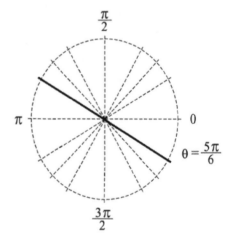

5. $r = 5$ for all θ. Graph is a circle with radius 5.

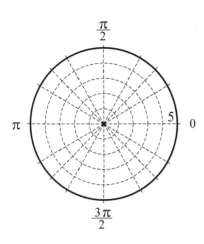

9. $r = 4\sec\theta = \dfrac{4}{\cos\theta}$; vertical line

θ	r
0	4
$\frac{\pi}{6}$	4.6
$\frac{\pi}{4}$	5.7
$\frac{\pi}{3}$	8
$\frac{\pi}{2}$	*
$\frac{2\pi}{3}$	-8
$\frac{3\pi}{4}$	-5.7
$\frac{5\pi}{6}$	4.6
π	-4
$\frac{5\pi}{4}$	-5.7
$\frac{3\pi}{2}$	*
$\frac{7\pi}{4}$	5.7
2π	4

* denotes undefined

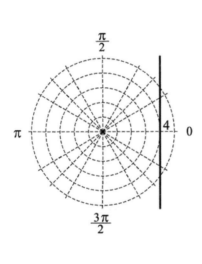

13. $r = 1 - \cos\theta$; cardioid

θ	r
0	0
$\frac{\pi}{4}$	0.3
$\frac{\pi}{2}$	1
$\frac{3\pi}{4}$	1.7
π	2
$\frac{5\pi}{4}$	1.7
$\frac{3\pi}{2}$	1
$\frac{7\pi}{4}$	0.3

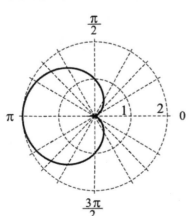

17. $r = 4\sin 2\theta$; rose (4 petals)

θ	r
0	0
$\frac{\pi}{8}$	2.8
$\frac{\pi}{4}$	4
$\frac{3\pi}{8}$	-2.8
$\frac{\pi}{2}$	0
$\frac{5\pi}{8}$	2.8
$\frac{3\pi}{4}$	-4
$\frac{7\pi}{8}$	-2.8
π	0
$\frac{9\pi}{8}$	2.8
$\frac{5\pi}{4}$	4
$\frac{11\pi}{8}$	2.8
$\frac{3\pi}{2}$	0
$\frac{13\pi}{8}$	-2.8
$\frac{7\pi}{4}$	-4
2π	-2.8

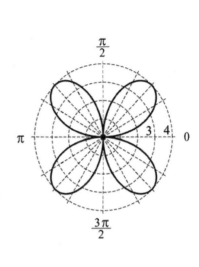

21. $r = 2^{\theta}$; spiral

θ	r
0	1
$\frac{\pi}{4}$	1.7
$\frac{\pi}{2}$	3.0
$\frac{3\pi}{4}$	5.1
π	8.8
$\frac{5\pi}{4}$	15.2
$\frac{3\pi}{2}$	26.2
$\frac{7\pi}{4}$	45.2
2π	77.9

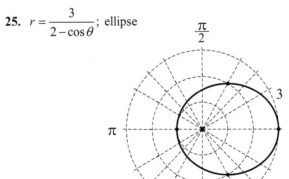

25. $r = \dfrac{3}{2 - \cos\theta}$; ellipse

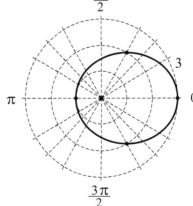

θ	r
0	3
$\frac{\pi}{4}$	2.32
$\frac{\pi}{2}$	1.5
$\frac{3\pi}{4}$	1.11
π	1
$\frac{5\pi}{4}$	1.11
$\frac{3\pi}{2}$	1.5
$\frac{7\pi}{4}$	2.32
2π	3

29. $r = 4\cos\dfrac{1}{2}\theta$

θ	r	θ	r
0	4.0	$\frac{13\pi}{6}$	−3.9
$\frac{\pi}{6}$	3.9	$\frac{9\pi}{4}$	−3.7
$\frac{\pi}{4}$	3.7	$\frac{7\pi}{3}$	−3.5
$\frac{\pi}{3}$	3.5	$\frac{5\pi}{2}$	−2.8
$\frac{\pi}{2}$	2.8	$\frac{8\pi}{3}$	−2.0
$\frac{2\pi}{3}$	2.0	$\frac{11\pi}{4}$	−1.5
$\frac{3\pi}{4}$	1.5	$\frac{17\pi}{6}$	−1.0
$\frac{5\pi}{6}$	1.0	3π	0
π	0	$\frac{19\pi}{6}$	1.0
$\frac{7\pi}{6}$	−1.0	$\frac{13\pi}{4}$	1.5
$\frac{5\pi}{4}$	−1.5	$\frac{10\pi}{3}$	2.0
$\frac{4\pi}{3}$	−2.0	$\frac{7\pi}{2}$	2.8
$\frac{3\pi}{2}$	−2.8	$\frac{11\pi}{3}$	3.5
$\frac{5\pi}{3}$	−3.5	$\frac{15\pi}{4}$	3.7
$\frac{7\pi}{4}$	−3.7	$\frac{23\pi}{6}$	3.9
$\frac{11\pi}{6}$	−3.9	4π	4.0
2π	−4.0		

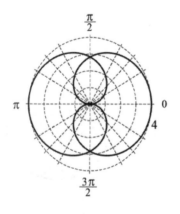

33. $r = \theta \quad (-20 \le \theta \le 20)$

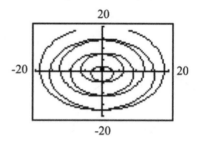

37. $r = 3\cos 4\theta$

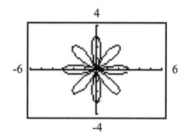

41. $r = \cos\theta + \sin 2\theta$

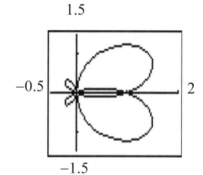

45. From the calculator screen the curves intersect at $(0, 0)$ and $(1, 1)$ where the tangent lines are horizontal and vertical showing the curves intersect at right angles.

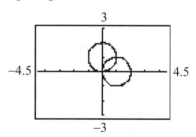

49. $r = 4.0 - \sin\theta$

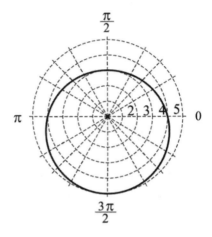

53. $R = \dfrac{\sin^2\theta}{\left(1 - 0.5\cos\theta\right)^2}$

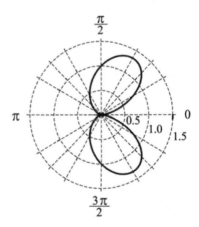

Chapter 21   Review Exercises

1. Given straight line; (x_1, y_1) is $(1, -7)$; $m = 4$

$y - y_1 = m(x - x_1);\ y - (-7) = 4(x - 1)$

$y + y = 4x - 4;\ y = 4x - 4 - 7$

$y = 4x - 11\ \text{or}\ 4x - y - 11 = 0$

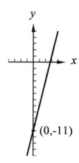

5. $x^2 + y^2 = 6x$

$x^2 - 6x + 9 + y^2 = 9$

$(x - 3)^2 + y^2 = 3^2;\ \text{circle, center } (3, 0),\ r = 3$

The concentric circle has equation

$(x - 3)^2 + y^2 = r^2\ \text{and passes through } (4, -3)$

$(4 - 3)^2 + (-3)^2 = r^2$

$\qquad\qquad r^2 = 10$

$\quad (x - 3)^2 + y^2 = 10$

$x^2 - 6x + 9 + y^2 = 10$

$x^2 - 6x + y^2 - 1 = 0$

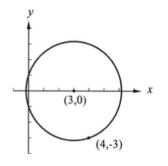

9. $a = 10,\ c = 8$

$a^2 = b^2 + c^2$

$100 = b^2 + 8^2 \Rightarrow b^2 = 36$

$\dfrac{x^2}{100} + \dfrac{y^2}{36} = 1\ \text{or}\ 9x^2 + 25y^2 = 900$

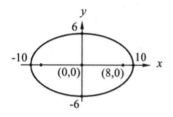

13. Given: $x^2 + y^2 + 6x - 7 = 0$

$(x^2 + 6x) + (y^2) = 7;\ (x^2 + 6x + 9) + y^2 = 7 + 9$

$(x + 3)^2 + (y + 0)^2 = 16$

$[x - (-3)]^2 + (y - 0)^2 = 4^2$

$C(h, k) = (-3, 0);\ r = 4$

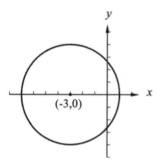

17. $8x^2 + 2y^2 = 2$

Given: $4x^2 + y^2 = 1$

$\dfrac{x^2}{\frac{1}{4}} + \dfrac{y^2}{1} = 1$

$a = 1,\ b = \frac{1}{4},\ c = \sqrt{1 - \frac{1}{4}} = \dfrac{\sqrt{3}}{2}$

vertices $(0, 1),\ (0, -1)$

foci: $\left(0, \dfrac{\sqrt{3}}{2}\right),\ \left(0, \dfrac{-\sqrt{3}}{2}\right)$

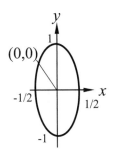

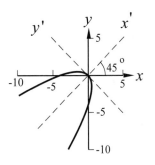

21. Given: $x^2 - 8x - 4y - 16 = 0$

$x^2 - 8x = 4y + 16$; $x^2 - 8x + 16 = 4y + 16 + 16$

$(x-4)^2 = 4y + 32$; $(x-4)^2 = 4(+8)$

$(x-4)^2 = 4(1)(y+8)$; $p = 1$

vertex (h, k) is $(4, -8)$; focus is $(4, -7)$

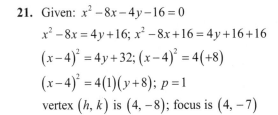

(4,-8)

29. $r = 4\cos 3\theta$

Let $\theta = 0$ to π in steps of $\dfrac{\pi}{48}$.

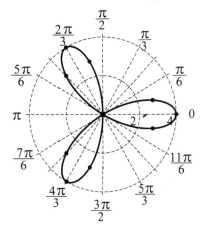

25. $x^2 - 2xy + y^2 + 4x + 4y = 0$

$B^2 - 4AC = (-2)^2 - 4(1)(1) = 0$, parabola

$\tan 2\theta = \dfrac{B}{A-C} = \dfrac{-2}{1-0} \Rightarrow \theta = 45°$

$\left(x'\cdot\dfrac{1}{\sqrt{2}} - y'\cdot\dfrac{1}{\sqrt{2}}\right)^2 - 2\left(x'\cdot\dfrac{1}{\sqrt{2}} - y'\cdot\dfrac{1}{\sqrt{2}}\right)$

$\left(x'\cdot\dfrac{1}{\sqrt{2}} + y'\cdot\dfrac{1}{\sqrt{2}}\right) + \left(x'\cdot\dfrac{1}{\sqrt{2}} + y'\cdot\dfrac{1}{\sqrt{2}}\right)^2$

$+ 4\left(x'\cdot\dfrac{1}{\sqrt{2}} - y'\cdot\dfrac{1}{\sqrt{2}}\right) + 4\left(x'\cdot\dfrac{1}{\sqrt{2}} + y'\cdot\dfrac{1}{\sqrt{2}}\right) = 0$

$y'^2 = -2\sqrt{2}x' \Rightarrow V(0, 0)$

33. $r = 2\sin\dfrac{\theta}{2}$

Let $\theta = 0$ to 4π in steps of $\dfrac{\pi}{24}$.

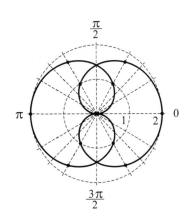

37. $x^2 + xy + y^2 = 2$

$x^2 + y^2 + xy = 2$

$r^2 + (r\cos\theta)(r\sin\theta) = 2$

$r^2 = \dfrac{2}{1 + \sin\theta\cos\theta}$

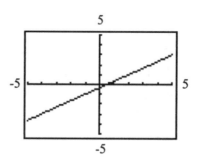

41. $r = \dfrac{4}{2 - \cos\theta}$

$2r - r\cos\theta = 4$

$2r = 4 + r\cos\theta = 4 + x$

$4r^2 = 16 + 8x + x^2$

$4(x^2 + y^2) = 4x^2 + 4y^2 = x^2 + 8x + 16$

$3x^2 + 4y^2 - 8x - 16 = 0$

45. $x^2 + y^2 - 4y - 5 = 0$

$y^2 - 4x^2 - 4 = 0$

From the graph, two real solutions.

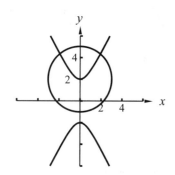

49. $x^2 + 3y + 2 - (1 + x)^2 = 0$

$3y = (1 + x)^2 - x^2 - 2$

$y = \dfrac{(1 + x)^2 - x^2 - 2}{3} - \dfrac{1 + 2x + x^2 - x^2 - 2}{3} = \dfrac{2x - 1}{3}$

$= \dfrac{2}{3}x - \dfrac{1}{3}$

Graph $y_1 = \dfrac{2}{3}x - \dfrac{1}{3}$

53. $x^2 - 4y^2 + 4x + 24y - 48 = 0$. Solve for y by completing the square.

$y^2 - 6y = 0.25x^2 + x - 12$

$y^2 - 6y + 9 = 0.25x^2 + x - 3$

$(y - 3)^2 = 0.25x^2 + x - 3$

$y = \pm\sqrt{0.25x^2 + x - 3} + 3$

Graph $y_1 = \sqrt{0.25x^2 + x - 3} + 3$

$y_2 = -\sqrt{0.25x^2 + x - 3} + 3$

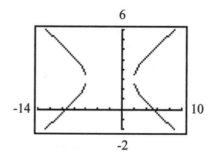

57. $r = 2 - 3\csc\theta = 2 - \dfrac{3}{\sin\theta}$

Graph $r_1 = 2 - \dfrac{3}{\sin\theta}$

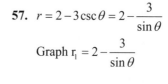

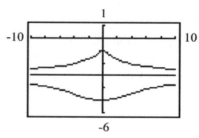

61.

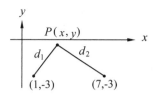

$d_1 + d_2 = 8$ describes an ellipse with center $(4, -3)$,

$a = 4 \Rightarrow a^2 = 16,\ c = 3 \Rightarrow c^2 = 9.$

$a^2 = b^2 + c^2 \Rightarrow 16 = b^2 + 9 \Rightarrow b^2 = 7$

$\dfrac{(x-4)^2}{16} + \dfrac{(y+3)^2}{7} = 1$

65. $\dfrac{(x-h)^2}{a^2} + \dfrac{(y-k)}{b^2} = 1$, for $a = b$

$(x-h)^2 + (y-k)^2 = a^2$, a circle

with center (h, k) and radius a.

69.

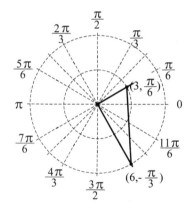

$\dfrac{\pi}{3} + \dfrac{\pi}{6} = \dfrac{\pi}{2} \Rightarrow$ right triangle

$d^2 = 3^2 + 6^2$

$d = \sqrt{45}$

73. The boundary line is the dashed graph of the

parabola $y = (x+2)^2$. Use $(0, 0)$ as a test point in

$$y > 4(x+2)^2$$
$$0 > 4(0+2)^2$$
$$0 > 16,\ F$$

Solution region does not contain $(0, 0)$. Shade region inside parabola.

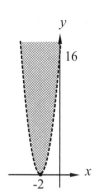

77.

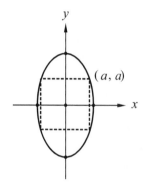

$7a^2 + 2a^2 = 18$

$9a^2 = 18$

$a = \sqrt{2}$

Square has side $= 2\sqrt{2}$

Area of square $= \left(2\sqrt{2}\right)^2 = 8$

81. $R_T = R + 2.5$; linear function with slope 1 and y-intercept $= 2.5$

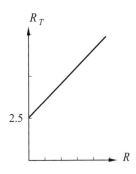

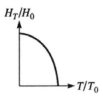

101. $a = \dfrac{120}{2} = 60; \; c = 60 - 15 = 45;$

$b = \sqrt{a^2 - b^2} = \sqrt{60^2 - 45^2} = \sqrt{1575}$

$A = \pi ab = \pi \cdot 60 \cdot \sqrt{1575} = 7500 \text{ ft}^2$

85. $y = 50.00 \text{ kg} \dfrac{2.010 \text{ kJ}}{1 \text{ kg} \cdot 1^\circ\text{C}} \left(T - 100^\circ\text{C} \right)$

$y = 100.5T - 10,050$

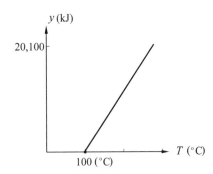

105.

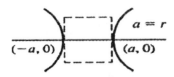

$\dfrac{x^2}{a^2} - \dfrac{y^2}{b^2} = 1$

$y = 40$ when $x = 40$

$y = 100$ when $x = 50$

$\left\{\begin{array}{l} \dfrac{40^2}{a^2} - \dfrac{40^2}{b^2} = 1 \text{ or } 40^2 b^2 - 40^2 a^2 = a^2 b^2 \\[2mm] \dfrac{50^2}{a^2} - \dfrac{100^2}{b^2} = 1 \text{ or } 50^2 b^2 - 100^2 a^2 = a^2 b^2 \end{array}\right\}$

Multiply by 100^2

Multiply by -40^2

$\left\{\begin{array}{l} 40^2 b^2 - 40^2 a^2 = a^2 b^2 \\ 50^2 - 100^2 a^2 = a^2 b^2 \end{array}\right\}$

$\left\{\begin{array}{l} 100^2 40^2 b^2 - 100^2 40^2 a^2 = 100^2 a^2 b^2 \\ -40^2 50^2 b^2 + 40^2 100^2 a^2 = -40^2 a^2 b^2 \end{array}\right\}$

Add

$\left\{\begin{array}{l} 16 \times 10^6 b^2 - 1.6 \times 10^7 a^2 = 10 \times 10^3 a^2 b^2 \\ 4 \times 10^6 b^2 + 1.6 \times 10^7 a^2 = -1.6 \times 10^3 a^2 b^2 \end{array}\right\}$

$12 \times 10^6 b^2 = 8.4 \times 10^3 a^2 b^2$

$12 \times 10^6 = 8.4 \times 10^3 a^2$

$a^2 = \dfrac{12 \times 10^6}{8.4 \times 10^3} = 1.42 \times 10^3$

$a = 37.8 \text{ ft}$

89. $A = \pi r^2 = \pi \left(490 \tan 7^\circ \right)$

$A = 11,000 \text{ ft}^2$

93.

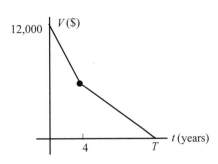

$V = -1250t + 12,000, \qquad 0 \le t \le 4$

$V - 7000 = -1000(t - 4) \qquad 4 \le t \le T$

$0 - 7000 = -1000(T - 4) \Rightarrow T = 11$

97. Graph is parabolic in shape

109. $r^2 = R^2 \cos 2\left(\theta + \dfrac{\pi}{2}\right)$

$r = R\sqrt{\cos 2\left(\theta + \dfrac{\pi}{2}\right)} = R\sqrt{\cos\left(2\theta + \pi\right)}$

The square root requires $\cos 2\left(\theta + \frac{\pi}{2}\right) \geq 0$. The

domain is $\dfrac{\pi}{4} \leq \theta \leq \dfrac{3\pi}{4}$ or $\dfrac{5\pi}{4} \leq \theta \leq \dfrac{7\pi}{4}$.

θ	r
$\frac{\pi}{4}$	0
$\frac{\pi}{3}$	$0.5R$
$\frac{\pi}{2}$	$1R$
$\frac{2\pi}{3}$	$0.5R$
$\frac{3\pi}{4}$	0
$\frac{5\pi}{4}$	0
$\frac{4\pi}{3}$	$0.5R$
$\frac{3\pi}{2}$	$1R$
$\frac{5\pi}{3}$	$0.5R$
$\frac{7\pi}{4}$	0

$$\dfrac{\left(x + \frac{b}{a^2 - b^2}\right)^2}{\dfrac{a^2}{\left(a^2 - b^2\right)^2}} + \dfrac{y^2}{\frac{1}{a^2 - b^2}} = 1$$

which is the equation of a conic.

113. $\dfrac{1}{r} = a + b\cos\theta$

$1 = ar + br\cos\theta$

$1 = a\sqrt{x^2 + y^2} + bx$

$a\sqrt{x^2 + y^2} = 1 - bx$

$a^2\left(x^2 + y^2\right) = 1 - 2bx + b^2 x^2$

$a^2 x^2 + a^2 y^2 + 2bx - b^2 x^2 = 1$

$\left(a^2 - b^2\right)x^2 + 2bx + a^2 y^2 = 1$

$x^2 + \dfrac{2b}{a^2 - b^2}\cdot x + \dfrac{a^2}{a^2 - b^2}\cdot y^2 = \dfrac{1}{a^2 - b^2}$

$x^2 + \dfrac{2b}{a^2 - b^2}\cdot x + \dfrac{b^2}{\left(a^2 - b^2\right)^2} + \dfrac{a^2}{a^2 - b^2}\cdot y^2$

$\qquad = \dfrac{1}{a^2 - b^2} + \dfrac{b^2}{\left(a^2 - b^2\right)^2}$

$\left(x + \dfrac{b}{a^2 - b^2}\right)^2 + \dfrac{a^2}{a^2 - b^2}\cdot y^2 = \dfrac{a^2 - b^2 + b^2}{\left(a^2 - b^2\right)^2}$

$\qquad = \dfrac{a^2}{\left(a^2 - b^2\right)^2}$

Chapter 22

INTRODUCTION TO STATISTICS

22.1   Frequency Distributions

1.

Est. (hrs)	0 − 5	6 − 11	12 − 17	18 − 23	24 − 29
Freq.	5	12	19	9	5

5.

Number	Frequency
143	1
144	3
145	1
146	3
147	2
148	4
149	3
150	1
151	1
152	0
153	1

9.

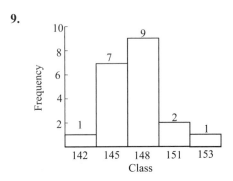

13.

Number	18	19	20	21	22	23	24	25
Frequency	1	3	2	4	3	1	0	1

17.

Messages	Frequency
13	2
38	7
63	18
88	41
113	56
138	32
163	8
188	3
213	3

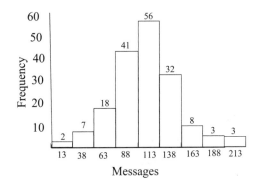

21.

Dist. (ft)	f (%)
155 − 159	1.7
160 − 164	12.5
165 − 169	26.7
170 − 174	30.0
175 − 179	20.0
180 − 184	8.3
185 − 189	0.8

25.

Dosage (mR)	Frequency
3.73 – 3.87	1
3.88 – 4.02	2
4.03 – 4.17	2
4.18 – 4.32	7
4.33 – 4.47	7
4.48 – 4.62	1

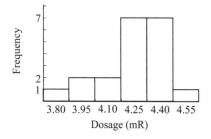

29.

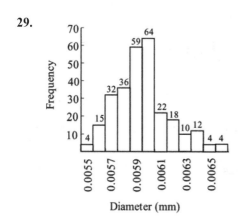

22.2 Measures of Central Tendency

1. 1, 2, 2, 3, 4, 4, 4, 6, 7, 7, 8, 9, 9, 11

There are 14 numbers. The median is halfway between the seventh and eighth numbers.

$$\text{median} = \frac{4+6}{2} = 5$$

5. Arrange the numbers in numerical order:

2, 3, 3, 3, 4, 4, 4, 4, 5, 5, 6, 6, 6, 7, 7

There are 15 numbers. The middle number is eighth. Since the eighth number is 4, the median is 4.

9. The arithmetic mean is:

$$\bar{x}$$

$$= \frac{2+3+3+3+4+4+4+4+5+5+6+6+6+7+7}{`5}$$

$$= \frac{69}{15} = 4.6$$

13. The mode is the number that occurs most frequently, which is 4 since it occurs 4 times.

17. Arrange in ascending order; $n = 20 \left(10^{\text{th}}, 11^{\text{th}}\right)$,

$M = 147.5$ mi

21. $\bar{x} = \dfrac{\sum xf}{\sum f} = \dfrac{313}{15} = 20.9$

25. 60th-61st, $M = 172$ ft

29. $M = 4.36$ mR; $f = 2$

33. $\bar{x} = \dfrac{\sum xf}{\sum f} = \dfrac{1.6673}{280} = 0.00595$ mm

37. $\bar{x} = \dfrac{\sum xf}{\sum f} = \dfrac{(862)(200)}{1000} = 862$ kW·h

41. Lowest = 525; highest = 800; midrange $= \dfrac{1325}{2}$

$= \$663$

45. $\bar{x} = \dfrac{\sum xf}{\sum f} = \dfrac{12,375}{14} = \$884$

The mean increased by 235 from 648 to 884.

22.3   Standard Deviation

1.

x	$x - \bar{x}$	$\left(x - \bar{x}\right)^2$
6	2	4
5	1	1
4	0	0
7	3	9
6	2	4
2	−2	4
1	−3	9
1	−3	9
5	1	1
3	−1	1
40		42

$$\bar{x} = 40 / 10 = 4$$

$$\frac{\sum\left(x - \bar{x}\right)^2}{n-1} = \frac{42}{10-1} = \frac{42}{9}$$

$$s = \sqrt{\frac{42}{9}} = 2.2$$

5.

x	$x - \bar{x}$	$\left(x - \bar{x}\right)^2$	x^2
0.45	−0.055	0.00303	0.2025
0.46	−0.045	0.00203	0.2116
0.47	−0.035	0.00123	0.2209
0.48	−0.025	6.3×10^{-4}	0.2304
0.48	−0.025	6.3×10^{-4}	0.2304
0.49	−0.015	2.3×10^{-4}	0.2401
0.49	−0.015	2.3×10^{-4}	0.2401
0.49	−0.015	2.3×10^{-4}	0.2401
0.50	−0.005	2.5×10^{-5}	0.25
0.51	0.005	2.5×10^{-5}	0.2601
0.53	0.025	6.3×10^{-4}	0.2809
0.53	0.025	6.3×10^{-4}	0.2809
0.53	0.025	6.3×10^{-4}	0.2809
0.55	0.045	0.00203	0.3025
0.55	0.045	0.00203	0.3025
0.57	0.045	0.00203	0.3249

$$\sum x = 8.08, \sum\left(x - \bar{x}\right)^2 = 0.0184, \sum x^2 = 4.0988$$

$$\bar{x} = \frac{\sum x}{n} = \frac{8.08}{16} = 0.505$$

$$s = \sqrt{\frac{\sum\left(x - \bar{x}\right)^2}{n-1}} = \sqrt{\frac{0.0184}{15}} = 0.035$$

9. Using equation 22.3

$$s = \sqrt{\frac{n\sum x^2 - \left(\sum x\right)^2}{n\left(n-1\right)}} = \sqrt{\frac{16\left(4.0988\right) - \left(8.08\right)^2}{16\left(15\right)}}$$

$$= 0.035$$

13. $\bar{x} = 0.505$, $s = 0.035$

```
1-Var Stats
x̄=.505
Σx=8.08
Σx²=4.0988
Sx=.0350238014
σx=.0339116499
↓n=16
■
```

17. Computer Instruction, exercise 13, section 22.1

x	x^2
18	324
19	361
19	361
19	361
20	400
20	400
21	441
21	441
21	441
21	441
22	484
22	484
22	484
23	529
25	625
313	6577

$$s = \sqrt{\frac{15\left(6577\right) - \left(313\right)^2}{15\left(14\right)}}$$

$$s = 1.8$$

21. Text messages in Exercise 17, Section 22.1

$$s = \sqrt{\frac{n\sum x^2 - \left(\sum x\right)^2}{n(n-1)}}$$

$$s = \sqrt{\frac{170(2,178,555) - (18,285)^2}{170(169)}}$$

$$s = 35.4$$

25. Fiber-optic cable diameters, Exercise 29, Section 22.1

$s = 0.00022$ mm

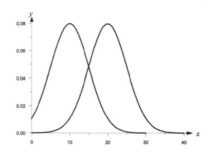

```
1-Var Stats
x̄=.0059546429
Σx=1.6673
Σx²=.00994211
Sx=2.234782E-4
σx=2.230788E-4
↓n=280
■
```

22.4   Normal Distributions

1. $\mu = 10$, $\alpha = 5$ and $\mu = 20$, $\alpha = 5$ are the same height with the first centered at $x = 10$ and the second centered at $x = 20$.

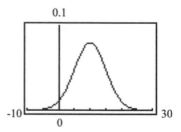

5. $\mu = 10$, $\sigma = 5$ is the graph in Fig. 22.9.

9. $200(0.68) = 136$ bags

13. $\mu = 1.50$, $\sigma = 0.05$. x between 1.45 and 1.55
$$z = \frac{1.45 - 1.50}{0.05} = -1, \; z = \frac{1.55 - 1.50}{0.05} = 1$$
$$(0.3413 + 0.3413)(500) = 341.3$$
341 batteries will have a voltage between 1.45 V and 1.55 V.

17. $\mu = 100,000$, $\sigma = 10,000$. x between 85,000 and 100,000
$$z = \frac{85,000 - 100,000}{10,000} = -1.5$$
$$(0.4332)(5000) = 2166$$
2166 tires will last between 85,000 km and 100,000 km

21. $\sigma_{\bar{x}} = \dfrac{\sigma}{\sqrt{n}} = \dfrac{10,000}{\sqrt{5000}} = 141$. About 68% have a mean lifetime from 99,859 km to 100,141 km.

25. Since 50% of area is to right of $z = 0$, 25.8% of area will be to the left of $z = 0$ which gives a z – value of -0.7.

29. 34% are between 12 and 16, 18% are between 16 and 18 for a total of 53% between 12 and 18.

33. $\bar{x} - \sigma = 13.7 - 6.1 = 7.6$

34 of 50 or 68% are within $\bar{x} \pm \sigma$ as compared with 68% for a normal distribution.

22.5   Statistical Process Control

1. | Subgroup | Amount of Drug (mg) of five capsules | | | | |
|---|---|---|---|---|---|
| 1 | 497 | 499 | 502 | 493 | 498 |
| 2 | 497 | 499 | 500 | 495 | 502 |
| 3 | 496 | 500 | 507 | 503 | 502 |
| 4 | 512 | 503 | 488 | 500 | 497 |
| 5 | 504 | 505 | 500 | 508 | 502 |
| 6 | 495 | 495 | 501 | 497 | 497 |
| 7 | 503 | 500 | 507 | 499 | 498 |
| 8 | 494 | 498 | 497 | 501 | 496 |
| 9 | 502 | 504 | 505 | 500 | 502 |
| 10 | 500 | 502 | 500 | 496 | 497 |
| 11 | 502 | 498 | 510 | 503 | 497 |
| 12 | 497 | 498 | 496 | 502 | 500 |
| 13 | 504 | 500 | 495 | 498 | 501 |
| 14 | 500 | 499 | 498 | 501 | 494 |
| 15 | 498 | 496 | 502 | 501 | 505 |
| 16 | 500 | 503 | 504 | 499 | 505 |
| 17 | 487 | 496 | 499 | 498 | 494 |
| 18 | 498 | 497 | 497 | 502 | 497 |
| 19 | 503 | 501 | 500 | 498 | 504 |
| 20 | 496 | 494 | 503 | 502 | 501 |

Subgroup	Mean x	Range R
1	497.8	9
2	498.6	7
3	501.6	11
4	500.0	24
5	503.8	8
6	497.0	6
7	501.4	9
8	497.2	7
9	502.6	5
10	499.0	6
11	502.0	13
12	498.6	6
13	499.6	9
14	498.4	7
15	500.4	9
16	502.2	6
17	494.8	12
18	498.2	5
19	501.2	6
20	499.2	9
Sums	9993.6	174
Means	499.7	8.7

$$UCL\left(\overline{x}\right) = \overline{\overline{x}} + A_2 R = 499.7 + 0.577\left(8.7\right) = 504.7 \text{ mg}$$

$$LCL\left(\overline{x}\right) = \overline{\overline{x}} - A_2 R = 499.7 - 0.577\left(8.7\right) = 494.7 \text{ mg}$$

5.

Hour	Torques	(N·m)	of five	engines	
1	366	352	354	360	362
2	370	374	362	366	356
3	358	357	365	372	361
4	360	368	367	359	363
5	352	356	354	348	350
6	366	361	372	370	363
7	365	366	361	370	362
8	354	363	360	361	364
9	361	358	356	364	364
10	368	366	368	358	360
11	355	360	359	362	353
12	365	364	357	367	370
13	360	364	372	358	365
14	348	360	352	360	354
15	358	364	362	372	361
16	360	361	371	366	346
17	354	359	358	366	366
18	362	366	367	361	357
19	363	373	364	360	358
20	372	362	360	365	367

Subgroup	Mean x	Range R
1	358.8	14
2	365.6	18
3	362.6	15
4	363.4	9
5	352.0	8
6	366.4	11
7	364.8	9
8	360.4	10
9	360.6	8
10	364.0	10
11	357.8	9
12	364.6	13
13	363.8	14
14	354.8	12
15	363.4	14
16	360.8	25
17	360.6	12
18	362.6	10
19	363.6	15
20	365.2	12
Sum	7235.8	248
Mean	361.79	12.4

$$CL: \quad \overline{x} = 361.79 \text{ N·m}$$

$$UCL(\overline{x}) = \overline{x} + A_2\overline{R} = 361.79 + 0.577(12.4)$$
$$= 368.9 \text{ N·m}$$

$$LCL(\overline{x}) = \overline{x} - A_2\overline{R} = 361.79 - 0.577(12.4)$$
$$= 354.6 \text{ N·m}$$

9.

Subgrp	Output	voltages	five	adapters	
1	9.03	9.08	8.85	8.92	8.90
2	9.05	8.98	9.20	9.04	9.12
3	8.93	8.96	9.14	9.06	9.00
4	9.16	9.08	9.04	9.07	8.97
5	9.03	9.08	8.93	8.88	8.95
6	8.92	9.07	8.86	8.96	9.04
7	9.00	9.05	8.90	8.94	8.93
8	8.87	8.99	8.96	9.02	9.03
9	8.89	8.92	9.05	9.10	8.93
10	9.01	9.00	9.09	8.96	8.98
11	8.90	8.97	8.92	8.98	9.03
12	9.04	9.06	8.94	8.93	8.92
13	8.94	8.99	8.93	9.05	9.10
14	9.07	9.01	9.05	8.96	9.02
15	9.01	8.82	8.95	8.99	9.04
16	8.93	8.91	9.04	9.05	8.90
17	9.08	9.03	8.91	8.92	8.96
18	8.94	8.90	9.05	8.93	9.01
19	8.88	8.82	8.89	8.94	8.88
20	9.04	9.00	8.98	8.93	9.05
21	9.00	9.03	8.94	8.92	9.05
22	8.95	8.95	8.91	8.90	9.03
23	9.12	9.04	9.01	8.94	9.02
24	8.94	8.99	8.93	9.05	9.07

Subgroup	Mean x	Range R
1	8.956	0.23
2	9.078	0.22
3	9.018	0.21
4	9.064	0.19
5	8.974	0.20
6	8.970	0.21
7	8.964	0.15
8	8.974	0.16
9	8.978	0.21
10	9.008	0.13
11	8.960	0.13
12	8.978	0.14
13	9.002	0.17
14	9.022	0.11
15	8.962	0.22
16	8.966	0.15
17	8.980	0.17
18	8.966	0.15
19	8.882	0.12
20	9.000	0.12
21	8.988	0.13
22	8.948	0.13
23	9.026	0.18
24	8.996	0.14
Sum	215.66	3.97
Mean	8.986	0.1654

$$CL:\ \overline{\overline{x}} = 8.986 \text{ V}$$

$$UCL\left(\overline{x}\right) = \overline{\overline{x}} + A_2\overline{R} = 8.986 + 0.577\left(0.1654\right)$$

$$= 9.081 \text{ V}$$

$$LCL\left(\overline{x}\right) = \overline{\overline{x}} - A_2\overline{R} = 8.986 - 0.577\left(0.1654\right)$$

$$= 8.891 \text{ V}$$

13. $CL:$ $\mu = \overline{\overline{x}} = 2.725$ in.

$$UCL(\overline{x}) = \mu + A\sigma = 2.725 + 1.342(0.0032)$$
$$= 2.729 \text{ in.}$$
$$LCL(\overline{x}) = \mu - A\sigma = 2.725 - 1.342(0.0032)$$
$$= 2.721 \text{ in.}$$

17.

Week	Accounts with errors	Proportion with errors
1	52	0.052
2	36	0.036
3	27	0.027
4	58	0.058
5	44	0.044
6	21	0.021
7	48	0.048
8	63	0.063
9	32	0.032
10	38	0.038
11	27	0.027
12	43	0.043
13	22	0.022
14	35	0.035
15	41	0.041
16	20	0.020
17	28	0.028
18	37	0.037
19	24	0.024
20	42	0.042
Total	738	

$CL:$ $\overline{p} = \dfrac{738}{1000(20)} = 0.0369$

$$\sigma_p = \sqrt{\dfrac{\overline{p}(1-\overline{p})}{n}}$$
$$= \sqrt{\dfrac{0.0369(1-0.0369)}{1000}} = 0.00596$$

$UCL(p) = 0.0369 + 3(0.00596) = 0.0548$
$LCL(p) = 0.0369 - 3(0.00596) = 0.0190$

22.6 Linear Regression

1.

x	y	xy	x^2
1	3	3	1
2	7	14	4
3	9	27	9
4	9	36	16
5	12	60	25
15	40	140	55

$n = 5$

$$m = \frac{n\sum xy - \sum x \sum y}{n\sum x^2 - \left(\sum x\right)^2}$$
$$= \frac{5(140) - 15(40)}{5(55) - 15^2} = 2$$

$$b = \frac{\sum x^2 \sum y - \sum xy \sum x}{n\sum x^2 - \left(\sum x\right)^2}$$
$$= \frac{55(40) - 140(15)}{5(55) - 15^2} = 2$$

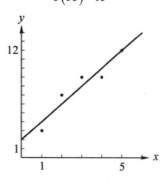

The equation of the least square line is $y = 2x + 2$.

5.

$t(h)$	1.0	2.0	4.0	8.0	10.0	12.0
$y(mg/dL)$	8.7	8.4	7.7	7.3	5.7	5.2

t	y	ty	t^2
1.0	8.7	8.7	1.0
2.0	8.4	16.8	4.0
4.0	7.7	30.8	16.0
8.0	7.3	58.4	64.0
10.0	5.7	57	100
12.0	5.2	62.4	144
37.0	43	234.1	329

$n = 6$

$$m = \frac{6(234.1) - 37.0(43)}{6(329) - 37^2} = -0.308$$

$$b = \frac{329(43) - 234.1(37)}{6(329) - 37^2} = 9.07$$

$$y = -0.308t + 9.07$$

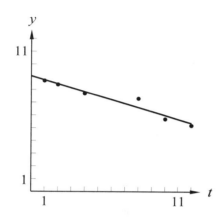

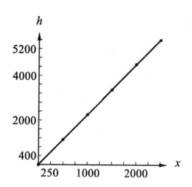

9.

x	h	$xh \cdot (10)^3$	$x^2 \cdot (10)^3$
0	0	0	0
500	1130	565	250
1000	2250	2250	1000
1500	3360	5040	2250
2000	4500	9000	4000
2500	5600	14,000	6250
7500	16,840	30,855	13,750

$n = 6$

$$m = \frac{6(30,855 \times 10^3) - (7500)(16,840)}{6(13,750 \times 10^3) - (7500)^2} = 2.24$$

$$b = \frac{(13,750 \times 10^3)(16,840) - (30,855 \times 10^3)(7500)}{6(13,750 \times 10^3) - (7500)^2}$$

$$= 5.24$$

$h = mx + b; \ h = 2.24x + 5.24$

Plot points:

x	y
750	1690
2250	5045

13.

f	V	fV	f^2
0.550	0.350	0.19250	0.302500
0.605	0.600	0.36300	0.366025
0.660	0.850	0.56100	0.435600
0.735	1.10	0.80850	0.540225
0.805	1.45	1.16725	0.648025
0.880	1.80	1.58400	0.774400
4.235	6.15	4.67625	3.066775

$n = 6$

$$m = \frac{6(4.67625) - (4.235)(6.15)}{6(3.066775) - (4.235)^2} = 1.432$$

$$b = \frac{(3.066775)(6.15) - (4.67625)(4.235)}{6(3.066775) - (4.235)^2} = -2.03$$

Therefore, since f is in PHz,

$$V = 4.32 \times 10^{-15} f - 2.03$$

$$f_0 = \frac{2.03}{4.32 \times 10^{-15}} = 0.470 \times 10^{15} \, \text{Hz} = 0.470 \, \text{PHz}$$

Plot points:

f (PHz)	V
0.600	0.562
0.800	1.426

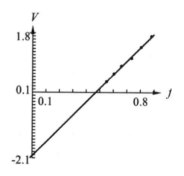

17.

x	1	3	6	5	8	10	4	7	3	8
y	15	12	10	8	9	2	11	9	11	7

$n = 10$

$$s = \sqrt{\overline{x^2} - \left(\overline{x}\right)^2}$$

$$s_x = \sqrt{\frac{373}{10} - \left(\frac{55}{10}\right)^2} = 2.655$$

$$s_y = \sqrt{\frac{990}{10} - \left(\frac{94}{10}\right)^2} = 3.262$$

$$m = -1.1064$$

$$r = -1.1064\left(\frac{2.655}{3.262}\right) = -0.901$$

22.7 Nonlinear Regression

1.

x	$f(x) = x^2$	y	$x^2 y$	$\left(x^2\right)^2$
0	0	2	0	0
1	1	3	3	1
2	4	10	40	16
3	9	25	225	81
4	16	44	704	256
5	25	65	1625	625
	55	149	2597	979

$n = 6$

$$m = \frac{6(2597) - 55(149)}{6(979) - 55^2} = 2.59$$

$$b = \frac{979(149) - 2597(55)}{6(979) - 55^2} = 1.07$$

$$y = 2.59x^2 + 1.07$$

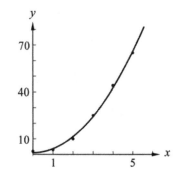

5. $y = mt^2 + b$

t	y	t^2	yt^2	$\left(t^2\right)^2$
1.0	6.0	1.0	6.0	1.0
2.0	23	4.0	92	16.0
3.0	55	9.0	495	81.0
4.0	98	16.0	1568	256
5.0	148	25.0	3700	625
	330	55.0	5861.0	979.0

$n = 5$

$$m = \frac{5(5861) - (55.0)(330)}{5(979.0) - (55.0)^2} = 5.97$$

$$b = \frac{(979.0)(330) - (5861)(55.0)}{5(979.0) - (55.0)^2} = 0.38$$

$$y = 5.97t^2 + 0.38$$

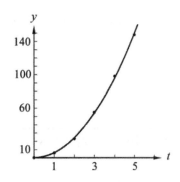

9.

f	$\frac{1}{f}$	T	$\frac{1}{f} \cdot T$	$\left(\frac{1}{f}\right)^2$
500	500^{-1}	220	0.44	4×10^{-6}
1000	1000^{-1}	102	0.102	1×10^{-6}
1500	1500^{-1}	77	$0.05\overline{13}$	$4.\overline{4} \times 10^{-7}$
2000	2000^{-1}	50	0.025	2.5×10^{-7}
2500	2500^{-1}	43	0.0172	1.6×10^{-7}
3000	3000^{-1}	30	0.01	$1.\overline{1} \times 10^{-7}$
10,500	0.0049	522	$0.64\overline{553}$	$5.96\overline{5} \times 10^{-6}$

$$m = \frac{6(0.6455\overline{3}) - 0.0049(522)}{6(5.96\overline{5} \times 10^{-6}) - 0.0049^2} = 1.11 \times 10^5$$

$$b = \frac{5.96\overline{5} \times 10^{-6}(522) - 0.6455\overline{3}(0.0049)}{6(5.96\overline{5} \times 10^{-6}) - 0.0049^2}$$

$$T = \frac{1.11 \times 10^5}{f} - 4$$

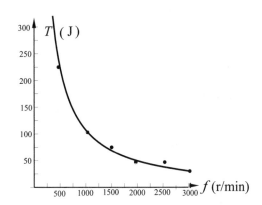

5.

Percent of on-time flights	Frequency	Relative Frequency (%)
67 – 70	3	15
71 – 74	4	20
75 – 78	8	40
79 – 82	3	15
83 – 86	2	10
Total	20	100

9.

Percent of on-time flights	Cumulative Frequency
less than 71	3
less than 75	7
less than 79	15
less than 83	18
less than 87	20

13. See 22.R.11. 0.014 mL/L is the standard deviation

17. Enter the 9 power numbers in the calculator as list L_1 and the corresponding frequencies as list L_2. 700 W is the median

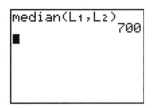

Chapter 22   Review Exercises

1. Enter the 20 numbers as list L_1 in the calculator. Then 2nd LIST ← to obtain.

```
NAMES OPS MATH
1:min(
2:max(
3:mean(
4:median(
5:sum(
6:prod(
7↓stdDev(
```

From which

```
median(L₁)
           76.5
■
```

21. Using L_1 and L_2 from problem 17, 17.3 W is the standard deviation

25. Enter counts as L_1 and intervals as L_2.

The median is 4.

Counts	0	1	2	3	4	5	6	7	8	9	10
Intervals	3	10	25	45	29	39	26	11	7	2	3

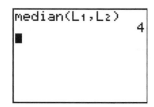

29. Enter speeds in L_1 and number cars in L_2, then

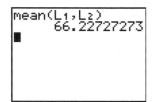

66.2 mi/h is the mean.

33. CL: $\bar{p} = \dfrac{540}{500 \cdot 20} = 0.0540$

$\sigma_p = \sqrt{\dfrac{\bar{p}(1-\bar{p})}{n}} = \sqrt{\dfrac{0.054(1-0.054)}{500}} = 0.01011$

$UCL(p) = 0.054 + 3(0.01011) = 0.0843$

$LCL(p) = 0.054 - 3(0.01011) = 0.0237$

37.

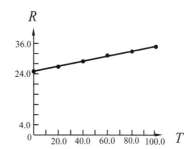

There are about 322 readings between 1.5 and 2.5 $\mu g/m^3$.

41.

T	R	TR	T^2
0.0	25.0	0	0
20.0	26.8	536	400
40.0	28.9	1156	1600
60.0	31.2	1872	3600
80.0	32.8	2624	6400
100	34.7	3470	10,000
300	179.4	9658	22,000

$\bar{T} = \dfrac{300}{6} = 50.0$ $\qquad (\bar{T})^2 = 2500$

$\bar{R} = \dfrac{179.4}{6} = 29.90$ $\qquad \bar{T}\,\bar{R} = 50.0(29.9) = 1495$

$\overline{TR} = \dfrac{9658}{6}$ $\qquad\qquad \overline{T^2} = \dfrac{22{,}000}{6}$

$s_T^2 = \overline{T^2} - (\bar{T})^2 = \dfrac{3500}{3}$

$m = \dfrac{\overline{TR} - \bar{T}\,\bar{R}}{s_T^2} = 0.0983$

$b = \bar{R} - m\bar{T} = 24.9857$

$R = mT + b;\ R = 0.0983T + 25.0$

(Answers may vary due to rounding.)

R

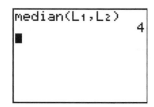

45.

x-Load (lb)	y-length (in.)	xy	x^2
0.0	10.0	0.0	0.0
1.0	11.2	11.2	1.0
2.0	12.3	24.6	4.0
3.0	13.4	40.2	9.0
4.0	14.6	58.4	16.0
5.0	15.9	79.5	25.0
15.0	77.4	213.9	55.0

$$m = \frac{n\sum xy - \left(\sum x\right)\left(\sum y\right)}{n\sum x^2 - \left(\sum x\right)^2}$$

$$= \frac{6(213.9) - 15(77.4)}{6(55.0) - (15.0)^2} = 1.17$$

$$b = \frac{\left(\sum x^2\right)\left(\sum y\right) - \left(\sum xy\right)\left(\sum x\right)}{n\sum x^2 - \left(\sum x\right)^2}$$

$$= \frac{55.0(77.4) - 213.9(15.0)}{6(55.0) - (15.0)^2} = 9.99$$

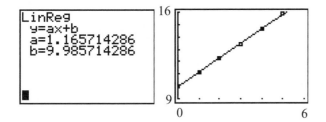

$$y = 0.0002x^2 + 15$$

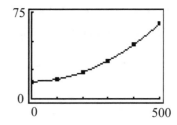

53. Enter t in L_1 as x and A in L_2 as y and use PwrReg.

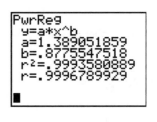

$$y = 1.39x^{0.878}$$

57. There are 7 numbers. The geometric mean is

$$\sqrt[7]{(8.0)(8.2)(8.8)(9.5)(9.7)(10.0)(10.7)}$$
$$= \sqrt[7]{5692009.7} = 5692009.7^{1/7} = 9.2 \text{ ppm}$$

61. Collect data in a chart of income versus education. Find equation of the least squares curve and use it to predict income based on education.

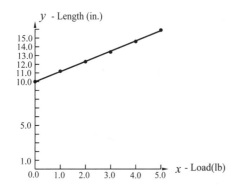

49.

$x(m)$	x^2	$y(m)$
0	0	15
100	10,000	17
200	40,000	23
300	90,000	33
400	160,000	47
500	250,000	65

Use LinReg for x^2 and y.

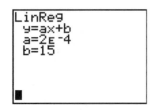

THE DERIVATIVE

23.1 Limits

1. $f(x) = \dfrac{1}{x+2}$ is not continuous at $x = -2$ because

$f(-2) = \dfrac{1}{-2+2}$ is a division by zero and the

function is not defined. The condition that the

function must exist is not satisfied.

5. $f(x) = 3x^2 - 98x$ is continuous for all real x since it

is defined for all x, and any small change in x will

produce only a small change in $f(x)$.

9. $f(x) = \sqrt{\dfrac{x}{x-2}}$ is continuous for $x \le 0$ and $x > 2$.

The function is not defined for $0 < x \le 2$. $0 < x < 2$

gives the square root of a negative and $x = 2$ gives

division by zero.

13. The graph is not continuous at $x = 2$. A small

change in x does not produce a small change in

y at $x = 2$. The function is continuous for $x \le 2$,

and continuous for $x > 2$.

17. (a) $f(2) = -1$

    (b) $\lim\limits_{x \to 2} f(x)$ does not exist

21. $f(x) = \begin{cases} x^2 & \text{for } x < 2 \\ 5 & \text{for } x \ge 2 \end{cases}$

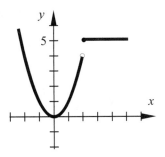

Not continuous at $x = 2$. Small change in x around

$x = 2$ produces a large change in $f(x)$.

25. $f(x) = \dfrac{x^3 - x}{x - 1}$

x	0.900	0.990	0.999
$f(x)$	1.7100	1.9701	1.9970

x	1.001	1.010	1.100
$f(x)$	2.0030	2.0301	2.3100

Therefore, $\lim\limits_{x \to 1} f(x) = 2$

29. $f(x) = \dfrac{2x+1}{5x-3}$

x	10	100	1000
$f(x)$	0.4468	0.4044	0.4004

Therefore, $\lim\limits_{x \to \infty} f(x) = 0.4$

33. $\lim\limits_{x \to 0} \dfrac{6x^2 + x}{x} = \lim\limits_{x \to 0} \dfrac{x(6x+1)}{x}$

$$= \lim\limits_{x \to 0} (6x+1)$$
$$= 0 + 1 = 1$$

37. $\lim\limits_{h \to 3} \dfrac{h^3 - 27}{h - 3} = \lim\limits_{h \to 3} \dfrac{(h-3)(h^2 + 3h + 9)}{h - 3}$

$$= \lim\limits_{h \to 3} (h^2 + 3h + 9)$$
$$= 3^2 + 3(3) + 9 = 27$$

41. For $p = -1$, $\sqrt{p} = \sqrt{-1} = i$. $\lim\limits_{p \to -1} \sqrt{p}(p+1.3)$

does not exist.

45. $\lim\limits_{h \to 0} \dfrac{\sqrt{9+h}-3}{h} \cdot \dfrac{\sqrt{9+h}+3}{\sqrt{9+h}+3} = \lim\limits_{h \to 0} \dfrac{9+h-9}{h\left(\sqrt{9+h}+3\right)}$

$= \lim\limits_{h \to 0} \dfrac{1}{\sqrt{9+h}+3} = \dfrac{1}{6}$

49. $\lim\limits_{x \to \infty} \dfrac{\sqrt{t^2+16}}{t+1} \cdot \dfrac{\frac{1}{t}}{\frac{1}{t}} = \lim\limits_{x \to \infty} \dfrac{\sqrt{\frac{t^2+16}{t^2}}}{1+\frac{1}{t}} = \lim\limits_{x \to \infty} \dfrac{\sqrt{1+\frac{16}{t^2}}}{1+\frac{1}{t}} = 1$

53. $\lim\limits_{x \to \infty} \dfrac{2x^2+x}{x^2-3} = \lim\limits_{x \to \infty} \dfrac{2+\frac{1}{x}}{1-\frac{3}{x^2}} = \dfrac{2}{1} = 2$

x	10	100	1000
$f(x)$	2.1649	2.0106	2.0010

57.

t	T
0	100
1	90
2	81

This is a geometric progression

$a_n = a_1 r^{n-1}$, $r = 0.9$, $a_1 = 100$

Therefore, $T = 100(0.9)^t$

$\lim\limits_{t \to 10} 100(0.9)^t = 34.9°\,C$

$\lim\limits_{t \to \infty} 100(0.9)^t = 100\lim\limits_{t \to \infty}100(0.9)^t = 100(0) = 0°\,C$

61. Let $y_1 = \dfrac{2^x-4}{x-2}$

```
Y₁(1.99)
        2.763001825
Y₁(1.999)
        2.771628038
Y₁(1.9999)
        2.772492634
■
```

```
Y₁(2.01)
        2.782220023
Y₁(2.001)
        2.77354985
Y₁(2.0001)
        2.772684815
■
```

$\lim\limits_{x \to 2} \dfrac{2^x-4}{x-2} = 2.77$

65. (a) $\lim\limits_{x \to 2^-} f(x) = -1$

(b) $\lim\limits_{x \to 2^+} f(x) = 2$

(c) $\lim\limits_{x \to 2} f(x)$ does not exist

69. For $x > 0$, $f(x) = \dfrac{x}{|x|} = \dfrac{x}{x} = 1$ from which

$\lim\limits_{x \to 0^+} f(x) = 1$

For $x < 0$, $f(x) = \dfrac{x}{|x|} = \dfrac{x}{-x} = -1$ from which

$\lim\limits_{x \to 0^-} f(x) = 1$

Since $RHL \neq LHL$, limit D.N.E. which means $f(x)$ is not continuous.

23.2 The Slope of a Tangent to a Curve

1. Find the slope of a tangent line to $y = x^2 + 3x$ at $(3, 18)$

$m_{PQ} = \dfrac{f(3+h)-f(3)}{h}$

$= \dfrac{\left[(3+h)^2+3(3+h)\right]-\left[3^2+3(3)\right]}{h}$

$m_{PQ} = \dfrac{9+6h+h^2+9+3h-9-9}{h} = \dfrac{h^2+9h}{h}$

$m_{PQ} = h+9$

$m_{tan} = \lim\limits_{h \to 0} m_{PQ} = \lim\limits_{h \to 0}(h+9) = 9$

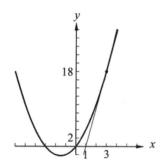

For $x_1 = -1$, $m_{tan} = 3 - 2(-1) = 5$.

For $x_1 = 3$, $m_{tan} = 3 - 2(3) = -3$.

5. $y = 2x^2 + 5x; \; P = (-2, -2)$

	Q_1	Q_2	Q_3	Q_4	P
x_2	-1.5	1.9	1.99	1.999	-2
y_2	-3	-2.28	-2.0298	-2.00299	-2
$y_2 - (-2)$	-1	-0.28	-0.0298	-0.00299	
$x_2 - (-2)$	0.5	0.1	0.01	0.001	
$m = \frac{y_2-(-2)}{x_2-(-2)}$	-2	-2.8	-2.98	-2.998	
$m_{\tan} = -3$					

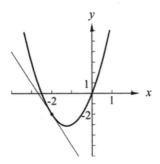

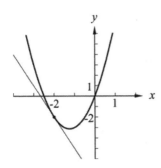

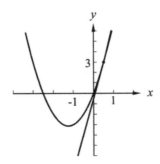

For $x_1 = -2$, $m_{\tan} = 4(-2) + 5 = -3$
For $x_1 = 0.5$, $m_{\tan} = 4(0.5) + 5 = 7$

9. $y = 2x^2 + 5x; \; P(-2, 2)$

$$m_{PQ} = \frac{f(-2+h) - f(-2)}{h}$$

$$= \frac{2(-2+h)^2 + 5(-2+h) - \left[2(-2)^2 + 5(-2)\right]}{h}$$

$$m_{PQ} = \frac{8 - 8h + 2h^2 - 10 + 5h - 8 + 10}{h} = \frac{-3h + 2h^2}{h}$$

$$m_{PQ} = -3 + 2h$$

$$m_{\tan} = \lim_{h \to 0}(-3 + 2h) = -3$$

17. $y = 6x - x^2; \; x = -1, \; x = 3, \; x = 1$

$$m_{PQ} = \frac{f(x_1 + h) - f(x_1)}{h}$$

$$= \frac{6(x_1 + h)^2 - (x_1 + h)^2 - \left(6x_1 - x_1^2\right)}{h}$$

$$= \frac{6x_1 + 6h - x_1^2 - 2x_1 h + h^2 - 6x_1 + x_1^2}{h}$$

$$m_{PQ} = \frac{6h - 2x_1 h + h^2}{h} = 6 - 2x_1 + h$$

$$m_{\tan} = \lim_{h \to 0}(6 - 2x_1 + h) = 6 - 2x_1$$

13. $y = 2x^2 + 5x; \; x = -2, \; x = 0.5$

$$m_{PQ} = \frac{f(x_1 + h) - f(x_1)}{h}$$

$$= \frac{2(x_1 + h)^2 + 5(x_1 + h) - \left(2x_1^2 + 5x_1\right)}{h}$$

$$m_{PQ} = \frac{2x_1^2 + 4x_1 h + 2h^2 + 5x_1 + 5h - 2x_1^2 - 5x_1}{h}$$

$$m_{PQ} = \frac{4x_1 h + 2h^2 + 5h}{h} = 4x_1 + 2h + 5$$

$$m_{\tan} = \lim_{h \to 0}(4x_1 + 2h + 5) = 4x_1 + 5$$

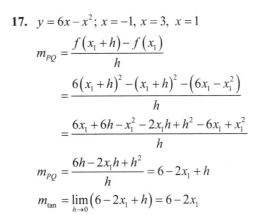

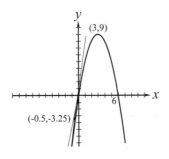

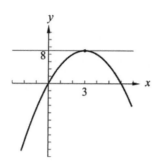

For $x_1 = -1$, $m_{\tan} = 6 - 2(-1) = 8$
For $x_1 = 3$, $m_{\tan} = 6 - 2(3) = 0$
For $x_1 = 1$, $m_{\tan} = 6 - 2(1) = 4$

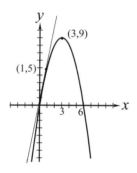

21. $y = x^5$; $x = -1$, $x = -0.5$, $x = 0.5$, $x = 1$

$$m_{PQ} = \frac{f(x_1 + h) - f(x_1)}{h} = \frac{(x_1 + h)^5 - x_1^5}{h}$$
$$= \frac{\left(x_1^5 + 5x_1^4 h + 10x_1^3 h^2 + 10x_1^2 h^3 + 5x_1 h^4 + h^5 - x_1^5\right)}{h}$$
$$m_{PQ} = 5x_1^4 + 10x_1^3 h + 10x_1^2 h^2 + h^4$$
$$m_{\tan} = \lim_{h \to 0}\left(5x_1^4 + 10x_1^3 h + 10x_1^2 h^2 + 5x_1 h^3 + h^4\right) = 5x_1^4$$

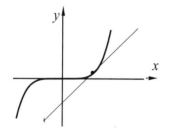

For $x_1 = -0.5$, $m_{\tan} = 5(-0.5)^4 = \dfrac{5}{16}$

For $x_1 = 0.5$, $m_{\tan} = 5(0.5)^4 = \dfrac{5}{16}$

For $x_1 = 1$, $m_{\tan} = 5(1)^4 = 5$

25. $y = \dfrac{1}{3}x^6$; $x = -0.5$, $x = 0$, $x = 0.5$

Graph $y_1 = \dfrac{1}{3}x^6$. Use DRAW feature to draw tangent line and dy/dx feature to find $m_{\tan}$.

$x_1 = -0.5$, $m_{\tan} = -0.0625$
$x_1 = 0$, $m_{\tan} = 0$
$x_1 = 0.5$, $m_{\tan} = 0.0625$

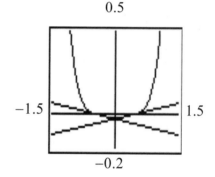

29. $y = 9 - x^3$; $P(2, 1)$, $Q(2.1, -0.261)$

From P to Q, x changes by 0.1 unit and $f(x)$ by -1.261 units. The average change in $f(x)$ for one unit change in x is $\dfrac{-1.261}{0.1} = -12.61$.

$$m_{PQ} = \frac{f(x_1 + h) - f(x_1)}{h} = \frac{9 - (x_1 + h)^3 - (9 - x_1^3)}{h}$$
$$m_{PQ} = -3x_1^2 - 3x_1 h - h^2$$
$$m_{\tan} = \lim_{h \to 0}\left(-3x_1^2 - 3x_1 h - h^2\right) = -3x_1^2$$
$$x = 2, \; m_{\tan} = -12 \text{ (Instantaneous rate of change)}$$
$$m_{PQ} = -12.61 \text{ (Average rate of change)}$$

33. $y = 12x - \dfrac{1}{3}x^3$

$m_{\tan} = 12 - x^2 = -4 \Rightarrow x = \pm 4$,

$$\left(4, \frac{80}{3}\right), \left(-4, -\frac{80}{3}\right)$$

37. $y = -2x^2 + 10$, $m_{\tan} = -4x_1 \big|_{x_1 = 1.5} = -6$

$\theta + \tan^{-1}(-6) = 180° \Rightarrow \theta = 180° - 99.46° = 80.54°$

$\alpha = $ angle of dive; $\alpha + \theta = 90° \Rightarrow \alpha = 90° - \theta$

$$\alpha = 9.46°$$

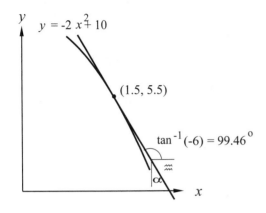

23.3   The Derivative

1. $y = 4x^2 + 3x$

$f(x+h) = 4(x+h)^2 + 3(x+h)$

$\qquad = 4(x^2 + 2xh + h^2) + 3x + 3h$

$f(x+h) = 4x^2 + 8xh + 4h^2 + 3x + 3h$

$f(x+h) - f(x)$

$\qquad = 4x^2 + 8xh + 4h^2 + 3x + 3h - 4x^2 - 3x$

$\qquad = 8xh + 4h^2 + 3h$

$\dfrac{f(x+h) - f(x)}{h} = \dfrac{8xh + 4h^2 + 3h}{h} = 8x + 4h + 3$

$\displaystyle\lim_{h \to 0} \frac{f(x+h) - f(x)}{h} = \lim_{h \to 0}(8x + 4h + 3) = 8x + 3$

$f'(x) = 8x + 3$

5. $y = 1 - 7x$

$f'(x) = \displaystyle\lim_{h \to 0} \frac{f(x+h) - f(x)}{h}$

$\qquad = \displaystyle\lim_{h \to 0} \frac{1 - 7(x+h) - (1 - 7x)}{h}$

$f'(x) = \displaystyle\lim_{h \to 0} \frac{1 - 7x - 7h - 1 + 7x}{h} = \lim_{h \to 0} \frac{-7h}{h}$

$\qquad = \displaystyle\lim_{h \to 0} -7 = -7$

9. $y = \pi x^2$

$f'(x) = \displaystyle\lim_{h \to 0} \frac{f(x+h) - f(x)}{h}$

$\qquad = \displaystyle\lim_{h \to 0} \frac{\pi(x+h)^2 - \pi x^2}{h}$

$\qquad = \displaystyle\lim_{h \to 0} \frac{\pi x^2 + 2\pi xh + h^2 - \pi x^2}{h}$

$f'(x) = \displaystyle\lim_{h \to 0} \frac{2\pi xh + h^2}{h} = \lim_{h \to 0}(2\pi x + h)$

$f'(x) = 2\pi x$

13. $y = 8x - 2x^2$

$f'(x) = \displaystyle\lim_{h \to 0} \frac{f(x+h) - f(x)}{h}$

$\qquad = \displaystyle\lim_{h \to 0} \frac{8(x+h)^2 - 2(x+h)^2 - (8x - 2x^2)}{h}$

$\qquad = \displaystyle\lim_{h \to 0} \frac{8x + 8h - 2x^2 - 4xh - 2h^2 - 8x + 2x^2}{h}$

$f'(x) = \displaystyle\lim_{h \to 0} \frac{8h - 4xh - 2h^2}{h} = \lim_{h \to 0}(8 - 4x - 2h)$

$f'(x) = 8 - 4x$

17. $y = \dfrac{\sqrt{3}}{x+2}$

$f'(x) = \displaystyle\lim_{h \to 0} \frac{f(x+h) - f(x)}{h} = \lim_{h \to 0} \frac{\frac{\sqrt{3}}{x+h+2} - \left(\frac{\sqrt{3}}{x+2}\right)}{h}$

$f'(x) = \displaystyle\lim_{h \to 0} \frac{\frac{\sqrt{3}}{x+h+2} - \left(\frac{\sqrt{3}}{x+2}\right)}{h} = \lim_{h \to 0} \frac{\frac{\sqrt{3}(x+2) - \sqrt{3}(x+h+2)}{(x+h+2)(x+2)}}{h}$

$f'(x) = \displaystyle\lim_{h \to 0} \frac{\frac{\sqrt{3}(x+2) - \sqrt{3}(x+h+2)}{(x+h+2)(x+2)}}{h} = \lim_{h \to 0} \frac{\frac{\sqrt{3}x + 2\sqrt{3} - \sqrt{3}x - \sqrt{3}h - 2\sqrt{3}}{(x+h+2)(x+2)}}{h}$

$f'(x) = \displaystyle\lim_{h \to 0} \frac{\frac{-\sqrt{3}h}{(x+h+2)(x+2)}}{h} = \lim_{h \to 0} \frac{-\sqrt{3}}{(x+h+2)(x+2)}$

$f'(x) = \dfrac{-\sqrt{3}}{(x+2)^2}$

21. $y = \dfrac{2}{x^2}$

$$f'(x) = \lim_{h \to 0} \frac{f(x+h) - f(x)}{h} = \lim_{h \to 0} \frac{\frac{2}{(x+h)^2} - \frac{2}{x^2}}{h}$$

$$f'(x) = \lim_{h \to 0} \frac{2x^2 - 2\left(x^2 + 2xh + h^2\right)}{hx^2(x+h)^2}$$

$$= \lim_{h \to 0} \frac{2x^2 - 2x^2 - 4xh - 2h^2}{hx^2(x+h)^2}$$

$$f'(x) = \lim_{h \to 0} \frac{-4x - 2h}{x^2(x+h)^2} = \frac{-4x}{x^4}$$

$$f'(x) = \frac{-4}{x^3}$$

25. $y = x^4 - \dfrac{8}{x}$

$$f(x+h) = (x+h)^4 - \frac{8}{x+h}$$

$$= x^4 + 4x^3 h + 6x^2 h^2 + 4xh^2 + 4xh^3 + h^4 - \frac{8}{x+h}$$

$$f(x+h) - f(x)$$

$$= 4x^3 h + 6x^2 h^2 + 4xh^3 + h^4 - \frac{8x - 8(x+h)}{x(x+h)}$$

$$f(x+h) - f(x)$$

$$= 4x^3 h + 6x^2 h^2 + 4xh^3 + h^4 - \frac{-8h}{x(x+h)}$$

$$\frac{f(x+h) - f(x)}{h}$$

$$= 4x^3 + 6x^2 h + 4xh^2 + h^3 + \frac{8}{x(x+h)}$$

$$f'(x) = \lim_{h \to 0} \frac{f(x+h) - f(x)}{h} = 4x^3 + \frac{8}{x^2}$$

$$f'(x) = 4x^3 + \frac{8}{x^2}$$

29. $y = \dfrac{11}{3x+2}; \ (3, 1)$

$$f(x+h) - f(x) = \frac{11}{3(x+h)} - \frac{11}{3x+2}$$

$$= \frac{-33h}{(3x+2)(3(x+h)+2)}$$

$$\frac{f(x+h) - f(x)}{h} = \frac{-33}{(3x+2)(3(x+h)+2)}$$

$$f'(x) = \lim_{h \to 0} \frac{-33}{(3x+2)(3(x+h)+2)} = \frac{-33}{(3x+2)^2}$$

$$\left. \frac{dy}{dx} \right|_{(3,1)} = \frac{-33}{(3(3)+2)^2} = \frac{-3}{11}$$

33. $y = \dfrac{3}{x^2 - 1}$

$$\frac{f(x+h) - f(x)}{h} = \frac{\frac{3}{(x+h)^2 - 1} - \frac{3}{x^2 - 1}}{h}$$

$$= \frac{-3(2x+h)}{(x^2 - 1)(x^2 + 2xh + h^2 - 1)}$$

$$f'(x) = \lim_{h \to 0} \frac{-3(2x+h)}{(x^2 - 1)(x^2 + 2xh + h^2 - 1)}$$

$$= \frac{-6x}{(x^2 - 1)^2}, \ x \neq \pm 1$$

Function is differentiable for all $x \neq \pm 1$.

37. $y = f(x) = 2x^2 - 16x$

$f'(x) = 4x - 16 = 0$ for $x = 4$

$f(4) = 2(4)^2 - 16(4) = -32$

At $(4, -32)$ on curve $y = 2x^2 - 16x$ the tangent line is horizontal.

41. For $y = x^4 + x^3 + x^2 + x, \ \dfrac{dy}{dx} = 4x^3 + 3x^2 + 2x + 1$

from which, as a guess, $y = x^n$, $n > 0$, would have

$\dfrac{dy}{dx} = nx^{n-1}$.

23.4 The Derivative as an Instantaneous Rate of Change

1. $s = 48t - 16t^2$

$$v = \lim_{h \to 0} \frac{48(t+h) - 16(t+h)^2 - 48t + 16t^2}{h}$$

$$= \lim_{h \to 0} \left(-16(h + 2t - 3) \right) = -32t + 48$$

$$\left. \frac{ds}{dt} \right|_{t=2} = -32(2) + 48 = -16 \text{ ft/s}$$

$$\left. \frac{ds}{dt} \right|_{t=4} = -32(4) + 48 = -80 \text{ ft/s}$$

5. $y = \dfrac{6}{3x+1}; (-3, -2)$

$$m_{\tan} = \lim_{h \to 0} \frac{\frac{16}{3(x+h)+1} - \frac{16}{3x+1}}{h}$$

$$m_{\tan} = \lim_{h \to 0} \frac{\frac{48x+16-48x-48h-16}{(3(x+h)+1)(3x+1)}}{h}$$

$$m_{\tan} = \lim_{h \to 0} \frac{-48}{(3(x+h)+1)(3x+1)} = \frac{-48}{(3x+1)^2}$$

$$m_{\tan} = \left. \frac{dy}{dx} \right|_{(-3, -2)} = \frac{-48}{(3(-3)+1)^2} = -\frac{3}{4}$$

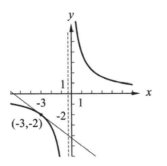

9. $s = 3t^2 - 4t; t = 2$

$$s = 3(2)^2 - 4(2) = 4$$

t (s)	1.0	1.5	1.9	1.99	1.999
s (ft)	−1.0	0.75	3.23	3.9203	3.9092003
$4 - s$ (ft)	5.0	3.25	0.77	0.0797	0.007997
$h = 2 - t$ (s)	1.0	0.5	0.1	0.01	0.001
$v = \frac{4-s}{h}$ (ft/s)	5.0	6.5	7.7	7.97	7.997

$v = 8$ ft/s when $t = 2$ s

13. $s = 3t^2 - 4t; t = 2$

$$v = \lim_{h \to 0} \frac{3(t+h)^2 - 4(t+h) - 3t^2 + 4t}{h}$$

$$= \lim_{h \to 0} \frac{3t^2 + 6th + 3h^2 - 4t - 4h - 3t^2 + 4t}{h}$$

$$v = \lim_{h \to 0} \frac{6th + 3h^2 - 4h}{h} = \lim_{h \to 0} (6t + 3 - 4) = 6t - 4$$

$$v = \left. \frac{ds}{dt} \right|_{t=2} = 6(2) - 4 = 8 \text{ ft/s}$$

17. $s = 12t^2 - t^4$

$$v = \frac{ds}{dt} = \lim_{h \to 0} \frac{12(t+h)^2 - (t+h)^4 - 12t^2 + t^4}{h}$$

$$v =$$

$$\lim_{h \to 0} \frac{12(t^2 + 2th + h^2) - t^4 - 4t^3h - 6t^2h^2 - 4th^3 - t^4 - 12t^2 + t^4}{h}$$

$$v = \lim_{h \to 0} \frac{24th + 12h^2 - 4t^3h - 6t^2h^2 - 4th^3}{h}$$

$$v = \lim_{h \to 0} \left(24t + 12h - 4t^3 - 6t^2h - 4th^2 \right) = 24 - 4t^3$$

21. $s = 7t^2 - \dfrac{2}{t+1}$

$$v = \frac{ds}{dt} = \lim_{h \to 0} \frac{7(t+h)^2 - \frac{2}{t+h+1} - 7t^2 + \frac{2}{t+1}}{h}$$

$$v = \lim_{h \to 0} \frac{14th + 7t^2h - \frac{2}{t+h+1} + \frac{2}{t+1}}{h}$$

$$v = \lim_{h \to 0} 14t + \frac{2}{(t+h+1)(t+1)}$$

$$v = 14t + \frac{2}{(t+1)^2}$$

25. $s = t^3 + 15t$

$$v = \frac{ds}{dt} = \lim_{h \to 0} \frac{(t+h)^3 + 15(t+h) - t^3 - 15t}{h}$$

$$v = \lim_{h \to 0} \frac{t^3 + 3t^2h + 3th^2 + h^3 = 15t + 15h - t^3 - 15t}{h}$$

$$v = \lim_{h \to 0} (3t^2 + 3th + h^2 + 15) = 3t^2 + 15$$

$$a = \frac{dv}{dt} = \lim_{h \to 0} \frac{3(t+h)^2 = 15 - 3t^2 - 15}{h}$$

$$= \lim_{h \to 0} \frac{3t^2 + 6th + 3h^2 + 15 - 3t^2 - 15}{h}$$

$$a = \lim_{h \to 0} (6t + 3h)$$

$$a = 6t$$

29. $c = 2\pi r$

$$\frac{dc}{dt} = 2\pi \frac{dr}{dt} = 2\pi(-0.0015)$$

$$\frac{dc}{dt} = -0.0094 \text{ cm/min}$$

33. $q = 30 - 2t$

$$i = \frac{dq}{dt} = \lim_{h \to 0} \frac{30 - 2(t+h) - 30 + 2t}{h} = \lim_{h \to 0} \frac{-2h}{h}$$

$$i = -2$$

37. $P = 500 + 250 \text{ m}^2$

$$\frac{dP}{dm} = \lim_{h \to 0} \frac{500 + 250(m+h)^2 - 500 - 250m^2}{h}$$

$$\frac{dP}{dm} = \lim_{h \to 0} \frac{250m^2 + 500mh + 250h^2 - 250m^2}{h}$$

$$\frac{dP}{dm} = \lim_{h \to 0} (500 + 250h) = 500m$$

$$\frac{dP}{dm}\bigg|_{m=0.92} = 500(0.92) = 460 \text{ W}$$

41. $V = \frac{48}{t+3}$

$$\frac{dV}{dt} = \lim_{h \to 0} \frac{\frac{48}{t+h+3} - \frac{48}{t+3}}{h} = \frac{-48}{(t+3)^2}$$

$$\frac{dV}{dt}\bigg|_{t=3} = \frac{-48}{(3+3)^2} = -\$1300/\text{year}$$

45. $r = k\sqrt{\lambda}$, $r = 3.72 \times 10^{-2}$ m, $\lambda = 59.2 \times 10^{-8}$ m

$$k = \frac{3.72 \times 10^{-2}}{\sqrt{59.2 \times 10^{-8}}} = 48.35$$

$$\frac{dr}{d\lambda} = \lim_{h \to 0} \frac{k\sqrt{\lambda + h} - k\sqrt{\lambda}}{h} = \frac{k}{2\sqrt{\lambda}} = \frac{48.35}{2\sqrt{\lambda}}$$

$$\frac{dr}{d\lambda} = \frac{24.2}{\sqrt{\lambda}}$$

23.5 Derivatives of Polynomials

1. $v = r^9$

$$\frac{dv}{dr} = 9r^{9-1}$$

$$\frac{dv}{dr} = 9r^8$$

5. $y = x^5$; $\dfrac{dy}{dx} = 5x^{5-1} = 5x^4$

9. $y = 5x^4 - 3\pi$; $\dfrac{dy}{dx} = 5\left(4x^{4-1}\right) - 0 = 20x^3$

13. $p = 5r^3 - 2r + 12$; $\dfrac{dp}{dr} = 5\left(3r^2\right) - 2 + 0$

$$= 15r^2 - 2$$

17. $f(x) = -6x^7 + 5x^3 + \pi^2$

$$\frac{f(x)}{dx} = -6\left(7x^6\right) + 5\left(3x^2\right) + 0 = -42x^6 + 15x^2$$

21. $y = 6x^2 - 8x + 1$;

$$\frac{dy}{dx} = \frac{d\left(6x^2\right)}{dx} - \frac{d\left(8x\right)}{dx} + \frac{d\left(1\right)}{dx} = 12x - 8 + 0$$

Since the derivative is a function of only x, we now evaluate it for $x = 2$.

$$\left.\frac{dy}{dx}\right|_{x=2} = 12\left(2\right) - 8 = 24 - 8 = 16$$

25. $y = 2x^6 - 4x^2$; $m_{\tan} = \dfrac{dy}{dx} = 12x^5 - 8x$

$$\left.\frac{dy}{dx}\right|_{x=-1} = m_{\tan} = 12\left(-1\right)^5 - 8\left(-1\right) = -12 + 8 = -4$$

Move the trace to $x = -1$ and observe that the function is decreasing and that the slope is neg

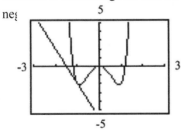

29. $s = 6t^5 - 5t + 2$; $v = \dfrac{ds}{dt} = 30t^4 - 5$

33. $s = 2t^3 - 4t^2$; $t = 4$

$$v = \frac{ds}{dt} = 2\left(3t^2\right) - 4\left(2t\right) = 6t^2 - 8t$$

$$\left.v\right|_{t=4} = 6\left(4^2\right) - 8\left(4\right) = 64$$

37. $y = 3x^2 - 6x$; $m_{\tan} = \dfrac{dy}{dx} = 6x - 6$

Tangent is parallel where slope is zero.

Therefore, $6x - 6 = 0$; $x = 1$

41. $x - 3y = 16 \Rightarrow y = \dfrac{1}{3}x - \dfrac{16}{3}$; $m_\perp = -3$

$$y = 2x^2 - 7x$$

$$\frac{dy}{dx} = 4x - 7 = -3 \Rightarrow x = 1$$

$$y = 2\left(1\right)^2 - 7\left(1\right) = -5$$

The tangent line to $y = 2x - 7x$ at $\left(1, -5\right)$ is perpendicular to the line $x - 3y = 16$.

45. $V = \pi r^2 \cdot h = \pi r^2 \cdot 20r = 20\pi r^3$

$$\left.\frac{dV}{dr} = 60\pi r^2\right|_{r=3.0} = 1700 \text{ mm}^2$$

49. $R = 16.0 + 0.450T + 0.0125T^2$

$$\left.\frac{dR}{dT} = 0.450 + 0.0250T\right|_{T=115} = 3.33 \ \Omega/^\circ\text{C}$$

53. $h = 0.000104x^4 - 0.0417x^3 + 4.21x^2 - 8.33x$

$$\left.\frac{dh}{dx} = 0.000416x^3 - 0.1251x^2 + 8.42x - 8.33\right|_{x=120}$$

$$\frac{dh}{dx} = -80.5 \text{ m/km}$$

$$\left.\frac{dF}{dx}\right|_{x=4} = 4\left(4\right)^3 - 36\left(4^2\right) + 92\left(4\right) - 60$$

$$= 256 - 576 + 368 - 60 = -12 \text{ N/cm}$$

23.6 Derivatives of Products and Quotients of Functions

1. $p(x) = (5 - 3x^2)(3 - 2x), u = 5 - 3x^2, v = 3 - 2x$

$\dfrac{d(uv)}{dx} = u\dfrac{dv}{dx} + v\dfrac{du}{dx}$

$p'(x) = (5 - 3x^2)(-2) + (3 - 2x)(-6x)$

$p'(x) = 18x^2 - 18x - 10$

5. $s = (3t + 2)(2t - 5)$

$u = 3t + 2, v = 2t - 5$

$\dfrac{du}{dt} = 3 \quad \dfrac{dv}{dt} = 2$

$\dfrac{ds}{dt} = (3t + 2)(2) + (2t - 5)(3)$

$\quad = 6t + 4 + 6t - 15 = 12t - 11$

9. $y = (2x - 7)(5 - 2x); u = (2x - 7); v = (5 - 2x)$

$\dfrac{dy}{dx} = (2x - 7)(-2) + (5 - 2x)(2)$

$\quad = -4x + 14 + 10 - 4x = -8x + 24$

$y = (2x - 7)(5 - 2x)$

$\quad = 10x - 4x^2 - 35 + 14x = -4x^2 + 24x - 35$

$\dfrac{dy}{dx} = -8x + 24$

13. $y = \dfrac{x}{8x + 3}; u = x; \dfrac{du}{dx} = 1; y = 8x + 3; \dfrac{dy}{dx} = 8$

$\dfrac{dy}{dx} = \dfrac{(8x + 3)(1) - x(8)}{(8x + 3)^2}$

$\quad = \dfrac{8x + 3 - 8x}{(8x + 3)^2} = \dfrac{3}{(8x + 3)^2}$

17. $y = \dfrac{6x^2}{3 - 2x}; u = 6x^2;$

$\dfrac{du}{dx} = 12x; v = 3 - 2x; \dfrac{dv}{dx} = -2$

$\dfrac{dy}{dx} = \dfrac{(3 - 2x)(12x) - (6x^2)(-2)}{(3 - 2x)^2}$

$\quad = \dfrac{36x - 24x^2 + 12x^2}{(3 - 2x)^2} = \dfrac{36x - 12x^2}{(3 - 2x)^2}$

21. $f(x) = \dfrac{3x + 8}{x^2 + 4x + 2}$

$\dfrac{df(x)}{dx} = \dfrac{(x^2 + 4x + 2)(3) - (3x + 8)(2x + 4)}{(x^2 + 4x + 2)^2}$

$\quad = \dfrac{-3x^2 - 16x - 26}{(x^2 + 4x + 2)^2}$

25. $y = (3x - 1)(4 - 7x)$

$\dfrac{dy}{dx} = (3x - 1)(-7) + (4 - 7x)(3)$

$\quad = -21x + 7 + 12 - 21x = -42x + 19$

$\dfrac{dy}{dx}\Big|_{x=3} = -42(3) + 19 = -126 + 19 = -107$

29. $y = \dfrac{3x - 5}{2x + 3}; y = 3x - 5; v = 2x + 3; du = 3dx;$

$dv = 2dx$

$\dfrac{dy}{dx} = \dfrac{(2x + 3)(3) - (3x - 5)(2)}{(2x + 3)^2} = \dfrac{6x + 9 - 6x + 10}{(2x + 3)^2}$

$\quad = \dfrac{19}{(2x + 3)^2}$

$\dfrac{dy}{dx}\Big|_{x=-2} = \dfrac{19}{(2(-2) + 3)^2} = \dfrac{19}{1} = 19$

33. For $u = c$ the product rule $\dfrac{d(u \cdot v)}{dx} = u \cdot \dfrac{dv}{dx} + v \cdot \dfrac{du}{dx}$

becomes $\dfrac{d(cv)}{dx} = c\dfrac{dv}{dx} + v\dfrac{dc}{dx} = c\dfrac{dv}{dx} + v(0)$

$\dfrac{d(cv)}{dx} = c\dfrac{dv}{dx}$, Equation 23.10

37. $\dfrac{d}{dx}(x^2 f(x)) = x^2 f'(x) + 2xf(x)$

41. (1) $y = \dfrac{x^2(1-2x)}{3x-7}$

$\dfrac{dy}{dx} = \dfrac{(3x-7)\left[x^2(-2)+(1-2x)(2x)\right]-x^2(1-2x)(3)}{(3x-7)^2}$

$= \dfrac{(3x-7)\left(-6x^2+2x\right)-3x^2+6x^3}{(3x-7)^2}$

$= \dfrac{-18x^3+6x^2+42x^2-14x-3x^2+6x^3}{(3x-7)^2}$

$= \dfrac{-12x^3+45x^2-14x}{(3x-7)^2}$

(2) $y = \dfrac{x^2-2x^3}{3x-7}$

$\dfrac{dy}{dx} = \dfrac{(3x-7)\left(2x-6x^2\right)-\left(x^2-2x^3\right)(3)}{(3x-7)^2}$

$= \dfrac{-12x^3+45x^2-14x}{(3x-7)^2}$

45. $y = \dfrac{x}{x^2+1};\ y = x;\ \dfrac{du}{dx} = 1;$

$v = x^2+1;\ \dfrac{dv}{dx} = 2x$

$\dfrac{dy}{dx} = \dfrac{(x^2+1)(1)-(x)(2x)}{(x^2+1)^2} = \dfrac{x^2+1-2x^2}{(x^2+1)^2}$

$= \dfrac{-x^2+1}{(x^2+1)^2}$

Therefore, $m_{\tan} = 0$ when $\dfrac{-x^2+1}{(x^2+1)^2} = 0;$

$-x^2+1 = 0;\ x^2 = 1;\ x = 1,\ -1$

49. $i = \dfrac{8R}{7R+12}$

$\dfrac{di}{dR} = \dfrac{(7R+12)(8)-8R(7)}{(7R+12)^2} = \dfrac{56R+96-56R}{(7R+12)^2}$

$\dfrac{di}{dR} = \dfrac{96}{(7R+12)^2}$

53. $T = \dfrac{2t}{0.05t+1}-20;\ t = 6\text{ h}$

$\dfrac{dT}{dt} = \dfrac{(0.05t+1)2-2t(0.05)}{(0.05t+1)^2}-0$

$= \dfrac{0.1t+2-0.1t}{(0.05t+1)^2}$

$\left.\dfrac{dT}{dt}\right|_{t=6.0\text{ h}} = 1.2°\text{C/h}$

57. $P = \dfrac{E^2 r}{R^2+2Rr+r^2}$

$\dfrac{dP}{dr} = \dfrac{\left(R^2+2Rr+r^2\right)E^2-E^2 r(0+2R+2r)}{\left(R^2+2Rr+r^2\right)^2}$

$\dfrac{dP}{dr} = \dfrac{E^2 R^2+2E^2 Rr+E^2 r^2-2E^2 Rr-2E^2 r^2}{\left(R^2+2Rr+r^2\right)^2}$

$\dfrac{dP}{dr} = \dfrac{E^2 R^2-E^2 r^2}{\left(R^2+2Rr+r^2\right)^2} = \dfrac{E^2\left(R^2-r^2\right)}{\left[(R+r)^2\right]^2}$

$\dfrac{dP}{dr} = \dfrac{E^2(R-r)(R+r)}{(R+r)^4} = \dfrac{E^2(R-r)}{(R+r)^3}$

23.7 The Derivative of a Power of a Function

1. $p(x) = (2 + 3x^3)^4$

$p'(x) = 4(2 + 3x^3)^3 (9x^2)$

$p'(x) = 36x^2 (2 + 3x^3)^3$

5. $y = 4\sqrt{x} = x^{1/2}$

$\dfrac{dy}{dx} = \dfrac{4}{2}x^{1/2-1} = \dfrac{4}{2}x^{-1/2} = \dfrac{4}{2}\left(\dfrac{1}{x^{1/2}}\right) = \dfrac{2}{2x^{1/2}}$

9. $y = \dfrac{3}{\sqrt[3]{x}} + 4\sqrt[3]{\pi} = 3x^{-1/3} + 4\sqrt[3]{\pi}$;

$\dfrac{dy}{dx} = 3\left(-\dfrac{1}{3}x^{-1/3-1}\right)$

$\dfrac{dy}{dx} = -1x^{-4/3} = -1\left(\dfrac{1}{x^{4/3}}\right) = -\dfrac{1}{x^{4/3}}$

13. $y = (4x^2 + 3)^5$

$\dfrac{dy}{dx} = 5(4x^2 + 3)^4 (8x) = 40x(4x^2 + 3)^4$

17. $y = (2x^3 - 3)^{1/3}$;

$\dfrac{dy}{dx} = \dfrac{1}{3}(2x^3 - 3)^{-2/3}(6x^2) = \dfrac{2x^2}{(2x^3 = 3)^{2/3}}$

21. $y = 4(2x^4 - 5)^{0.75}$; $u = 2x^4 - 5$; $\dfrac{du}{dx} = 8x^3$

$\dfrac{dy}{dx} = 4\left[0.75(2x^4 - 5)^{-0.25}(8x^3)\right] = \dfrac{24x^3}{(2x^4 - 5)^{0.25}}$

25. $u = v\sqrt{8v + 5} = v(8v + 5)^{1/2}$

$\dfrac{du}{dv} = v\left(\dfrac{1}{2}\right)(8v + 5)^{-1/2}(8) + (8v + 5)^{1/2}(1)$

$= 4v(8v + 5)^{-1/2} + (8v + 5)^{1/2}(1)$

$= \dfrac{4v}{(8v + 5)^{1/2}} + \dfrac{(8v + 5)}{(8v + 5)^{1/2}} = \dfrac{12v + 5}{(8v + 5)^{1/2}}$

29. $y = \dfrac{6x\sqrt{x + 2}}{x + 4}$

$\dfrac{dy}{dx} = \dfrac{(x + 4)\left[6x \cdot \frac{1}{2\sqrt{x+2}} + 6\sqrt{x + 2}\right] - 6x\sqrt{x + 2}\,(1)}{(x + 4)^2}$

$\dfrac{dy}{dx} = \dfrac{3x(x + 4) + 6(x + 4)(x + 2) - 6x(x + 2)}{(x + 4)^2 \sqrt{x + 2}}$

$\dfrac{dy}{dx} = \dfrac{3x^2 + 12x + 6x^2 + 36x + 48 - 6x^2 - 12x}{(x + 4)^2 \sqrt{x + 2}}$

$\dfrac{dy}{dx} = \dfrac{3(x^2 + 12x + 16)}{(x + 4)^2 \sqrt{x + 2}}$

33. $y = \sqrt{3x + 4}$; $x = 7$

$y = (3x + 4)^{1/2}$; $u = 3x + 4$; $n = \dfrac{1}{2}$; $\dfrac{du}{dx} = 3$

$\dfrac{dy}{dx} = \dfrac{1}{2}(3x + 4)^{-1/2}(3) = \dfrac{3}{2}(3x + 4)^{-1/2}$

$= \dfrac{3}{2\sqrt{3x + 4}}$

$\dfrac{dy}{dx}\Big|_{x=7} = \dfrac{3}{2\sqrt{3(7) + 4}} = \dfrac{3}{2\sqrt{25}} = \dfrac{3}{2(5)} = \dfrac{3}{10}$

37. $\dfrac{d}{dx}(x^{3/2}) = \dfrac{d}{dx}(x(x^{1/2})) = x\left(\dfrac{1}{2}x^{-1/2}\right) + x^{1/2}(1)$

$= \dfrac{1}{2}x^{1/2} + x^{1/2} = \dfrac{3}{2}x^{1/2}$

41. $y = \dfrac{x^2}{\sqrt{x^2 + 1}} = \dfrac{x^2}{(x^2 + 1)^{1/2}}$

$\dfrac{dy}{dx} = \dfrac{(x^2 + 1)^{1/2}\,2x - x^2\,\frac{1}{2}(x^2 + 1)^{-1/2}(2x)}{(x^2 + 1)}$

$\dfrac{dy}{dx} = \dfrac{2x(x^2 + 1)^{1/2} - \frac{x^3}{(x^2+1)^{1/2}}}{(x^2 + 1)}$

$$= \frac{\frac{2x\left(x^2+1\right)-x^3}{\left(x^2+1\right)^{1/2}}}{\left(x^2+1\right)}$$

$$\frac{dy}{dx} = \frac{x^3+2x}{\left(x^2+1\right)^{3/2}} = 0$$

$x^3 + 2x = 0;\ x\left(x^2+2\right) = 0$

$x = 0$ or $x = \pm\sqrt{-2}$ (imaginary). Therefore,

$$\frac{dy}{dx} = 0 \text{ for } x = 0$$

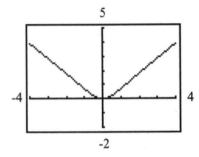

45. $y^2 = 4x;\ y = \sqrt{4x} = 2\sqrt{x} = 2x^{1/2}$

$$m_{\tan} = \frac{dy}{dx} = \frac{d\left(2x^{1/2}\right)}{dx} = 2\left(\frac{1}{2}\right)x^{-1/2} = \frac{1}{\sqrt{x}}$$

$$m_{\tan}\Big|_{x=1} = \frac{1}{\sqrt{1}} = \frac{1}{1} = 1$$

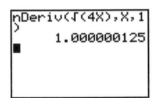

49. $s = \left(8t - t^2\right)^{2/3};\ t = 6.25 s$

$$\frac{ds}{dt} = \frac{2}{3}\left(8t - t^2\right)^{-1/3}\left(8 - 2t\right)$$

$$v = \frac{2\left(8 - 2t\right)}{3\sqrt[3]{8t - t^2}};\ v\Big|_{t=6.25} = -1.35 \text{ cm/s}$$

53. $v = \sqrt{\dfrac{l}{a} + \dfrac{a}{l}} = \left(\dfrac{l}{a} + \dfrac{a}{l}\right)^{1/2}$

$$\frac{dv}{dl} = \frac{1}{2}\left(\frac{1}{a} + \frac{a}{l}\right)^{-1/2}\left(\frac{1}{a} - \frac{a}{l^2}\right) = 0$$

$$\frac{1}{a} - \frac{a}{l^2} = 0$$

$$\frac{1}{a} = \frac{a}{l^2}$$

$$l^2 = a^2$$

$$l = a$$

57. $\lambda_r = \dfrac{2a\lambda}{\sqrt{4a^2 - \lambda^2}} = \dfrac{2a\lambda}{\left(4a^2 - \lambda^2\right)^{1/2}}$

$$\frac{d\lambda_r}{d\lambda}$$

$$= \frac{\left(4a^2 - \lambda^2\right)^{1/2}(2a) - 2a\lambda\left(\frac{1}{2}\right)\left(4a^2 - \lambda^2\right)^{-1/2}(-2\lambda)}{\left(4a^2 - \lambda^2\right)}$$

$$= \frac{2a\left(4a^2 - \lambda^2\right)^{1/2} + 2a\lambda^2\left(4a^2 - \lambda^2\right)^{-1/2}}{\left(4a^2 - \lambda^2\right)}$$

$$= \frac{\left(4a^2 - \lambda^2\right)^{-1/2}\left[(2a)\left(4a^2 - \lambda^2\right) + 2a\lambda^2\right]}{\left(4a^2 - \lambda^2\right)}$$

$$= \frac{8a^3}{\left(4a^2 - \lambda^2\right)^{3/2}}$$

23.8   Differentiation of Implicit Functions

1. $y^3 + 2x^2 = 5$

$$3y^2 \cdot \frac{dy}{dx} + 4x = 0$$

$$\frac{dy}{dx} = \frac{-4x}{3y^2}$$

5. $\dfrac{d}{dx}\left(\dfrac{3}{x^4 y}\right) = \dfrac{x^4 y(0) - 3\left(x^4 y' + 4x^3 y\right)}{\left(x^4 y\right)^2}$

$$= \frac{-3\left(xy' + 4y\right)}{x^5 y^2}$$

9. $4y - 3x^2 = x;\ \dfrac{d}{dx}(4y) - \dfrac{d}{dx}\left(3x^2\right) = \dfrac{d}{dx}(x)$

$$4\frac{dy}{dx} - 6x = 1;\ 4\frac{dy}{dx} = 1 + 6x;\ \frac{dy}{dx} = \frac{1 + 6x}{4}$$

13. $y^5 = x^2 - 1;\ \dfrac{d}{dx}\left(y^5\right) = 2x;\ 5x\dfrac{dy}{dx} = 2x;\ \dfrac{dy}{dx} = \dfrac{2x}{5y^4}$

17. $y + 3xy - 4 = 0$

$$\frac{dy}{dx} + 3\frac{d}{dx}(xy) = 0;\ \frac{dy}{dx} + 3\left(x\frac{dy}{dx} + y(1)\right) = 0$$

$$\frac{dy}{dx} + 3x\frac{dy}{dx} + 3y = 0;\ \frac{dy}{dx}(1 + 3x) = -3y$$

$$\frac{dy}{dx} = \frac{-3y}{1 + 3x}$$

21. $\dfrac{3x^2}{y^2 + 1} + y = 3x + 1$

$$\frac{\left(y^2 + 1\right)6x - 3x^2\left(2\frac{dy}{dx}\right)}{\left(y^2 + 1\right)^2} + \frac{dy}{dx} = 3$$

$$\frac{6x\left(y^2 + 1\right) - 6x^2 y \frac{dy}{dx} + \left(y^2 + 1\right)^2 \frac{dy}{dx}}{\left(y^2 + 1\right)^2} = 3$$

$$\left(y^2 + 1\right)^2 \frac{dy}{dx} - 6x^2 y\frac{dy}{dx} = 3\left(y^2 + 1\right)^2 - 6x\left(y^2 + 1\right)$$

$$= 3\left(y^2 + 1\right)\left(y^2 + 1 - 2x\right)$$

$$\frac{dy}{dx}\left[\left(y^2 + 1\right)^2 - 6x^2 y\right] = 3\left(y^2 + 1\right)\left(y^2 - 2x + 1\right)$$

$$\frac{dy}{dx} = \frac{3\left(y^2 + 1\right)\left(y^2 - 2x + 1\right)}{\left(y^2 + 1\right)^2 - 6x^2 y}$$

25. $2\left(x^2 + 1\right)^3 + \sqrt{y^2 + 1} = 17$

$$6\left(x^2 + 1\right)^2 (2x) + \frac{1}{2\sqrt{y^2 + 1}}2y\frac{dy}{dx} = 0$$

$$\frac{dy}{dx} = \frac{-12x\left(x^2 + 1\right)^2 \sqrt{y^2 + 1}}{y}$$

29. $5y^4 + 7 = x^4 - 3y;\ (3, -2)$

$$20y^3 \frac{dy}{dx} = 4x^3 - 3\frac{dy}{dx};\ 20y^3\frac{dy}{dx} + 3\frac{dy}{dx} = 4x^3$$

$$\frac{dy}{dx}\left(20y^3 + 3\right) = 4x^3;\ \frac{dy}{dx} = \frac{4x^3}{20y^3 + 3};$$

$$\left.\frac{dy}{dx}\right|_{(3,\,-2)} = -\frac{108}{157}$$

33. $x^2 + y^2 = 4x$

$$2x + 2y\frac{dy}{dx} = 4$$

$$\frac{dy}{dx} = \frac{2 - x}{y} = 0$$

$$x = 2$$

$$2^2 + y^2 = 4(2)$$

$$y^2 = 4$$

$$y = \pm 2$$

The graph of $x^2 + y^2 = 4x$ has a horizontal tangent line at $(2, 2)$ and $(2, -2)$.

37. $\omega^2 = \dfrac{1}{LC} - \dfrac{R^2}{L^2}$

$$2\omega\frac{d\omega}{dL} = \frac{-1}{L^2 C} + \frac{2R^2}{L^3}$$

$$\frac{d\omega}{dL} = \frac{-1}{2\omega L^2 C} + \frac{R^2}{\omega L^3}$$

41. $PV = n\left(RT + aP - \dfrac{bP}{T}\right)$

$PV = nRT + naP - \dfrac{nbP}{T}$

$V\dfrac{dP}{dT} = nR + na\dfrac{dP}{dT} - \dfrac{nbT\frac{dP}{dT} - nbP}{T^2}$

$VT^2\dfrac{dP}{dT} = nRT^2 + naT^2\dfrac{dP}{dT} - nbT\dfrac{dP}{dT} + nbP$

$\dfrac{dP}{dT}\left(VT^2 - naT^2 + nbT\right) = nRT^2 + nbP$

$\dfrac{dP}{dT} = \dfrac{nRT^2 + nbP}{VT^2 - naT^2 + nbT}$

45. $r^2 = 2rR + 2R - 2r;\ 2r = 2R + \dfrac{dR}{dr}(2r) + 2\dfrac{dR}{dr} - 2$

$2r - 2R + 2 = 2r\dfrac{dR}{dr} + 2\dfrac{dR}{dr} = \dfrac{dR}{dr}(2r+2)$

$\dfrac{dR}{dr} = \dfrac{2(r - R + 1)}{2(r+1)} = \dfrac{r - R + 1}{r+1}$

23.9   Higher Derivatives

1. $y = 5x^3 - 2x^2$

$y' = 15x^2 - 4x$

$y'' = 30x - 4$

$y''' = 30$

$y^{(4)} = 0$

$y^{(n)}(x) = 0,\ n \ge 4$

5. $f(x) = x^3 - 6x^4;\ f'(x) = 3x^2 - 24x^3;$

$f''(x) = 6x - 72x^2;\ f'''(x) = 6 - 144x;$

$f^{(4)}(x) = -144;\ f^{(n)}(x) = 0,\ n \ge 5$

9. $f(r) = r(4r+9)^3 = r\left(64r^3 + 432r^2 + 972r + 729\right)$

$f(r) = 64r^4 + 432r^3 + 972r^2 + 729r$

$f'(r) = 256r^3 + 1296r^2 + 1944r + 729$

$f''(r) = 768r^2 + 2592r + 1944$

$f'''(r) = 1536r + 2592$

$f^{(4)}(r) = 1536$

$f^{(n)}(r)0,\ n \ge 5$

13. $y = 5x + 8\sqrt{x} = 5x + 8x^{1/2};\ y' = 5 + \dfrac{8}{2}x^{-1/2}$

$y'' = 4\left(-\dfrac{1}{2}\right)x^{-3/2} = \dfrac{-2}{x^{3/2}}$

17. $f(p) = \dfrac{4.8\pi}{\sqrt{1+2p}} = 4.8\pi(1+2p)^{-1/2}$

$f'(p) = -2.4\pi(1+2p)^{-3/2}(2) = -4.8\pi(1+2p)^{-3/2}$

$f''(p) = 7.2\pi(1+2p)^{-5/2}(2) = \dfrac{14.4\pi}{(1+2p)^{5/2}}$

21. $y = (3x^2 - 1)^5;\ y' = 5(3x^2 - 1)^4(6x)$

$= 30x(3x^2 - 1)^4;$

$y'' = 30x(4)(3x^2 - 1)^3(6x) + (3x^2 - 1)^4(30)$

$= 30(3x^2 - 1)^3(27x^2 - 1)$

25.

$u = \dfrac{v^2}{4v+15}$

$u' = \dfrac{(4v+15)2v - v^2(4)}{(4v+15)^2} = \dfrac{4v^2 + 30v}{(4v+15)^2}$

$u'' = \dfrac{(4v+15)^2(8v+30) - (4v^2 + 30v)(2)(4v+15)(4)}{(4v+15)^4}$

$u'' = \dfrac{450}{(4v+15)^3}$

29. $x^2 - xy = 1 - y^2;\ 2x - (xy' + y) = -2yy'$

$2x - xy' - y = -2yy';\ 2yy' - xy' = y - 2x$

$y' = \dfrac{y-2x}{2y-x}$;

$y'' = \dfrac{(2y-x)(y'-2)=(y-2x)(2y'-1)}{(2y-x)^2}$

$y'' = \left[(2y-x)\left(\dfrac{y-2x}{2y-x}-2\right)-(y-2x)\left(2\dfrac{y-2x}{2y-x}-1\right)\right]$

$\quad \div (2y-x)^2$

$y'' = \left[y-2x-2(2-y)-\dfrac{2(y-2x)^2}{2y-x}+y-2x\right]$

$\quad \div (2y-x)^2$

$y'' = \left[-2(y+x)-\dfrac{2(y-2x)^2}{(2y-x)}\right]\div(2y-x)^2$

$y'' = \left(-4y^2-2xy+2x^2-2y^2+8xy-8x^2\right)\div(2y-x)^3$

$y'' = \dfrac{-6\left(y^2-xy+x^2\right)}{(2y-x)^3}$

33. $y = 3x^{2/3}-\dfrac{2}{x}=3x^{2/3}-2x^{-1}$

$y' = 2x^{-1/3}+2x^{-2}$

$y'' = -\dfrac{2}{3}x^{-4/3}-4x^{-3}=\dfrac{-2}{3x^{4/3}}-\dfrac{4}{x^3}$;

$y''\Big|_{x=-8} = \dfrac{-2}{48}+\dfrac{4}{512}$

$y''\Big|_{x=-8} = \dfrac{-2}{6\times8}+\dfrac{4}{8\times64}=-\dfrac{13}{384}$

37. $s = 26t-4.9t^2$, $t = 3.0$ s

$v = 26-9.8t$

$a = -9.8$ m/s^2

41. $\dfrac{d(uv)}{dx} = u\dfrac{dv}{dx}+v\dfrac{du}{dx}$

$\dfrac{d^2(uv)}{dx^2} = u\dfrac{d^2v}{dx^2}+\dfrac{du}{dx}\dfrac{dv}{dx}+v\dfrac{d^2u}{dx^2}+\dfrac{du}{dx}\dfrac{dv}{dx}$

$\qquad = u\dfrac{d^2v}{dx^2}+2\dfrac{du}{dx}\dfrac{dv}{dx}+v\dfrac{d^2u}{dx^2}$

45. $P(t) = 8000\left(1+0.02t+0.005t^2\right)$

$P'(t) = 8000(0.02+0.01t)$

$P''(t) = 8000(0.01) = 80$

49. $s = 2250t-16.1t^2$; $s' = 2250-32.2t$

$s'' = a = -32.2$ ft/s^2

53. $y = 0.0001\left(x^5-25x^2\right)$

$\dfrac{dy}{dx} = 0.0001\left(5x^4-50x\right)$

$\dfrac{d^2y}{dx^2}\Big|_{x=3.00} = 0.0001\left(20x^3-50\right)\Big|_{x=3.00}$

$\dfrac{d^2y}{dx^2}\Big|_{x=3.00} = 0.0001\left(20(3.00)^3-50\right)$

$\dfrac{d^2y}{dx^2}\Big|_{x=3.00} = 0.049$

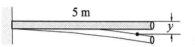

Chapter 23   Review Exercises

1. $\lim\limits_{x \to 4}(8-3x) = 8-3(4) = -4$

5. $\lim\limits_{x \to 2}\dfrac{4x-8}{x^2-4} = \lim\limits_{x \to 2}\dfrac{4(x-2)}{(x-2)(x+2)}$

$\qquad = \lim\limits_{x \to 2}\dfrac{4}{x+2} = \dfrac{4}{2+2} = 1$

9. $\lim\limits_{x \to \infty}\dfrac{2+\frac{3x}{x+4}}{3-\frac{1}{x^2}} = \lim\limits_{x \to \infty}\dfrac{2+\frac{3}{1+\frac{4}{x}}}{3-\frac{1}{x^2}} = \dfrac{5}{3}$

13. $y = 7+5x$

$\dfrac{dy}{dx} = \lim\limits_{h \to 0}\dfrac{f(x+h)-f(x)}{h}$

$\qquad = \lim\limits_{h \to 0}\dfrac{7+5(x+h)-7-5x}{h}$

$\dfrac{dy}{dx} = \lim\limits_{h \to 0}\dfrac{5h}{h} = \lim\limits_{h \to 0}5 = 5$

17. $y = \dfrac{2}{x^2}$

$\dfrac{dy}{dx} = \lim\limits_{h \to 0}\dfrac{\frac{2}{(x+h)^2}-\frac{2}{x^2}}{h} = \lim\limits_{h \to 0}\dfrac{2x^2-2(x+h)^2}{hx^2(x+h)^2}$

$\qquad = \lim\limits_{h \to 0}\dfrac{2x^2-2x^2-4xh-2h^2}{hx^2(x+h)^2}$

$\qquad = \lim\limits_{h \to 0}\dfrac{-4xh-2h^2}{hx^2(x+h)^2} = \lim\limits_{h \to 0}\dfrac{-4x-2h}{x^2(x+h)^2}$

$\qquad = \dfrac{-4x}{x^4} = \dfrac{-4}{x^3}$

21. $y = 2x^7 - 3x^2 + 5$

$\dfrac{dy}{dx} = 2(7x^6) - 3(2x) + 0 = 14x^6 - 6x$

25. $f(y) = \dfrac{12y}{1-5y}$

$\dfrac{df(y)}{dy} = \dfrac{(1-5y)(12)-12y(-5)}{(1-5y)^2} = \dfrac{12-60y+60y}{(1-5y)^2}$

$\dfrac{df(y)}{dy} = \dfrac{12}{(1-5y)^2}$

29. $y = \dfrac{3\pi}{(5-2x^2)^{3/4}}$

$\dfrac{dy}{dx} = \dfrac{(5-2x^2)^{3/4}(0)-3\pi\left[-3x(5-2x^2)^{-1/4}\right]}{(5-2x^2)^{3/2}}$

$\qquad = \dfrac{9\pi x(5-2x^2)^{-1/4}}{(5-2x^2)^{3/2}}$

$\qquad = \dfrac{9\pi x}{(5-2x^2)^{3/2}(5-2x^2)^{1/4}} = \dfrac{9\pi x}{(5-2x^2)^{7/4}}$

33. $y = \dfrac{\sqrt{4x+3}}{2x} = 2(4x+3)^{-1/2}$

$\dfrac{dy}{dx} = \dfrac{2x(2)(4x+3)^{-1/2}-(4x+3)^{1/2}(2)}{(2x)^2}$

$\qquad = \dfrac{(4x+3)^{-1/2}\left[4x-(4x+3)(2)\right]}{4x^2}$

$\qquad = \dfrac{-4x-6}{(4x+3)^{1/2}(4x^2)} = \dfrac{2(-2x-3)}{2(2x^2)(4x+3)^{1/2}}$

$\qquad = \dfrac{-2x-3}{2x^2(4x+3)^{1/2}}$

37. $y = \dfrac{4}{x} + 2\sqrt[3]{x},\ x = 8$

$y = 4x^{-1} + 2x^{1/3};\ \dfrac{dy}{dx} = 4(-1)x^{-2} + 2\left(\dfrac{1}{3}x^{-2/3}\right)$

$\qquad = \dfrac{-4}{x^2} + \dfrac{2}{3x^{2/3}}$

$\dfrac{dy}{dx}\Big|_{x=8} = \dfrac{-4}{8^2} + \dfrac{2}{3(8)^{2/3}} = \dfrac{-4}{64} + \dfrac{2}{3(4)} = \dfrac{-1}{16} + \dfrac{1}{6}$

$\qquad = \dfrac{-3}{48} + \dfrac{8}{48} = \dfrac{5}{48}$

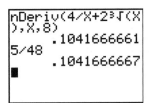

41. $y = 3x^4 - \dfrac{1}{x} = 3x^4 - x^{-1}$

$y' = 12x^3 + x^{-2}$

$y'' = 36x^2 - 2x^{-3}$

45. From graph, as $x \to 0^+$

$1/x^2$ increases most rapidly.

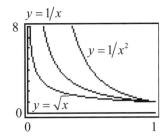

49. $y = \dfrac{2(x^2 - 4)}{x - 2}, \ x \neq 2$

$y = \dfrac{2(x + 2)(x - 2)}{(x - 2)}$

$= 2(x + 2)\big|_{x=2}$

$= 2(2 + 2) = 8$

Just to the left of $x = 2$, the trace feature gives $x = 1.9787234$, $y = 7.9574468$ and just to the right of $x = 2$, the trace feature gives $x = 2.0319149$, $y = 8.0638298$ which would appear to give $y = 8$ for $x = 2$. However, using the value feature shows there is no y-value for $x = 2$.

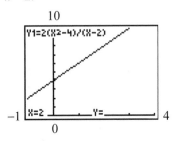

53. $y = 7x^4 - x^3$

$\dfrac{dy}{dx} = 28x^3 - 3x^2 \Big|_{(-1,\, 8)}$

$= 28(-1)^3 - 3(-1)^2$

$= -31 = m_{\text{tan}} \text{ at } (-1,\, 8)$

The nDeriv feature of calculator gives -31.000029

```
nDeriv(7X^4-X^3,
X,-1)
            -31.000029
■
```

57. $A = 5000(1 + 0.25i)^8$

$\dfrac{dA}{di} = 40,000(1 + 0.25i)^7 (0.25)$

$= 10,000(1 + 0.25i)^7$

61. $y = 0.0015x^2 + C$

$\dfrac{dy}{dx} = 2(0.0015)x = 0.3$

$x = \dfrac{0.3}{2(0.0015)}$

$y = 0.3\left(\dfrac{0.3}{2(0.0015)}\right) - 10$

$= 0.0015\left(\dfrac{0.3}{2(0.0015)}\right)^2 + C$

$C = 5$

65. $F = \dfrac{Gm_1 m_2}{r^2}$

$\dfrac{dF}{dr} = \dfrac{-2Gm_1 m_2}{r^3}$

69. $r_f = \dfrac{2(R^3 - r^3)}{3(R^2 - r^2)}$

$\dfrac{dr_f}{dR} = \dfrac{3(R^2 - r^2)(6R^2) - 2(R^3 - r^3)(6R)}{9(R^2 - r^2)^2}$

$$\frac{dr_f}{dR} = \frac{18R^2(R+r)(R-r)-12R(R-r)(R^2+Rr+r^2)}{9(R+r)^2(R-r)^2}$$

$$\frac{dr_f}{dR} = \frac{6R^2(R+r)-4R(R^2+Rr+r^2)}{3(R+r)^2(R-r)}$$

$$\frac{dr_f}{dR} = \frac{6R^3+6R^2r-4R^3-4R^2r-4Rr^2}{3(R+r)^2(R-r)}$$

$$\frac{dr_f}{dR} = \frac{2R^3+2R^2r-4Rr^2}{3(R+r)^2(R-r)}$$

$$= \frac{2R(R^2+Rr-2r^2)}{3(R+r)^2(R-r)}$$

$$\frac{dr_f}{dR} = \frac{2R(R+2r)(R-r)}{3(R+r)^2(R-r)}$$

$$= \frac{2R(R+2r)}{3(R+r)^2}$$

73. $y = kx(x^4+450x^2-950) = kx^5+450kx^3-950kx$

$$\frac{dy}{dx} = 5kx^4+1350kx^2-950k$$

77. $T = \dfrac{10(1-t)}{0.5t+1}$

$$\frac{dT}{dt} = \frac{(0.5t+1)(-10)-10(1-t)(0.5)}{(0.5t+1)^2}$$

$$= \frac{-5t-10-5+5t}{(0.5t+1)^2}$$

$$\frac{dT}{dt} = \frac{-15}{(0.5t+1)^2}$$

81. $A = lw = 75 \Rightarrow l = \dfrac{75}{w}$

$$p = 2l+2w = \frac{150}{w}+2w$$

$$\frac{dp}{dw} = \frac{-150}{w^2}+2$$

85. $A = \pi r^2$

$$\frac{dA}{dr} = 2\pi r \big|_{r=1.8} = 11.3 \text{ km}^2/\text{km}$$

89. $x^2 = h^2+(vt)^2$, $h = \dfrac{2640}{5280}$, $v = 400$

$$2x\frac{dx}{dt} = 2v^2t$$

$$\frac{dx}{dt} = \frac{v^2t}{x}\Bigg|_{v=400,\ x=\sqrt{\left(\frac{2640}{5280}\right)^2+\left(400\left(\frac{0.600}{60}\right)\right)^2},\ t=\frac{0.600}{60}}$$

$$\frac{dx}{dt} = 397 \text{ mi/h}$$

Chapter 24

APPLICATIONS OF THE DERIVATIVE

24.1   Tangents and Normals

1. $x^2 + 4y^2 = 17, (1, 2)$

$2x + 8yy' = 0$

$y' = \dfrac{-x}{4y}\Big|_{(1, 2)} = \dfrac{-1}{4(2)} = -\dfrac{1}{8}$

$y - 2 = -\dfrac{1}{8}(x-1)$

$8y - 16 = -x + 1$

$x + 8y - 17 = 0$

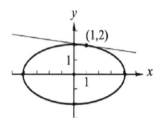

5. $y = \dfrac{1}{x^2+1}; \left(-1, \dfrac{1}{2}\right); y = (x^2+1)^{-1}$

$\dfrac{dy}{dx} = m_{\tan} = -(x^2+1)^{-2}(2x); m_{\tan} = \dfrac{-2x}{(x^2+1)^2};$

$m_{\tan} = \dfrac{1}{2}$ for $x = -1$

Eq. T.L.: $y - \dfrac{1}{2} = \dfrac{1}{2}(x+1); 2y - 1 = x + 1$

$2y - x - 2 = 0$

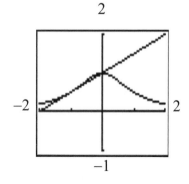

9. $y^2(2-x) = x^3 \Rightarrow y^2 = \dfrac{x^3}{2-x}$

$\dfrac{dy}{dx} = m_{\tan} = \dfrac{3x^2 - x^3}{y(2-x)^2}\Big|_{(1, 1)} = 2$

$m_{\text{normal}} = \dfrac{-1}{2}$

Eq. of normal: $y - 1 = \dfrac{-1}{2}(x-1) \Rightarrow x + 2y - 3 = 0$

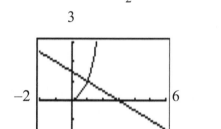

13. $y = (2x-1)^3$; normal line $m = -\dfrac{1}{24}, x > 0$

Therefore, $m_{\tan} = 24$

$m_{\tan} = 3(2x-1)^2(2) = 6(x-1)^2$

$6(2x-1)^2 = 24; (2x-1)^2 = 4; 2x - 1 = \pm 2$

$2x = \pm 2 + 1 = 3$ and -1

$x = \dfrac{3}{2}, y = 8$

Eq. of N.L.: $y - 8 = -\dfrac{1}{24}\left(x - \dfrac{3}{2}\right);$

$24y - 192 = -x + \dfrac{3}{2}$

$48y - 384 = -2x + 3; 2x + 48y - 387 = 0$

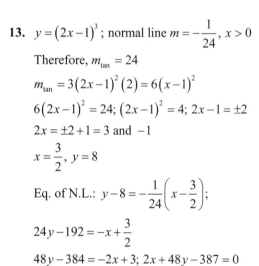

17. $y = x + 2x^2 - x^4$

$y' = 1 + 4x - 4x^3$

At $(1, 2)$, $y' = 1 + 4(1) - 4(1)^3 = 1$

Eq. of T.L.: at $(1, 2)$: $y - 2 = 1(x - 1)$

$$y = x + 1$$

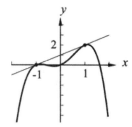

At $(-1, 0)$, $y' = 1 + 4(1) - 4(-1)^3 = 1$

Eq. of TL at $(-1, 0)$:

$y - 0 = 1(x - (-1))$

$$y = x + 1$$

The tangent lines are the same: $y = x + 1$

21. $x^2 + y^2 = a^2$

$2x + 2yy' = 0$

$$y' = \frac{-x}{y}\Big|_{(x_1, y_1)} = \frac{-x_1}{y_1}$$

$$y - y_1 = \frac{-x_1}{y_1}(x - x_1)$$

$$y_1 y - y_1^2 = -x_1 x + x_1^2$$

$$x_1 x + y_1 y = x_1^2 + y_1^2 = a^2$$

25. Equation of a curve:

$x^2 = 4py$

$$p = \frac{x^2}{4y} = \frac{(100)^2}{4(30)}$$

$$p = \frac{250}{3}$$

Therefore, $x^2 = \frac{1000}{3}y$

29. $y = \frac{4}{x^2 + 1}$; $(-2 < x < 2)$

Supports at $(-1, 2)$, $(0, 4)$, $(1, 2)$

$y = 4(x^2 + 1)^{-1}$

$$\frac{dy}{dx} = m_{\tan} = -4(x^2 + 1)^{-2}(2x)$$

$$m_{\tan} = \frac{-8x}{(x^2 + 1)^2}$$

$m_{\tan}\big|_{x=-1} = 2$; $m_{\tan}\big|_{x=0} = 0$;

$m_{\tan}\big|_{x=1} = -2$

Eq. of N.L.: at $(-1, 2)$:

$$m_{NL} = -\frac{1}{2}$$

$y - 2 = -\frac{1}{2}(x + 1)$; $2y - 4 = -x + 1$;

$x + 2y - 3 = 0$

Eq. of N.L.: at $(0, 4)$

$x = 0$; m_{NL} is undefined (vertical).

Eq. of N.L.: at $(1, 2)$:

$$m_{NL} = \frac{1}{2}$$

$y - 2 = \frac{1}{2}(x - 1)$; $2y - 4 = x - 1$;

$x - 2y + 3 = 0$

24.2 Newton's Method for Solving Equations

1. $x^2 - 5x + 1 = 0$, $0 < x < 1$, $f(x) = x^2 - 5x + 1$

$f(0) = 1$, $f(1) = -3$, choose $x_1 = 0.5$

$f'(x) = 2x - 5$

$x_2 = x_1 - \dfrac{f(x_1)}{f'(x_1)} = 0.5 - \dfrac{0.5^2 - 5(0.5) + 1}{2(0.5) - 5}$

$\quad = 0.1875$

$x_3 = 0.2086148649$

$x_4 = 0.2087121505$

$x_5 = 0.2087121525$

$x_6 = 0.2087121525$

Using the quadratic formula,

$x = \dfrac{(-5) - \sqrt{(-5)^2 - 4(1)(1)}}{2(1)}$

$\quad = 0.2087121525$

5. $x^3 - 6x^2 + 10x - 4 = 0$ (between 0 and 1)

$f(x) = x^3 - 6x^2 + 10x - 4$; $f'(x) = 3x^2 - 12x + 10$;

$f(0) = -4$; $f(1) = 1$

Let $x_1 = 0.7$

n	x_n	$f(x_n)$	$f'(x_n)$	$x_n - \dfrac{f(x_n)}{f'(x_n)}$
1	0.7	0.403	3.07	0.5687296
2	0.5687296	−0.0694666	4.1456049	0.5854863
3	0.5854863	−0.0012009	4.002547	0.5857863
4	0.5857863	−0.0000005	4.0000012	0.5857864

$x_4 = x_3 = 0.5857864$ to seven decimal places;

$x = 0.58578644$, calculator

9. $x^4 - x^3 - 3x^2 - x - 4 = 0$; (between 2 and 3)

$f(x) = x^4 - x^3 - 3x^2 - x - 4$;

$f'(x) = 4x^3 - 3x^2 - 6x - 1$

$f(2) = -10$; $f(3) = 20$.

Let $x_1 = 2.3$

n	x_n	$f(x_n)$	$f'(x_n)$	$x_n - \dfrac{f(x_n)}{f'(x_n)}$
1	2.3	−6.3529	17.998	2.6529781
2	2.6529781	3.0972725	36.657001	2.5684848
3	2.5684848	0.2175007	31.576097	2.5615967
4	2.5615967	0.0013683	−31.179599	2.5615528

$x_3 = x_4 = 2.5615528$ to seven decimal places;

$x = 2.5615528$, calculator

13. $2x^2 = \sqrt{2x + 1}$ or $2x^2 - \sqrt{2x + 1} = 0$, (the positive real solution)

$4x^4 - 2x - 1 = 0$ (Square both sides)

$f(x) = 4x^4 - 2x - 1$; $f'(x) = 16x^3 - 2$

Let $x_1 = 0.8$

n	x_n	$f(x_n)$	$f'(x_n)$	$x_n - \dfrac{f(x_n)}{f'(x_n)}$
1	0.8	−0.9616	6.192	0.955297
2	0.955297	0.420707	11.94875	0.920087
3	0.920087	0.026491	10.46257	0.917555
4	0.917555	1.30109×10^{-4}	10.35997	0.917543
5	0.917543	3.186×10^{-9}	10.359467	0.917543

The positive root is approximately 0.9175433; to seven decimal places. $x = 0.91754334$, calculator

17. $f(x) = x^3 - 2x^2 - 5x + 4$. From graph one, root lies between -1 and -2, a second between 0 and 1 and a third between 3 and 4.

$f'(x) = 3x^2 - 4x - 5$

Let $x_1 = -1.7$

n	x_n	$f(x_n)$	$f'(x_n)$	$x_n - \dfrac{f(x_n)}{f'(x_n)}$
1	−1.7	1.807	10.47	−1.8725883
2	−1.872588	−0.2166267	13.010114	−1.8559377
3	−1.855937	−0.0021074	12.757265	−1.8557725
4	−1.855772	-2.065005×10^{-7}	12.754765	−1.8557725

Let $x_1 = 0.7$

1	0.7	-0.137	-6.33	0.6783570
2	0.6783570	3.6703855×10^{-5}	-6.3329233	0.6783628

Let $x_1 = 3.1$

1	3.1	-0.929	11.43	3.1812773
2	3.11812773	0.0487608	12.6364672	3.1774186
3	3.1774186	1.1226889×10^{-4}	12.5782926	3.1774097

The roots are $-1.8557725, 0.6783628$, and 3.1774097.

21. $f(x) = x^2 - a$

$f'(x) = 2x$

$$x_2 = x_1 - \frac{f(x_1)}{f'(x_1)} = x_1 - \frac{x_1^2}{2x_1} = x_1 - \frac{x_1}{2} + \frac{a}{2x_1}$$

$x_2 = \frac{x_1}{2} + \frac{a}{2x_1}$. Similarly, $x_3 = \frac{x^2}{2} + \frac{a}{2x_2}$ which generalizes to

$$x_{n+1} = \frac{x_n}{2} + \frac{a}{2x_n}.$$

25. $x_1 = -1.4$

$x_2 = -1.407172738$

$x_3 = -1.41070945$

$x_4 = -1.412465769$

$x_5 = -1.413340965$

$x_6 = -1.413777828$

$x_7 = -1.413996076$

$x_8 = -1.414105155$

$x_9 = -1.414159685$

$x_{10} = -1.414186948$

$x = -1.4142$ to four decimal places

calculator: -1.414213

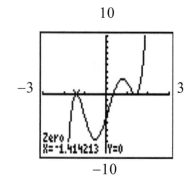

29. $V = \dfrac{1}{6}\pi h\left(h^2 + 3r^2\right) = \dfrac{1}{6}\pi h^3 + \dfrac{1}{2}\pi r^2 h$

$180{,}000 = \dfrac{1}{6}\pi h\left[h^2 + 3(60)^2\right] = \dfrac{1}{6}\pi h^3 + 1800\pi h$

$f(h) = \dfrac{1}{6}\pi h^3 + \dfrac{1}{2}\pi r^2 = \dfrac{1}{2}\pi h^2 + 1800\pi$

$f'(h) = \dfrac{1}{2}\pi h^2 + \dfrac{1}{2}\pi r^2 = \dfrac{1}{2}\pi h^2 + 1800\pi$

n	h_n	$f(h_n)$	$f'(h_n)$	$h_n - \dfrac{f(h_n)}{f'(h_n)}$
1	29	−3238.813	6975.906	29.464286
2	29.464286	9.874670	7018.544	29.462879
3	29.462879			

$h_3 = 29.462879;\ h = 29.5$ m

24.3   Curvilinear Motion

1. $x = 4t^2$ $\qquad\qquad$ $y = 1 - t^2$

$v_x = \dfrac{dx}{dt} = 8t\big|_{t=2} = 16$ $\qquad$ $\dfrac{dy}{dt} = -2t\big|_{t=2} = -4$

$v = \sqrt{16^2 + (-4)^2} = 16.5$

$\tan\theta = \dfrac{-4}{16},\ \theta = -14.0°$

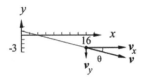

5. $x = t\left(2t + 1\right)^2$

$\dfrac{dx}{dt} = t(2)(2t+1)(2) + (2t+1)^2(1)$

$\qquad = 12t^2 + 8t + 1 = v_x$

$y = 6\left(4t + 3\right)^{-1/2}$

$\dfrac{dy}{dt} = 6\left(-\dfrac{1}{2}\right)(4t+3)^{-3/2}(4) = \dfrac{-12}{(4t+3)^{3/2}} = v_y$

$v_x\big|_{t=0.5} = 8$

$v_y\big|_{t=0.5} = -1.0733$

$v = \sqrt{8^2 + (-1.0773)^2}$

$v = 8.07$

$\alpha = \tan^{-1}\dfrac{1.0733}{8} = 7.641°$

t	x	y
0	0	3.464
0.5	2	2.683
1	9	2.268

Therefore, $\theta = 360° - \alpha = 352.4°$

Therefore, v is 8.07 at $\theta = 352.34°$

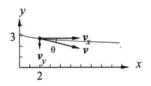

9. $x = t\left(2t + 1\right)^2$; $v_x = 12t^2 + 8t + 1$; $a_x = 24t + 8$;

$a_x\big|_{t=0.5} = 20.0$

$y = \dfrac{6}{\sqrt{4t + 3}}$; $v_y = -12\left(4t + 3\right)^{-3/2}$

$a_y = \dfrac{72}{\left(4t + 3\right)^{5/2}}$; $a_y\big|_{t=0.5} = 1.288$

$a = \sqrt{20.0^2 + 1.288^2}$

$a = 20.041$

$\theta = \tan^{-1}\dfrac{1.288}{20.0}$

$\theta = 3.68°$

Therefore, a is 20.0 at $\theta = 3.7°$.

13. $y = 2.0 + 0.80x - 0.20x^2$, $\dfrac{dy}{dt} = v_y$

$\qquad = 0.80\dfrac{dx}{dt} - 0.40x\dfrac{dx}{dt}$

$v_y = 0.80v_x = -0.40xv_x\big|_{(4.0,\ 2.0)}$

$\qquad = 0.80(5.0) - 0.40(4.0)(5.0) = -4.0$

$v = \sqrt{(5.0)^2 + (-4.0)^2} = 6.4$ m/s

$\tan\theta = \dfrac{-4.0}{5.0},\ \theta = -39°$

17. $x = 25t$, $v_x = \dfrac{dx}{dt} = 25$, $a_x = \dfrac{d^2x}{dt^2} = 0$

$y = 15t - 3.7t^2$, $v_y = \dfrac{dy}{dt} = 15 - 7.4t\Big|_{t=6.0}$

$\qquad\qquad\qquad\qquad = 15 - 7.4(6.0)$

$v = \sqrt{25^2 + (15 - 7.4(6.0))^2} = 39$ m/s

$\theta_v = \tan^{-1}\dfrac{15 - 7.4(6.0)}{25} = -50°$

$a_y = \dfrac{d^2y}{dt^2} = -7.4$

$a = \sqrt{a_x^2 + a_y^2} = \sqrt{0^2 + (-7.4)^2} = 7.4$ m/s²

$\theta_a = \tan^{-1}\dfrac{-7.4}{0} = -90°$

21. $y = 3.00 + x^{-1.50}$, at $t = 0$, $(x, y) = (1.00, 4.00)$.

$v_x = 1.20$ cm/s.

$\dfrac{dy}{dt} = -1.50x^{-2.50}\dfrac{dx}{dt} = -1.50x^{-2.50}(1.20)$

$\qquad = -1.80x^{-2.50}$

At $t = 0.500$ s, $x = 1.00 + 1.20(0.500) = 1.60$

$\dfrac{dy}{dt} = -1.8(1.60)^{-2.50} = -0.556$

$v = \sqrt{1.2^2 + (-0.556)^2} = 1.32$ cm/s

$\theta = \tan^{-1}\dfrac{-0.556}{1.20} = -24.9°$

25. $y = x - \dfrac{1}{90}x^3$; $v_x = x$; rocket hits the ground when

$y = 0$.

$v_y = v_x - \dfrac{1}{30}x^2 v_x$; $v_y = x - \dfrac{1}{30}x^2(x) = x - \dfrac{1}{30}x^3$

$y = 0 = x - \dfrac{1}{90}x^3 = x\left(x - \dfrac{1}{90}x^2\right)$

$x = 0$, $x = \sqrt{90} = 9.487$

$x = 0$; rocket leaves the ground

$x = 9.487$; rocket returns (crashed) to ground

$v_x = x$; $v_x\big|_{x=9.487} = 9.487$

$v_y = x - \dfrac{1}{30}x^3$; $v_y\big|_{x=9.487} = -18.97$

$v = \sqrt{90.0 + (18.97)^2} = 21.2$

$\alpha = \tan^{-1}\dfrac{18.97}{\sqrt{9.487}} = 63.4°$

Therefore, $\theta = 296.6°$

Therefore, v is 21.2 mi/min at $\theta = 296.6°$.

29. $h = k\sqrt{x} = kx^{1/2}$

$k = \dfrac{h}{\sqrt{x}} = \dfrac{280}{\sqrt{400}}$; $v_x = 350$ m/s

$\dfrac{dh}{dt} = k\cdot\dfrac{1}{2}x^{-1/2}v_x$

$\dfrac{dh}{dt} = \dfrac{kv_x}{2\sqrt{x}}$; $\dfrac{dh}{dt}\Big|_{x=400} = \dfrac{\frac{280}{\sqrt{400}}(350)}{2\sqrt{400}} = 122.5$ m/s

$v = \sqrt{350^2 + 122.5^2}$

$v = 370.8$ m/s

$\theta = \tan^{-1}\left(\dfrac{122.5}{350}\right) = 19.3°$

Therefore, v is 370 m/s at $\theta = 19°$.

24.4   Related Rates

1. $E = 2.800T + -0.012T^2$

$$\frac{dE}{dt} = 2.800\frac{dT}{dt} + 0.024T\frac{dT}{dt}$$

$$\left.\frac{dE}{dt}\right|_{T=100°\,\text{C}} = 2.800(1.00) + 0.024(100)(1.00)$$

$$\left.\frac{dE}{dt}\right|_{T=100°\,\text{C}} = 5.20 \text{ V/min}$$

5. $x^2 + 3y^2 + 2y = 10$

$$2x\frac{dx}{dt} + 6y\frac{dy}{dt} + 2\frac{dy}{dt} = 0$$

$$2(3)(2) + 6(-1)\frac{dy}{dt} + 2\frac{dy}{dt} = 0$$

$$\frac{dy}{dt} = 3$$

9. $v = 18\sqrt{T}$

$$\frac{dv}{dt} = \frac{18}{2\sqrt{T}}\frac{dT}{dt} = \frac{9}{\sqrt{25}}(0.2)$$

$$= 0.36 \text{ ft/s}^2$$

13. $T = \pi\sqrt{\dfrac{L}{96}} = \dfrac{\pi}{\sqrt{96}}L^{1/2}$

$$\frac{dT}{dt} = \frac{\pi}{\sqrt{96}}L^{-1/2}\frac{dL}{dT}$$

$$= \frac{\pi}{2\sqrt{96}}(16.0)^{-1/2}(0.100)$$

$$\frac{dT}{dt} = 0.00401$$

17. $r = \sqrt{0.4\lambda};\; \dfrac{d\lambda}{dt} = 0.10\times10^{-7};\; r = (0.4\lambda)^{1/2}$

$$\frac{dr}{dt} = \frac{1}{2}(0.4\lambda)^{-1/2}(0.4)\frac{d\lambda}{dt}$$

$$= 0.2(0.4\lambda)^{-1/2}\frac{d\lambda}{dt}$$

$$\left.\frac{dr}{dt}\right|_{\lambda=6.0\times10^{-7}} = 0.2\left[0.4\left(6.0\times10^{-7}\right)\right]^{-1/2}\left(0.10\times10^{-7}\right)$$

$$= \frac{2\times10^{-9}}{\sqrt{24}\times10^{-4}} = 4.1\times10^{-6} \text{ m/s}$$

21. $y = \dfrac{1.5}{18}x \Rightarrow 12 = x$

$$v = \frac{1}{2}xy(12) = 6xy, \quad 0 < x < 18,$$
$$0 < y < 1.5$$

$$v = 72y^2$$

$$\frac{dv}{dt} = 144y\frac{dy}{dt}$$

$$0.80 = 144(1.0)\frac{dy}{dt}$$

$$\frac{dy}{dt} = 0.0056 \text{ m/min}$$

25. $V = x^3;\; \dfrac{dx}{dt} = -0.50 \text{ mm/min}$

$$\frac{dV}{dt} = 3x^2\frac{dx}{dt};\; \left.\frac{dV}{dt}\right|_{x=8.20} = 3(8.20)^2(-0.50)$$

$$= -101 \text{ mm}^3 / \text{min}$$

29. $p = \dfrac{k}{v};\; \dfrac{dv}{dt} = 20 \text{ cm}^3 / \text{min}; \; v = 810 \text{ cm}^3$

$$230 = \frac{k}{650}; \; k = 1.495\times10^5 \text{ kPa}\times\text{cm}^3$$

$$p = \frac{149{,}500}{v} = 149{,}500v^{-1}; \; \frac{dp}{dt} = -149{,}500v^{-2}\frac{dv}{dt}$$

$$\left.\frac{dp}{dt}\right|_{v=810} = -149{,}500(810)^{-2}(20) = -4.6 \text{ kPa/min}$$

33.

$$I = \frac{8.00k}{x^2}$$

$$\frac{dI}{dt} = \frac{-2(8.00)k}{x^3}\frac{dx}{dt} = \frac{-2(8.00)k}{100^3}(-50.0)$$

$$\frac{dI}{dt} = 0.000800k \text{ units/s}$$

37.

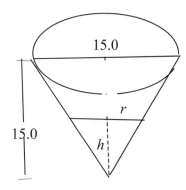

$$\frac{r}{h} = \frac{\frac{15.0}{2}}{15.0}$$

$$r = \frac{1}{2}h$$

$$V = \frac{1}{3}\pi r^2 h = \frac{1}{3}\pi\left(\frac{h}{2}\right)^2 h = \frac{\pi h^3}{12}$$

$$\frac{dV}{dt} = \frac{\pi h^2}{4} \cdot \frac{dh}{dt}$$

$$-18.0 = \frac{\pi(10^2)}{4}\frac{dh}{dt}$$

$$\frac{dh}{dt} = -0.23 \text{ cm/min}$$

41. $z^2 = 20^2 + x^2$; $\dfrac{dz}{dt} = -10.0$ ft/s; $z = 36.0$ ft,

$x = 29.9$ ft

$$2z\frac{dz}{dt} = 2x\frac{dx}{dt}$$

$$\frac{dx}{dt} = \frac{z}{x}\frac{dz}{dt}$$

$$\frac{dx}{dt}\bigg|_{x=36.0} = \frac{36.0}{29.9}(-10.0)$$

$$\frac{dx}{dt} = -12.0 \text{ ft/s}$$

The negative sign indicates the boat is approaching the wharf.

24.5 Using Derivatives in Curve Sketching

1. $f(x) = x^3 - 6x^2$

$f'(x) = 3x^2 - 12x = 3(x - 4)$ with $x = 0$, $x = 4$
as critical values.

If $x < 0$, $f'(x) = 3x(x - 4) > 0$. $f(x)$ increasing

If $0 < x < 4$, $f'(x) = 3x(x - 4) > 0$. $f(x)$
decreasing

If $x > 4$, $f'(x) = 3x(x - 4) > 0$. $f(x)$ increasing

inc. $x < 0$, $x > 4$; dec. $0 < x < 4$

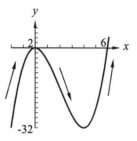

5. $y = x^2 + 2x$; $y' = 2x + 2$; $2x + 2 > 0$

$2x > -2$; $x > -1$; $f(x)$ increases.

$2x + 2 < 0$; $2x < -2$; $x < -1$; $f(x)$ decreases.

9. $y = x^2 + 2x$; $y' = 2x + 2$; $y' = 0$ at $x = 1$

$y'' = 2 > 0$ at $x = -1$ and $(-1, -1)$

is a relative minimum.

13. $y = x^2 + 2x$; $y' = 2x + 2$; $y'' = 2$

Thus, $y'' > 0$ for all x. The graph is concave up
for all x and has no points of inflection.

17. $y = x^2 + 2x$

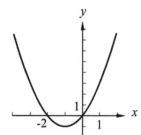

21. $y = 12x - 2x^2$; $y' = 12 - 4x$

$y' = 0$ at $x = 3$; for $x = 3$,

$y = 12(3) - 2(3)^2 = 18$ and $(3, 18)$ is a critical

point. $12 - 4x < 0$ for $x > 3$ and the function

decreases; $12 - 4x > 0$ for $x < 3$ and the function

increases; $y'' = -4$; thus $y'' < 0$ for all x. There

are no inflections; the graph is concave down for

all x, and $(3, 18)$ is a maximum point.

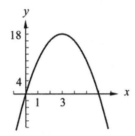

25. $y = x^3 + 3x^2 + 3x + 2$

$y' = 3x^2 + 6x = 3(x^2 + 2x + 1)$

   $= 3(x + 1)(x + 1)$

$3(x + 1)(x + 1) = 0$ for $x = -1$

$(-1, 1)$ is a critical point.

$3(x + 1)(x + 1) > 0$ for $x < -1$ and the slope is

positive.

$3(x + 1)(x + 1) > 0$ for $x > -1$ and the slope is

positive.

$y'' = 6x + 6$; $6x + 6 = 0$ for $x = 1$, and $(-1, 1)$ is

an inflection point.

$6x + 6 < 0$ for $x < -1$ and the graph is concave

down.

$6x + 6 > 0$ for $x > -1$ and the graph is concave up.

Since there is no change in slope from positive to

negative or vice versa, there are no maximum or

minimum points.

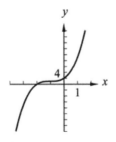

29. $y = 4x^3 - 3x^4 + 6$

$y' = 12x^2 - 12x^3 = 12x^2(1 - x) = 0$

$12x^2(1 - x) = 0$ for $x = 0$ and $x = 1$

$(0, 6)$ and $(1, 4)$ are critical points.

$12x^2 - 12x^3 > 0$ for $x < 0$ and the slope is positive.

$12x - 12x^3 > 0$ for $0 < x < 1$ and the slope is positive.

$12x - 12x^3 < 0$ for $x > 1$ and the slope is negative.

$y'' = 24x - 36x^2$; $24x - 36x^2 = 12x(2 - 3x) = 0$

for $x = 0$, $x = \frac{2}{3}$

$(0, 6)$ and $(\frac{2}{3}, \frac{178}{27})$ are possible inflection points.

$24x - 36x^2 < 0$ for $x < 0$ and the graph is concave

down.

$24x - 36x^2 > 0$ for $0 < x < \frac{2}{3}$ and the graph is

concave up.

$24x - 36x^2 < 0$ for $x > \frac{2}{3}$ and the graph is concave

down.

$(1, 7)$ is a relative maximum point since $y' = 0$

at $(1, 7)$ and the slope is positive for $x < 1$ and

negative for $x > 1$. $(0, 6)$ and $(\frac{2}{3}, \frac{178}{27})$ are

inflection pts since there is a concavity change.

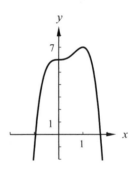

33. $y = x^3 = 12x$; $y' = 3x^2 - 12$; $y'' = 6x$. On graphing

calculator with $x_{min} = -5$, $x_{max} = 5$, $y_{min} = -20$,

$y_{max} = 20$, enter $y_1 = x^3 - 12x$; $y_2 = 3x^2 - 12$;

$y_3 = 6x$. From the graph is observed that the

maximum minimum values of y occur when y' is

zero. A maximum value for y occurs when $x = -2$,

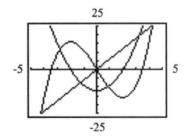

37. The left relative max is above the left
relative min and below the right relative min.

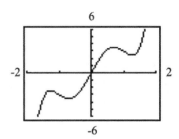

41. $(0, 0)$ is a point on a curve.

$f'(x) > 0$ for all $x \Rightarrow f$ increases everywhere

$\Rightarrow$ no max or min points

$f''(x) < 0$ for $x < 0 \Rightarrow f$ is concave down

for $x < 0$

$f''(x) > 0$ for $x > 0 \Rightarrow f$ is concave up

for $x > 0$

$\Rightarrow (0, 0)$ is an inflection point

one possibility is $y = f(x) = x^3 + x$

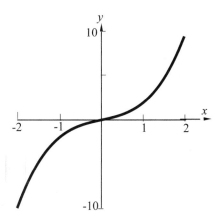

45. $P = 4i - 0.5i^2$

$P' = 4 - 1.0i = 0$ for $i = 4.0$

$P'' = -1.0 < 0$, conc. down everywhere

$P' > 0$ for $x < 4$, P inc.

$P' < 0$ for $x > 4$, P dec.

$P(4) = 8$, $(4, 8)$ max.

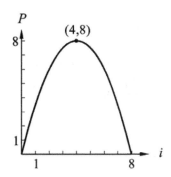

49. $R = 75 - 18i^2 + 8i^3 - i^4$

$R' = -36i + 24i^2 - 4i^3 = -4i(i^2 - 6i + 9)$

$= -4i(i - 3)^2$

$R' = 0$ for $i = 0$ and $i = 3$

$(0, 75)$ and $(3, 48)$ are critical points.

$R' > 0$ for $i < 0$, $R' < 0$ for $0 < i < 3$

$R' < 0$ for $i > 3$

Max. at $(0, 75)$, no max. or min. at $(3, 48)$

$R'' = -36 + 48i - 12i^2 = -12(i - 1)(i - 3)$

$(1, 64)$ and $(3, 48)$ are possible inflection points.

$R'' < 0$ for $i < 1$, concave down

$R'' > 0$ for $1 < i < 3$, concave up

$R'' < 0$ for $i > 3$, concave down

$(1, 64)$ and $(3, 48)$ are inflection points.

(From calculator graph, $R = 0$ for $i = -1.5$

and $i = 5.0$)

53. $V = x(8 - 2x)(12 - 2x)$

$V = 4(24x - 10x^2 + x^3)$

$= 4x^3 - 40x^2 + 96x$

$f'(x) = 4(24 - 20x + 3x^2) = 0$

By quadratic solution:

$x = 1.57$ and $x = 5.10$ (reject)

$f''(x) = 4(-20 + 6x)$

$f''(1.57) < 0$, rel max. $(1.57, 67.6)$

$f''(5.10) > 0$, rel min. $(5.10, -20.2)$

$f''(x) = 0$; $x = \dfrac{20}{6}$, infl. $\left(\dfrac{10}{3}, 23.7\right)$

57. $f(-1) = 0$, root at $(-1, 0)$

$f(2) = 2$, point on curve $(2, 2)$

$f'(x) = < 0$ for $x < -1$

Therefore, $f(x)$ decreasing for $x < -1$.

$f'(x) > 0$ for $x > -1$

Therefore, $f(x)$ increasing for $x > -1$.

$f''(x) < 0$ for $0 < x < 2$

Therefore, $f(x)$ concave down for $0 < x < 2$.

$f''(x) > 0$ for $x < 0$ or $x > 2$

Therefore, $f(x)$ concave up for $x < 0$.

Therefore, $f(x)$ concave up for $x > 2$.

Summary: $(-1, 0)$ min. point

Inflection point at y-intercept

$(2, 2)$ inflection point

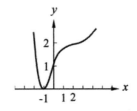

24.6   More on Curve Sketching

1. $y = x - \dfrac{4}{x}$

(1) Intercepts:

For $x = 0$, y is undefined which means curve is not continuous at $x = 0$ and there are no y-intercepts.

For $y = 0$, $0 = x - \dfrac{4}{x}$, $x^2 - 4 = 0$, $x = \pm 2$ are the x-intercepts.

(2) Symmetry: none

(3) Behavior as x becomes large:

As $x \to \pm\infty$, $\dfrac{4}{x} \to 0$ and $y \to x$. $y = x$ is a slant asymptote.

(4) Vertical asymptotes:

y is undefined for $x = 0$. As $x \to 0^-$, $y \to \infty$ and as $x \to 0^+$, $y \to -\infty$. The x-axis is a vertical asymptote.

(5) Domain and range:

domain: $x \neq 0$, range: $-\infty < y < \infty$

(6) Derivatives:

$y' = 1 + \dfrac{4}{x^2} > 0$, y inc. for $x \neq 0$

$y'' = -\dfrac{8}{x^3} > 0$ for $x < 0$, y conc. up

$y'' = -\dfrac{8}{x^3} < 0$ for $x > 0$, y conc. down

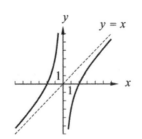

5. $y = x^2 + \dfrac{2}{x} = \dfrac{x^2 + 2}{x}$

(1) $\dfrac{2}{x}$ is undefined for $x = 0$, so the graph is not continuous at the y-axis; i.e., no y-intercept exists.

(2) $\dfrac{x^3 + 2}{x} = 0$ at $x = \sqrt[3]{-2} = -\sqrt[3]{2}$. There is an x-intercept at $\left(-\sqrt[3]{2}, 0\right)$.

(3) As $x \to \infty$, $x^2 \to \infty$ and $\dfrac{2}{x} \to 0$, so $x^2 + \dfrac{2}{x} \to \infty$.

(4) As $x \to 0$ through positive x, $x^2 \to 0$ and $\dfrac{2}{x} \to \infty$, so $x^2 + \dfrac{2}{x} \to \infty$.

(5) As $x \to -\infty$, $x^2 \to \infty$ and $\frac{2}{x} \to 0$ so $x^2 + \frac{2}{x} \to \infty$.

$x = 0$ is a vertical asymptote

(6) As $x \to 0$ through negative numbers, $x^2 \to 0$ and $\frac{2}{x} \to -\infty$, so $x^2 + \frac{2}{x} \to -\infty$.

(7) $y' = 2x - 2x^{-2} = 0$ at $x = 1$ and the slope is zero at $(1, 3)$.

(8) $y'' = 2 + 4x^{-3} = 0$ at $x = -\sqrt[3]{2}$ and $\left(-\sqrt[3]{2}, 0\right)$ is an inflection point.

(9) $y'' > 0$ at $x = 1$, so the graph is concave up and $(1, 3)$ is a relative minimum.

(10) Since $\left(-\sqrt[3]{2}, 0\right)$ is an inflection pt, $f''(-1) < 0$ and the graph is concave down. $f''(-2) > 0$ and the graph is concave up.

(11) Not symmetrical about the x or y-axis.

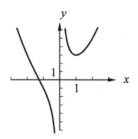

9. $y = \dfrac{x^2}{x+1} = x - 1 + \dfrac{1}{x+1}$

Intercepts:

(1) Function undefined at $x = -1$; not continuous at $x = -1$.

(2) At $x = 0$, $y = 0$. The origin is the only intercept.

(3) Behavior as x becomes large: As $x \to \pm\infty$, $y \to x - 1$, so $y = x - 1$ is a slant asymptote.

Vertical asymptotes:

(4) As $x \to -1$ from the left, $x + 1 \to 0$ through negative values and $\frac{x^2}{x+1} \to -\infty$ since $x^2 > 0$ for all x. As $x \to -1$ from the right, $x + 1 \to 0$ through positive values and $\frac{x^2}{x+1} \to +\infty$. $x = -1$ is an asymptote.

Symmetry:

(5) The graph is not symmetrical about the y-axis or the x-axis.

Derivatives:

(6) $y' = \frac{x^2 + 2x}{(x+1)^2}$; $y' = 0$ at $x = -2$, $x = 0$.

$(-2, -4)$ and $(0, 0)$ are critical points. Checking the derivative at $x = -3$, the slope is positive, and at $x = -1.5$ the slope is negative. $(-2, -4)$ is a relative maximum point. Checking the derivative at $x = -0.5$, the slope is negative and at $x = 1$ the slope is positive, so $(0, 0)$ is a relative minimum point.

Int. $(0, 0)$, max. $(-2, -4)$, min $(0, 0)$, asym. $x = -1$

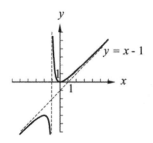

13. $y = \dfrac{4}{x} - \dfrac{4}{x^2}$

Intercepts:

(1) There are no y intercepts since $x = 0$ is undefined.

(2) $y = 0$ when $x = 1$ so $(1, 0)$ is an x-intercept,

Asymptotes:

(3) $x = 0$ is an asymptote; the denominator is 0.

   $y = 0$ is an asymptote

Symmetry:

(4) Not symmetrical about the y-axis since $\frac{4}{x} - \frac{4}{x^2}$ is different from $\frac{4}{(-x)} - \frac{4}{(-x)^2}$

(5) Not symmetrical about the x-axis since $y = \frac{4}{x} - \frac{4}{x^2}$ is different from $-y = \frac{4}{x} - \frac{4}{x^2}$

(6) Not symmetrical about the origin since $y = \frac{4}{x} - \frac{4}{x^2}$ is different from $-y = \frac{4}{-x} - \frac{4}{(-x)^2}$

Derivatives:

(7) $y' = -4x^{-2} + 8x^{-3} = 0$ at $x = 2$; $(2, 1)$ is a relative maximum.

(8) $y'' = 8x^{-3} - 24x^{-4} = 0$ at $x = 3$ so $\left(3, \frac{8}{9}\right)$ is a possible inflection.

$y'' < 0$ (concave down) for $x < 3$ and 0 (concave up) for $x > 3$ so $\left(3, \frac{8}{9}\right)$ is an inflection.

Behavior as x becomes large:

(9) As $x \to \infty$ or $-\infty$, $\frac{4}{x}$ and $-\frac{4}{x^2}$ each approach 0.

As $x \to 0$, $\frac{4}{x} - \frac{4}{x^2} = \frac{4x-4}{x^2}$ approaches $-\infty$, through positive or negative values of x.

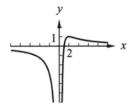

17. $y = \dfrac{9x}{9 - x^2}$

Intercept:

(1) Intercept, $(0, 0)$

(2) Asymptotes at $x = -3$, $x = 3$

Derivatives:

(3) $y' = \dfrac{\left(9 - x^2\right)(9) - (9)(-2x)}{\left(9 - x^2\right)^2}$

$81 + 9x^2 = 0;\ 9x^2 = -81;\ x^2 = \sqrt{-9}$ (imaginary)

No real value max. or min., ± 3 are critical values.

(4) y''

$= \dfrac{\left(9 - x^2\right)^2 (18x) - \left(81 + 9x^2\right)(2)\left(9 - x^2\right)(-2x)}{\left(9 - x^2\right)^2}$

$= \dfrac{-18x^5 - 324x^3 + 4374x}{\left(9 - x^2\right)^4}$

$-18x^5 - 324x^3 + 4374x = 0;$

$-18x\left(x^4 + 18x^2 - 243\right) = 0$

$-18x = 0;\ x = 0$

$x^4 + 18x^2 - 243 = 0;\ \left(x^2 + 27\right)\left(x^2 - 9\right) = 0$

(imaginary) $x^2 = 9;\ x = \pm 3$ (these are asymptotes)

$y\big|_{x=0} = 0$; a possible inflection is $(0, 0)$

at $\left(-1, -\frac{9}{8}\right)$, $y'' = -4032$; concave down at $\left(1, \frac{9}{8}\right)$, $y'' = 4032$; concave up, and $(0, 0)$ is an inflection point.

Symmetry:

(5) There is symmetry to the origin.

(6) As $x \to +\infty$ and as $x \to -\infty$, $y \to 0$. Therefore, $y = 0$ is an asymptote.

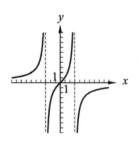

21.

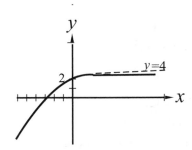

Use $y = -2e^{-x} + 4$

25. $C_T = \dfrac{6C}{6+C}; \dfrac{dC_T}{dC} = \dfrac{36}{(6+C)^2}; \dfrac{d^2C_T}{dC^2} = \dfrac{-72}{(6+C)^3}$

$f'(C) \neq 0; f''(C) = 0$, therefore no max, no min, no infl. points

$f'(C) > 0$ for all C, therefore C_T is increasing for all C.

$f''(C) < 0$ for $C > -6$, therefore C_T is concave down for $C > -6$ (only values of $C \geq 0$ have physical significance.)

$C = 0$, $C_T = 0$, $(0, 0)$ is the only intercept

No symmetry WRT axes or origin.

$\lim\limits_{x \to +\infty} \dfrac{6C}{6+C} = \lim\limits_{x \to +\infty} \dfrac{6}{\frac{6C}{C}+1} = 6$, therefore, horizontal asymptote at $C_T = 6$.

Vertical asymptote at $C = -6$ (capacitance > 0)

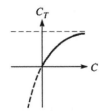

29. $v = k\left(r - \dfrac{1}{r^3}\right) = r - \dfrac{1}{r^3}$ for $k = 1$.

(1) Intercepts: For $r = 0$, v is undefined. There is no y-intercept. For $v = 0 = r - \frac{1}{r^3}$

$r^4 = 1$, $r = \pm 1$ are the x-intercepts.

(2) Symmetry: none

(3) Behavior as x becomes large:

As $x \to +\infty$, $v \to r$. $v = r$ is a slant asymptote.

(4) Vertical asymptotes: $r = 0$ is a vertical asymptote since v is undefined for $r = 0$.

(5) Domain and range: Domain is all $r \neq 0$ range is $-\infty < v < \infty$

(6) Derivatives:

$v' = 1 + \dfrac{3}{r^4} > 0$, v inc. $r \neq 0$

$v'' = -\dfrac{12}{r^5}$, $v'' < 0$, $r > 0$, v conc. down

$v'' = -\dfrac{12}{r^5}$, $v'' > 0$, $r < 0$, conc. up.

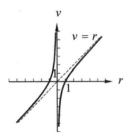

24.7 Applied Maximum and Minimum Problems

1. $A = xy$, $2x + 2y = 2400 \Rightarrow x + y = 1200$

$A = (1200 - y)y = 1200y - y^2$

$A' = 1200 - 2y = 0 \Rightarrow y = 600$

$A'' = -2 < 0$, $y = 600$ is max.

$x + 600 = 1200 \Rightarrow x = 600$

$A_{max} = 600(600) = 360,000 \text{ ft}^2$

5. $P = EI - RI^2$; $\dfrac{dP}{dI} = E - 2RI = 0$; $I = \dfrac{E}{2R}$

$\dfrac{d^2P}{dI^2} = -2R < 0$ for all I, therefore max. power

at $I = \dfrac{E}{2R}$

9. $S = 360A - 0.1A^2$, find maximum S.

$S' = 360 - 0.3A^2$; $A^2 = 1200$; $A = 35$ m^2

$S'' = -0.6A < 0$ for all valid (positive) A so the graph is concave down and $A = 35$ m^2 is a max. Maximum savings are

$S = 360(35) - 0.1(35)^2 = \$8300$.

13. $y = x^2 - 4$

$l = $ distance to origin $= \sqrt{x^2 + y^2}$

$l = \sqrt{x^2 + (x^2 - 4)^2}$

$L = l^2 = x^2 + x^4 - 8x^2 + 16$

$\quad = x^4 - 7x^2 + 16$

$L' = 4x^3 - 14x = 0$

$x = \sqrt{\dfrac{14}{4}} = 1.87$

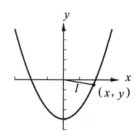

$L'' = 12x^2 - 14\big|_{x=1.87} > 0$, $x = 1.87$ is a minimum

$y = 1.87^2 - 4 = \dfrac{1}{2}$

$l = \sqrt{1.87^2 + \left(\dfrac{1}{2}\right)^2} = 1.94$ units is closest particle comes to origin.

17.

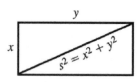

$P = 2x + 2y = 48$

$\quad x + y = 24$

Diagonal will be a minimum if $l = s^2$ is a minimum.

$l = x^2 + y^2$

$l = x^2 + (24 - x)^2$

$l = x^2 + 24x^2 - 48x + x^2$

$l = 2x^2 - 48x + 24x^2$

$\dfrac{dl}{dx} = 4x - 48 = 0$

$\qquad x = 12$

from which $y = 12$

Dimensions are 12 in. by 12 in., a square will minimize the diagonal.

21.

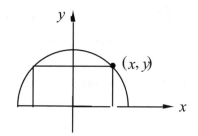

$x^2 + y^2 = 7.0^2$

$A = 2xy = 2x\sqrt{7.0^2 - x^2}$

$\dfrac{dA}{dx} = 2x\dfrac{-2x}{2\sqrt{7.0^2 - x^2}} + 2\sqrt{7.0^2 - x^2} = 0$

$x = \dfrac{7.0}{\sqrt{2}} = 4.95$, $y = 4.95$

dimensions: 4.95 cm by 9.90 cm

25. $A = \dfrac{1}{2}xy$; $12.0^2 = x^2 + y^2$

$y^2 = 144 - x^2$; $y = \sqrt{144 - x^2}$

$A = \dfrac{1}{2}x\sqrt{144 - x^2}$

$\dfrac{dA}{dx} = \dfrac{1}{2}\left[x \cdot \dfrac{1}{2}(144 - x^2)^{-1/2}(-2x) + \sqrt{144 - x^2}\right]$

$\quad = \dfrac{1}{2}\left[\dfrac{-x^2}{\sqrt{144 - x^2}} + \sqrt{144 - x^2}\right]$

$\quad = \dfrac{1}{2}\left[\dfrac{-x^2 + 144 - x^2}{\sqrt{144 - x^2}}\right] = \dfrac{1}{2}\dfrac{(144 - 2x^2)}{\sqrt{144 - x^2}}$

$\dfrac{dA}{dx} = 0 = 144 - 2x^2$; $x = \sqrt{72} = 8.49$

Test $\sqrt{72}$: $f'\sqrt{71} > 0; f'\sqrt{73} < 0$

Therefore, max. $\left(\sqrt{72}, \sqrt{72}\right)$.

Therefore, legs of triangle will be equal at 8.49 cm for max. area.

29. $lw = 384$

$A = 384 + 4(24) + 2(6.00w) + 2(4.00l)$

$A = 408 + 12.0w + 8.00\dfrac{384}{w}$

$\dfrac{dA}{dw} = 12.0 - \dfrac{8.00(384)}{w^2} = 0$

$w = 16.0, \quad l = \dfrac{384}{16.0} = 24.0$

Dimensions: $w + 8.00 = 24.0$ cm

$\qquad\qquad\quad l + 12.0 = 36.0$ cm

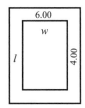

33. $V = (10 - 2x)(15 - 2x), \ 0 < x < 5$

$V = 4x^3 - 50x^2 + 150x$

$V' = 8x^2 - 100x = 150 = 0$

$x = 1.96, 6.37 > 5,$ reject

$V'' = 12x^2 - 100x + 150\big|_{x=1.96} > 0,$

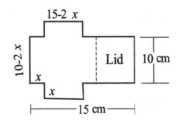

V is a maximum for $x = 2$ cm.

37. $y = k\left(x^4 - 5Lx^3 + 3L^2x^2\right)$

$= 2kx^4 - 5kLx^3 + 3kL^2x^2$

$y' = 8kx^3 - 15kLx^2 + 6kL^2x = 0$

$kx\left(8x^2 - 15Lx + 6L^2\right) = 0$

$kx = 0; \ x = 0$

$8x^2 - 15Lx + 6L^2 = 0$

$x = \dfrac{-(-15) \pm \sqrt{(-15)^2 - 4(8)(6)}}{2(8)}$

$= \dfrac{5 \pm \sqrt{33}}{16}$

$= \dfrac{15 \pm 5.75}{16}$

$x = 0.58L, 1.30L$ (not valid - this distance is greater than L, the length of the beam)

41. $l + w + h = 45, \ l = 2.4w \Rightarrow h = 45 - 3w$

$v = lwh = 2.4w(w)(45 - 3w) = 108w^2 - 8.16w^3$

$\dfrac{dv}{dw} = 216w - 24.48w^2 = 0 \Rightarrow w = \dfrac{150}{17},$

$l = 2.4w = \dfrac{360}{17}$

$h = 45 - 3.4w = 15.$

dimensions: 21 in. by 9 in. by 15 in.

45. $n(x) = \dfrac{k}{x^2} + \dfrac{8k}{(8-x)^2}$

$n'(x) = \dfrac{-2k}{x^3} + \dfrac{8k \cdot (-2)(-1)}{(8-x)^3} = 0 \Rightarrow 8x^3 = (8-x)^3$

$\qquad\qquad\qquad\qquad\qquad\qquad 2x = 8 - x$

$\qquad\qquad\qquad\qquad\qquad\qquad 3x = 8$

$\qquad\qquad\qquad\qquad\qquad\qquad x = \dfrac{8}{3}$

$n''(x) = \dfrac{6k}{x^4} + \dfrac{48k}{(8-x)^4}$

$n''\left(\dfrac{8}{3}\right) > 0 \Rightarrow n(x)$ is min at $x = \dfrac{8}{3}$ km from A.

49. $2x + \pi d = 400;\ \pi d = 400 - 2x;\ d = \dfrac{400 - 2x}{\pi}$

$A = x(d) = x\left(\dfrac{400 - 2x}{\pi}\right) = \dfrac{400x - 2x^2}{\pi}$

$A' = \dfrac{400 - 4x}{\pi} = 0;\ 400 - 4x = 0;\ x = 100 \text{ m}$

53. $xy = 7000 = A_B$ of building

$(x + 20.0)(y + 40.0) = A_L$ of lot $= A_L$

$A_L = (x + 20.0)(y + 40.0)$

$y = \dfrac{7000}{x}$

Therefore, $A_L = (x + 20.0)\left(\dfrac{7000}{x}\right)$

$\dfrac{dA_L}{dx} = (x + 20.0)\left(\dfrac{-7000}{x^2} + 40.0\right) + \left(\dfrac{7000}{x} + 40.0\right)1 = 0$

$\quad = \dfrac{-7000(x + 20.0)}{x^2} + \dfrac{7000 + 40.0x}{x} = 0$

$-7000(x + 20.0) + x(7000 + 40.0x) = 0$

$-7000x - 140,000 + 7000x + 40.0x^2 = 0$

$x = \sqrt{\dfrac{140,000}{40.0}} = 59.2 \text{ m}$

Test $x = 59.2$

$f'(59.1) < 0$

$f'(59.3) > 0$

Min at $x = 59.2$ m

Therefore, $x = 59.2$ m

Therefore, $y = 118$ m

Therefore, dimensions of building are 59.2 m by 118 m.

24.8 Differentials and Linear Approximations

1. $s = \dfrac{4t}{t^3 + 4}$

$ds = \dfrac{(t^3 + 4)(4) - 4t(3t^2)}{(t^3 + 4)^2}\,dt$

$ds = \dfrac{-8t^3 + 16}{(t^3 + 4)^2}\,dt$

5. $y = x^5 + 4x;$

$\dfrac{dy}{dx} = 5x^4 + 4$

$dy = (5x^4 + 4)\,dx$

9. $s = 2(3t^2 - 5)^4;$

$\dfrac{ds}{dt} = 8(3t^2 - 5)^3(6t)$

$ds = 8(3t^2 - 5)^3(6t)\,dt$

$ds = 48t(3t^2 - 5)^3\,dt$

13. $y = x^2(1 - x)^3$

$dy = \left[x^2 \cdot 3(1 - x)^2(-1) + (1 - x)^3 \cdot 2x\right]dx$

$dy = \left(-3x^2(1 - x)^2 + 2x(1 - x)^3\right)dx$

$dy = (1 - x)^2\left(-3x^2 + 2x(1 - x)\right)dx$

$dy = (1 - x)^2\left(-5x^2 + 2x\right)dx$

$dy = x(1 - x)^2(-5x + 2)\,dx$

17. $y = f(x) = 7x^2 + 4x,\ dy = f'(x)\,dx$

$\quad = (14x + 4)\,dx$

$\Delta y = f(x + \Delta x) - f(x)$

$\quad = 7(4.2)^2 + 4(4.2) - (7 \cdot 4^2 + 4 \cdot 4) = 12.28$

$dy = (14 \cdot 4 + 4)(0.2) = 12$

21. $f(x) = x^2 + 2x$; $f'(x) = 2x + 2$

$L(x) = f(a) + f'(a)(x-a)$

$\quad = f(0) + f'(0)(x-0)$

$L(x) = 0^2 + 2 \cdot 0 + (2 \cdot 0 + 2)(x-0)$

$L(x) = 2x$

25. $C = 2\pi r$, $r = 6370$, $dr = 250$

$dC = 2\pi dr = 2\pi(250) = 1570$ km

29. $\lambda = \dfrac{k}{f}$, $685 = \dfrac{k}{4.38 \times 10^{14}}$;

$k = 3.00 \times 10^{17}$ mm - H$_z$

$\dfrac{d\lambda}{df} = \dfrac{-k}{f^2}$

$d\lambda = \dfrac{-k}{f^2} df$

$d\lambda = \dfrac{-3.00 \times 10^{17}}{\left(4.38 \times 10^{14}\right)^2} \cdot \left(0.20 \times 10^{14}\right)$

$d\lambda = -31$ nm

33. $r = k\sqrt{\lambda}$

$\dfrac{dr}{d\lambda} = \dfrac{k}{2\sqrt{\lambda}}$; $\dfrac{dr}{r} = \dfrac{1}{r} \cdot \dfrac{kd\lambda}{2\sqrt{\lambda}}$

$\quad = \dfrac{1}{2} \cdot \dfrac{k}{k\sqrt{\lambda}} \cdot \dfrac{d\lambda}{\sqrt{\lambda}}$

$\dfrac{dr}{r} = \dfrac{1}{2} \cdot \dfrac{d\lambda}{\lambda}$

37. $y = f(x) = \sqrt{x}$, $x = 4$, $\Delta x = dx = 0.05$

$\Delta y = f(x + \Delta x) - f(x) \approx dy = \dfrac{1}{2\sqrt{x}} dx$

$\sqrt{4.05} \approx \sqrt{4} + \dfrac{1}{2\sqrt{4}}(0.05)$

$\sqrt{4.05} \approx 2.0125$

41. $f(x) = \sqrt{2-x}$; $f'(x) = \dfrac{-1}{2\sqrt{2-x}}$

$L(x) = f(a) + f'(a)(x-a)$

$L(x) = f(1) + f'(1)(x-1)$

$\quad = \sqrt{2-1} + \dfrac{-1}{2\sqrt{2-1}}(x-1)$

$L(x) = 1 - \dfrac{1}{2}(x-1) = -\dfrac{1}{2}x + \dfrac{3}{2}$

$\sqrt{1.9} = f(0.1) \approx L(0.1)$

$\quad = 1 - \dfrac{1}{2}(0.1-1) = 1.45$

Chapter 24   Review Exercises

1. $y = 3x - x^2$ at $(-1, -4)$; $y' = 3 - 2x$

$y'|_{x=-1} = 3 - 2(-1) = 5$

$m = 5$ for tangent line

$y - y_1 = 5(x - x_1)$

$y - (-4) = 5[x - (-1)]$; $y + 4 = 5x + 5$

$5x - y + 1 = 0$

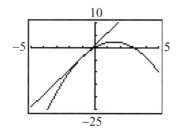

5. $y = \sqrt{x^2 + 3}$; $m = \dfrac{1}{2}$

$y = (x^2 + 3)^{1/2}$;

$\dfrac{dy}{dx} = \dfrac{1}{2}(x^2 + 3)^{-1/2}(2x)$

$\qquad = \dfrac{x}{\sqrt{x^2 + 3}}$

$m_{\tan} = \dfrac{dy}{dx}$

$2x = \dfrac{x}{\sqrt{x^2 + 3}} = \dfrac{1}{2}$

Squaring both sides, $4x^2 = x^2 + 3$; $3x^2 = 3$;

$x^2 = 1$; $x = 1$ and $x = -1$. Therefore, the abscissa of

the point at which $m = \dfrac{1}{2}$ is 1 or -1. If $x = 1$,

$y = \sqrt{1^2 + 3} = \sqrt{4} = 2$ or -2. If $x = -1$,

$y = \sqrt{(-1)^2 + 3} = 2$ or -2. The possible points

where the slope of the tangent line is $\frac{1}{2}$ are $(1, 2)$,

$(1, -2)$, $(-1, 2)$, $(-1, -2)$. A sketch of the curve

shows that the only relative maximum or minimum

point is at $m(0, 1.7)$. Therefore, the point is $(1, 2)$.

$y - 2 = \dfrac{1}{2}(x - 1)$; $y = \dfrac{1}{2}x + \dfrac{3}{2}$ is the equation of the

tangent line.

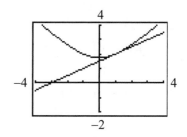

9. $y = 0.5x^2 + x$; $v_y = \dfrac{dy}{dt} = \dfrac{x\,dx}{dt} = +\dfrac{dx}{dt}$; $x = 0.5\sqrt{x}$

Substituting, $v_y = x(0.5\sqrt{x}) + 0.5\sqrt{x}$

Find v_y at $(2, 4)$:

$v_y|_{x=2} = 2(0.5\sqrt{2}) + 0.5\sqrt{2} = \sqrt{2} + 0.5\sqrt{2} =$

$1.5\sqrt{2} = 2.12$

13. $x^3 - 3x^2 - x + 2 = 0$ (between 0 and 1)

$f(x) = x^3 - 3x^2 - x + 2$; $f'(x) = 3x^2 - 6x - 1$

$f(0) = 0^3 - 3(0^2) - 0 + 2 = 2$; $f(1)$

$\qquad = 1^3 - 3(1^2) - 1 + 2 = -1$

The root is possibly closer to 1 than 0. Let $x_1 = 0.6$:

n	x_n	$f(x_n)$	$f'(x_n)$	$x_n - \dfrac{f(x_n)}{f'(x_n)}$
1	0.6	0.536	−3.52	0.7522727
2	0.7522727	−0.0242935	−3.8158936	0.7459063
3	0.7459063	−0.0000304	−3.8063092	0.7458983

$x_4 = x_3 = 0.7458983$

17. $y = 4x^2 + 16x$

(1) The graph is continuous for all x.

(2) The intercepts are $(0, 0)$ and $(-4, 0)$.

(3) As $x \to +\infty$ and $-\infty$, $y \to +\infty$.

(4) The graph is not symmetrical about either axis or the origin.

(5) $y' = 8x + 16$; $y' = 0$ at $x = -2$. $(-2, -16)$ is a critical point.

(6) $y'' = 8 > 0$ for all x; the graph is concave up and $(-2, -16)$ is a minimum.

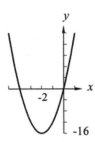

21. $y = x^4 - 32x$

(1) The graph is continuous for all x.

(2) The intercepts are $(0, 0)$ and $\left(2\sqrt[3]{4}, 0\right)$.

(3) As $x \to -\infty$, $y \to +\infty$; as $x \to +\infty$, $y \to +\infty$.

(4) The graph is not symmetrical about either axis or the origin.

(5) $y' = 4x^3 - 32 = 0$ for $x = 2$

(6) $y'' = 12x^2$; $y'' = 0$ at $x = 0$; $(0, 0)$ is a possible point of inflection. Since $f''(x) > 0$; the graph is concave up everywhere (except 0) and $(0, 0)$ is not an inflection point. $(2, -48)$ is a minimum

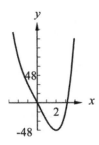

25. $y = f(x)$

$= 4x^3 + \dfrac{12}{x}$

$dy = f'(x)\, dx$

$= \left(12x^2 - \dfrac{12}{x^2}\right) dx$

29. $y = f(x) = 4x^3 - 12$, $x = 2$, $\Delta x = 0.1$

$\Delta y - dy = f(x + \Delta x) - f(x) - f'(x)\, dx$

$= 4(2.1)^3 - 12 - \left(4(2)^3 - 12\right) - 12(2)^2 (0.1)$

$= 0.244$

33. $V = f(r) = \dfrac{4}{3}\pi r^3$, $r = 3.500$, $\Delta r = 0.012$

$dV = f'(r)\, dr = 4\pi r^2\, dr = 4\pi (3.500)^2 (0.012)$

$dV = 1.85$ m^3

37. $Z = \sqrt{R^2 + X^2} = \left(R^2 + X^2\right)^{1/2}$

$dZ = \dfrac{1}{2}\left(R^2 + X^2\right)^{-1/2} (2R)\, dR$

$dZ = \dfrac{R\, dR}{\sqrt{R^2 + X^2}}$

relative error $= \dfrac{dZ}{Z} = \dfrac{R\, dR}{R^2 + X^2}$

41. $y = x^2 + 2$ and $y = 4x - x^2$

$y' = 2x$; $y' = 4 - 2x$

$2x = 4 - 2x$; $4x = 4$; $x = 1$

The point $(1, 3)$ belongs to both graphs; the slope of the tangent line is 2.

$y - y_1 = 2(x - x_1)$; $y - 3 = 2(x - 1)$; $y - 3 = 2x - 2$

$2x - y + 1 = 0$ is the equation of the tangent line.

45. (a) False, $f(x) = x^3$ is a counterexample.

(b) False, $f(x) = x^4$ is a counterexample.

49. $y = k\left(x^4 - 30x^3 + 1000x\right) = 0$. By inspection $x = 0.0$ m is the first value of x where the deflection is zero. From the graph of $y = x^4 - 30x^3 + 1000x$ there is a zero between $x = 6$ and $x = 7$. Letting $x_1 = 6$, successive iterations of Newton's method give

$x_1 = 6$

$x_2 = 6.593023253$

$x_3 = 6.527855923$

$x_4 = 6.527036576$

$x_5 = 6.527036447$

$x_6 = 6.527036447$

$x = 6.527$ m is the second value where the defection is zero.

53. $D^2 = (4.0t)^2 + 250^2$

$2D\dfrac{dD}{dt} = 32t$

$\dfrac{dD}{dt} = \dfrac{16(60)}{\sqrt{(4.0(60))^2 + 250^2}} = 2.77$ ft/s

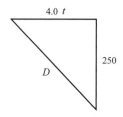

57. $y = x^2 - \dfrac{2}{x}$

(1) Intercepts: y is undefined for $x = 0 \Rightarrow$ no

y-intercept. $y = 0$ gives $0 = x^2 - \dfrac{2}{x}$; $x^3 = 2$;

$x = \sqrt[3]{2} \approx 1.26$

(2) Symmetry: none

(3) Behavior as x becomes large: $y \to \infty$ as

$x \to \pm\infty$

(4) Vertical asymptotes: $x = 0$ is a vertical asymptote

(5) Domain: $x \neq 0$; Range: $-\infty < y < \infty$

(6) Derivatives: $y' = 2x + \dfrac{2}{x^2} = 0$ for $x = -1$

$y'' = 2 - \dfrac{4}{x^3}\Big|_{x=-1} > 0 \Rightarrow (-1, 3)$ is a min

$y'' = 0 = 2 - \dfrac{4}{x^3} \Rightarrow x^3 = 2 \Rightarrow x = \sqrt[3]{2}$

$y'' < 0$ for $x < \sqrt[3]{2}$; $y'' > 0$ for $x > \sqrt[3]{2} \Rightarrow (\sqrt[3]{2}, 0)$

is an inflection point

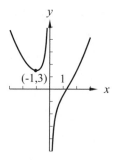

From the graph $y = x^2 - \dfrac{2}{x} > 0$ for $x < 0$ or $x > 1.26$

$x^2 - \dfrac{2}{x} > 0 \Rightarrow x^2 > \dfrac{2}{x}$ for $x < 0$ or $x > 1.26$

The graph of $x^2 > \dfrac{2}{x}$ is

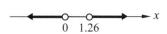

61. $f(0) = 2 \Rightarrow y$-intercept is 2

$\left.\begin{array}{l} f'(x) < 0 \text{ for } x < 0 \Rightarrow f \text{ is dec. for } x < 0 \\ f'(x) > 0 \text{ for } x > 0 \Rightarrow f \text{ is inc. for } x > 0 \end{array}\right\} (0, 2)$ min

$f''(x) > 0$ for all $x \Rightarrow f$ is conc. up for all x.

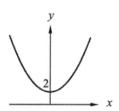

65.

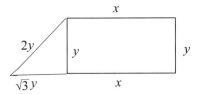

$$p = 2x + \left(3 + \sqrt{3}\right)y = 1500 \Rightarrow y = \frac{1500 - 2x}{3 + \sqrt{3}}$$

$$A = xy + \frac{1}{2}\sqrt{3}y^2$$

$$A = \frac{1500x - 2x^2}{3 + \sqrt{3}} + \frac{\sqrt{3}}{2} \cdot \frac{1500^2 - 6000x + 4x^2}{\left(3 + \sqrt{3}\right)^2}$$

$$\frac{dA}{dx} = \frac{1500 - 4x}{3 + \sqrt{3}} + \frac{\sqrt{3}}{2} \cdot \frac{-6000 + 8x}{\left(3 + \sqrt{3}\right)^2} = 0$$

$$x = 160 \text{ m}$$

$$y = \frac{1500 - 2x}{3 + \sqrt{3}} = 250 \text{ m}$$

69. $y = \dfrac{300}{0.0005x^2 + 2} - 50.0 < x < 100$

$$y' = 300\left(0.0005x^2 + 2\right)^{-2}\left(0.001x\right)\big|_{x=50} = -1.42$$

$y' < 0$ for $0 < x < 100 \Rightarrow y$ is dec for $0 < x < 100$

$$y' = \frac{-0.3x}{\left(0.0005x^2 + 2\right)^2}$$

y''

$$= \frac{\left(0.0005x^2 + 2\right)^2 \left(-0.3\right) + 0.3x\left(2\left(0.0005x^2 + 2\right)^1 \left(0.001x\right)\right)}{\left(0.0005x^2 + 2\right)^4}$$

$y'' = 0$ for $x = 37$, $y(37) = 63$

$y'' < 0$ for $x < 37$; $y'' > 0$ for $x > 37 \Rightarrow x = 37$ is infl.

y is conc. down for $x < 37$ and conc. up for $x > 37$.

$y(0) = 100$, y-intercept

$y = 0 \Rightarrow x = 89$, x-intercept

$L(x) = -1.42(x - 50) + y(50) = -1.42x + 71 + 42$

$L(x) = -1.42x + 113$

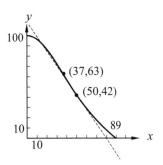

73. $\dfrac{1}{C} = \dfrac{1}{C_1} + \dfrac{1}{C_2}$

$$C_1 + C_2 = 12$$

$$C_T = \frac{C_1 + C_2}{C_1 C_2} = \frac{12}{C_1\left(12 - C_1\right)} = \frac{12}{12C_1 - C_1^2}$$

$$\frac{dC_T}{dC_1} = \frac{-12\left(12 - 2C_1\right)}{\left(12C_1 - C_1^2\right)^2} = 0, \; C_1 = 6$$

which is a minimum since

$$\frac{d^2 C_T}{dC_1^2}\bigg|_{C_1=6} > 0.$$

Each capacitor is 6 μF.

77. $t = \dfrac{\sqrt{16 + x^2}}{3} + \dfrac{5 - x}{5}$

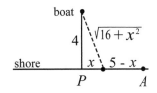

$$\frac{dt}{dx} = \frac{x}{3\sqrt{16 + x^2}} - \frac{1}{5} = 0$$

$$5x = 3\sqrt{16 + x^2}$$

$$25x^2 = 9\left(16 + x^2\right) = 144 + 9x^2$$

$$16x^2 = 144$$

$$x^2 = 9$$

$$x = 3, \text{ a minimum since}$$

$$\frac{d^2 t}{dx^2} = \frac{16}{3\left(16 + x^2\right)^{3/2}}\bigg|_{x=3} > 0$$

The boat should land 3 km from P toward A.

81. $V = \dfrac{1}{3}\pi r^2 h = \dfrac{1}{3}\pi r^2 \cdot r = \dfrac{1}{3}\pi r^3$

$\dfrac{dV}{dt} = \pi r^2 \dfrac{dr}{dt}$

$100 = \pi (10.0)^2 \dfrac{dr}{dt}$

$\dfrac{dr}{dt} = 0.318 \text{ ft/min}$

85. $V = 1350 = 0.75 l^2 h \Rightarrow h = \dfrac{1800}{l^2}$

$c = 0.75 l^2 (6) + 2(0.75 lh)(9)$
$\qquad\qquad + 2(lh)(9) + 0.75 l^2 h (4.5)$

$c = 4.5 l^2 + 13.5 l \left(\dfrac{1800}{l^2}\right) + 18 l \left(\dfrac{1800}{l^2}\right) + 3.375 l^2$

$c = 7.875 l^2 + \dfrac{56,700}{l}$

$\dfrac{dc}{dl} = 15.75 l - \dfrac{56,700}{l^2} = 0 \Rightarrow l = 15.326$

$\Rightarrow h = \dfrac{1800}{l^2} = 7.66$

$w = 0.75\, l = 11.49$

dimensions for minimum cost: 11 ft by 15 ft by 7.7 ft

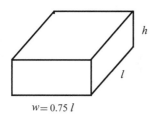

$w = 0.75\ l$

Chapter 25

INTEGRATION

25.1 Antiderivatives

1. $f(x) = 12x^3$; power of x required is 4, $F(x) = ax^4$
   $F'(x) = 4ax^3 = 12x^3$; $4a = 12$, $a = 3$
   $F(x) = 3x^4$

5. $3x^2$; the power of x required in the antiderivative is 3. Therefore, we must multiply by $\frac{1}{3}$. The antiderivative of $3x^2$ is $\frac{1}{3}(3x^3) = x^3$. $a = 1$.

9. The power of x required in the antiderivative of $f(x) = 9\sqrt{x}$ is $\frac{3}{2}$. Multiply by $\frac{2}{3}$. The anti-derivative of $9\sqrt{x}$ is $\frac{2}{3} \cdot 9x^{3/2}$. $a = 6$.

13. The power of x required in the antiderivative of $\frac{5}{2}x^{3/2}$ is $\frac{5}{2}$. Multiply by $\frac{2}{5}$. The antiderivative of $\frac{5}{2}x^{3/2}$ is $\frac{2}{5}\left(\frac{5}{2}\right)x^{5/2} = x^{5/2}$.

17. $f(x) = 2x^2 - 3x$; $2x^2 \rightarrow ax^3$; $\frac{d}{dx}(ax^3) = a3x^2$
    $3a = 2$; $a = \frac{2}{3}$, therefore, $\frac{2}{3}x^3$; $3x \rightarrow ax^2$;
    $\frac{d}{dx}(ax^2) = a2x = 3x$
    $2a = 3$; $a = \frac{3}{2}$, therefore, $\frac{3}{2}x^2$.
    antiderivative of $2x^2 - 3x$ is $\frac{2}{3}x^3 - \frac{3}{2}x^2$.

21. $f(x) = \frac{-7}{x^6}$; $+\frac{1}{3^2}$ $f(x) = -7x^{-6} + \frac{1}{9}$ power required is -5 therefore, x^{-5};
    $\frac{d}{dx}ax^{-5} = -5ax^{-6}$; $-5a = -7$; $a = \frac{7}{5}$ therefore, $\frac{7}{5}x^{-5}$.
    Therefore, antiderivative is $\frac{7}{5x^5} + \frac{x}{9}$.

25. The antiderivative of $f(x) = 50x^{99} - 39x^{-79}$ is
    $$F(x) = \frac{50}{99+1}x^{99+1} - \frac{39}{-79+1}x^{-79+1} = \frac{1}{2}x^{100} + \frac{1}{2}x^{-78}$$

29. $f(x) = 6(2x+1)^5(2)$; $u^n = (2x+1)^5$ therefore, power required is 6.
    $u = 2x+1$; $\frac{du}{dx} = 2$; $\frac{d}{dx}a(2x+1)^6 = 6a(2x+1)^5(2)$
    therefore, $a = 1$.
    Therefore, antiderivative is $(2x+1)^6$.

33. $f(x) = 4x^3(2x^4+1)^4$, $u^n = (2x^4+1)^4$
    $u = 2x^4+1$; $\frac{du}{dx} = 8x^3$
    Power needed is 5.
    $\frac{d}{dx}a(2x^4+1)^5 = 5a(2x^4+1)^4 8x^3$; $5a = \frac{1}{2}$; $a = \frac{1}{10}$.
    Therefore, antiderivative is
    $\frac{1}{5}(2x^4+1)^5\frac{4}{8}x^3 = \frac{1}{10}(2x^4+5)^5$.
    Check:
    $$\frac{d}{dx}\frac{1}{10}(2x^4+1)^5 = \frac{1}{10}\left[5(2x^4+1)^4 8x^3\right]$$
    $$= \frac{1}{10}\left[40(2x^4+1)^4 x^3\right]$$
    $$= (2x^4+1)^4(4x^3)$$

37. $f(x) = (3x+1)^{1/3}$; $u^n = (3x+1)^{1/3}$; $u = 3x+1$
    $\frac{du}{dx} = 3$ therefore, $\frac{1}{3}$ needed.
    Power of u needed is $\frac{4}{3}$.
    $\frac{d}{dx}a(3x+1)^{4/3} = a\frac{4}{3}(3x+1)^{1/3}(3)$; $\frac{4}{3}a = 1$; $a = \frac{3}{4}$

Therefore, $\dfrac{3}{4}(3x+1)^{4/3}\dfrac{1}{3}=\dfrac{1}{4}(3x+1)^{4/3}$

Check:

$\dfrac{d}{dx}\dfrac{1}{4}(3x+1)^{4/3}=\dfrac{1}{4}\cdot\dfrac{4}{3}(3x+1)^{1/3}(3)=(3x+1)^{1/3}$.

Therefore, the antiderivative is $\dfrac{1}{4}(3x+1)^{4/3}$.

41. $F(x)=(x+5)^3$ is the correct antiderivative of

$f(x)=3(x+5)^2$ since $F'(x)=3(x+5)(1)=f(x)$

$F(x)=(2x+5)^3$ is not the correct antiderivative of

$f(x)=3(2x+5)^2$ since

$F'(x)=3(2x+5)^2(2)=2\cdot3(2x+5)^2$

$=2f(x)\neq f(x)$

25.2   The Indefinite Integral

1. $\displaystyle\int 8x\,dx=8\int x^1\,dx=8\dfrac{x^{1+1}}{1+1}+C=4x^2+C$

5. $\displaystyle\int 2x\,dx=2\int x\,dx;\ u=x;\ du=dx;\ n=1$

$2\displaystyle\int x\,dx=2\left(\dfrac{x^{1+1}}{1+1}\right)+C=x^2+C$

9. $\displaystyle\int 8x^{3/2}\,dx;\ u=x;\ du=dx;\ n=\tfrac{3}{2}$

$\displaystyle\int 8x^{3/2}\,dx=\dfrac{8x^{(3/2)+1}}{\tfrac{3}{2}+1}+C$

$=\dfrac{8x^{5/2}}{\tfrac{5}{2}}+C$

$=\dfrac{16}{5}x^{5/2}+C$

13. $\displaystyle\int(x^2-3x^5)\,dx=\int x^2\,dx-\int 3x^5\,dx$

$=\dfrac{x^3}{3}-\dfrac{3x^6}{6}+C$

$=\dfrac{1}{3}x^3-\dfrac{1}{2}x^6+C$

17. $\displaystyle\int\left(\dfrac{t^2}{2}-\dfrac{2}{t^2}\right)dt=\dfrac{t^{2+1}}{2(2+1)}-\dfrac{2t^{-2+1}}{-2+1}+C$

$=\dfrac{t^3}{6}+\dfrac{2}{t}+C$

21. $\displaystyle\int(2x^{-2.3}+3^{-2})\,dx=\int 2x^{-2/3}\,dx+\int 3^{-2}x^0\,dx$

$=2\int x^{-2/3}\,dx+3^{-2}\int x^0\,dx$

$=2\cdot\dfrac{x^{1/3}}{\tfrac{1}{3}}(3x^{1/3})+3^{-2}(x^1)$

$=6x^{1/3}+\dfrac{1}{9}x+C$

25. $\displaystyle\int(x^2-1)^5(2x\,dx);\ u=x^2-1;\ du=2x\,dx;\ n=5$

$\displaystyle\int(x^2-1)^5(2x\,dx)=\dfrac{(x^2-1)^6}{6}+C$

$=\dfrac{1}{6}(x^2-1)^6+C$

29. $\displaystyle\int 40(2\theta^5+5)^7\theta^4\,d\theta=\dfrac{40}{10}\int(2\theta^5+5)^7\cdot(10\theta^4)\,d\theta$

$=\dfrac{40}{10}\cdot\dfrac{(2\theta^5+5)^8}{8}+C$

$=\dfrac{(2\theta^5+5)^8}{2}+C$

33. $\displaystyle\int\dfrac{4x\,dx}{\sqrt{6x^2+1}}=4\int(6x^2+1)^{-1/2}x\,dx$

$u=6x^2+1;\ du=12x;\ n=-\dfrac{1}{2}$

$\displaystyle\int(6x^2+1)^{-1/2}x\,dx=\dfrac{4}{12}\int(6x^2+1)^{-1/2}(12x\,dx)$

$=\dfrac{4}{12}\dfrac{(6x^2+1)^{1/2}}{1/2}+C$

$=\dfrac{2}{3}\sqrt{6x^2+1}+C$

37. $\dfrac{dy}{dx}=6x^2;\ dy=6x^2\,dx$

$y=\displaystyle\int 6x^2\,dx=6\int x^2\,dx=\dfrac{6x^3}{3}+C=2x^3+C$

The curve passes through $(0,2)$. $2=2(0^3)+C$,

$C=2;\ y=2x^3+2$

41. $\int 3x^2 dx = x^3 + C$

$\int 3x^2 dx \neq x^3$ since the constant of integration must be included.

45. $\int 3(2x+1)^2 dx = \frac{3}{2}\int(2x+1)^2 (2dx)$

$= \frac{3}{2}\frac{(2x+1)^3}{3} + C$

$= \frac{(2x+1)^3}{3} + C$

$\int 3(2x+1)^2 dx \neq (2x+1)^3 + C$ because the factor of $\frac{1}{2}$ is missing.

49. $f'(t) = 4x - 5$

$f(x) = 2x^2 - 5x + C$

$f(-1) = 10 = 2(-1)^2 - 5(-1) + C \Rightarrow C = 3$

$f(x) = 2x^2 - 5x + 3$

53. $V(t) = \int(5.00 + 0.01t)\,dt$

$V(t) = 5.00t + 0.005t^2 + C$ where $C = 0$

since $v(0) = 0$

$V(t) = 5.00t + 0.005t^2$

57. $\frac{dT}{dr} = -4500(r+1)^{-3}$, $T = 2500°C$ for $r = 0$.

$\int dT = \int -4500(r+1)^{-3}\,dr$

$T = 2250(r+1)^{-2} + C$

$2500 = 2250(0+1)^{-2} + C$

$C = 250$

$T = 2250(r+1)^{-2} + 250$

61. $\frac{d^2y}{dx^2} = \frac{d(y')}{dx} = 6$; $dy' = 6dx$

$y' = 6x + C$; $y' = 8$ for $x = 1$

$8 = 6(1) + C$; $C = 2$

$y' = \frac{dy}{dx} = 6x + 2$; $dy = (6x+2)\,dx$

$y = 3x^2 + 2x + C_1$; curve passes through $(1, 2)$

$2 = 3(1)^2 + 2(1) + C_1$; $C_1 = -3$

$y = 3x^2 + 2x - 3$

25.3   The Area Under a Curve

1. (a)

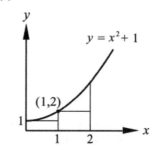

$A = 1(1) + 1(2)$

$A = 3$

(b)

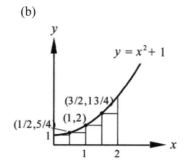

$A = \frac{1}{2}\left[1 + \frac{5}{4} + 2 + \frac{13}{4}\right]$

$A = \frac{15}{4}$

5. $y = 3x$, between $x = 0$ and $x = 3$

(a)

x	y
1	3
2	6
3	9

$n = 3; \Delta x = 1$

$A = 1(0 + 3 + 6) = 9$; (first rectangle has 0 height)

(b)

x	y
0	0
0.3	0.9
0.6	1.8
0.9	2.7
1.2	3.6
1.5	4.5
1.8	5.4
2.1	6.3
2.4	7.2
2.7	8.1
3.0	9.0

$n = 10; \Delta x = 0.3$

$A = 0.3(0 + 0.9 + 1.8 + 2.7 + 3.6 + 4.5 + 5.4$

$\quad + 6.3 + 7.2 + 8.1)$

$\quad = 0.3(40.5) = 12.15$

9. $y = 4x - x^2$, between $x = 1$ and $x = 4$

(a) $n = 6, \Delta x = 0.5$

$A = 0.5(3.00 + 3.75 + 3.75 + 3.00 + 1.75 + 0.00)$

$A = 7.625$

x	y
1.0	3.00
1.5	3.75
2.0	4.00
2.5	3.75
3.0	3.00
3.5	1.75
4.0	0.00

$(y = 4.00$ is not the height of any inscribed rectangle$)$

(b) $n = 10, \Delta x = 0.3$

$A = 0.3(3.00 + 3.51 + 3.84 + 3.96 + 3.75 + 3.36$

$\quad + 2.79 + 2.04 + 1.11)$

$A = 8.208$

x	y
1.0	3.00
1.3	3.51
1.6	3.84
1.9	3.99
2.2	3.96
2.5	3.75
2.8	3.36
3.1	2.79
3.4	2.04
3.7	1.11
4.0	0.00

$(y = 3.99$ is not the height of any inscribed rectangle$)$

13. $y = \dfrac{1}{\sqrt{x+1}}$, between $x = 3$ and $x = 8$

(a) $n = 5, \Delta x = \dfrac{8 - 3}{5} = 1$

$A = \displaystyle\sum_{i=1}^{5} A_i = \sum_{i=1}^{5} y_i \Delta x$

$y_1 = f(4)$

$A = (0.447 + 0.408 + \cdots + 0.354 + 0.333)(1)$

$A = 1.92$

x	y
3	0.5
4	0.447
5	0.408
6	0.378
7	0.355
8	0.333

(b) $n = 10$, $\Delta x = \dfrac{8-3}{10} = 0.5$

$$A = \sum_{i=1}^{10} A_i = \sum_{i=1}^{10} y_i \Delta x$$

$$y_1 = f(3.5)$$

$$A = (0.471 + 0.447 + \cdots + 0.343 + 0.333)(0.5)$$

$$A = 1.96$$

x	y
3	0.5
3.5	0.471
4	0.447
4.5	0.426
5	0.408
5.5	0.392
6	0.378
6.5	0.365
7	0.354
7.5	0.343
8	0.333

17. $y = x^2$, between $x = 0$ and $x = 2$

$$A_{0,2} = \left[\int x^2 dx \right]_0^2 = \frac{x^3}{3}\Big|_0^2 = \frac{8}{3} - 0 = \frac{8}{3}$$

21. $y = \dfrac{1}{x^2} = x^{-2}$, between $x = 1$ and $x = 5$

$$A_{1,5} = \left[\int x^{-2} dx \right]_1^5 = \frac{x^{-1}}{-1}\Big|_1^5 = \frac{-1}{x}\Big|_1^5$$

$$= -\frac{1}{5} - (-1) = \frac{4}{5} = 0.8$$

25. $y = 3x$, $x = 0$ to $x = 3$, $n = 10$, $\Delta x = 0.3$

Using the table in 5(b)

$$A = 0.3(0.9 + 1.8 + 2.7 + 3.6 + 4.5 + 5.4 + 6.3 + 7.2$$

$$+ 8.1 + 9.0)$$

$$A = 14.85$$

$$\begin{array}{ccccc} A_{\text{inscribed}} & < & A_{\text{exact}} & < & A_{\text{circumscribed}} \\ 12.15 & < & 13.5 & < & 14.85 \end{array}$$

$\frac{12.15 + 14.85}{2} = 13.5$ because the extra area above $y = 3x$ using circumscribed rectangles is the same as the omitted area under $y = 3x$ using inscribed rectangles.

25.4   The Definite Integral

1. $\displaystyle\int_1^4 (x^{-2} - 1)\,dx = -\frac{1}{x} - x\Big|_1^4 = -\frac{1}{4} - 4 - \left(-\frac{1}{1} - 1\right)$

$$\int_1^4 (x^{-2} - 1)\,dx = -\frac{9}{4}$$

5. $\displaystyle\int_1^4 x^{5/2}\,dx = \frac{2}{7} x^{7/2}\Big|_1^4 = \frac{256}{7} - \frac{2}{7} = \frac{254}{7}$

9. $u = 1 - x$; $du = -dx$

$$\int_{-1.6}^{0.7} (1-x)^{1/3}\,dx = -\int_{-1.6}^{0.7} (1-x)^{1/3}(-dx)$$

$$= -\frac{3}{4}(1-x)^{4/3}\Big|_{-1.6}^{0.7}$$

$$= -\frac{3}{4}(0.2008 - 3.5752) = 2.53$$

13. $\displaystyle\int_{0.5}^{2.2} (\sqrt[3]{x} - 2)\,dx = \int_{0.5}^{2.2} x^{1/3}\,dx - 2\int_{0.5}^{2.2} dx$

$$= \left(\frac{3}{4}x^{4/3} - 2x\right)\Big|_{0.5}^{2.2}$$

$$= (2.1460 - 4.4) - (0.2976 - 1)$$

$$= -1.5516 = -1.552$$

17. $u = 4 - x^2$; $du = -2x\,dx$

$$\int_{-2}^{-1} 12x(4-x^2)^3\,dx = -6\int_{-2}^{-1} (4-x^2)^3(-2x\,dx)$$

$$= -\frac{6(4-x^2)^4}{4}\Big|_{-2}^{-1}$$

$$= -\frac{243}{2}$$

21. $u = 6x + 1$; $du = 6\,dx$

$$\int_{2.75}^{3.25} \frac{dx}{\sqrt[3]{6x+1}} = \int (6x+1)^{-1/3}\,dx$$

$$= \frac{1}{6}\int (6x+1)^{-1/3}(6\,dx)$$

$$= \frac{1}{6} \cdot \frac{3}{2}(6x+1)^{2/3}\Big|_{2.75}^{3.25} = \frac{1}{4}(6x+1)^{2/3}\Big|_{2.75}^{3.25}$$

$$= \frac{1}{4}(7.4904 - 6.7405) = 0.1875$$

25. $u = 4t + 1;\ du = 4dt$

$$\int_3^7 \sqrt{16t^2 + 8t + 1}\,dt = \int_3^7 \sqrt{(4t+1)^2}\,dt$$

$$= \int_3^7 \sqrt{(4t+1)}\,dt$$

$$= \frac{1}{4}\int_3^7 (4t+1)^1\, 4dt = \frac{1}{4}\frac{(4t+1)^2}{2}$$

$$= \frac{1}{8}(4t+1)^2\Big|_3^7 = 84$$

29. $\displaystyle\int_{-1}^2 \frac{8x-2}{\left(2x^2 - x + 1\right)^3}\,dx$

$$= \int_{-1}^2 (8x-2)\left(2x^2 - x + 1\right)^{-1}\,dx$$

$$= 2\int_{-1}^2 (4x-1)\left(2x^2 - x + 1\right)^{-3}\,dx$$

$$= \frac{2\left(2x^2 - x + 1\right)^{-2}}{-2}\Bigg|_{-1}^2$$

$$= -\left(2x^2 - x + 1\right)^{-2}\Big|_{-1}^2$$

$$= -\frac{1}{7^2} - \left(-\frac{1}{4^2}\right) = 0.0421$$

33. $\displaystyle\int_{\sqrt5}^3 8z\sqrt[4]{z^4 + 8z^2 + 16}\ dz$

$$= \int_{\sqrt5}^3 8z\left(\left(z^2+4\right)^2\right)^{1/4}\,dz = \int_{\sqrt5}^3 8z\left(z^2+4\right)^{1/2}\,dz$$

$$= \frac{2}{3}\left(z^2+4\right)^{3/2}\Big|_{\sqrt5}^3 = 53.0$$

37. $y^2 = 4x,\quad y > 0$

For $x = 1,\quad y^2 = 4(1) \Rightarrow y = 2$

For $x = 4,\quad y^2 = 4(4) \Rightarrow y = 4$

$$\int_{x=1}^{x=4} y\,dx = \int_1^4 2\sqrt{x}\,dx = 2\cdot\frac{2}{3}x^{3/2}\Big|_1^4 = \frac{4}{3}\left(4^{3/2} - 1^{3/2}\right) = \frac{28}{3}$$

41. $\displaystyle\int_{-1}^1 t^{2k}\,dt = \frac{t^{2k+1}}{2k+1}\Bigg|_{-1}^1 = \frac{1^{2k+1}}{2k+1} - \frac{(-1)^{2k+1}}{2k+1}$

$$\int_{-1}^1 t^{2k}\,dt = \frac{1}{2k+1} - \frac{(-1)^{2k}(-1)^1}{2k+1}$$

$$= \frac{1}{2k+1} + \frac{1}{2k+4}$$

$$\int_{-1}^1 t^{2k}\,dt = \frac{2}{2k+1}$$

45.

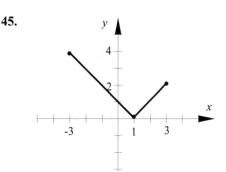

$$\int_{-3}^3 |x-1|\,dx = \frac{1}{2}(4)(4) + \frac{1}{2}(2)(2) = 10$$

49. $\displaystyle W = \int_0^{80}(1000 - 5x)\,dx$

$$= \left(1000x - \frac{5}{2}x^2\right)\Bigg|_0^{80} = 1000(80) - \frac{5}{2}(80)^2$$

$$ - [0 - 0]$$

$$= 80{,}000 - 16{,}000 = 64{,}000\ \text{ft}\cdot\text{lb}$$

53. $\displaystyle\frac{3N}{2E_F^{3/2}}\int_0^{E_F} E^{3/2}\,dE = \frac{3N}{2E_F^{3/2}}\cdot\frac{E^{5/2}}{\frac{5}{2}}\Bigg|_0^{E_F}$

$$\frac{3N}{2E_F^{3/2}}\int_0^{E_F} E^{3/2}\,dE = \frac{3N}{2E_F^{3/2}}\cdot\frac{2E^{5/2}}{5} = \frac{3NE_F}{5}$$

25.5 Numerical Integration: The Trapezoidal Rule

1. $\int_1^3 \frac{1}{x}\,dx,\ n=2,\ h=\dfrac{b-a}{n}=\dfrac{3-1}{2}=1$

x	y
1	1
2	$\frac{1}{2}$
3	$\frac{1}{3}$

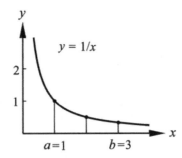

$$A=\frac{1}{2}\left(1+2\left(\frac{1}{2}\right)+\frac{1}{3}\right)=\frac{7}{6}$$

5. $\int_1^4 \left(1+\sqrt{x}\right)dx;\ n=6;\ h=\dfrac{4-1}{6}=\dfrac{1}{2};\dfrac{h}{2}=\dfrac{1}{4}$

n	x_n	y_n
0	1	2
1	1.5	2.22
2	2	2.41
3	2.5	2.58
4	3	2.73
5	3.5	2.87
6	4	3

$$A_T=\frac{1}{4}\Big[2+2(2.22)+2(2.41)+2(2.58)+2(2.73)$$
$$+2(2.87)+3\Big]$$
$$A=\frac{1}{4}(30.646)=7.661$$

$$A=\int_1^4\left(1+x^{1/2}\right)dx=\left(x+\frac{2}{3}x^{3/2}\right)\Big|_1^4$$

$$A=4+\frac{16}{3}-\left(1+\frac{2}{3}\right)=\frac{23}{3}=7.667$$

9. $\int_0^5 \sqrt{25-x^2}\,dx;\ n=5;\ x=\dfrac{5}{5}=1;\dfrac{h}{2}=\dfrac{1}{2}$

n	x_n	y_n
0	0	5
1	1	4.90
2	2	4.58
3	3	4
4	4	3
5	5	0

$$A_T=\frac{1}{2}\Big[5+2(4.90)+2(4.58)+2(4)+2(3)+0\Big]$$
$$=18.98$$

13. $\int_0^4 2^x\,dx;\ n=12;\ h=\dfrac{4}{12}=\dfrac{1}{3};\dfrac{h}{2}=\dfrac{1}{6}$

x	y
0	1
$\frac{1}{3}$	1.260
$\frac{2}{3}$	1.587
1	2
$1\frac{1}{3}$	2.520
$1\frac{2}{3}$	3.175
2	4
$2\frac{1}{3}$	5.040
$2\frac{2}{3}$	6.350
3	8
$3\frac{1}{3}$	10.080
$3\frac{2}{3}$	12.699
4	16

$$A=\frac{1}{6}\Big[1+2(1.260)+2(1.587)+2(2)+2(.520)$$
$$+2(3.175)+2(4)+2(5.040)+2(6.350)$$
$$+2(8)+2(10.080)+2(12.699)+16\Big]$$
$$A=21.74$$

17. The approximate value is less than the exact value because the tops of all the trapezoids are below the curve.

21. $L = 2\int_0^{50}\sqrt{6.4\times 10^{-7}x^2+1}\,dx$; $n = 10$; $h = \dfrac{50}{10} = 5$;

$\dfrac{h}{2} = \dfrac{5}{2}$

x	y
0	2
5	2.000016
10	2.000063999
15	2.000143955
20	2.000255984
25	2.00039996
30	2.000575917
35	2.000783846
40	2.00103738
45	2.00129558
50	2.001599361

$L = \dfrac{5}{2}[2 + 2(2.000016 + 2.000063999 + 2.000143995$

$\qquad + 2.000255984 + 2.00039996 + 2.000575917$

$\qquad + 2.000783846 + 2.001023838 + 2.00129558)$

$\qquad + 2.001599361]$

$L = 100.0267917$ ft

25.6 Simpson's Rule

1. $\displaystyle\int_0^1 \dfrac{dx}{x+2}$, $n = 2$, $h = \dfrac{1-0}{2} = \dfrac{1}{2}$

x	y
0	$\frac{1}{2}$
$\frac{1}{2}$	$\frac{2}{5}$
1	$\frac{1}{3}$

$\displaystyle\int_0^1 \dfrac{dx}{x+2} = \dfrac{\frac{1}{2}}{3}\left(\dfrac{1}{2} + 4\left(\dfrac{2}{5}\right) + \dfrac{1}{3}\right) = 0.40\overline{5}$

5. $\displaystyle\int_1^4\left(2x+\sqrt{x}\right)dx$; $n = 6$; $\Delta x = \dfrac{4-1}{6} = \dfrac{1}{2}$; $\dfrac{\Delta x}{3} = \dfrac{1}{6}$

$A_S = \dfrac{1}{6}\big[3 + 4(4.22) + 2(5.41) + 4(6.58) + 2(7.73)$

$\qquad + 4(8.87) + 10\big]$

$A_S = \dfrac{1}{6}(117.999) = 19.67$

n	x_n	y_n
1	1	3
2	1.5	4.22
3	2	5.41
4	2.5	6.58
5	3	7.73
6	3.5	8.87
7	4	10

$A = \displaystyle\int_1^4\left(2x+\sqrt{x}\right)dx = 2\int_1^4 x\,dx + \int_1^4 x^{1/2}\,dx$

$\quad = \left(x^2 + \dfrac{2}{3}x^{3/2}\right)\Big| = 16 + \dfrac{16}{3} + -\left(1 + \dfrac{2}{3}\right)$

$\quad = \dfrac{59}{3} = 19.67$

9. $\displaystyle\int_1^5 \dfrac{1}{x^2+x}\,dx$; $n = 10$; $\Delta x = 0.4$; $\dfrac{\Delta x}{3} = \dfrac{0.4}{3}$

$A_S = \dfrac{0.4}{3}\big[0.5000 + 4(0.2976) + 2(0.1984)$

$\qquad + 4(0.1420) + 2(0.1068) + 4(0.0833)$

$\qquad + 4(0.0388) + 0.0333\big]$

$\quad = \dfrac{0.4}{3}(3.8349) = 0.5114$

13. $\Delta x = 2$; $\dfrac{\Delta x}{3} = \dfrac{2}{3}$

x	y
2	0.67
4	2.34
6	4.56
8	3.67
10	3.56
12	4.78
14	6.87

$$\int_2^{14} y\,dx = \frac{2}{3}\big[0.67 + 4(2.34) + 2(4.56) + 4(3.67)$$
$$+ 2(3.56) + 4(4.78) + 6.87\big] = 44.63$$

17. $\bar{x} = 0.9129\int_0^3 x\sqrt{0.3 - 0.1x}\,dx;\ n = 12;\ \Delta x = \frac{3}{12} = \frac{1}{4};$

$$\frac{\Delta x}{3} = \frac{1}{12}$$

n	x_n	y_n
1	0	0
2	0.25	0.131
3	0.50	0.25
4	0.75	0.356
5	1	0.447
6	1.25	0.523
7	1.50	0.581

n	x_n	y_n
8	1.75	0.619
9	2	0.632
10	2.25	0.616
11	2.5	0.559
12	2.75	0.435
13	3	0

$$\bar{x} = A_S = \frac{1}{12}\big[0 + 4(0.131) + 2(0.25) + 4(0.356)$$
$$+ 2(0.447) + 4(0.523) + 2(0.581)$$
$$+ 4(0.619) + 2(0.632) + 4(0.616)$$
$$+ 2(0.559) + 4(0.435) + (0)\big](0.9129)$$
$$= \frac{1}{12}(15.6572)(0.9129) = 1.191$$
$$\bar{x} = 1.200 \text{ in.}$$

Chapter 25   Review Exercises

1. $\displaystyle\int (4x^3 - x)\,dx = \int 4x^3\,dx - \int x\,dx = \frac{4x^4}{4} - \frac{x^2}{2} + C$

$$= x^4 - \frac{1}{2}x^2 + C$$

5. $\displaystyle\int_1^4 \left(\frac{\sqrt{x}}{2} + \frac{2}{\sqrt{x}}\right) dx = \frac{1}{2}\int_1^4 x^{1/2}\,dx + 2\int_1^4 x^{-1/2}\,dx$

$$= \frac{1}{2}\frac{x^{3/2}}{\frac{3}{2}} + \frac{2x^{1/2}}{\frac{1}{2}}\Bigg|_1^4 = \frac{1}{3}x^{3/2} + 4x^{1/2}\Bigg|_1^4$$

$$= \left[\frac{1}{3}(4)^{3/2} + 4(4)^{1/2}\right]$$

$$- \left[\frac{1}{3}(1)^{3/2} + 4(1)^{1/2}\right] = \frac{19}{3}$$

9. $\displaystyle\int \left(5^2 + \frac{6}{x^3}\right) dx = \int 5^2\,dx + \int \frac{6}{x^3}\,dx = \int 5^2\,dx + \int 6x^{-3}\,dx$

$$= (25x) + \left(-\frac{6}{2}x^{-2}\right)$$

$$= 25x - 3x^{-2} = 25x - \frac{3}{x^2} + C$$

13. $\displaystyle\int \frac{10\,dn}{(9 - 5n)^3} = 10\int (9 - 5n)^{-3}\,dn$

$$-\frac{10}{5}\int (9 - 5n)^{-3}(-5\,dn) = -2 \cdot \frac{(9 - 5n)^{-2}}{-2} + C$$

$$= \frac{1}{(9 - 5n)^2} + C = \frac{1}{(9 - 5n)^2} + C$$

17. $\displaystyle\int_0^2 \frac{3x\,dx}{\sqrt[3]{1 + 2x^2}} = \int_0^2 (1 + 2x^2)^{-1/3}(3x)\,dx$

$$u = 1 + 2x^2;\ du = 4x\,dx;\ n = -\frac{1}{3}$$

$$\frac{3}{4}\int_0^2\left(1+2x^2\right)^{-1/3}(4x)\,dx=\frac{3}{4}\cdot\frac{\left(1+2x^2\right)^{2/3}}{\frac{2}{3}}\bigg|_0^2$$

$$=\frac{9}{8}\left(1+2x^2\right)^{2/3}\bigg|_0^2$$

$$=\frac{9}{8}\left[1+2\left(2^2\right)\right]^{2/3}$$

$$-\frac{9}{8}\left[1+2(0)^2\right]^{2/3}$$

$$=\frac{9}{8}(9)^{2/3}-\frac{9}{8}(1)^{2/3}$$

$$=\frac{9}{8}\left(\sqrt[3]{81}-1\right)=\frac{9}{8}\left(3\sqrt[3]{3}-1\right)$$

21. $\displaystyle\int\frac{\left(2-3x^2\right)dx}{\left(2x-x^3\right)^2}=\int\left(2x-x^3\right)^{-2}\left(2-3x^2\right)dx;$

$u=2x-x^3;\ du=2-3x^2;\ n=-2$

$$\int\left(2x-x^3\right)^{-2}\left(2-3x^2\right)dx=\frac{\left(2x-x^3\right)^{-1}}{-1}+C$$

$$=-\frac{1}{\left(2x-x^3\right)}+C$$

25. $\displaystyle\frac{dy}{dx}=3-x^2$

$$y=3x-\frac{x^3}{3}+C$$

$$3=3(-1)-\frac{(-1)^3}{3}+C$$

$$C=\frac{17}{3}$$

$$y=3x-\frac{x^3}{3}+\frac{17}{3}$$

29. $\displaystyle\int_3^8 F(v)\,dv-\int_4^8 F(v)\,dv=\int_3^8 F(v)\,dv+\int_8^4 F(v)\,dv$

$$=\int_3^4 F(v)\,dv$$

33. $f''(x)=\left(6x+5\right)^{-1/2}$

$$f'(x)=\frac{1}{6}\int\left(6x+5\right)^{-1/2}(6dx)=\frac{1}{3}\left(6x+5\right)^{1/2}+C_1$$

$$f(x)=\frac{1}{18}\int\left(6x+5\right)^{1/2}(6dx)+\int C_1 dx$$

$$f(x)=\frac{1}{27}\left(6x+5\right)^{3/2}+C_1 x+C_2$$

37. $A=\displaystyle\int_1^3(6x-1)\,dx$

$$A=3x^2-x\big|_1^3$$

$$A=3\left(3^2\right)-3-\left[3\left(1^2\right)-1\right]$$

$$A=22$$

41. $\Delta x=\dfrac{b-a}{n}=\dfrac{3-1}{4}=\dfrac{1}{2},\ \dfrac{\Delta x}{2}=\dfrac{1}{4}$

$y_0=1,\ f(1)=\dfrac{1}{2(1)-1}=1,\ y_1=1.5,$

$f(1.5)=\dfrac{1}{2(1.5)-1}=\dfrac{1}{2},\ y_2=2.0,$

$f(2.0)=\dfrac{1}{2(2.0)-1}=\dfrac{1}{3},\ y_3=2.5,$

$f(2.5)=\dfrac{1}{2(2.5)-1}=\dfrac{1}{4},\ y_4=3.0,$

$f(3.0)=\dfrac{1}{2(3.0)-1}=\dfrac{1}{5}$

$$\int_1^3\frac{dx}{2x-1}\approx\frac{1}{4}\left[1+2\left(\frac{1}{2}\right)+2\left(\frac{1}{3}\right)+2\left(\frac{1}{4}\right)+\frac{1}{5}\right]=0.842$$

45. $\Delta x=\dfrac{b-a}{n}=\dfrac{4-1}{3}=1$

$$\int_1^4 x\sqrt[3]{2x^2+1}\,dx\approx\Delta x\left[f(1)+f(2)+f(3)\right]$$

$$=\left[1\sqrt[3]{2(1)^2+1}+2\sqrt[3]{2(2)^2+1}+3\sqrt[3]{2(3)^2+1}\right]\approx13.6$$

49.

Simpson with $n=2$. $\Delta x=\dfrac{4-1}{2}=\dfrac{3}{2},\ \dfrac{\Delta x}{3}=\dfrac{1}{2}$

$$\int_1^4 x\sqrt[3]{2x^2+1}\,dx$$

$$\approx\frac{1}{2}\left[1\sqrt[3]{2(1)^2+1}+4(2.5)\sqrt[3]{2(2.5)^2+1}+4\sqrt[3]{2(4)^2+1}\right]$$

$$\int_1^4 x\sqrt[3]{2x^2+1}\,dx\approx19.0417$$

53. $y(x)=4+\sqrt{1+8x-2x^2}.$

$$A=\frac{1}{2}\left[y(0)+y\left(\frac{1}{2}\right)+y(1)+y\left(\frac{3}{2}\right)+y(2)\left(\frac{5}{2}\right)\right.$$

$$\left.+y(3)+y\left(\frac{7}{2}\right)+y(4)\right]$$

$$A=24.68\ \text{m}^2$$

57. $Q = \int_0^3 6t^2 \, dt = 2t^3 \Big|_0^3 = 2(3)^3 = 54 \text{ C}$

61. $A = 2\int_0^5 \sqrt{5-y} \, dy = -2\int_5^0 \sqrt{u} \, du$

$\qquad = -2\left(\dfrac{2}{3}\right) u^{3/2} \Big|_5^0$

$\quad u = 5 - y, \; du = -dy$

$\quad A = -\dfrac{4}{3}\left[0 - 5^{3/2}\right]$

$\quad A = 14.9 \text{ m}^2$

Chapter 26

Applications of Integration

26.1 Applications of the Indefinite Integral

1. $s = \int (v_0 - 32t)\, dt = v_0 t - 16t^2 + C$

$200 = v_0(0) - 16(0)^2 + C$

$200 = C$

$s = v_0 t - 16t^2 + 200$

$0 = v_0(2.5) - 16(2.5)^2 + 200$

$v_0 = -40 \text{ ft/s}$

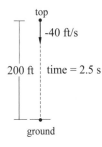

5. $\dfrac{ds}{dt} = -0.25 \text{ m/s}$

$ds = -0.25\, dt$

$s = -0.25 \int dt = -0.25t + C_1$

$t = 0,\ s = 8; \text{therefore},\ C_1 = 8$

$s = -0.25t + 8.00 = 8.00 - 0.25t$

9. $\dfrac{d^2 s}{dt^2} = -8.0t$

$v = \dfrac{ds}{dt} = -4.0t^2 + 64 = 0 \Rightarrow t = 4.0$

$s = \dfrac{-4.0t^3}{3} + 64t$

$s = \dfrac{-4.0}{3}(4.0)^3 + 64(4.0) = 171 \text{ ft}$

13. At impact: $t = 0,\ a = -250 \text{ m/s}^2$,

$v_{\text{impact}} = 96 \text{ km/h} = 26.\overline{6} \text{ m/s},\ s_{\text{impact}} = 0$

At stop: $t = t_{\text{stop}},\ v = 0,\ s = 1.4 \text{ m}$

$a = -250$

$v = -250t + v_{\text{impact}} = -250t + 26.\overline{6}$

$s = -125t^2 + 26.\overline{6}t + s_{\text{impact}},\ s_{\text{impact}} = 0$

$s = -125t^2 + 26.\overline{6}$

$v = 0 = -250t_{\text{stop}} + 26.\overline{6},\ t_{\text{stop}} = 0.10\overline{6}$

$s = -125(0.10\overline{6})^2 + 26.\overline{6} = 1.4 \text{ m}$

17. $v = \int -32\, dt = -32t + C;\ v = v_0;\ t = 0,$

$C = v_0;\ v = -32t + v_0;$

$s = \int (-32t + v_0)\, dt = -16t^2 + v_0 t + C_1$

$s = 0,\ t = 0;\ C_1 = 0;\ s = -16t^2 + v_0;$

$s = 90 \text{ ft when } v = 0$

$90 = 16t^2 + v_0 t;\ 0 = -32t + v_0;\ t = \dfrac{v_0}{32};$

$90 = -16\left(\dfrac{v_0}{32}\right)^2 + v_0\left(\dfrac{v_0}{32}\right);\ \dfrac{v_0^2}{64} = 90$

$v_0 = \sqrt{64(90)} = 76 \text{ ft/s}$

21. $q = \int i\, dt = \int 0.230 \times 10^{-6}\, dt = 0.230 \times 10^{-6} t + C;$

$q = 0,\ t = 0;\ C = 0;\ q = 0.230 \times 10^{-6} t$

Find q for $t = 1.50 \times 10^{-3}\, s$:

$q = 0.230 \times 10^{-6}(1.50 \times 10^{-3})$

$= 0.345 \times 10^{-9} = 0.345 \text{ nC}$

25. $V_c = \dfrac{1}{C} \int i\, dt = \dfrac{1}{2.5 \times 10^{-6}} \int 0.025\, dt$

$= \dfrac{1}{2.5 \times 10^{-6}}(0.025)t = 1.0 \times 10^4 t + C$

$v_c = 0,\ t = 0;\ C = 0;\ v_c = 1.0 \times 10^4 t$

Find v_c for $t = 0.012 \text{ s}:\ = 1.0 \times 10^4 t(0.012)$

$= 120 \text{ V}$

29. $\omega = \dfrac{d\theta}{dt} = 16t + 0.5t^2$; $d\theta = \left(6t + 0.50t^2\right)dt$

$\theta = 8t^2 + \dfrac{0.50t^3}{3} + C$; $\theta = 0$, $t = 0$; $C = 0$;

$\theta = 8t^2 + \dfrac{0.50t^3}{3}$; find θ for $t = 10.0$ s

$\theta = 8(10.0)^2 + \dfrac{0.50}{3}(10.0)^3 = 970$ rad

33. $\dfrac{dV}{dx} = \dfrac{-k}{x^2}$; $V = -\displaystyle\int \dfrac{k}{x^2}dx = kx^{-1} + C = \dfrac{k}{x} + C$

$\lim_{V \to 0} V = \lim_{x \to \infty} \dfrac{k}{x} + C$; $0 = 0 + C$; $C = 0$; therefore,

$V\big|_{x = x_1} = \dfrac{k}{x_1}$

26.2   Areas by Integration

1. $A = \displaystyle\int_1^3 y\,dx = \int_1^3 x^2\,dx$

$= \dfrac{x^3}{3}\bigg|_1^3$

$A = \dfrac{(3)^3}{3} - \dfrac{(1)^3}{3}$

$A = \dfrac{26}{3}$

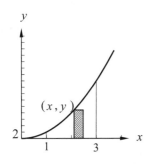

5. $y = 8 - 2x^2$, $y = 0$

$A = \displaystyle\int_{-2}^{2} \left(8 - 2x^2\right)dx = \dfrac{64}{3}$

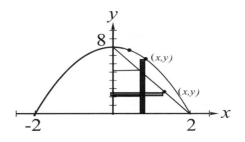

$A = \displaystyle\int_0^3 x\,dy = \int_0^3 \left(-\dfrac{1}{4}y + \dfrac{3}{2}\right)dy$

$= -\dfrac{1}{8}y^2 + \dfrac{3}{2}y\bigg|_0^3$

$= -\dfrac{1}{8}(3)^2 + \dfrac{3}{2}(3) + \dfrac{1}{8}(0)^2 - \dfrac{3}{2}(0)$

$= -\dfrac{9}{8} + \dfrac{9}{2} = -\dfrac{9}{8} + \dfrac{36}{8} = \dfrac{27}{8}$

9. $y = 3x^{-2}$; $y = 0$, $x = 2$, $x = 3$

$A = \displaystyle\int_2^3 3x^{-2}\,dy = -3x^{-1}\bigg|_2^3 = -\dfrac{3}{x}\bigg|_2^3 = \dfrac{1}{2}$

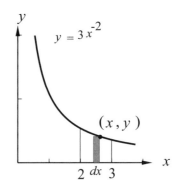

13. $y = \dfrac{2}{\sqrt{x}}$; $x = 0$, $y = 1$, $y = 4$

$\sqrt{x} = \dfrac{2}{y}$; $x = \dfrac{4}{y^2}$

$A = \displaystyle\int_1^4 x\,dy;\ A = 4\int_1^4 y^{-2}\,dy = -4\,y^{-1}\Big|_1^4$

$= -\dfrac{4}{y}\Big|_1^4 = -\dfrac{4}{4} - \left(-\dfrac{4}{1}\right)$

$= -1 + 4 = 3$

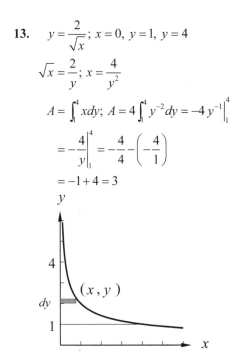

17. $A = \displaystyle\int_0^{16}\left(0 - \left(x - 4\sqrt{x}\right)\right)dx$

$A = -\dfrac{x^2}{2} + \dfrac{4x^{3/2}}{\frac{3}{2}}\Big|_0^{16}$

$A = \dfrac{128}{3}$

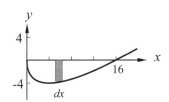

21. $y = x^4 - 8x^2 + 16$, $y = 16 - x^4$

$A = 2\displaystyle\int_0^2\left(16 - x^4 - \left(x^4 - 8x^2 + 16\right)\right)dx$

$A = 2\displaystyle\int_0^2\left(8x^2 - 2x^4\right)dx = 2\left(\dfrac{8}{3}x^3 - \dfrac{2}{5}x^5\right)\Big|_0^2 = \dfrac{256}{15}$

$\displaystyle\int_{-2}^2 = 25_0^2$ because of symmetry with respect to y-axis.

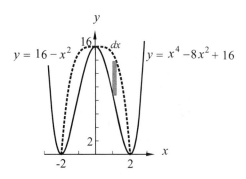

$y = 16 - x^2$ $y = x^4 - 8x^2 + 16$

25. $y = \dfrac{1}{2}x^5$; $x = -1$, $x = 2$, $y = 0$

$A_T = A_{(-1 \to 0) + A_{(0 \to 2)}}$

$= \text{negative} + \text{positive}$

$= -\left(A\big|_{-1}^0\right) + A\big|_0^2$

$A = -\displaystyle\int_{-1}^0 y\,dx + \int_0^2 y\,dx$

$= -\displaystyle\int_{-1}^0 \dfrac{x^5}{2}\,dx + \int_0^2 \dfrac{x^5}{2}\,dx = -\dfrac{x^6}{12}\Big|_{-1}^0 + \dfrac{x^6}{12}\Big|_0^2$

$= 0 - \left(-\dfrac{1}{12}\right) + \dfrac{64}{12} - 0 = \dfrac{65}{12}$

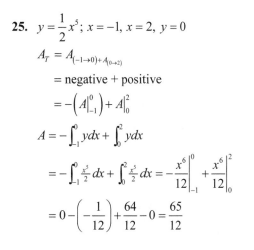

29. $A = 2\displaystyle\int_1^2\left(4x - x^2 - \left(4 - x^2\right)\right)dx$

$= 2\displaystyle\int_1^2\left(4x - 4\right)dx = 2\left(2x^2 - 4x\right)\Big|_1^2 = 4$

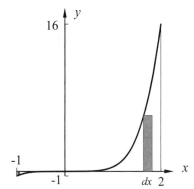

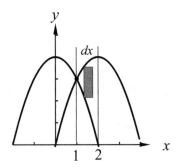

33.

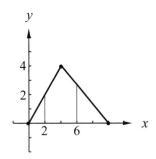

$$A = \int_0^4 x\,dx + \int_4^{10}\left(-\frac{2}{3}(x-4)+4\right)dx = \int_4^{10}\left(-\frac{2}{3}x+\frac{20}{3}\right)dx$$

$$A = \frac{x^2}{2}\Big|_0^4 + \left(-\frac{x^2}{3}+\frac{20x}{3}\right)\Big|_4^{10}$$

$$= \frac{4^2}{2}+\left(-\frac{10^2}{3}+\frac{20(10)}{3}+\frac{4^2}{3}-\frac{20(4)}{3}\right)$$

$$A = 20$$

37.

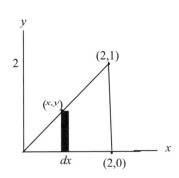

$$A = 2\int_0^2 y\,dx = 2\int_0^2 \frac{1}{2}x\,dx = \frac{1}{2}\Big|_0^2 = 2$$

41. $y = 8x;\ x = 0,\ y = 4$

(a) Using horizontal elements,

$$A = \int_0^4 x\,dy = \frac{1}{8}\int_0^4 y\,dy = \frac{1}{8}\frac{y^2}{2}\Big|_0^4$$

$$= 16\,y^2\Big|_0^4 = 1-0 = 1$$

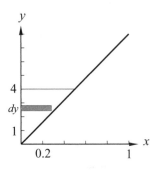

(b) Using vertical elements,

$$A = \int_0^{1/2}(4-8x)\,dx = 4x-4x^2\Big|_0^{1/2} = 1$$

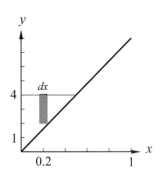

45. $dw = p\,dt;\ p = 12t - 4t^2$

$$w = \int_0^3 \left(12t-4t^2\right)dt = 6t^2 - \frac{4}{3}t^3\Big|_0^3$$

$$= 6(3)^2 - \frac{4}{3}(3)^2 - 0 = 54-36 = 18.0\text{ J}$$

49. $y = x^3 - 2x^2 - x + 2$ and $y = x^2 - 1$

Find the points of intersection:

$$x^3 - 2x^2 - x + 2 = x^2 - 1$$

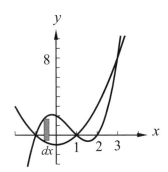

$x^3 - 3x^2 - x + 3 = 0; \ (x^2 - 1)(x - 3) = 0$

$x^2 - 1; \ x = 3; \ x = \pm 1$

$A = \int_{-1}^{1} \left[(x^3 - 2x^2 - x + 2) - (x^2 - 1) \right] dx$

$= \int_{-1}^{1} (x^3 - 3x^2 - x + 3) dx$

$= \frac{1}{4}x^4 - x^3 - \frac{1}{2}x^2 + 3x \Big|_{-1}^{1}$

$= \left(\frac{1}{4} - 1 - \frac{1}{2} + 3 \right) - \left(\frac{1}{4} + 1 - \frac{1}{2} - 3 \right)$

$= 4 \text{ cm}^2$

26.3 Volumes by Integration

1. $y = x^3, x = 2, y = 0$ about x-axis.

$$V = \int_0^2 \pi y^2 dx = \pi \int_0^2 (x^3)^2 dx$$

$$V = \pi \frac{x^7}{7} \Big|_0^2 = \frac{128\pi}{7}$$

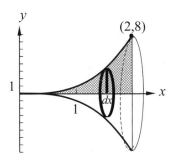

5. $V = \int_0^2 2\pi x(4 - 2x)dx = \frac{16\pi}{3}$

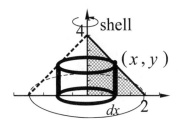

9. $y = 3\sqrt{x}, y = 0, x = 4$

Disk: $dV = \pi y^2 dx$

$$V = \pi \int_0^4 y^2 dx = \int_0^4 9x\, dx$$

$$= \pi \left(\frac{9}{2}x^2 \right) \Big|_0^4$$

$$= \pi \left[\frac{9}{2}(4)^2 - 0 \right] = 72\pi$$

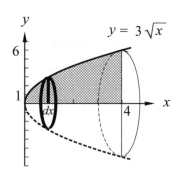

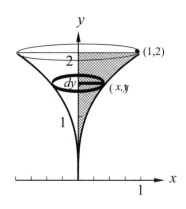

13. $y = x^2 + 1, x = 0, , x = 3, y = 0$
Disk: $dV = \pi y^2 dx$

$$V = \pi \int_0^3 (x^2 + 1)^2 dx$$

$$= \pi \int_0^3 (x^4 + 2x^2 + 1) dx$$

$$= \pi \left(\frac{1}{5}x^5 + \frac{2}{3}x^3 + x \right) \Big|_0^3$$

$$= \pi \left[\frac{1}{5}(3)^5 + \frac{2}{3}(3)^3 + 3 - 0 \right]$$

$$= \frac{348}{5}\pi$$

21. $x^2 - 4y^2 = 4, x = 3, h = 2y$
Shell: $dV = 2\pi x(2y) dx$

$$V = 4\pi \int_2^3 x \sqrt{\frac{x^2 - 4}{4}} dx$$

$$= \frac{2\pi}{2} \int_2^3 (x^2 - 4)^{1/2} 2x\, dx$$

$u = x^2 - 4, du = 2x\, dx$

$$V = \pi \cdot \frac{2}{3}(x^2 - 4)^{3/2} \Big|_2^3 = \frac{2\pi}{3}(5^{3/2}) - 0$$

$$= \frac{10\sqrt{5}}{3}\pi$$

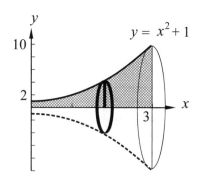

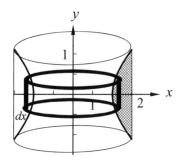

17. $y = 2x^{1/3}, \ x = 0, \ y = 2$

$$\frac{y^3}{8} = x \Rightarrow \frac{y^6}{64} = x^2$$

$$V = \pi \int_0^2 \frac{y^6}{64} = \frac{2\pi}{7}$$

25. $y = \sqrt{4 - x^2}$, Quad I
Shell: $dV = 2\pi xy\, dx$

$$V = 2\pi \int_0^2 x\sqrt{4 - x^2}\, dx$$

$$u = 4 - x^2, du = -2x\, dx$$

$$V = -\pi \int_0^2 (4 - x^2)^{1/2}(-2x\, dx)$$

$$= -\pi \frac{2}{3}(4 - x^2)^{3/2}\Big|_0^2$$

$$= -\frac{2\pi}{3}(0 - 8) = \frac{16\pi}{3}$$

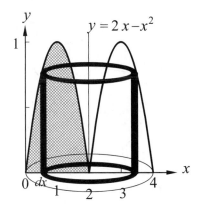

29. $y = 2x - x^2$, $y = 0$, rotated around
$x = 2$, using shells $r = 2 - x$
$h = y, t = dx$
$dV = 2\pi(2 - x)y\, dx$

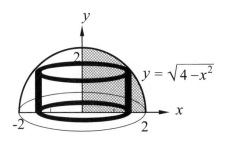

$$V = 2\pi \int_0^2 (2 - x)(2 - x^2)\, dx$$

$$= 2\pi \int_0^2 (4x - 4x^2 + x^3)\, dx$$

$$= 2\pi \left[2x^2 - \frac{4}{3}x^3 + \frac{1}{4}x^4\right]\Big|_0^2$$

$$= 2\pi \left[8 - \frac{32}{3} + 4\right] = \frac{8}{3}\pi$$

33.

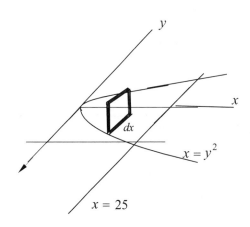

$$V = \int_0^{25} (2y)^2\, dx = \int_0^{25} 4y^2\, dx$$

$$V = \int_0^{25} 4x\, dx = \frac{4x^2}{2}\Big|_0^{25}$$

$$V = 1250 \text{ cm}^3$$

37.

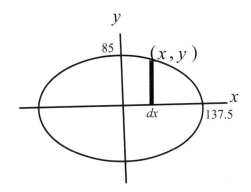

$$\frac{x^2}{a^2} + \frac{y^2}{b^2} = 1$$

$$y^2 = b^2 - \frac{b^2 x^2}{a^2} = \frac{a^2 b^2 - b^2 x^2}{a^2}$$

$$V = 2 \int_0^a \pi \cdot \frac{a^2 b^2 - b^2 x^2}{a^2} \, dx$$

$$V = \frac{2\pi}{a^2} \left(a^2 b^2 x - b^2 \frac{x^3}{3} \right)\Big|_0^a = \frac{2\pi}{a^2} \left(a^3 b^2 - \frac{a^3 b^2}{3} \right)$$

$$= \frac{2\pi}{a^2} \left(\frac{2a^3 b^2}{3} \right)$$

$$V = \frac{4\pi}{3} ab^2 = \frac{4\pi}{3} \left(\frac{275}{2} \right) \left(\frac{170}{2} \right)^2 = 4.16 \times 10^6 \text{ mm}^3$$

26.4   Centroids

1. $y = |x|$ is symmetric with respect to

y-axis $\Rightarrow \bar{x} = 0$

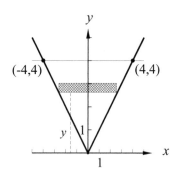

$$\bar{y} = \frac{\int_0^4 y(2x)\,dy}{\int_0^4 2x\,dy}$$

$$\bar{y} = \frac{\int_0^4 2y^2\,dy}{\int_0^4 2y\,dy} = \frac{\frac{y^3}{3}\Big|_0^4}{\frac{y^2}{2}\Big|_0^4}$$

$$= \frac{\frac{4^3}{3}}{\frac{4^2}{2}} = \frac{8}{3}$$

$$(\bar{x}, \bar{y}) = \left(0, \frac{8}{3} \right)$$

5. $M\bar{x} = m_1 x_1 + m_2 x_2 + m_3 x_3 + m_4 x_4$

$$(42 + 24 + 15 + 84)\bar{x} = 42(-3.5) + 24(0) + 15(2.6)$$
$$+ 84(3.7)$$
$$\bar{x} = 1.2 \text{ cm}$$

9. Break area into three rectangles.

First: center $(-1, -1)$; $A_1 = 4$

Second: center $(0, \frac{1}{2})$; $A_2 = 4$

Third: center $(2\frac{1}{2}, 1\frac{1}{2})$; $A_3 = 3$

Therefore,

$$4(-1) + 4(0) + 3\left(2\frac{1}{2}\right) = (4 + 4 + 3)\bar{x}$$

$$\bar{x} = 0.32$$

Therefore, $4(-1) + 4\left(\frac{1}{2}\right) + 3\left(1\frac{1}{2}\right) = 11\bar{y}$

$$\bar{y} = 0.23$$

Therefore, $(0.32 \text{ in}, 0.23 \text{ in})$ is the center of mass.

13. $y = 4 - x$, and axes

$$\bar{x} = \frac{\int_0^4 xy\,dx}{\int_0^4 y\,dx} = \frac{\int_0^4 x(4-x)\,dx}{\int_0^4 (4-x)\,dx} = \frac{\int_0^4 (4x - x^2)\,dx}{\int_0^4 (4-x)\,dx}$$

$$= \frac{\left(2x^2 - \frac{1}{3}x^3\right)\Big|_0^4}{\left(4x - \frac{1}{2}x^2\right)\Big|_0^4} = \frac{\frac{32}{3}}{8} = \frac{32}{24} = \frac{4}{3}$$

$$\bar{y} = \frac{\int_0^4 y(x)\,dy}{\int_0^4 x\,dy} = \frac{\int_0^4 y(4-y)\,dy}{\int_0^4 (4-y)\,dy} = \frac{\int_0^4 (4y - y^2)\,dy}{\int_0^4 y(4-y)\,dy}$$

$$= \frac{\left(2y^2 - \frac{1}{3}y^3\right)\Big|_0^4}{\left(4y - \frac{1}{2}y^2\right)\Big|_0^4} = \frac{4}{3}$$

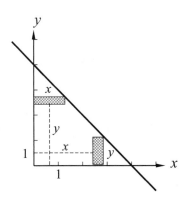

$$\left(\overline{x},\, \overline{y}\right) = \left(\frac{4}{3},\, \frac{4}{3}\right).$$

17. $A = \int_0^2 \left(3x + 2 - 2(x+1)\right) dx + \int_2^3 \left(8 - (2x+1)\right) dx$

$= 3$

$\overline{x} = \dfrac{\int_0^2 \left(3x + 2 - 2(x+1)\right) dx + \int_2^3 \left(8 - (2x+1)\right) dx}{3}$

$= \dfrac{5}{3}$

$\overline{y} = \dfrac{\int_2^8 y\left(\frac{y}{2} - 1 - \frac{y-2}{3}\right) dy}{3} = 6$

$$\left(\overline{x},\, \overline{y}\right) = \left(\frac{5}{3},\, 6\right)$$

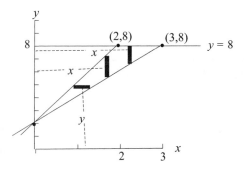

21. Curves intersect when $\dfrac{x^2}{4p} = a$

$x = \pm 2\sqrt{pa}$

$y = a$

Region is symmetric with respect to y-axis, $\overline{x} = 0$.

$\overline{y} = \dfrac{\int y(2x)\, dy}{\int 2x\, dy}$

$= \dfrac{\int_0^a 2y(2\sqrt{py})\, dy}{\int_0^a 2(2\sqrt{py})\, dy}$

$\overline{y} = \dfrac{4\sqrt{p}\int_0^a y^{3/2}\, dy}{4\sqrt{p}\int_0^a y^{1/2}\, dy}$

$= \dfrac{\frac{2}{5}y^{5/2}\Big|_0^a}{\frac{2}{3}y^{3/2}\Big|_0^a}$

$= \dfrac{3}{5}\dfrac{a^{5/2}}{a^{3/2}} = \dfrac{3}{5}a$

$$\left(\overline{x},\, \overline{y}\right) = \left(0,\, \frac{3}{5}a\right)$$

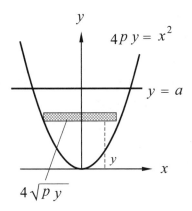

25. $y^2 = 4x,\ y = 0,\ x = 1,$

$y^2 = 4x,\ x = \dfrac{y^2}{4}$

Rotate about y-axis, $\overline{x} = 0$

$\overline{y} = \dfrac{\int_0^2 y\left(-\left(\frac{y^2}{4}\right)^2 + 1\right) dy}{\int_0^2 \left(\left(-\frac{y^2}{4}\right)^2 + 1\right) dy}$

$\overline{y} = \dfrac{\int_0^2 \left(-\frac{1}{16}y^5 + y\right) dy}{\int_0^2 \left(-\frac{1}{16}y^4 + 1\right) dy} = \dfrac{-\frac{1}{96}y^6 + \frac{1}{2}y^2\Big|_0^2}{-\frac{1}{80}y^5 + y\Big|_0^2}$

$= \dfrac{-\frac{64}{96} + 2}{-\frac{32}{80} + 2} = \dfrac{\frac{128}{96}}{\frac{128}{80}} = \dfrac{5}{6}$

$$\left(\overline{x},\, \overline{y}\right) = \left(0,\, \frac{5}{6}\right)$$

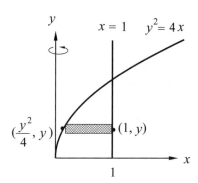

29. Triangle is area bounded by $y = \dfrac{a}{b}x$, x-axis, $x = b$

$$\bar{x} = \frac{\int_0^b xy\, dx}{\int_0^b y\, dx} = \frac{\int_0^b x\frac{a}{b}x\, dx}{\int_0^b \frac{a}{b}x\, dx}$$

$$= \frac{\frac{a}{b}\int_0^b x^2\, dx}{\frac{a}{b}\int_0^b x\, dx} = \frac{\frac{x^3}{3}\Big|_0^b}{\frac{x^2}{2}\Big|_0^b} = \frac{\frac{b^3}{3}}{\frac{b^2}{2}} = \frac{2}{3}b$$

$$\bar{y} = \frac{\int_0^b y(b-x)\, dy}{\int_0^b (b-x)\, dy} = \frac{\int_0^b y\left(b-\frac{b}{a}y\right)\, dy}{\int_0^b \left(b-\frac{b}{a}y\right)\, dy}$$

$$= \frac{\int_0^b \left(by-\frac{b}{a}y^2\right)\, dy}{\int_0^b \left(b-\frac{b}{a}y\right)\, dy} = \frac{\frac{b}{2}y^2 - \frac{b}{3a}y^3\Big|_0^a}{by - \frac{b}{2a}y^2\Big|_0^a}$$

$$= \frac{\frac{a^2 b}{2} - \frac{a^2 b}{3}}{ab - \frac{ab}{2}} = \frac{a}{3}$$

$$\left(\bar{x}, \bar{y}\right) = \left(\frac{2}{3}b, \frac{1}{3}a\right)\Bigg|_{a=4.5,\ b=3.0} = (2.0\text{ m}, 1.5\text{ m})$$

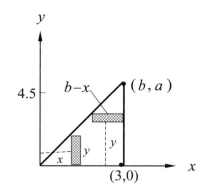

33. Bounded area: $y = -4x + 80, y = 60, y = 0,$
$x = 0, \bar{x} = 0$

$y = -4x + 80;\ 4x = 80 - y;\ x = 20 - \dfrac{1}{4}y$

$$\bar{y} = \frac{\int_0^{60} yx^2\, dy}{\int_0^{60} x^2\, dy} = \frac{\int_0^{60} y\left(20 - \frac{1}{4}y\right)^2 dy}{\int_0^{60} \left(20 - \frac{1}{4}y\right)^2 dy}$$

$$\bar{y} = \frac{\int_0^{60} \left(400y - 10y^2 + \frac{1}{16}y^3\right) dy}{\int_0^{60} \left(400 - 10y + \frac{1}{16}y^2\right) dy}$$

$$= \frac{200y^2 - \frac{10}{3}y^3 + \frac{1}{64}y^4\Big|_0^{60}}{400y - 5y^2 + \frac{1}{48}y^3\Big|_0^{60}}$$

$$= \frac{720\,000 - 720\,000 + 202\,500}{24\,000 - 18\,000 + 4\,500}$$

$= 19.3$ cm from larger base

$$\left(\bar{x}, \bar{y}\right) = (0, 19.3)$$

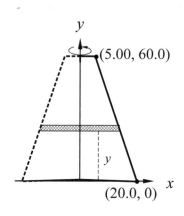

26.5   Moments of Inertia

1. $I_y = k\displaystyle\int_0^1 x^2 y\, dx$

$I_y = k\displaystyle\int_0^1 x^2 (4x)\, dx$

$I_y = 4k\dfrac{x^4}{4}\Big|_0^1$

$I_y = k$

$m = k\displaystyle\int_0^1 y\, dx = k\displaystyle\int_0^1 4x\, dx$

$m = 4k\dfrac{x^2}{2}\Big|_0^1 = 2k$

$$R_y^2 = \frac{I_y}{m} = \frac{k}{2k}$$

$$R_y = \frac{\sqrt{2}}{2}$$

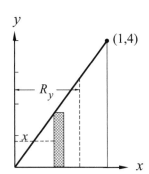

5. $I = m_1 x_1^2 + m_2 x_2^2 + m_3 x_3^2$
$I = 45.0(-3.80)^2 + 90.0(0.00)^2 + 62.0(5.50)^2$
$I = 2530 \text{ g} \cdot \text{cm}^2$

$I = MR^2$
$2530 = (45.0 + 9.0 + 62.0)R^2$
$R = 3.58 \text{ cm}$

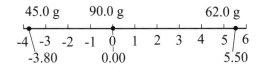

9. $y^2 = x$, $x = 9$, x-axis, with respect to the x-axis

$$I_x = k \int_0^3 y^2 \left(9 - y^2 \right) dy = k \int_0^3 \left(9y^2 - y^4 \right) dy$$

$$= k \left(3y^3 - \frac{1}{5} y^5 \right) \Big|_0^3 = k \left(81 - \frac{243}{5} \right) = \frac{162}{5} k$$

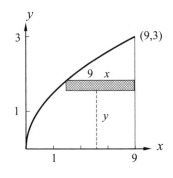

13. $y = x^2$, $x = 3$, x-axis, with respect to the x-axis

$$I_x = k \int_0^9 y^2 \left(3 - \sqrt{y} \right) dy$$

$$= k \int_0^9 \left(3y^2 - y^{5/2} \right) dy = k \left(y^3 - \frac{2}{7} y^{7/2} \right) \Big|_0^9$$

$$= \frac{729}{7}$$

$$m = k \int_0^9 \left(3 - \sqrt{y} \right) dy$$

$$= k \left(3y - \frac{2}{3} y^{3/2} \right) \Big|_0^9 = 9k$$

$$R_x^2 = \frac{\frac{729}{7} k}{9k}; \quad R_x = \frac{9}{\sqrt{7}}$$

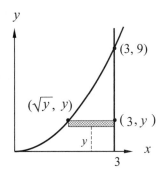

17. $y^2 = 4x$, $y = 2$, y-axis, rotated about x-axis

$$\frac{y^2}{4} = x$$

$$I_x = 2\pi k \int_0^2 xy^3 dy = 2\pi k \int_0^2 \frac{y^2}{4} y^3 dy = \frac{2\pi k}{4} \int_0^2 y^5 dy$$

$$= \frac{\pi k}{2} \frac{y^6}{6} \Big|_0^2 = \frac{16\pi k}{3}$$

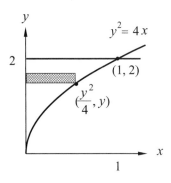

21. $I_y = \dfrac{m}{\frac{1}{2}(3.0)(4.5)} \displaystyle\int_0^3 x^2\left(-1.5x+4.5\right)dx$

$I_y = 1.5m$

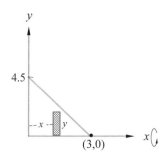

25. $r = 0.600$ cm, $h = 0.800$ cm, $m = 3.00$ g

$y = \dfrac{0.600}{0.800}x = 0.750x; \; x = 1.333y$

$I_x = 2\pi k \displaystyle\int_0^{0.600}(0.800-1.333y)y^3\,dy$

$\quad = 2\pi k(0.200y^4 - 0.2667y^5)\Big|_0^{0.600}$

$\quad = 2\pi k(0.005\,181)$

$m = \dfrac{k}{3}\pi r^2 h, \; 2\pi k = \dfrac{6m}{r^2 h} = \dfrac{6(3.00)}{(0.600^2)(0.800)}$

$\quad = 62.5$ g/cm^3

$I_x = (62.5)(0.005\,181) = 0.324$ g · cm^2

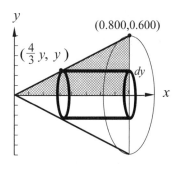

26.6   Other Applications

1. $f(x) = kx$
$\quad 6.0 = k(2.0)$
$\quad\quad k = 3.0$ lb/in.

$W = \displaystyle\int_{3.0}^{6.0} 3.0x\,dx - 1.5x^2\Big|_{3.0}^{6.0}$

$W = 41$ lb · in.

5. $f(x) = kx; \; 6.0 = k(1.5); \; k = 4.0$ lb/in

$w = \displaystyle\int_0^{2.0} 4.0x\,dx = 2.0x^2\Big|_0^{2.0} = 2.0(2.0^2) - 0$

$\quad = 8.0$ lb · in

9. $f(x) = \dfrac{kq_1 q_2}{x^2}, \; 1.0$ pm $= 1.0 \times 10^{-12}$,

4.0 pm $= 4.0 \times 10^{-12}$

$W = \displaystyle\int_{1.0\times10^{-12}}^{4.0\times10^{-12}} \dfrac{9.0 \times 10^9 (1.6 \times 10^{-19})^2}{x^2}dx$

$\quad = 9.0 \times 10^9 (1.6 \times 10^{-19})^2 \left(-\dfrac{1}{x}\right)\Bigg|_{1.0\times10^{-12}}^{4.0\times10^{-12}}$

$\quad = -23.04 \times 10^{-29}[0.25 \times 10^{12} - 10^{12}]$
$\quad = -23.04 \times 10^{-29}(-0.75 \times 10^{12})$
$\quad = 1.7 \times 10^{-16}$ J

13. $W = 1500(24) + 2\displaystyle\int_0^{24} 12(24-x)\,dx$

$W = 36,000 + 24\left(24x - \dfrac{x^2}{2}\right)\Bigg|_0^{24}$

$W = 36,000 + 6912$
$W = 43,000$ lb · ft

17.

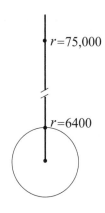

$$F = \frac{Gm_1m_2}{r^2}$$

$$160 = \frac{Gm_1m_2}{6400^2} \Rightarrow Gm_1m_2 = 160\left(6400^2\right)$$

$$W = \int_{75,000}^{6400} -\frac{Gm_1m_2}{r^2}\, dr = \frac{Gm_1m_2}{r}\Bigg|_{75,000}^{6400}$$

$$W = 160\left(6400^2\right)\left(\frac{1}{6400} - \frac{1}{75,000}\right)$$

$$W = 940,000 \text{ N}\cdot\text{km}$$

21. $W = \displaystyle\int_0^{0.100} 9800\pi\,(0.050)^2\,dy\,(0.150 - y)$

$\qquad = 0.770 \text{ N}\cdot\text{m}$

25. $F = w\displaystyle\int_a^b lh\,dh = 64.0\int_{4.00}^{9.00} 10.0h\,dh$

$\qquad = 64.0(10.0)\dfrac{h^2}{2}\Bigg|_{4.00}^{9.00}$

$\qquad = 640(40.5 - 8.0) = 20,800 \text{ lb}$

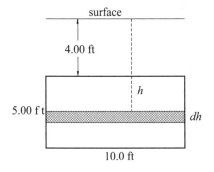

29. $y = x^2, y = 20$

$$F = 62.4\int_0^4 2xy\,dy$$

$$= 62.4(2)\int_0^4 \sqrt{y}(20 - y)dy$$

$$= 62.4(2)\int_0^4 (20\sqrt{y} - y\sqrt{y})dy$$

$$= 124.8\left[20\cdot\frac{2}{3}y^{3/2} - \frac{2}{5}y^{5/2}\right]\Bigg|_0^4$$

$$= 124.8\left[\frac{40}{3}(8) - \frac{2}{5}(32)\right]$$

$$= 11,700 \text{ lb}$$

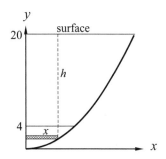

33. $i = 0.4t - 0.1t^2$. Find i_{av} with respect to time for $t = 0$ to $t = 4$.

$$i_{av} = \frac{\displaystyle\int_0^{4.0} i\,dt}{4.0 - 0} = \frac{\displaystyle\int_0^{4.0}\left(0.4t - 0.1t^2\right)}{4.0} = 0.27\,\mu\text{A}$$

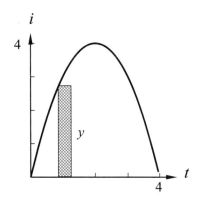

37. $s = \int_a^b \sqrt{1 + \left(\dfrac{dy}{dx}\right)^2}\, dx;$

$y = 0.4x^{3/2};$

$\dfrac{dy}{dx} = 0.06x^{1/2}$

$s = \int_0^{100} \sqrt{1 + (0.06x^{1/2})^2}\, dx$

$\quad = \int_0^{100} \sqrt{(1 + 0.0036x)}\, dx$

$\quad = \int_0^{100} (1 + 0.0036x)^{1/2}\, dx$

$= \dfrac{1}{0.0036} \int_0^{100} (1 + 0.0036x)^{1/2}(0.0036)(dx)$

$= \dfrac{1}{0.0036} \left[\dfrac{2}{3}(1 + 0.0036x)^{3/2}\right]\Big|_0^{100}$

$= \dfrac{1}{0.0036} \left[\dfrac{2}{3}(1.586) - \dfrac{2}{3}(1)^{3/2}\right]$

$= \dfrac{1}{0.0036} \left[\dfrac{2}{3}(0.586)\right]$

$= \dfrac{1}{0.0036}[0.391] = 109 \text{ ft}$

Chapter 26   Review Exercises

1. $t = \dfrac{s}{v} = \dfrac{17.1}{45.0}$

$\text{drop} = \dfrac{1}{2}(9.8)\left(\dfrac{17.1}{45.0}\right)^2 = 0.708 \text{ m}$

5. $a = -0.750$

$v = -0.750t + 2.50 = 0 \Rightarrow t = \dfrac{2.50}{0.750}$

$s = -\dfrac{0.750}{2}t^2 + 2.50t\Big|_{t=\frac{2.50}{0.750}} = 4.17 < 4.20$

The ball does not make it to the hole.

9. $i = 0.25(2\sqrt{t} - t);\ q = \int i\, dt$

$q = \int_0^2 0.25(2\sqrt{t} - t)\, dt$

$\quad = \int_0^2 0.25(2t^{1/2} - t)\, dt$

$\quad = \int_0^2 (0.50t^{1/2} - 0.25t)\, dt$

$\quad = \left(\dfrac{1}{3}t^{3/2} - \dfrac{1}{8}t^2\right)\Big|_0^2$

$\quad = \dfrac{\sqrt{8}}{3} - \dfrac{1}{2} = 0.44 \text{ C}$

13. $dy = \left(20 + \dfrac{1}{40}x^2\right) dx;$

$y = \int \left(20 + \dfrac{1}{40}x^2\right) dx = 20x + \dfrac{1}{120}x^3 + C$

$y = 0 \text{ when } x = 0$

$20(0) + \dfrac{1}{120}(0)^3 + C = 0;\ C = 0$

$y = 20x + \dfrac{1}{120}x^3$

17. $y^2 = 2x, y = x - 4$

$x_1 = \frac{1}{2}y^2$; $x_2 = y + 4$;

$A = \int_{-2}^{4} (x_2 - x_1)\,dy$

$= \int_{-2}^{4} \left(y + 4 - \frac{1}{2}y^2 \right) dy$

$= -\frac{1}{6}y^3 + \frac{1}{2}y^2 + 4y \Big|_{-2}^{4}$

$= -\frac{1}{6}(64) + \frac{1}{2}(16) + 4(4) + \frac{1}{6}(-8) - \frac{1}{2}(4) - 4(-2)$

$= 18$

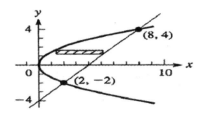

21. $\dfrac{A_I}{A_{II}} = \dfrac{\int_0^1 \left(1 - x^{2n}\right) dx}{\int_0^1 x^{2n}\,dx} = \dfrac{x - \frac{x^{2n+1}}{2n+1}\Big|_0^1}{\dfrac{x^{2n+1}}{2n+1}\Big|_0^1}$

$= \dfrac{1 - \frac{1}{2n+1}}{\frac{1}{2n+1}} = \dfrac{2n+1-1}{1}$

$\dfrac{A_I}{A_{II}} = \dfrac{2n}{1}$

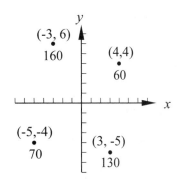

25. $y = x^3 - 4x^2$ has $(0,0)$ and $(4,0)$ as intercepts. Using shells,

$V = \int_0^4 2\pi(-y\,dy) = \int_0^4 -2\pi x(x^3 - 4x^2)\,dx$

$V = -2\pi \int_0^4 (x^4 - 4x^3)\,dx = -2\pi \left(\frac{x^5}{5} - x^4 \right)\Big|_0^4$

$V = -2\pi \left(\frac{4^5}{5} - 4^4 \right) = \frac{512\pi}{5}$

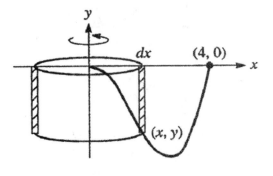

29.

$\bar{x} = \dfrac{70(-5) + 160(-3) + 60(4) + 130(3)}{70 + 160 + 60 + 130} = -\dfrac{10}{21}$

$\bar{y} = \dfrac{70(-4) + 160(6) + 60(4) + 130(-5)}{420} = \dfrac{9}{14}$

$(\bar{x}, \bar{y}) = (-0.5, 0.6)$

33. $y = \sqrt{x}$, $x = 1$, $x = 4$, $y = 0$

Volume symmetric to x-axis: $\overline{y} = 0$

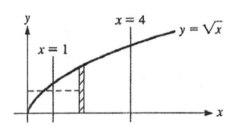

$$\overline{x} = \frac{\int_1^4 x\,y^2\,dx}{\int_1^4 y^2\,dx} = \frac{\int_1^4 x(x)\,dx}{\int_1^4 x\,dx}$$

$$= \frac{\int_1^4 x^2\,dx}{\int_1^4 x\,dx}$$

$$\overline{x} = \frac{\frac{x^3}{3}\Big|_1^4}{\frac{x^2}{2}\Big|_1^4} = \frac{\frac{4^3}{3} - \frac{1^3}{3}}{\frac{4^2}{2} - \frac{1^2}{2}} = \frac{\frac{63}{3}}{\frac{15}{2}} = \frac{14}{5}$$

$$(\overline{x},\, \overline{y}) = \left(\frac{14}{5},\, 0\right)$$

37. $I_x = 2\pi k \int_c^d (x_2 - x_1)y^3\,dy$

$$I_x = 2\pi(0.0114) \int_0^{3.00(20.0)^{0.10}}$$

$$\left(20.0 - \left(\frac{y}{3.00}\right)^{10}\right) y^3\,dy$$

$$I_x = 68.7 \text{ g} \cdot \text{mm}^2$$

41.

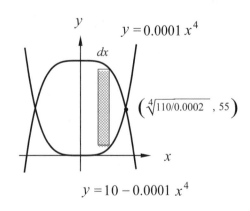

$y = 0.0001\,x^4$

dx

$\left(\sqrt[4]{110/0.0002}\;,\, 55\right)$

x

$y = 10 - 0.0001\,x^4$

$$A = 2\int_0^{\sqrt[4]{\frac{110}{0.0002}}} \left(110 - 0.0001x^4 - 0.0001x^4\right)dx$$

$$A = 220x - 0.004\frac{x^5}{5}\Big|_0^{\sqrt[4]{\frac{110}{0.0002}}} = 4790 \text{ cm}^2$$

45. Disks: $V = \int_0^4 \pi y^2\,dx = \int_0^4 \pi x^3\,dx = 200 \text{ cm}^2$

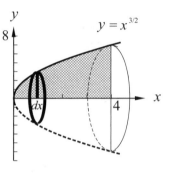

49. Using disks and $\dfrac{x^2}{1.5^2} + \dfrac{y^2}{10^2} = 1$ as equation of

ellipse.

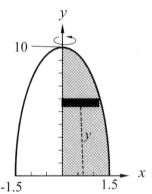

$$V = \int_0^{10} \pi x^2\,dy$$

$$V = \pi \int_0^{10} 1.5^2\left(1 - \frac{y^2}{10^2}\right)dy$$

$$V = \pi(1.5)^2\left(y - \frac{1}{3}\frac{y^3}{10^2}\right)\Big|_0^{10}$$

$$V = \pi(1.5)^2\left(10 - \frac{1}{3}\frac{10^3}{10^2}\right) = 47 \text{ m}^3$$

53. The circumference of the bottom, $c = 2\pi r = 9\pi$, equates to l, the length of the vertical surface area.

$$F = 68.0 \int_0^{3.25} (9\pi h)\, dh = 68.0 \left[4.50\pi h^2 \right]_0^{3.25}$$
$$= 68.0 \left[4.50\pi (3.25)^2 - 0 \right] = 10,200 \text{ lb}$$

57. $a = 5 \times 10^{14}$

$v = 5 \times 10^{14} t$

$s = 2.5 \times 10^{14} t^2 = 0.025 \Rightarrow t = 1 \times 10^{-8}$

$v = 5 \times 10^{14} \left(1 \times 10^{-8} \right) = 5 \times 10^6 \text{ m/s}$

Chapter 27

DIFFERENTIATION OF TRANSCENDENTAL FUNCTIONS

27.1 Derivatives of the Sine and Cosine Functions

1. $r = \sin^2 2\theta^2$

$$\frac{dr}{d\theta} = 2\sin 2\theta^2 \cos 2\theta^2 (2(2\theta))$$

$$\frac{dr}{d\theta} = 8\theta \sin 2\theta^2 \cos 2\theta^2$$

$$\frac{dr}{d\theta} = 4\theta \sin 4\theta^2$$

5. $y = 2\sin(2x^3 - 1)$

$$\frac{dy}{dx} = 2\cos(2x^3 - 1)(6x^2) = 12x^2 \cos(2x^3 - 1)$$

9. $y = 3x + 2\cos(3x - \pi)$

$$\frac{dy}{dx} = 3 + 2\left[-\sin(3x - \pi)(3)\right] = 3 - 6\sin(3x - \pi)$$

13. $y = 3\cos^3(5x + 2)$;

$$\frac{dy}{dx} = 3\cdot 3\cos^2(5x + 2)[-\sin(5x + 2)(5)]$$

$$\frac{dy}{dx} = -45\cos^2(5x + 2)\sin(5x + 2)$$

17. $y = 3x^3 \cos 5x$;

$$\frac{dy}{dx} = 3[x^3(-5\sin 5x) + \cos 5x(3x^2)]$$

$$\frac{dy}{dx} = 9x^2 \cos 5x - 15x^3 \sin 5x$$

21. $y = \sqrt{1 + \sin 4x} = (1 + \sin 4x)^{1/2}$

$$\frac{dy}{dx} = \frac{1}{2}(1 + \sin 4x)^{-1/2}(4\cos 4x)$$

$$\frac{dy}{dx} = \frac{2\cos 4x}{\sqrt{1 + \sin 4x}}$$

25. $y = \dfrac{2\cos x^2}{3x - 1}$

$$\frac{dy}{dx} = \frac{(3x - 1)(2)(-\sin x^2)(2x) - 2\cos x^2(3)}{(3x - 1)^2}$$

$$\frac{dy}{dx} = \frac{-4x(3x - 1)\sin x^2 - 6\cos x^2}{(3x - 1)^2}$$

$$\frac{dy}{dx} = \frac{4x(1 - 3x)\sin x^2 - 6\cos x^2}{(3x - 1)^2}$$

29. $s = \sin(3\sin 2t)$

$$\frac{ds}{dt} = \cos(3\sin 2t)(3\cos 2t(2))$$

$$\frac{ds}{dt} = 6\cos 2t \cos(3\sin 2t)$$

33. $p = \dfrac{1}{\sin s} + \dfrac{1}{\cos s}$

$$\frac{dp}{ds} = \frac{\sin s(0) - \cos s}{\sin^2 s} + \frac{\cos s(0) - (-\sin s)}{\cos^2 s}$$

$$\frac{dp}{ds} = \frac{-\cos s}{\sin^2 s} + \frac{\sin s}{\cos^2 s}$$

37. (a) Set mode to radian, and graph $y_1 = \dfrac{\sin x}{x}$.

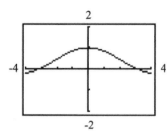

(b) Check the values, using a calculator. When $x = 0.5$ $y = 0.95885108$; when $x = 0.1$, $y = 0.99833417$; when $x = 0.05$, $y = 0.99958339$; when $x = 0.01$, $y = 0.99998333$; when $x = 0.001$, $y = 0.99999983$.

41. $y = \sin x$

$m_{\tan}\big|_{x=0} = 1.0$ $m_{\tan}\big|_{x=\pi/4} = 0.7$ $m_{\tan}\big|_{x=\pi/2} = 0.0$

$m_{\tan}\big|_{x=3\pi/4} = -0.7$ $m_{\tan}\big|_{x=\pi} = -1.0$ $m_{\tan}\big|_{x=5\pi/4} = -0.7$

$m_{\tan}\big|_{x=3\pi/2} = 0.0$ $m_{\tan}\big|_{x=7\pi/4} = 0.7$ $m_{\tan}\big|_{x=2\pi} = 1.0$

Plot points: $\left(0, 1.0\right), \left(\frac{\pi}{4}, 0.7\right), \left(\frac{\pi}{2}, 0.0\right), \left(\frac{3\pi}{4}, -0.7\right),$

$\left(\pi, -1.0\right), \left(\frac{5\pi}{4}, -0.7\right), \left(\frac{3\pi}{2}, 0.0\right), \left(\frac{7\pi}{4}, 0.7\right), \left(2\pi, 1.0\right).$

Resulting curve is $y = \cos x$.

45. $\dfrac{d}{dx}\sin x = \cos x;$

$\dfrac{d^2}{dx^2}\sin x = -\sin x$

$\dfrac{d^3}{dx^3}\sin x = -\cos x;$

$\dfrac{d^4}{dx^4}\sin x = \sin x$

49. $x = 30°,\ dx = 1° = \dfrac{\pi}{180},\ y = f(x) = \sin x$

$\Delta y = f(x + \Delta x) - f(x) \approx dy = f'(x)\,dx = \cos x\,dx$

$\sin 31° - \sin 30° \approx \cos 30°\left(\dfrac{\pi}{180}\right)$

$\sin 31° \approx \sin 30° + \cos 30°\left(\dfrac{\pi}{180}\right)$

$\sin 31° \approx 0.5151$

53. $d = 3.0\sin 188t \cos 188t$

$\dfrac{dd}{dt} = 3.0\left[\sin 188t(-\sin 188t)(188) + \cos 188t(\sin 188t)(188)\right]$

$\dfrac{dd}{dt} = 3.0(188)\left[-\sin^2 188t + \cos^2 188t\right]\big|_{t=2\times10^{-3}}$

$\dfrac{dd}{dt} = 3.0(188)\left[-\sin^2 188(2\times10^{-3}) + \cos^2 188(2\times10^{-3})\right]$

$\dfrac{dd}{dt} = 410 \text{ mm/s}$

57. $r = \dfrac{100}{1 - \cos\theta} = 100(1 - \cos\theta)^{-1}$

$\dfrac{dr}{d\theta} = -100(1 - \cos\theta)^{-2}(\sin\theta)$

$\dfrac{dr}{d\theta} = \dfrac{-100\sin\theta}{(1 - \cos\theta)^2}$

$\theta = 120° = \dfrac{2\pi}{3}$

$\dfrac{dr}{d\theta}\bigg|_{\theta=120°} = \dfrac{-100\sin\frac{2\pi}{3}}{\left(1 - \cos\frac{2\pi}{3}\right)^2} = -38.5 \text{ km}$

27.2 Derivatives of the Other Trigonometric Functions

1. $y = 3\sec^2 x^2$

$\dfrac{dy}{dx} = 3(2)(\sec x^2)\dfrac{d}{dx}(\sec x^2)$

$\dfrac{dy}{dx} = 6\sec x^2 \sec x^2 \tan x^2(2x)$

$\dfrac{dy}{dx} = 12x\sec^2 x^2 \tan x^2$

5. $y = 5\cot(0.25\pi - 2\theta)$

$\dfrac{dy}{d\theta} = -5\csc^2(0.25\pi - 2\theta)\cdot(-2)$

$= 10\csc^2(0.25\pi - 2\theta)$

9.

$y = 4x^3 - 3\csc\sqrt{2x+3}$

$\dfrac{dy}{dx} = 12x^2 - 3\left[-\csc\sqrt{2x+3}\cot\sqrt{2x+3}\cdot\dfrac{1}{2}(2x+3)^{-1/2}(2)\right]$

$\dfrac{dy}{dx} = 12x^2 + \dfrac{3\csc\sqrt{2x+3}\cot\sqrt{2x+3}}{\sqrt{2x+3}}$

13. $y = 2\cot^4\left(\dfrac{1}{2}x + \pi\right)$

$\dfrac{dy}{dx} = 2(4)\cot^3\left(\dfrac{1}{2}x + \pi\right)\left[-\csc^2\left(\dfrac{1}{2}x + \pi\right)\left(\dfrac{1}{2}\right)\right]$

$= -4\cot^3\left(\dfrac{1}{2}x + \pi\right)\csc^2\left(\dfrac{1}{2}x + \pi\right)$

17. $y = 3\csc^4\left(7x - \dfrac{\pi}{2}\right)$

$\dfrac{dy}{dx} = 3 \cdot 4\csc^3\left(7x - \dfrac{\pi}{2}\right)$

$\qquad -\csc\left(7x - \dfrac{\pi}{2}\right)\cot\left(7x - \dfrac{\pi}{2}\right)(7)$

$\qquad = -84\csc^4\left(7x - \dfrac{\pi}{2}\right)\cot\left(7x - \dfrac{\pi}{2}\right)$

21. $y = 4\cos x \csc x^2$

$\dfrac{dy}{dx} = 4[\cos x(-\csc x^2 \cot x^2 \cdot 2x) + \csc x^2(-\sin x)]$

$\dfrac{dy}{dx} = -4\csc x^2(2x\cos x \cot x^2 + \sin x)$

25. $y = \dfrac{2\cos 4x}{1 + \cot 3x}$

$\dfrac{dy}{dx} = \dfrac{(1 + \cot 3x)[-2\sin 4x(4)] - 2\cos 4x(-\csc^2 3x)(3)}{(1 + \cot 3x)^2}$

$\dfrac{dy}{dx} = \dfrac{-8\sin 4x(1 + \cot 3x) + 6\cos 4x \csc^2 3x}{(1 + \cot 3x)^2}$

$\dfrac{dy}{dx} = \dfrac{2(-4\sin 4x - 4\sin 4x \cot 3x + 3\cos 4x \csc^2 3x)}{(1 + \cot 3x)^2}$

29. $r = \tan(\sin 2\pi\theta)$

$\dfrac{dr}{d\theta} = \sec^2(\sin 2\pi\theta)\cos(2\pi\theta)(2\pi)$

$\qquad = 2\pi\cos 2\pi\theta \sec^2(\sin 2\pi\theta)$

33. $x\sec y - 2y = \sin 2x$

$x\sec y \tan y \dfrac{dy}{dx} + \sec y - 2\dfrac{dy}{dx} = 2\cos 2x$

$x\sec y \tan y \dfrac{dy}{dx} - 2\dfrac{dy}{dx} = 2\cos 2x - \sec y$

37. $y = \tan 4x \sec 4x$

$\dfrac{dy}{dx} = \tan 4x \cdot 4\sec 4x \tan 4x + \sec 4x \cdot 4\sec^2 4x$

$dy = 4\sec 4x(\tan^2 4x + \sec^2 4x)dx$

41. (a)

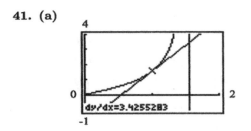

(b)

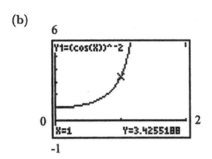

The values are the same in the first four decimal places.

45. $y = 2\cot 3x;\; x = \dfrac{\pi}{12};$

$\dfrac{dy}{dx} = 2(-\csc^2 3x)(3) = -6\csc^2 3x;$

$\dfrac{dy}{dx}\bigg|_{x=\pi/12} = -6\csc^2\dfrac{\pi}{4} = -6(\sqrt{2})^2 = -12$

49. $x = d(\sec kL - 1)$

$x' = kd\sec(kL)\tan(kL)$

53. $\theta = \dfrac{3t}{(2t + 10)}$

$h = 1000\tan\dfrac{3t}{2t + 10};\; t = 5.0\,\text{s}$

$\dfrac{dh}{dt} = 1000\left[\sec^2\left(\dfrac{3t}{2t + 10}\right)\right]\left(\dfrac{(2t + 10)3 - 3t(2)}{(2t + 10)^2}\right)$

$\dfrac{dh}{dt}\bigg|_{t=5.0} = \left(\sec^2\dfrac{15}{20}\right)\left(\dfrac{30{,}000}{400}\right) = 140\,\text{ft/s}$

27.3 Derivatives of the Inverse Trigonometric Functions

1. $y = \sin^{-1} x^2$

$$\frac{dy}{dx} = \frac{1}{\sqrt{1-(x^2)^2}}(2x) = \frac{2x}{\sqrt{1-x^4}}$$

5. $y = 2\sin^{-1} 3x^3$

$$\frac{dy}{dx} = 2\frac{1}{\sqrt{1-9x^6}}(9x^2) = \frac{18x^2}{\sqrt{1-9x^6}}$$

9. $y = 6\cos^{-1}\sqrt{2-x}$

$$\frac{dy}{dx} = 6 \cdot \frac{-1}{\sqrt{1-(2-x)}} \cdot \frac{1}{2}(2-x)^{-1/2}(-1)$$

$$\frac{dy}{dx} = \frac{3}{\sqrt{x-1}\sqrt{2-x}} = \frac{3}{\sqrt{(x-1)(2-x)}}$$

13. $y = 6\tan^{-1}\left(\frac{1}{x}\right)$

$$\frac{dy}{dx} = 6x\frac{1}{1+\frac{1}{x^2}}\left(-\frac{1}{x^2}\right) + 6\tan^{-1}\frac{1}{x}$$

$$\frac{dy}{dx} = \frac{-6x}{x^2+1} + 6\tan^{-1}\frac{1}{x}$$

17. $y = 0.4u\tan^{-1} 2u$

$$\frac{dy}{du} = 0.4u \cdot \frac{2}{1+4u^2} + 0.4\tan^{-1} 2u$$

$$\frac{dy}{du} = \frac{0.8u}{1+4u^2} + 0.4\tan^{-1} 2u$$

21. $y = \dfrac{\sin^{-1} 2x}{\cos^{-1} 2x}$

$$\frac{dy}{dx} = \frac{\cos^{-1} 2x\frac{1}{\sqrt{1-4x^2}}(2) - \sin^{-1} 2x\frac{1}{\sqrt{1-4x^2}}(2)}{(\cos^{-1} 2x)^2}$$

$$\frac{dy}{dx} = \frac{2}{\sqrt{1-4x^2}}\left[\frac{\cos^{-1} 2x + \sin^{-1} 2x}{(\cos^{-1} 2x)^2}\right]$$

$$\frac{dy}{dx} = \frac{2(\cos^{-1} 2x + \sin^{-1} 2x)}{\sqrt{1-4x^2}(\cos^{-1} 2x)^2}$$

25. $u = \left[\sin^{-1}(4t+3)\right]^2$

$$\frac{du}{dt} = 2\left[\sin^{-1}(4t+3)\right]\frac{4}{\sqrt{1-(4t+3)^2}}$$

$$= \frac{2\sqrt{2}\sin^{-1}(4t+3)}{\sqrt{-2t^2-3t-1}}$$

29. $y = \dfrac{1}{1+4x^2} - \tan^{-1} 2x$

$$= (1+4x^2)^{-1} - \tan^{-1} 2x$$

$$\frac{dy}{dx} = -(1+4x^2)^{-2}(8x) - \frac{1}{1+4x^2}(2)$$

$$\frac{dy}{dx} = \frac{-8x}{(1+4x^2)^2} - \frac{2}{1+4x^2} = \frac{-8x-2(1+4x^2)}{(1+4x^2)^2}$$

$$\frac{dy}{dx} = \frac{-8x-2-8x^2}{(1+4x^2)^2} = \frac{-2(1+4x+4x^2)}{(1+4x^2)^2}$$

$$= \frac{-2(1+2x)^2}{(1+4x^2)^2}$$

33. $2\tan^{-1} xy + x = 3$

$$2\frac{1}{1+x^2y^2}\left(x\frac{dy}{dx}+y\right)+1=0$$

$$\frac{2x}{1+x^2y^2}\frac{dy}{dx} + \frac{2y}{1+x^2y^2} = -1$$

$$\frac{2x}{1+x^2y^2}\frac{dy}{dx} = -1 - \frac{2y}{1+x^2y^2}$$

$$\frac{2x}{1+x^2y^2}\frac{dy}{dx} = \frac{-1-x^2y^2-2y}{1+x^2y^2}$$

$$2x\frac{dy}{dx} = -1 - x^2y^2 - 2y; \quad \frac{dy}{dx} = \frac{-(x^2y^2+2y+1)}{2x}$$

37. $y = (\sin^{-1} x)^3$

$$\frac{dy}{dx} = 3(\sin^{-1} x)^2 \frac{1}{\sqrt{1 - x^2}}$$

$$dy = \frac{3(\sin^{-1} x)^2 dx}{\sqrt{1 - x^2}}$$

41. $y = x \tan^{-1} x$

$$\frac{dy}{dx} = \frac{x}{1 + x^2} + \tan^{-1} x$$

$$\frac{d^2 y}{dx^2} = \frac{-x}{\left(1 + x^2\right)^2}(2x) + \frac{1}{1 + x^2} + \frac{1}{1 + x^2}$$

$$\frac{d^2 y}{dx^2} = \frac{2}{\left(1 + x^2\right)^2}$$

45. $y = \tan^{-1} 2x$; $\dfrac{dy}{dx} = \dfrac{1}{1 + (2x)^2}(2) = \dfrac{2}{1 + 4x^2}$;

$$\frac{d^2 y}{dx^2} = \frac{(1 + 4x^2)(0) - 2(8x)}{(1 + 4x^2)^2} = \frac{-16x}{(1 + 4x^2)^2}$$

49. $t = \dfrac{1}{\omega} \sin^{-1} \dfrac{A - E}{mE}$

$$= \frac{1}{\omega} \sin^{-1} \left(\frac{A - E}{E}\right)\left(\frac{1}{m}\right)$$

$$= \frac{1}{\omega} \sin^{-1} \left(\frac{A - E}{E}\right) m^{-1}$$

$$u = \left(\frac{A - E}{E}\right) m^{-1}; \ \frac{du}{dm} = -\left(\frac{A - E}{E}\right) m^{-2}$$

$$\frac{dt}{dm} = \frac{1}{\omega \sqrt{1 - \left(\frac{-A + E}{E}\right)^2 m^{-2}}} \left(\frac{-A + E}{Em^2}\right)$$

$$= \frac{E - A}{\omega E m^2 \sqrt{1 - \frac{(A - E)^2}{E^2 m^2}}}$$

$$= \frac{E - A}{\omega E m^2 \sqrt{\frac{E^2 m^2 - (A - E)^2}{E^2 m^2}}}$$

$$\frac{dt}{dm} = \frac{E - A}{\omega m \sqrt{E^2 m^2 - (A - E)^2}}$$

53. $\tan \theta = \dfrac{h}{x}$

$$\theta = \tan^{-1} \frac{h}{x}$$

$$\frac{d}{dx} \frac{h}{x} = \frac{d}{dx} h x^{-1} = -h x^{-2} = -\frac{h}{x^2}$$

$$\frac{d\theta}{dx} = \frac{1}{1 + \frac{h^2}{x^2}} - \frac{h}{x^2}$$

$$\frac{d\theta}{dx} = \frac{-h}{\frac{x^2 + h^2}{x^2}\left(x^2\right)}$$

$$\frac{d\theta}{dx} = \frac{-h}{h^2 + x^2}$$

27.4   Applications

1. Sketch the curve $y = \sin x - \dfrac{x}{2}, 0 \le x \le 2\pi$.

$x = 0 \Rightarrow y = 0, (0, 0)$ is both x-intercept and y-intercept. Using Newton's method or zero feature on graphing calculator $(1.90, 0)$ is the only x-intercept for $0 \le x \le 2\pi$. For $x = 2\pi$

$y = \sin 2\pi - \dfrac{2\pi}{2} = -\pi \Rightarrow (2\pi, -\pi)$ is right hand end point.

$$\frac{dy}{dx} = \cos x - \frac{1}{2} = 0 \text{ for } x = \frac{\pi}{3}, \frac{5\pi}{3}$$

$$\frac{d^2 y}{dx^2} = -\sin x = \frac{-\sqrt{3}}{2} \text{ for } x = \frac{\pi}{3} \Rightarrow \max\left(\frac{\pi}{3}, 0.34\right)$$

$$\frac{d^2 y}{dx^2} = -\sin x = \frac{\sqrt{3}}{2} \text{ for } x = \frac{5\pi}{3} \Rightarrow \min\left(\frac{5\pi}{3}, -3.48\right)$$

$$\frac{d^2 y}{dx^2} = -\sin x = 0 \text{ for } x = 0, \pi$$

$$\frac{d^2 y}{dx^2} < 0 \text{ for } 0 < x < \pi \Rightarrow \text{infl } (0, 0)$$

$$\frac{d^2 y}{dx^2} > 0 \text{ for } \pi < x < \frac{\pi}{2} \Rightarrow \text{infl } \left(\pi, -\frac{\pi}{2}\right)$$

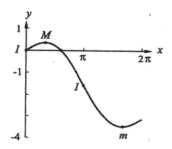

5. $y = \tan^{-1} x$

$\dfrac{dy}{dx} = \dfrac{1}{1+x^2} > 0$ for all x.

9. $y = x\sin^{-1} x\big|_{x=0.5} = 0.26179939$

$\dfrac{dy}{dx} = \dfrac{x}{\sqrt{1-x^2}} + \sin^{-1} x\big|_{x=0.5} = 1.1009497$

$y - 0.26179939 = 1.1009497(x - 0.5)$

$\qquad\qquad y = 1.1x - 0.29$

13. $y = 6\cos x - 8\sin x$; minimum value occurs when

$f'(x) = 0$.

$f'(x) = -6\sin x - 8\cos x = 0$

$\sin x = -\dfrac{8}{6}\cos x$

$\tan x = -\dfrac{8}{6} = -\dfrac{4}{3}$

$\alpha = 0.927\,(\text{a 3, 4, 5 triangle})$

$Q_1 = 2.214,\ Q_2 = 5.356$

$f''(x) = -6\cos x + 8\sin$

$f''(2.214) > 0$, min; $f''(5.356) < 0$, max.

Minimum occurs where $x = 2.214$ rad.

$f(2.214) = 6\left(\dfrac{-3}{5}\right) - 8\left(\dfrac{4}{5}\right)$

$\qquad = \dfrac{-18}{5} + \dfrac{-32}{5} = \dfrac{-50}{5} = -10$

Minimum value of $y = -10$.

17. $y = 0.50\sin 2t + 0.30\cos t$;

$v = \dfrac{dy}{dt} = 1.00\cos 2t - 0.30\sin t$

$v\big|_{t=0.40s} = 1.00\cos 0.80 - 0.30\sin 0.40$

$\qquad = 0.58$ ft/s

$a = \dfrac{d^2y}{dt^2} = -2.00\sin 2t - 0.30\cos t$

$a\big|_{t=0.40s} = -200\sin 0.80 - 0.30\cos 0.40$

$\qquad = -1.7$ ft/s^2

21.

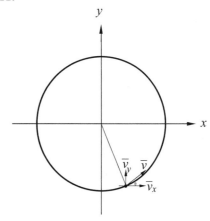

$v = \sqrt{v_x^2 + v_y^2} = \sqrt{\left(\dfrac{dx}{dt}\right)^2 + \left(\dfrac{dy}{dt}\right)^2}$

$v = \sqrt{\left(-19(6\pi)\sin 6\pi t\right)^2 + \left(19(6\pi)\cos 6\pi t\right)^2}\Big|_{t=0.600}$

$v = 358$ cm/s

$\tan\theta = \dfrac{v_y}{v_x} = \dfrac{19(6\pi)\cos 6\pi t}{19(6\pi)\sin 6\pi t}\Big|_{t=0.600}$

$\theta = 18.0°$

25. $\qquad s = 16t^2$

$\tan\theta = \dfrac{200 - s}{100} = \dfrac{200 - 16t^2}{100}$

$\qquad = 2 - 0.16t^2$

$\theta = \tan^{-1}(2 - 0.16t^2)$

$\dfrac{d\theta}{dt} = \dfrac{1}{1 + (2 - 0.16t^2)^2}(-0.32t)$

$\qquad = \dfrac{-0.32t}{5 - 0.64(t)^2 + 0.0256(t)^4}$

$\dfrac{d\theta}{dt}\Big|_{t=1.0} = \dfrac{-0.32(1.0)}{5 - 0.64(1.0)^2 + 0.0256(1.0)^4}$

$\qquad = -0.073$ rad/s

29. $F = \dfrac{0.25w}{0.25\sin\theta + \cos\theta}$

$\dfrac{dF}{d\theta}$

$= \dfrac{(0.25\sin\theta + \cos\theta)(0) - 0.25w(0.25\cos\theta - \sin\theta)}{(0.25\sin\theta + \cos\theta)^2}$

$= 0$

$\tan\theta = 0.25 \Rightarrow \theta = 14°$

33. $V = 0.48(1.2 - \cos 1.26t)$

$\dfrac{dV}{dt} = 0.48(1.26)\sin 1.26t = 0$

$\dfrac{d^2V}{dt^2} = 0.48(1.26)^2\cos 1.26t = 0 \Rightarrow 1.26t = \dfrac{\pi}{2}$

$t = \dfrac{\pi}{2.52}$

$\dfrac{dV_{max}}{dt} = 0.48(1.26)\sin 1.26\left(\dfrac{\pi}{2.52}\right)$

$V_{max} = 0.60$ L/s

37. $\theta = \tan^{-1}\dfrac{x}{15.0}$

$\dfrac{d\theta}{dt} = \dfrac{1}{1 + \frac{x^2}{225}} \cdot \dfrac{1}{15.0} \cdot \dfrac{dx}{dt}$

$\dfrac{dx}{dt} = 15.0\left(1 + \dfrac{x^2}{225}\right)\dfrac{d\theta}{dt} = 15.0\left(1 + \dfrac{30.0^2}{225}\right)(0.75)$

$\dfrac{dx}{dt} = 56$ m/s

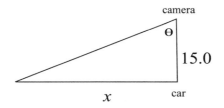

27.5   Derivative of the Logarithmic Function

1. $y = \ln\cos 4x$

$\dfrac{dy}{dx} = \dfrac{1}{\cos 4x}(-\sin 4x)(4)$

$\dfrac{dy}{dx} = -4\tan 4x$

5. $y = 4\log_5(3 - x)$

$\dfrac{dy}{dx} = 4\dfrac{1}{3 - x}\log_5 e(-1) = \dfrac{4}{x - 3}\log_5 e$

9. $y = 2\ln\tan 2x$

$\dfrac{dy}{dx} = 2\dfrac{1}{\tan 2x}\sec^2 2x(2) = \dfrac{4\sec^2 2x}{\tan 2x} = \dfrac{4\sec^2 2x}{\frac{\sec 2x}{\csc 2x}}$

$\dfrac{dy}{dx} = 4\sec 2x\csc 2x$

13. $y = \ln\left(x - x^2\right)^2$

$\dfrac{dy}{dx} = \dfrac{1}{\left(x - x^2\right)^3} \cdot 3\left(x - x^2\right)^2(1 - 2x)$

$= \dfrac{-6x + 3}{x - x^2}$

17. $y = 3x\ln(6 - x)^2 = 6x\ln(6 - x)$

$\dfrac{dy}{dx} = 6x \cdot \dfrac{-1}{6 - x} + 6\ln(6 - x)$

$= \dfrac{6x}{x - 6} + 6\ln(6 - x)$

21. $r = 0.5\ln\cos\left(\pi\theta^2\right)$

$\dfrac{dr}{d\theta} = 0.5\dfrac{1}{\cos\left(\pi\theta^2\right)}\left(-\sin\left(\pi\theta^2\right)\right)(2\pi\theta)$

$\dfrac{dr}{d\theta} = -\pi\theta\tan\left(\pi\theta^2\right)$

25. $u = 3v \ln^2 2v$

$$\frac{du}{dv} = 3v \cdot 2 \ln 2v \cdot \frac{1}{2v} \cdot 2 + 3 \ln^2 2v$$

$$\frac{du}{dv} = 6 \ln 2v + 3 \ln^2 2v$$

29. $r = \ln \dfrac{v^2}{v+2}$

$$= \ln v^2 - \ln(v+2)$$

$$\frac{dr}{dv} = \frac{1}{v^2}(2v) - \frac{1}{v+2}$$

$$= \frac{2}{v} - \frac{1}{v+2}$$

$$= \frac{2v+4-v}{v(v+2)}$$

$$\frac{dr}{dv} = \frac{v+4}{v(v+2)}$$

33. $y = x - \ln^2(x+y)$

$$\frac{dy}{dx} = 1 - 2\ln(x+y)\frac{1}{x+y}\left(1 + \frac{dy}{dx}\right)$$

$$\frac{dy}{dx} = 1 - \frac{2\ln(x+y)}{x+y} - \frac{2\ln(x+y)}{x+y}\frac{dy}{dx}$$

$$\frac{dy}{dx} + \frac{2\ln(x+y)}{x+1} = 1 - \frac{2\ln(x+y)}{x+1}$$

$$= \frac{x+y-2\ln(x+y)}{x+y}$$

$$\frac{dy}{dx}\left[\frac{x+y+2\ln(x+y)}{x+y}\right] = \frac{x+y-2\ln(x+y)}{x+y}$$

$$\frac{dy}{dx} = \frac{x+y-2\ln(x+y)}{x+y+2\ln(x+y)}$$

37. $\dfrac{\ln 2.0001 - \ln 2.0000}{0.0001} = 0.4999875006$, calculator,

which is the slope of secant line through

$(2.0000, \ln 2.0000)$ and $(2.0001, \ln 2.0001)$

$0.5 = \dfrac{d \ln x}{dx}$ for $x = 2$ and is the slope of tangent

line through $(2.0000, \ln 2.0000)$

41. $\ln \sin 45° = -0.3466, x = 45° = \dfrac{\pi}{4}$,

$$44° = 44° \left(\frac{\pi}{180°}\right), \Delta x = -1° \left(\frac{\pi}{180°}\right)$$

$$y = \ln \sin x$$

$$\frac{dy}{dx} = \frac{1}{\sin x}(\cos x)$$

$$\Delta y = f(x + \Delta x) - f(x) \approx dy = f'(x)dx$$

$$\ln \sin 44° - \ln \sin 45° \approx \frac{\cos \frac{\pi}{4}}{\sin \frac{\pi}{4}}\left(-\frac{\pi}{180}\right)$$

$$\ln \sin 44° \approx -0.3466 - \frac{\pi}{180}$$

$$\ln \sin 44° \approx -0.3641$$

45. $f(x) = 2\ln(\tan x)$, $f\left(\dfrac{\pi}{4}\right) = 0$

$$f'(x) = 2\cot x + 2\tan x, \quad f'\left(\frac{\pi}{4}\right) = 4$$

$$L(x) = 4\left(x - \frac{\pi}{4}\right) = 4x - \pi$$

49. $y = x^x$; $\ln y = x \ln x$

$$\frac{1}{y}\frac{dy}{dx} = x\frac{1}{x} + \ln x = 1 + \ln x$$

$$\frac{dy}{dx} = (1 + \ln x)y = x^x(1 + \ln x)$$

Eq. (23-15); $\dfrac{d}{dx}u^n$, n is a constant. In x^x

both the base and the exponent are variable.

53. $b = 10\log\left(\dfrac{I}{I_0}\right) = 10\log\left(I_0^{-1}I\right); u = \left(I_0^{-1}I\right);$

$$\frac{db}{dt} = 10\left(\frac{I_0}{I}\right)(\log e)\left(\frac{I}{I_0}\right)\frac{dI}{dt} = \frac{10}{I}\log e \frac{dI}{dt}$$

57. $t = 5\ln \dfrac{16}{16 - 0.1v}$; $v = 100$ ft/s

$$t = 5[\ln 16 - \ln(16 - 0.1v)]$$

$$\frac{dt}{dv} = 5\left[0 - \frac{1}{16 - 0.1v}(-0.1)\right]$$

$$\frac{dt}{dv} = \frac{0.5}{16 - 0.1v}; \quad \frac{dt}{dv}\bigg|_{v=100} = 0.083 \text{ s}^2/\text{ft}$$

27.6   Derivative of the Exponential Function

1. $y = \ln \sin e^{2x}$

$$\frac{dy}{dx} = \frac{1}{\sin e^{2x}}(\cos e^{2x})(2e^{2x})$$

$$\frac{dy}{dx} = 2e^{2x} \cot e^{2x}$$

5. $y = 6e^{\sqrt{x}}$

$$\frac{dy}{dx} = 6e^{\sqrt{x}} \cdot \frac{1}{2\sqrt{x}} = \frac{3e^{\sqrt{x}}}{\sqrt{x}}$$

9. $R = Te^{-3T}$

$$\frac{dR}{dT} = T\left(e^{-3T}\right)(-3) + (1)\left(e^{-3T}\right)$$

$$= e^{-3T} - 3Te^{-3T} = e^{-3T}\left(1 - 3T\right)$$

13. $r = \dfrac{2\left(e^{2s} - e^{-2s}\right)}{e^{2s}}$

$$r = 2\left(1 - e^{-4s}\right)$$

$$\frac{dr}{ds} = 2\left(4e^{-4s}\right)$$

$$\frac{dr}{ds} = 8e^{-4s}$$

17. $y = \dfrac{2e^{3x}}{4x+3}$;

$$\frac{dy}{dx} = \frac{(4x+3)(2e^{3x})(3) - (2e^{3x})(4)}{(4x+3)^2}$$

$$\frac{dy}{dx} = \frac{(12x+9)(2e^{3x}) - 8e^{3x}}{(4x+3)^2}$$

$$= \frac{2e^{3x}(12x+5)}{(4x+3)^2}$$

21. $y = \left(2e^{2x}\right)^3 \sin x^2$
       $= 8e^{6x} \sin x^2$

$$\frac{dy}{dx} = 8e^{6x}(\cos x^2)(2x) + \sin x^2 (8e^{6x})(6)$$

$$= 16e^{6x}(x \cos x^2 + 3 \sin x^2)$$

25. $y = xe^{xy} + \sin y$

$$\frac{dy}{dx} = x\left(e^{xy}\right)\left(x\frac{dy}{dx} + y\right) + (1)e^{xy} - \csc^2 y \frac{dy}{dx}$$

$$= x\left(e^{xy}\right)\left(x\frac{dy}{dx}\right) + x\left(e^{xy}\right)(y) + e^{xy} - \csc^2 y \frac{dy}{dx}$$

$$\frac{dy}{dx} - x\left(e^{xy}\right)\left(x\frac{dy}{dx}\right) + \csc^2 y \frac{dy}{dx} = x\left(e^{xy}\right)y + e^{xy}$$

$$\frac{dy}{dx}\left(1 - x\left(e^{xy}\right)(x) + \csc^2 y\right) = x\left(e^{xy}\right)y + e^{xy}$$

$$\frac{dy}{dx} = \frac{xy\left(e^{xy}\right) + e^{xy}}{1 - x^2 e^{xy} + \csc^2 y} = \frac{e^{xy}\left(xy + 1\right)}{1 - x^2 e^{xy} + \csc^2 y}$$

29. $I = \ln \sin 2e^{6t}$

$$\frac{dI}{dt} = \frac{1}{\sin 2e^{6t}} \cdot \cos 2e^{6t} \cdot 12e^{6t}$$

33. (a) $e = e^x = 2.7182818$ when $x = 1.0000$. This is the slope of a tangent line to the curve $f(x) = e^x$ when $x = 1.0000$. It is the value of $f'(x) = e^x$, since $\frac{d e^x}{dx} = e^x$.

(b) $\dfrac{e^{1.0001} - e^{1.0000}}{0.0001} = 2.7184178$ This is the slope of a secant line through the curve

$f(x) = e^x$ at $x = 1.0000$, where $\Delta x = 0.0001$.

$$\lim_{\Delta x \to 0} \frac{e^{(x + \Delta x)} - e^x}{\Delta x} = \frac{d e^x}{dx} = e^x$$

For $\Delta x = 0.0001$, the slope of the tangent line is approximately equal to the slope of the secant line.

37. $y = e^{-x/2} \cos 4x$

$$\frac{dy}{dx} = -4 \sin 4x e^{-x/2} - \frac{1}{2}e^{-x/2} \cos 4x \Big|_{x=0.625}$$

$$= -1.458341701$$

numerical derivative feature gives -1.458339582

41.

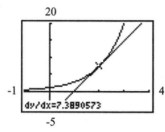

$$\frac{d}{dx}\sinh x = \frac{d}{dx}\frac{1}{2}(e^x - e^{-x}) = \frac{1}{2}(e^x + e^{-x})$$

$$\frac{d^2}{dx^2}\sinh x = \frac{d}{dx}\frac{1}{2}(e^x + e^{-x}) = \frac{1}{2}(e^x - e^{-x})$$

$$= \sinh x$$

$$\frac{d}{dx}\cosh x = \frac{d}{dx}\frac{1}{2}(e^x + e^{-x}) = \frac{1}{2}(e^x - e^{-x})$$

45. $y = \dfrac{e^{2x} - 1}{e^{2x} + 1}$

$$\frac{dy}{dx} = 1 - y^2$$

$$\frac{dy}{dx} = \frac{(e^{2x} + 1)2e^{2x} - (e^{2x} - 1)2e^{2x}}{(e^{2x} + 1)^2}$$

$$\frac{dy}{dx} = \frac{2e^{2x}(e^{2x} + 1 - e^{2x} + 1)}{(e^{2x} + 1)^2} = \frac{4e^{2x}}{(e^{2x} + 1)^2}$$

$$1 - y^2 = 1 - \frac{(e^{2x} - 1)^2}{(e^{2x} + 1)^2}$$

$$= \frac{(e^{2x} + 1)^2 - (e^{2x} - 1)^2}{(e^{2x} + 1)^2}$$

$$1 - y^2 = \frac{e^{4x} + 2e^{2x} + 1 - e^{4x} + 2e^{2x} - 1}{(e^{2x} + 1)^2}$$

$$= \frac{4e^{2x}}{(e^{2x} + 1)^2}$$

49. $y = Ae^{kx} + Be^{-kx}$

$$y' = Ake^{kx} - Bke^{-kx}$$

$$y'' = Ak^2 e^{kx} + Bk^2 e^{-kx}$$

$$= k^2\left(Ae^{kx} + Be^{-kx}\right)$$

$$= k^2 y$$

53. $i = 4.42e^{-66.7t}\sin(226t)$

$$\frac{di}{dt} = 4.42e^{-66.7t}\cos(226t)\cdot 226 - 294.8e^{-66.7t}\sin(226t)$$

$$\frac{di}{dt} = e^{-66.7t}(999\cos(226t) - 295\sin(226t))$$

57. $\dfrac{d}{dx}\sinh u = \dfrac{d}{dx}\dfrac{1}{2}\left(e^u - e^{-u}\right) = \dfrac{1}{2}\left(e^u\dfrac{du}{dx} - e^{-u}\cdot -\dfrac{du}{dx}\right)$

$$\frac{d}{dx}\sinh u = \frac{1}{2}e^u\frac{du}{dx} + \frac{1}{2}e^{-u}\frac{du}{dx} = \left[\frac{1}{2}\left(e^u + e^{-u}\right)\right]\frac{du}{dx}$$

$$\frac{d}{dx}\sinh u = \cosh\frac{du}{dx}$$

$$\frac{d}{dx}\cosh u = \frac{d}{dx}\frac{1}{2}\left(e^u + e^{-u}\right) = \frac{1}{2}\left(e^u\frac{du}{dx} + e^{-u}\cdot -\frac{du}{dx}\right)$$

$$\frac{d}{dx}\cosh u = \frac{1}{2}e^u\frac{du}{dx} - \frac{1}{2}e^{-u}\frac{du}{dx} = \left[\frac{1}{2}\left(e^u - e^{-u}\right)\right]\frac{du}{dx}$$

$$\frac{d}{dx} = \cosh u = \sinh u\frac{du}{dx}$$

27.7   L'Hospital's Rule

1. $\lim\limits_{x\to\infty}\dfrac{3e^{2x}}{5\ln x}\overset{\text{LH}}{=}\lim\limits_{x\to\infty}\dfrac{\frac{d}{dx}3e^{2x}}{\frac{d}{dx}5\ln x}$

$$=\lim\limits_{x\to\infty}\dfrac{6e^{2x}}{\frac{5}{x}}=\lim\limits_{x\to\infty}\dfrac{6xe^{2x}}{5}=\infty$$

5. $\lim\limits_{\theta\to0}\dfrac{\tan\theta}{\theta}\overset{\text{LH}}{=}\lim\limits_{\theta\to0}\dfrac{\frac{d}{d\theta}\tan\theta}{\frac{d}{d\theta}\theta}$

$$=\lim\limits_{\theta\to0}\dfrac{\sec^2\theta}{1}=1$$

9. $\lim\limits_{t\to\frac{\pi}{4}}\dfrac{1-\sin 2t}{\frac{\pi}{4}-t}\overset{\text{LH}}{=}\lim\limits_{t\to\frac{\pi}{4}}\dfrac{-2\cos 2t}{-1}=0$

13. $\lim\limits_{x\to0}\dfrac{\sin x-x}{x^3}\overset{\text{LH}}{=}\lim\limits_{x\to0}\dfrac{\cos x-1}{3x^2}$

$$\overset{\text{LH}}{=}\lim\limits_{x\to0}\dfrac{-\sin x}{6x}$$

$$=-\dfrac{1}{6}\lim\limits_{x\to0}\dfrac{\sin x}{x}=-\dfrac{1}{6}$$

17. $\lim\limits_{x\to0}x\cot x=\lim\limits_{x\to0}\dfrac{x}{\tan x}\overset{\text{LH}}{=}\lim\limits_{x\to0}\dfrac{1}{\sec^2 x}$

$$=\lim\limits_{x\to0}\cos^2 x=1$$

21. $\lim\limits_{x\to0}\dfrac{2\sin x}{5e^x}=\dfrac{2}{5}\dfrac{\lim\limits_{x\to0}\sin x}{\lim\limits_{x\to0}e^x}=\dfrac{2}{5}\cdot\dfrac{0}{1}=0$

25. $\lim\limits_{x\to\frac{\pi}{2}}\dfrac{4\tan x}{\sec^2 x}=\lim\limits_{x\to\frac{\pi}{2}}\dfrac{\frac{4\sin x}{\cos x}}{\frac{1}{\cos^2 x}}$

$$=\lim\limits_{x\to\frac{\pi}{2}}4\sin x\cos x$$

$$=0$$

29. $\lim\limits_{x\to0}\dfrac{\ln\sin x}{\ln\cos x}\overset{\text{LH}}{=}\lim\limits_{x\to0}\dfrac{\frac{1}{\sin x}\cdot\cos x}{\frac{1}{\tan x}\cdot\sec^2 x}$

$$=\lim\limits_{x\to0}\dfrac{\frac{\cos x}{\sin x}}{\frac{\cos x}{\sin x}\cdot\frac{1}{\cos^2 x}}$$

$$=\lim\limits_{x\to0}\cos^2 x=1$$

33. $\lim\limits_{x\to2}\dfrac{2x^3-7x^2+11x-10}{x^3-5x^2+7x-2}$

$$\overset{\text{LH}}{=}\lim\limits_{x\to2}\dfrac{6x^2-14x+11}{3x^2-10x+7}=-7$$

37. $\lim\limits_{\theta\to\frac{\pi}{2}^-}\left(\sec\theta-\tan\theta\right)$

$$=\lim\limits_{\theta\to\frac{\pi}{2}^-}\left(\dfrac{1}{\cos\theta}-\dfrac{\sin\theta}{\cos\theta}\right)$$

$$=\lim\limits_{\theta\to\frac{\pi}{2}^-}\left(\dfrac{1-\sin\theta}{\cos\theta}\right)\overset{\text{LH}}{=}\lim\limits_{\theta\to\frac{\pi}{2}^-}\dfrac{-\cos\theta}{\sin\theta}=0$$

41. $y=\left(1+x^2\right)^{\frac{1}{x}}\Rightarrow\ln y=\dfrac{1}{x}\ln\left(1+x^2\right)$

$$\lim\limits_{x\to+\infty}\ln y=\lim\limits_{x\to+\infty}\dfrac{\ln\left(1+x^2\right)}{x}$$

$$\lim\limits_{x\to+\infty}\ln y\overset{\text{LH}}{=}\lim\limits_{x\to+\infty}\dfrac{\frac{2x}{1+x^2}}{1}\overset{\text{LH}}{=}\lim\limits_{x\to+\infty}\dfrac{2}{2x}=0$$

$$e^{\lim\limits_{x\to+\infty}\ln y}=e^0=1$$

$$\lim\limits_{x\to+\infty}\left(1+x^2\right)^{\frac{1}{x}}=1$$

45. $\sin x$ varies between -1 and 1 as $x\to\infty$.

27.8 Applications

1. $y = e^{-x}\sin x, 0 \le x \le 2\pi$

$y = 0$ for $x = 0, \pi, 2\pi$. Intercepts: $(0,0)$, $(\pi, 0), (2\pi, 0)$

$\dfrac{dy}{dx} = e^{-x}\cos x - e^{-x}\sin x = 0$ for $\sin x = \cos x$

$$x = \frac{\pi}{4}, \frac{5\pi}{4}$$

$\dfrac{d^2 y}{dx^2} = e^{-x}(-\sin x - \cos x) - e^{-x}(\cos x - \sin x)$

$\dfrac{d^2 y}{dx^2} = e^{-x}(-2\cos x) = 0$ for $x = \dfrac{\pi}{2}, \dfrac{3\pi}{2}$

$\dfrac{d^2 y}{dx^2}\Big|_{x=\frac{\pi}{4}} = -2e^{-\pi/4}\cos\dfrac{\pi}{4}$

$\qquad = -0.64 < 0 \Rightarrow \left(\dfrac{\pi}{4}, 0.322\right)$ is a max

$\dfrac{d^2 y}{dx^2}\Big|_{\frac{5\pi}{4}} = -2e^{-5\pi/4}\cos\dfrac{5\pi}{4}$

$\qquad = 0.03 > 0 \Rightarrow \left(\dfrac{5\pi}{4}, -0.014\right)$ is a min

$\dfrac{d^2 y}{dx^2}$ changes from negative to positive at $x = \dfrac{\pi}{2}$

$\left(\dfrac{\pi}{2}, 0.208)\right)$ is an infl. point

$\dfrac{d^2 y}{dx^2}$ changes from positive to negative at $x = \dfrac{3\pi}{2}$

$\left(\dfrac{3\pi}{2}, -0.009\right)$ is an infl. point

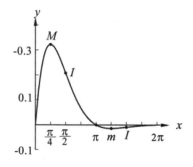

5. $y = 3xe^{-x} = \dfrac{3x}{e^x}$

$\dfrac{dy}{dx} = 3x\left(-e^{-x}\right) + 3e^{-x} = 3e^{-x}\left(1 - x\right)$

$\dfrac{d^2 y}{dx^2} = 3e^{-x}\left(-1\right) + 3\left(1 - x\right)\left(-e^{-x}\right)$

$\qquad = -3e^{-x} - 3e^{-x} + 3xe^{-x} = 3e^{-x}\left(x - 2\right)$

(1) Intercepts: $x = 0$, $y = 0$ (origin)

(2) Symmetry: none

(3) As $x \to +\infty$, $y \to 0$ positively. Horizontal asymptote $y = 0$. As $x \to -\infty$, $y \to -\infty$.

(4) Vertical asymptote: none

(5) Domain: all x, range to be determined.

(6) $\dfrac{dy}{dx} = 0$; $e^{-x}\left(1 - x\right) = 0$; $x = 1$; $f''(1) < 0$,

$\qquad$ max. $\left(1, \dfrac{3}{e}\right)$. Range: $y \le \dfrac{3}{e}$

$\dfrac{d^2 y}{dx^2} = 0$; $e^{-x}\left(x - 2\right) = 0$; $x = 2$; infl. $\left(2, \dfrac{6}{e^2}\right)$

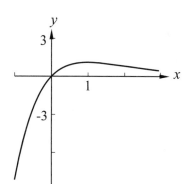

9. $y = 4e^{-x^2}$; $\dfrac{dy}{dx} = 4e^{-x^2}(-2x) = \dfrac{-8x}{e^{x^2}}$

$\dfrac{d^2 y}{dx^2} = \dfrac{e^{x^2}(-8) + 8x(2xe^{x^2})}{e^{2x^2}} = 8e^{x^2}(2x^2 - 1)$

(1) Intercepts: $x = 0, y = 4, (0,4)$ intercept.

(2) Symmetry: yes, with respect to y-axis.

(3) As $x \to \pm\infty, y \to 0$ positively; x-axis is a horizontal asymptote.

(4) No vertical asymptote.

(5) Domain: all x; range: to be determined.

(6) $\dfrac{dy}{dx} = 0$; $\dfrac{-8x}{e^{x^2}} = 0$; $f''(0) < 0$, max. $(0, 4)$

$\dfrac{d^2y}{dx^2} = 0$; $2x^2 - 1 = 0$; $x = \pm\sqrt{\dfrac{1}{2}} = \pm\dfrac{\sqrt{2}}{2}$;

$\left(-\dfrac{\sqrt{2}}{2}, \dfrac{4}{\sqrt{e}}\right)$, $\left(\dfrac{\sqrt{2}}{2}, \dfrac{4}{\sqrt{e}}\right)$ are inflection points.

Range: $0 < y \le 4$

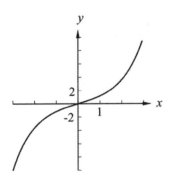

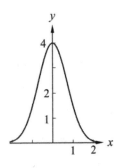

13. $y = \dfrac{1}{2}(e^x - e^{-x})$; $\dfrac{dy}{dx} = \dfrac{1}{2}(e^x + e^{-x})$

$\dfrac{d^2y}{dx^2} = \dfrac{1}{2}(e^x - e^{-x})$

(1) Intercepts: $x = 0, y = 0, (0, 0)$

(2) No symmetry.

(3) As $x \to +\infty, y \to +\infty, x \to -\infty, y \to -\infty$

(4) No vertical asymptote.

(5) Domain: all x; range: all y.

(6) $\dfrac{dy}{dx} = 0$; $e^x = -e^{-x} = -\dfrac{1}{e^x}$; $(e^x)^2 = -1$;

$e^x = \sqrt{-1}$. $\dfrac{dy}{dx} > 0$, inc for all x.

(imaginary), no max,. no min.

$\dfrac{d^2y}{dx^2} = 0$, $e^x = e^{-x} = \dfrac{1}{e^x}$; $(e^x)^2 = 1$; $e^x = \pm 1$

$e^x \ne -1$; $e^x = 1$; $x = 0$, therefore infl. at $(0, 0)$

17. $y = x^2 \ln x$;

$\dfrac{dy}{dx} = x^2 \left(\dfrac{1}{x}\right) + (\ln 2)(2x) = x + 2x \ln x$;

$\dfrac{dy}{dx}\bigg|_{x=1} = 1 + 2\ln 1 = 1 + 2(0) = 1$

Slope is 1, $x = 1, y = 0$; using slope intercept form of the equation and substituting gives $0 = 1(1) + b$ or $b = 1$. The equation is $y = (1)x - 1$ or $y = x - 1$.

21. $f(x) = x^2 - 3 + \ln 4x$

pick $x_1 = 1$ and use NEWTON calculator program

$x_2 = 1.204568524$

$x_3 = 1.197344417$

$x_4 = 1.197333849$

$x_5 = 1.197333849$

zero feature gives 1.1973338

25. $P = 100e^{-0.005t}$; $t = 100$ days;

$\dfrac{dP}{dt} = 100e^{-0.005t}(-0.005)$

$\dfrac{dP}{dt} = -0.5e^{-0.005t}\big|_{t=100} = -0.303$ W/day

29. $\ln p = \dfrac{a}{T} + b \ln T + c$; $p = e^{(a/T + b \ln T + c)}$

$\dfrac{dp}{dT} = e^{(a/T + b \ln T + c)} \left(-aT^{-2} + \dfrac{b}{T}\right)$

$= p\left(\dfrac{-a + bT}{T^2}\right)$

$= e^{(a/T + b \ln T + c)} \left(\dfrac{-a}{T^2} + \dfrac{b}{T}\right)$

$= \dfrac{p(-a + bT)}{T^2}$

33. $y = \ln \sec x;\ -1.5 \le x \le 1.5;\ u = \sec x;$

$\dfrac{dy}{dx} = \dfrac{1}{\sec x} \cdot \sec x \tan x = \tan x = 0$ at $x = 0$;
$x = 0$ is a critical value; also, multiples of 2π.
$\dfrac{d^2y}{dx^2} = \sec^2 x;\ \sec^2(0) = 1$ so the curve is concave up and there is a minimum point at $x = 0$,
$y = \ln \sec 0 = \ln 1 = 0$ recurring at multiples of
$x = 2\pi;\ (2\pi, 0), (4\pi, 0), \ldots (0, 0)$ is an intercept.

Asymptotes occur where $\dfrac{dy}{dx} = \tan x$ is undefined.

These values are odd multiples of $\dfrac{\pi}{2};\ -\dfrac{\pi}{2}, \dfrac{\pi}{2},$
$\dfrac{3\pi}{2} \ldots$

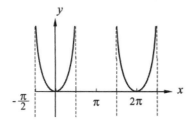

37. $y = 6.0e^{-0.020x} \sin(0.20x), 0 \le x \le 60$

$\dfrac{dy}{dx} = e^{-0.020x} \left[\dfrac{6\cos(0.2x)}{5} - \dfrac{3\sin(0.2x)}{25} \right] = 0$

$\tan 0.2x = 10$
$\quad 0.2x = \tan^{-1} 10 + k\pi$
$\qquad x = 5\tan^{-1} 10 + 5k\pi$

$k = 0 \quad x_1 = 7.355638372, y_1 = 5.153476505$
$k = 1 \quad x_2 = 23.06360164, y_2 = -3.764113107$
$k = 2 \quad x_3 = 38.77156491, y_3 = 2.749318343$
$k = 3 \quad x_4 = 54.47952818, y_4 = -2.008109516$

$(x_1, y_1) = 117.6°\text{ W } 50.2°\text{N, maximum}$
$(x_2, y_2) = 101.9°\text{ W } 41.2°\text{ N, minimum}$
$(x_3, y_3) = 86.2°\text{ W } 47.7°\text{ N, maximum}$
$(x_4, y_4) = 70.5°\text{ W } 43.0°\text{ N, minimum}$

41. $s = kx^2 \ln \dfrac{1}{x} = k[x^2(\ln 1 - \ln x)] = -kx^2 \ln x$

$\dfrac{ds}{dx} = -k\left(x^2 \dfrac{1}{x} + \ln x\, 2x \right) = -k(x + 2x \ln x)$

$\dfrac{ds}{dx} = -kx(1 + 2\ln x) = 0$

For max., min.:

$x = 0;\ \ln x = -\dfrac{1}{2};\ x = e^{-1/2} = \dfrac{1}{\sqrt{e}} = 0.607$

45. $a = 2w^2 \left(\cos\theta + 0.50\cos 2\theta \right)$

$\dfrac{da}{d\theta} = 2w^2 \left(-\sin\theta - 0.50\sin(2\theta)(2) \right) = 0$

$\sin\theta + \sin 2\theta = 0$

$\sin\theta(1 + 2\cos\theta = 0)$

$\sin\theta = 0 \quad , \qquad \cos\theta = -\dfrac{1}{2}$

$\theta = 0°, 180°, 360° \qquad \theta = 120°,\ 240°$

θ is a max at $0°, 360°$ and min at $120°, 240°$.

Chapter 27　Review Exercises

1. $y = y\cos(4x-1);\ \dfrac{dy}{dx} = \left[-3\sin(4x-1)\right][4]$

$= -12\sin(4x-1)$

5. $y = \csc^2(3x+2);\ \dfrac{dy}{dx}$

$= 2\csc(3x+2)\left[-\csc(3x+2)\cot(3x+2)\right](3)$

$= -6\csc^2(3x+2)\cot(3x+2)$

9. $y = \left(e^{x-3}\right)^2;\ \dfrac{dy}{dx} = \left(e^{x-3}\right)\left(e^{x-3}\right)(1) = 2e^{2(x-3)}$

13. $y = 10\tan^{-1}\left(\dfrac{x}{5}\right);\ \dfrac{dy}{dx} = 10\left[\dfrac{1}{1+\left(\frac{x}{5}\right)^2}\right]\dfrac{1}{5}$

$= \dfrac{2}{1+\left(\frac{x}{5}\right)^2} = \dfrac{2}{1+\frac{x^2}{25}} = \dfrac{50}{25+x^2}$

17. $y = \sqrt{\csc 4x + \cot 4x} = \left(\csc 4x + \cot 4x\right)^{1/2}$

$\dfrac{dy}{dx} = \dfrac{1}{2}\left(\csc 4x + \cot 4x\right)^{-1/2}$

$\left(-4\csc 4x\cot 4x - 4\csc^2 4x\right)$

$= \dfrac{1}{2}\left(\csc 4x + \cot 4x\right)^{-1/2}$

$\left(-4\csc 4x\cot 4x - 4\csc^2 4x\right)$

$= \dfrac{1}{2}\left(\csc 4x + \cot 4x\right)^{-1/2}\left(-4\csc 4x\right)$

$\left(\csc 4x + \cot 4x\right)$

$= -2\csc 4x\left(\csc 4x + \cot 4x\right)^{1/2}$

$= \left(-2\csc 4x\right)\sqrt{\csc 4x + \cot 4x}$

21. $y = \dfrac{\cos^2 x}{e^{3x} + \pi^2}$

$\dfrac{dy}{dx} = \dfrac{\left(e^{3x}+\pi^2\right)\left[2\cos x(-\sin x)\right] - \left(\cos^2 x\right)\left(e^{3x}\right)(3)}{\left(e^{3x}+\pi^2\right)^2}$

$= \dfrac{\left(e^{3x}+\pi^2\right)\left[-2\sin x\cos x\right] - 3e^{3x}\cos^2 x}{\left(e^{3x}+\pi^2\right)^2}$

$= \dfrac{-2\sin x\cos x}{e^{3x}+\pi^2} - \dfrac{3e^{3x}\cos^2 x}{\left(e^{3x}+\pi^2\right)^2}$

25. $y = \dfrac{\ln\left(\csc x^2\right)}{x}$

$\dfrac{dy}{dx} = \dfrac{\frac{x}{\csc x^2}\left(-\csc x^2\cot x^2\right)(2x) - \ln\csc x^2}{x^2}$

$\dfrac{dy}{dx} = \dfrac{-2x\cot x^2 - \ln\csc x^2}{x^2}$

29. $L = 0.1e^{-2t}\sec(\pi t)$

$\dfrac{dL}{dt} = 0.1e^{-2t}\sec(\pi t)\cdot\pi\tan(\pi t)$

$\qquad + \sec(\pi t)\cdot 0.1(-2)e^{-2t}$

$\dfrac{dL}{dt} = 0.1e^{-2t}\sec(\pi t)\left(\pi\tan(\pi t) - 2\right)$

33. $\tan^{-1}\dfrac{y}{x} = x^2 e^y;\ u = \dfrac{y}{x} = yx^{-1};\ \dfrac{du}{dx} = -yx^{-2} + x^{-1}\dfrac{dy}{dx}$

$\dfrac{1}{1+\left(yx^{-1}\right)^2}\left(-yx^{-2} + x^{-1}\right)\dfrac{dy}{dx} = x^2 e^y\dfrac{dy}{dx} + 2xe^y$

$\dfrac{\frac{-y}{x^2} + \frac{1}{x}\frac{dy}{dx}}{1+y^2 x^{-2}} = x^2 e^y\dfrac{dy}{dx} + 2xe^y$

$\dfrac{-y}{x^2} + \dfrac{1}{x}\dfrac{dy}{dx} = \left(x^2 e^y\dfrac{dy}{dx} + 2xe^y\right)\left(1+y^2 x^{-2}\right)$

$\dfrac{-y}{x^2} + \dfrac{1}{x}\dfrac{dy}{dx} = x^2 e^y\dfrac{dy}{dx} + 2xe^y + y^2 e^y\dfrac{dy}{dx}$

$\qquad\qquad + 2x^{-1}y^2 e^y$

$\dfrac{1}{x}\dfrac{dy}{dx} - x^2 e^y\dfrac{dy}{dx} - y^2 e^y\dfrac{dy}{dx} = 2xe^y + 2x^{-1}y^2 e^y + \dfrac{y}{x^2}$

$\dfrac{dy}{dx}\left(\dfrac{1}{x} - x^2 e^y - y^2 e^y\right) = 2xe^y + \dfrac{2y^2 ey}{x} + \dfrac{y}{x^2}$

$\dfrac{dy}{dx}\left(\dfrac{1 - x^3 e^y - xy^2 e^y}{x}\right) = \dfrac{2xe^y + 2xy^2 e^y + y}{x^2}$

$$\frac{dy}{dx} = \frac{2x^3e^y + 2xy^2e^y + y}{x^2}$$

$$\cdot \frac{x}{1 - x^3e^y - xy^2e^y}$$

$$\frac{dy}{dx} = \frac{2x^3e^y + 2xy^2e^y + y}{x - x^4e^y - x^2y^2e^y}$$

37. $e^x \ln xy + y = e^x$

$$e^x \cdot \frac{1}{xy}\left(x\frac{dy}{dx} + y\right) + e^x \cdot \ln xy + \frac{dy}{dx} = e^x$$

$$e^x\left(x\frac{dy}{dx} + y\right) + xye^x \ln xy + x\frac{dy}{dx} = xye^x$$

$$\left(xe^x + xy\right)\frac{dy}{dx} = xye^x - xye^x \ln xy - ye^x$$

$$\frac{dy}{dx} = \frac{ye^x\left(x - x\ln xy - 1\right)}{x\left(y + e^x\right)}$$

41. $y = x - \cos 0.5x\big|_{x=0} = -1 \Rightarrow y\text{-int: }(0, -1)$

$0 = x - \cos 0.5x \Rightarrow x = 0.9 \Rightarrow x\text{-int: }(0, 0.9)$

$$\frac{dy}{dx} = 1 + 0.5\sin 0.5x = 0 \Rightarrow \sin x = -2, \text{ no solution}$$

$\Rightarrow$ no critical points

$$\frac{d^2y}{dx^2} = 0.25\cos 0.5x = 0 \Rightarrow x = \pi + k(2\pi)$$

$$\frac{d^2y}{dx^2} = \text{changes sign at } x = \pi + k(2\pi)$$

$\Rightarrow x = \pi + k(2\pi)$ are inflection points

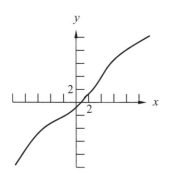

45. $y = 4\cos^2\left(x^2\right);$ slope $= \dfrac{dy}{dx}$

$$= 2\left[4\cos\left(x^2\right)\right]\left[-\sin\left(x^2\right)\right](2x)$$

$$= -16x\cos x^2 \sin x^2$$

$$\frac{dy}{dx}\bigg|_{x=1} = -16\cos\left(1^2\right)\sin\left(1^2\right)$$

$$= -16(0.5403)(0.8415)$$

$$= -7.27$$

$$f(1) = 4\cos^2\left(1^2\right) = 4(0.5403)^2 = 1.168$$

$$y = -7.27x + b; 1.168 = -7.27(1) + b, b = 8.44;$$

$$y = -7.27x + 8.44; 7.27x + y - 8.44 = 0$$

49. $\displaystyle\lim_{x\to 0}\frac{\sin 2x}{\sin 3x} = \lim_{x\to 0}\frac{\frac{1}{3}\frac{\sin 2x}{2x}}{\frac{1}{2}\frac{\sin 3x}{3x}} = \frac{2}{3}$

53. $\displaystyle\lim_{x\to\infty}\frac{\ln x}{\sqrt[3]{x}} \overset{\text{LH}}{=} \lim_{x\to\infty}\frac{\frac{1}{x}}{\frac{1}{3}x^{-\frac{2}{3}}} = 3\lim_{x\to\infty}\frac{1}{\sqrt[3]{x}} = 0$

57. $y = \sin 3x$

$$\frac{dy}{dx} = 3\cos 3x$$

$$\frac{d^2y}{dx^2} = -9\sin 3x = -9y$$

61. $y = e^x$

$$\frac{dy}{dx} = e^x + 2e^{-x}$$

$$\frac{d^2y}{dx^2} = e^x - 2e^{-x} > 0$$

$$e^{2x} > 2$$

$$2x > \ln 2$$

$$x > \frac{1}{2}\ln 2 \approx 0.3466$$

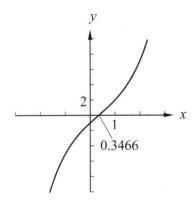

$y = e^x - 2e^{-x}$ is concave up for $x > \dfrac{1}{2}\ln 2$

65. $\ln r = \ln a - \ln \cos\theta - \dfrac{v}{w}\ln\left(\sec\theta + \tan\theta\right)$

$$\frac{1}{r}\frac{dr}{d\theta} = -\frac{1}{\cos\theta}\left(-\sin\theta\right) - \frac{v}{w}\frac{\sec\theta\tan\theta + \sec^2\theta}{\sec\theta + \tan\theta}$$

$$\frac{dr}{d\theta} = r\left(\tan\theta - \frac{v}{w}\sec\theta\right)$$

69. $\mu = \tan\theta,\ 18°\left(\dfrac{\pi}{180°}\right) = \dfrac{\pi}{10},\ \Delta\theta = 2°\left(\dfrac{\pi}{180°}\right) = \dfrac{\pi}{90}$

$$\Delta\mu \approx d\mu = \sec^2\theta\, d\theta = \sec^2\frac{\pi}{10}\left(\frac{\pi}{90}\right) = 0.03859$$

73. $y = 0.75\left(\sec\sqrt{0.15t} - 1\right)$

$$\frac{dy}{dt} = 0.75\sec\sqrt{0.15t}\,\tan\sqrt{0.15t}\,\frac{0.15}{2\sqrt{0.15t}}\bigg|_{t=5.0}$$

$$\frac{dy}{dt}\bigg|_{t=5.0} = 0.12 \text{ cm/s}$$

77. $W = 25\sin^2 2t$

$$P = \frac{dW}{dt} = 50\sin 2t\cos 2t\,(2)$$

$$P = 100\sin 2t\cos 2t$$

81. $I = kE_0^2\cos^2\dfrac{1}{2}\theta$

$$\frac{dI}{d\theta} = 2kE_0^2\cos^2\frac{1}{2}\theta\left(-\sin\frac{1}{2}\theta\right)\left(\frac{1}{2}\right)$$

$$\frac{dI}{d\theta} = -kE_0^2\cos\frac{1}{2}\theta\sin\frac{1}{2}\theta$$

85. $w = \sqrt{\dfrac{g}{l\cos\theta}},\ g = 9.800 \text{ m/s}^2,\ l = 0.6375 \text{ m}$

$\theta = 32.50°,\ d\theta = 0.25°$

$$\Delta w \approx dw = \frac{1}{2}\sqrt{\frac{g}{l\cos\theta}}\tan\theta\, d\theta$$

$$\Delta w \approx \frac{1}{2}\sqrt{\frac{9.800}{0.6375\cos 32.5°}}\tan 32.5°\left(\frac{\pi}{180°}\right)$$

$\Delta w \approx 0.005934 \text{ rad/s}$

89. $n = 2601 - 2290e^{-0.126t}$

$$\frac{dn}{dt} = 228.54\,e^{-0.126t}\bigg|_{t=15}$$

$$\frac{dn}{dt} = 44 \text{ million/year}$$

93.

jet ⟶ 880 ft/s

6800 ft

tower θ

x

$$\theta = \tan^{-1}\frac{6800}{x},\ \frac{d\theta}{dt} = \frac{d\theta}{dx}\frac{dx}{dt}$$

$$\frac{d\theta}{dt} = \frac{1}{1+\left(\frac{6800}{x}\right)^2}\left(\frac{-6800}{x^2}\right)(880)\bigg|_{x=\frac{6800}{\tan 13°}}$$

$$\frac{d\theta}{dt} = -0.0065486205$$

When the angle of elevation of the jet from the contral tower is $13°$, the angle of elevation is changing -0.0065 rad/s.

97. $x = r\left(\theta - \sin\theta\right);\ y = r\left(1 - \cos\theta\right);\ r = 5.00 \text{ cm};$

$$\frac{d\theta}{dt} = 0.12 \text{ rad/s};\ \theta = 35°$$

horizontal component of velocity,

$$v_x = \frac{dx}{dt} = \frac{d\left[5.5(\theta - \sin\theta)\right]}{dt} = 5.5\left(\frac{d\theta}{dt} - \cos\frac{d\theta}{dt}\right)$$

$$= 5.5\left(0.12 - 0.12\cos 35°\right)$$

$$= 0.119 \text{ cm/s}$$

vertical component of velocity,

$$v_y = \frac{dy}{dt} = \frac{d\left[5.5(1 - \cos\theta)\right]}{dt} = 5.5\sin\theta\frac{d\theta}{dt}$$

$$= (5.5)(0.12)\sin 35°$$

$$= 0.379$$

$$v = \sqrt{0.119^2 + 0.379^2} = 0.4 \text{ cm/s}.$$

$$\theta = \tan^{-1}\frac{0.379}{0.119} = 72.5°$$

101a.

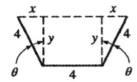

$$\cos\theta = \frac{y}{4}; \ y = 4\cos\theta$$

$$\sin\theta = \frac{y}{4}; \ x = 4\sin\theta$$

$$A = (4 + x)y = 4y + xy = 16\cos\theta + 16\sin\theta\cos\theta$$

$$= 16\cos\theta(1 + \sin\theta)$$

$$\frac{dA}{d\theta} = -16\sin\theta + 16\sin\theta(-\sin\theta) + 16\cos\theta(\cos\theta)$$

$$= -16\sin\theta - 16\sin^2\theta + 16\cos^2\theta$$

$$= 16\left(\cos^2\theta - \sin^2\theta - \sin\theta\right)$$

$$= 16\left(1 - 2\sin^2\theta - \sin\theta\right)$$

(1) Domain: $0 < \theta < \dfrac{\pi}{2}$

(2) A-intercept at $\theta = 0$, $(0, 16)$. To find θ-intercept,
$\quad A = 0$

$$16\cos\theta(1 + \sin\theta) = 0$$

$$16\cos\theta = 0 \qquad\qquad 1 + \sin\theta = 0$$

$$\theta = \frac{\pi}{2} \qquad\qquad\qquad \sin\theta = -1,$$

$$\theta = -\frac{\pi}{2}$$

θ-intercept: $\left(\dfrac{\pi}{2}, 0\right)$

(3) Critical value is

$$16\left(1 - 2\sin^2\theta - \sin\theta\right) = 0$$

$$1 - 2\sin^2\theta - \sin\theta = 0$$

$$-1 + 2\sin^2\theta + \sin\theta = 0$$

$$(2\sin\theta - 1)(\sin\theta + 1) = 0$$

$$2\sin\theta = 1 \qquad\qquad \sin\theta = -1$$

$$\sin\theta = \frac{1}{2} \qquad\qquad \theta = -\frac{\pi}{2} \text{ (not in domain)}$$

$$\theta = \frac{\pi}{6}$$

$$A\left(\frac{\pi}{6}\right) = 20.8$$

(4) $\dfrac{d^2A}{d\theta^2} = 16\left(-4\sin\theta\cos\theta - \cos\theta\right)$

$$= -16\cos\theta(4\sin\theta + 1)$$

$$= -16(4\sin\theta + 1)\big|_{\theta = \pi/6}$$

$$= -16\left[4(0.5) + 1\right] = -16(3) = -48 < 0.$$

Curve is concave down and at $\theta = \pi/6$; A is a maximum of 20.8.

101 b.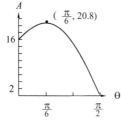

105.

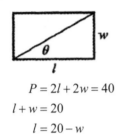

$$P = 2l + 2w = 40$$
$$l + w = 20$$
$$l = 20 - w$$

(a) Using algebra, $A = lw = (20 - w)w = 20w - w^2$

$$\frac{dA}{dw} = 20 - 2w = 0 \Rightarrow w = 10 \text{ and } l = 10$$

The area is maximum when the rectangle is a square.

(b) Using trigonometry $\tan \theta = \dfrac{w}{l} = \dfrac{w}{20 - x}$ from which

$w = \dfrac{20 \tan \theta}{1 + \tan \theta}$, then $A = \dfrac{20^2 \tan \theta}{1 + \tan \theta} - \dfrac{20^2 \tan^2 \theta}{(1 + \tan \theta)^2}$ and

taking $\dfrac{dA}{d\theta} = 0$ gives a maximum for $\theta = 45°$; again,

a square.

Chapter 28

METHODS OF INTEGRATION

28.1   The General Power Formula

1. Change $\cos x$ to $-\sin x$, then

$$\int \cos^3 x(-\sin x \, dx) = \frac{1}{4}\cos^4 x + C$$

5. $u = \cos x;\ n = \dfrac{1}{2};\ du = -\sin x \, dx$

$$0.4\int \sqrt{\cos x}\,\sin x \, dx = 0.4\int (\cos x)^{1/2}(-\sin x \, dx)$$

$$= -\frac{0.8}{3}(\cos x)^{3/2} + C$$

9. Let $u = \cos 2x;\ n = 1;\ du = -2\sin 2x \, dx$

$$\int_0^{\pi/8} \frac{\cos 2x}{\csc 2x}\,dx = \int_0^{\pi/8} \cos 2x \sin 2x \, dx$$

$$= -\frac{1}{2}\int_0^{\pi/8}(\cos 2x)^{-1}(2\sin 2x \, dx)$$

$$= -\frac{1}{2}\frac{(\cos 2x)^2}{2}\Big|_0^{\pi/8}$$

$$= -\frac{1}{4}\cos^2 2x\Big|_0^{\pi/8}$$

$$= -\frac{1}{4}\left(\cos^2 \frac{\pi}{4} - \cos^2 0\right)$$

$$= -\frac{1}{4}\left(\frac{1}{2} - 1\right) = \frac{1}{8}$$

13. $u = \tan^{-1} 5x;\ du = \dfrac{1}{1+25x^2}(5\,dx);\ n = 1$

$$\int \frac{5\tan^{-1} 5x}{1+25x^2}\,dx = \int (\tan^{-1} 5x)^1 \frac{5\,dx}{1+25x^2}$$

$$= \frac{1}{2}(\tan^{-1} 5x)^2 + C$$

17. $u = \ln(2x+3);\ du = \dfrac{1}{2x+3}(2\,dx);\ n = 1$

$$\frac{1}{4}\int_0^{1/2}\left[\ln(2x+3)\right]\frac{2\,dx}{2x+3} = \frac{1}{4}\frac{\left[\ln(2x+3)\right]^2}{2}\Bigg|_0^{1/2}$$

$$= \frac{1}{8}\ln^2(2x+3)\Big|_0^{1/2}$$

$$= \frac{1}{8}\left(\ln^2 4 - \ln^2 3\right) = 0.0894$$

21. $\displaystyle\int \frac{4e^{2t}}{\left(1-e^{2t}\right)^3}\,dt.$ Let $u = 1-e^{2t},\ du = -2e^{2t}\,dt.$

$$\int \frac{4e^{2t}}{\left(1-e^{2t}\right)^3}\,dt = \frac{4}{2}\int\left(1-e^{2t}\right)^{-3}\left(2e^{2t}\,dt\right)$$

$$-2\int u^{-3}\,du = -2\frac{u^{-3+1}}{-3+1} + C$$

$$= -2\frac{1}{-2\left(u^2\right)} + C$$

$$= \frac{1}{\left(1-e^{2t}\right)^2} + C$$

25. $\displaystyle\int_{\pi/6}^{\pi/4}(1+\cot x)^2 \csc^2 x \, dx$

$$= \int_{\pi/6}^{\pi/4}(1+\cot x)^2(-\csc^2 x \, dx)$$

$$= -\frac{(1+\cot x)^3}{3}\Bigg|_{\pi/6}^{\pi/4}$$

$$= -\frac{(1+\cot \frac{\pi}{4})^3}{3} + \frac{(1+\cot \frac{\pi}{6})^3}{3}$$

$$= 2\sqrt{3} + \frac{2}{3}$$

$$= 4.1308$$

29. $\int \dfrac{dx}{x \ln^2 x} = \int \ln^{-2} x \left(\dfrac{1}{x} dx\right) = \int u^{-2} du$ with

$u = \ln x,\ du = \dfrac{1}{x} dx,\ n = -2$

33. $V = \int_0^2 \pi y^2 dx = \int_0^2 \pi \left(e^x\right)^2 dx$

$V = \dfrac{\pi}{2} \int_0^2 e^{2x} \left(2\,dx\right)$

$V = \dfrac{\pi}{2} e^{2x} \Big|_0^2$

$V = \dfrac{\pi}{2} \left(e^4 - 1\right)$

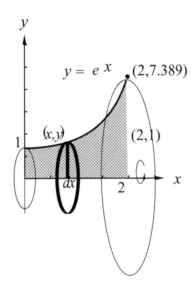

37. $\dfrac{dy}{dx} = m = \dfrac{(\ln x)^2}{x}$; passes through $(1,2)$

$dy = \dfrac{(\ln x)^2}{x} dx;\ y = \int (\ln x)^2 \dfrac{dx}{x};\ u = \ln x;$

$du = \dfrac{dx}{x};\ n = 2$

$y = \dfrac{1}{3}(\ln x)^3 + C;\ x = 1, y = 2$

$2 = \dfrac{1}{3}(\ln 1)^3 + C;$ therefore, $C = 2$

Therefore, $y = \dfrac{1}{3}(\ln x)^3 + 2$

41. $i = 3(1 - e^{-t})^2 e^{-t};\ \dfrac{dq}{dt} = 3(1 - e^{-t})^2 e^{-t};$

$t = 0, q = 0$

$dq = 3(1 - e^{-t})^2 e^{-t} dt;\ q = 3 \int (1 - e^{-t})^2 e^{-t} dt$

$n = 2;\ \mu = 1 - e^{-t};\ du = e^{-t} dt$

$q = \dfrac{3(1 - e^{-t})^3}{3} = (1 - e^{-t})^3 + C;\ t = 0, q = 0$

$0 = (1 - 1)^3 + C;$ therefore, $C = 0$

Therefore, $q = (1 - e^{-t})^3$

28.2   The Basic Logarithmic Form

1. Add $2x$ to the integrand, then

$\int \dfrac{2x\,dx}{x^2 + 1} = \ln \left|x^2 + 1\right| + C$

5. $\int \dfrac{2x\,dx}{4 - 3x^2};\ u = 4 - 3x^2;\ du = -6x\,dx$

$-\dfrac{1}{3} \int \dfrac{-6x\,dx}{4 - 3x^2} = -\dfrac{1}{3} \ln \left|4 - 3x^2\right| + C$

9. $0.4 \int \dfrac{\csc^2 2\theta\,d\theta}{\cot 2\theta};\ u = \cot 2\theta;\ du = -2\csc^2 2\theta\,d\theta$

$\dfrac{0.4}{2} \int \dfrac{2\csc^2 2\theta\,d\theta}{\cot 2\theta} = -0.2 \ln \left|\cot 2\theta\right| + C$

13. $u = 1 - e^{-x};\ du = e^{-x} dx$

$\int \dfrac{e^{-x}\,dx}{1 - e^{-x}} = \ln \left|1 - e^{-x}\right| + C$

17. $u = 1 + 4\sec x;\ du = 4\sec x \tan x\,dx$

$\int \dfrac{8\sec x \tan x\,dx}{1 + 4\sec x} = 2 \int \dfrac{4\sec x \tan x\,dx}{1 + 4\sec x}$

$\hspace{3cm} = 2 \ln \left|1 + 4\sec x\right| + C$

21. $u = \ln r;\ du = \dfrac{dr}{r}$

$0.5 \int \dfrac{dr}{r \ln r} = 0.5 \int \dfrac{\frac{dr}{r}}{\ln r} = 0.5 \ln \left|\ln r\right| + C$

25. $n = -\dfrac{1}{2}$; $u = 1 - 2x$; $du = -2dx$

$$\int \frac{16\,dx}{\sqrt{1-2x}} = -8\int (1-2x)^{-1/2}\,(-2\,dx)$$

$$= -8(1-2x)^{1/2}\,(2) + C$$

$$= -16\sqrt{1-2x} + C$$

29. $u = 4 + \tan 3x$; $du = 3\sec^2 3x\,dx$

$$\int_0^{\pi/12} \frac{\sec^2 3x}{4 + \tan 3x}\,dx = \frac{1}{3}\int_0^{\pi/12} \frac{3\sec^2 3x\,dx}{(4 + \tan 3x)}$$

$$= \frac{1}{3}\ln|4 + \tan 3x|\Big|_0^{\pi/12}$$

$$= \frac{1}{3}(\ln 5 - \ln 4) = \frac{1}{3}\ln\frac{5}{4}$$

$$= 0.0744$$

33.

$$\begin{array}{r} 1 \\ x + 4\,\overline{)\,x - 4} \\ \underline{x + 4} \\ -8 \end{array}$$

$$\int \frac{x-4}{x+4}\,dx = \int dx - \int \frac{8}{x+4}\,dx$$

$$= x - 8\ln|x+4| + C$$

37. $m = \dfrac{dy}{dx} = \dfrac{\sin x}{3 + \cos x}$; $y = \displaystyle\int \frac{1}{3 + \cos x} \times \sin x\,dx$

$$y = -\int \frac{1}{3 + \cos x}(-\sin x)\,dx;$$

let $u = 3 + \cos x$; $du = -\sin x\,dx$

$$y = -\int \frac{1}{u}\,du = -\ln|u| + C$$

$$= -\ln(3 + \cos x) + C$$

$$2 = -\ln\left(3 + \cos\frac{\pi}{3}\right) + C;\ \text{substitute values of } x$$

and y

$$2 = -\ln(3 + 0.5) + C;\ C = 2 + \ln 3.5$$

$$y = -\ln(3 + \cos x) + \ln 3.5 + 2;\ \text{substituting for } C$$

$$y = \ln\frac{3.5}{3 + \cos x} + 2$$

41. $v = \dfrac{ds}{dt} = \dfrac{12.0}{0.200t + 1}$

$$s = \int_0^{3.00} \frac{12.0}{0.200t + 1}\,dt = 60\ln|t + 5|\Big|_0^{3.00} = 28.2 \text{ km}$$

45. $t = \dfrac{1}{k}\displaystyle\int \frac{dP}{P} = \dfrac{1}{k}\ln P + C;\ \dfrac{dP}{dt} = kP$

$t = 0, P = 249$ M; $t = 10, P = 275$ M;
$t = 30, P = ?$

(a) $0 = \dfrac{1}{k}\ln 249 + C$; therefore, $C = -\dfrac{1}{k}\ln 249$

(b) $10 = \dfrac{1}{k}\ln 275 - \dfrac{1}{k}\ln 249 = \dfrac{1}{k}\ln\dfrac{275}{249}$

Therefore, $k = \dfrac{1}{10}\ln\dfrac{275}{249}$

Therefore, $t = \dfrac{1}{\frac{1}{10}\ln\frac{275}{249}}\ln P - \dfrac{1}{\frac{1}{10}\ln\frac{275}{249}}\ln 249$

$$t = \frac{10}{\ln\frac{275}{249}}(\ln P - \ln 249)$$

$$\ln P|_{t=30} = t\left(\frac{\ln\frac{275}{249}}{10}\right) + \ln 249 = 5.815$$

$$P = e^{5.815} = 335.4 \text{ M} = 335 \text{ million people}$$

49. $y = \dfrac{50}{x^2 + 20}$; $x = 3.00$; $A = 6.61 \text{ m}^2$

$u = x^2 + 20$; $du = 2x\,dx$

$$\bar{x} = \frac{\int_0^3 x(y)\,dx}{6.61}$$

$$\bar{x} = \frac{1}{6.61}\int_0^3 x\frac{50}{x^2 + 20}\,dx$$

$$\bar{x} = \frac{25}{6.61}\int_0^3 \frac{2x\,dx}{x^2 + 20} = \frac{25}{6.61}\ln|x^2 + 20|\Big|_0^3$$

$$\bar{x} = \frac{25}{6.61}(\ln 29 - \ln 20) = 1.41 \text{ m}$$

28.3 The Basic Exponential Form

1. $\int 3x^2 e^{x^3} dx = \int e^{x^3}(3x^2 dx)$

$$= e^{x^3} + C$$

5. $u = 2x + 5;\ du = 2\,dx$

$$\int 4e^{2x+5} dx = 2\int e^{2x+5}(2\,dx) = 2e^{2x+5} + C$$

9. $y = x^3;\ du = 3x^2 dx$

$$\int 6x^2 e^{x^3} dx = 6\int e^{x^3}(x^2 dx) = \frac{6}{3}\int e^{x^3}(3x^2 dx)$$

$$= 2e^{x^3} + C$$

13. $u = 2\sec\theta;\ du = 2\sec\theta\tan\theta\,d\theta$

$$\int 14(\sec\theta\tan\theta)e^{2\sec\theta}\,d\theta = 7\int e^{2\sec\theta}\,2\sec\theta\tan\theta\,d\theta$$

$$= 7e^{2\sec\theta} + C$$

17. $u = -2x;\ du = -2\,dx$

$$\int_1^3 3e^{2x}(e^{-2x} - 1)dx = 3\int(e^0 - e^{2x})dx$$

$$= 3\int dx - \frac{3}{2}\int e^{2x}(2\,dx)$$

$$= 3x - \frac{3}{2}e^{2x}\Big|_1^3$$

$$= \left[9 - \frac{3}{2}e^6 - \left(3 - \frac{3}{2}e^2\right)\right]$$

$$= 9 - 3 - \frac{3}{2}e^6 + \frac{3}{2}e^2$$

$$= 6 - \frac{3}{2}(e^6 - e^2) = -588.06$$

21. $\int \frac{e^{\tan^{-1} 2x}}{8x^2 + 2}dx = \frac{1}{2}\int \frac{e^{\tan^{-1} 2x}(2dx)}{2(4x^2 + 1)}$

$$= \frac{1}{4}e^{\tan^{-1} 2x} + C$$

25. $u = \cos^2 x;$

$du = 2\cos x(-\sin x)dx = -2\sin x\cos x\,dx$

$$= -2\sin 2x\,dx$$

$$\int_0^\pi (\sin 2x)e^{\cos^2 x}dx = -\frac{1}{2}\int_0^\pi e^{\cos^2 x}(-2\sin 2x\,dx)$$

$$= -\frac{1}{2}e^{\cos^2 x}\Big|_0^\pi$$

$$= -\frac{1}{2}[(\cos\pi)^2 - (\cos 0)^2]$$

$$= -\frac{1}{2}[(-1)^2 - 1^2] = 0$$

29. $A = \int_0^2 3e^x dx = 3e^x\Big|_0^2 = 3e^2 - 3e^0 = 19.2$

33. $1 + e^x \overline{\smash{\big)}1} \Rightarrow \frac{1}{1 + e^2}\, 1 - \frac{e^x}{1 + e^x}$

$\qquad \dfrac{1 + e^x}{-e^x}$

$$\int \frac{dx}{1 + e^x} = \int dx - \int \frac{e^x}{1 + e^x}dx$$

$$= x - \ln\left|1 + e^x\right| + C \text{ and since } 1 + e^x > 0$$

$$= x - \ln\left(1 + e^x\right) + C$$

37. $y_{av} = \dfrac{\int_0^4 4e^{x/2}dx}{4 - 0} = \dfrac{8\int e^{x/2}\left(\frac{1}{2}dx\right)}{4}$

$$= \dfrac{8e^{x/2}\Big|_0^4}{4} = 2e^{x/2}\Big|_0^4$$

$$= 2(e^2 - 1) = 12.8$$

41. $s = \int 12.0e^{-t/6.00}\,dt = -72.0e^{-t/6.00} + C$

$$s = 0 = -72 + C \Rightarrow C = 72$$

$$s = -72.0e^{-t/6.00} + 72$$

time	1st runner	2nd runner
3.00	28.2	28.3
4.00	35.3	35.0

a) At 3.00 h second runner is ahead by 0.1 km

b) At 4.00 h first runner is ahead by 0.3 km

45. $A = 2\int_0^{100}\left(-19.46(e^{x/38.92}+e^{-x/38.92})+230.9\right)dx$

$\quad\quad = 26,520 \text{ m}^2$

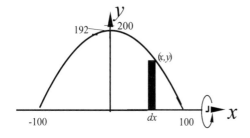

28.4 Basic Trigonometric Forms

1. Change $x\,dx$ to $3x^2dx$ then

$$\int \sec^2 x^3(3x^2dx) = \tan x^3 + C$$

5. $u = 3\theta;\ du = 3\,d\theta$

$$\int 0.3\sec^2 3\theta\,d\theta = 0.3\int \sec^2 3\theta(3\,d\theta)$$

$$= 0.1\tan 3\theta + C$$

9. $u = x^3;\ du = 3x^2dx$

$$\int_{0.5}^1 x^2\cot x^3 dx = \frac{1}{3}\int_{0.5}^1 \cot x^3(3x^2dx)$$

$$= \frac{1}{3}\ln|\sin x^3|\Big|_{0.5}^1$$

$$= \frac{1}{3}\left(\ln|\sin 1| - \ln\left|\sin\frac{1}{8}\right|\right)$$

$$= 0.6365$$

13. $u = \dfrac{1}{x} = x^{-1};$

$$du = -1x^{-2}dx = -\frac{dx}{x^2}$$

$$\int \frac{\sin\left(\frac{1}{x}\right)}{2x^2}dx = -\frac{1}{2}\int \sin\left(\frac{1}{x}\right)\left(-\frac{dx}{x^2}\right)$$

$$= -\frac{1}{2}\left[-\cos\left(\frac{1}{x}\right)\right] + C$$

$$= \frac{1}{2}\cos\left(\frac{1}{x}\right) + C$$

17. $\displaystyle\int\sqrt{\frac{1-\cos x}{8}}\,dx = \frac{1}{2}\int\sqrt{\sin^2\frac{x}{2}}\,dx = \frac{1}{2}\int\left|\sin\frac{x}{2}\right|dx$

$$= \int\sin\frac{x}{2}\left(\frac{1}{2}dx\right)\text{ for } 0\le x < 2\pi$$

$$= -\cos\frac{x}{2} + C$$

21. $\displaystyle\int\frac{2\tan T}{1-\tan^2 T}dT = \frac{1}{2}\int\tan 2T\,(2dT)$

$$= -\frac{1}{2}\ln|\cos 2T| + C$$

25. $\sin 3x\left(\dfrac{1}{\sin 3x} + \dfrac{1}{\cos 3x}\right) = 1 + \tan 3x;$

$u = 3x;\ du = 3\,dx$

$\displaystyle\int_0^{\pi/9}\sin 3x(\csc 3x + \sec 3x)dx$

$\quad = \displaystyle\int_0^{\pi/9}(1+\tan 3x)dx$

$\quad = \displaystyle\int_0^{\pi/9}dx + \int_0^{\pi/9}\tan 3x\,dx$

$\quad = \displaystyle\int_0^{\pi/9}dx + \frac{1}{3}\int_0^{\pi/9}\tan 3x(3\,dx)$

$\quad = \left(x - \dfrac{1}{3}\ln|\cos 3x|\right)\Big|_0^{\pi/9}$

$\quad = \dfrac{\pi}{9} - \dfrac{1}{3}\ln\left|\cos\dfrac{\pi}{3}\right| - \left(0 - \dfrac{1}{3}\ln|\cos 0|\right)$

$\quad = \dfrac{\pi}{9} - \dfrac{1}{3}\ln\left(\dfrac{1}{2}\right) = \dfrac{\pi}{9} + \dfrac{1}{3}\ln 2$

$\quad = 0.580$

29. $\displaystyle\int \frac{dx}{1+\sin x} = \int \frac{dx}{1+\sin x} \cdot \frac{1-\sin x}{1-\sin x}$

$\displaystyle\qquad = \int \frac{(1-\sin x)}{1-\sin^2 x}\, dx$

$\displaystyle\qquad = \int \frac{1-\sin x}{\cos^2 x}\, dx$

$\displaystyle\qquad = \int \frac{1}{\cos^2 x}\, dx + \int \frac{-\sin x\, dx}{\cos^2 x}$

$\displaystyle\qquad = \int (\sec^2 x\, dx - \sec x \tan x)dx$

$\displaystyle\qquad = \tan x - \sec x + C$

33. $y = \sec x;\ x = 0;\ x = \dfrac{\pi}{3}$

$y = 0$; rotated about x-axis, disks.

$dV = \pi r^2 dx;\ r = y$

$\displaystyle V = \int_0^{\pi/3} \pi y^2 dx = \pi \int_0^{\pi/3} \sec^2 x\, dx$

$\displaystyle V = \pi \tan x \Big|_0^{\pi/3} = \pi\sqrt{3} = 5.44$

37. $y = \tan x^2;\ y = 0, x = 1$

$\displaystyle \overline{x} = \frac{\displaystyle\int_0^1 xy\, dx}{\displaystyle\int_0^1 y\, dx} = \frac{\displaystyle\int_0^1 (\tan x^2)x\, dx}{\displaystyle\int_0^1 \tan x^2 dx}$

$u = x^2;\ du = 2x\, dx$

$\displaystyle \overline{x} = \frac{\dfrac{1}{2}\displaystyle\int_0^1 \tan x^2 (2x\, dx)}{0.3984} = \frac{-\dfrac{1}{2}\ln\left|\cos x^2\right|\Big|_0^1}{0.3984}$

$\displaystyle \overline{x} = \frac{-\dfrac{1}{2}(\ln\cos 1 - \ln\cos 0)}{0.3984} = \frac{0.3078}{0.3984}$

$\displaystyle \qquad = 0.7726 \text{ m}$

28.5 Other Trigonometric Forms

1. Change dx to $\cos 2x\, dx$, then

$$\int \sin^2 2x(\cos 2x\, dx) = \frac{1}{2}\int \sin^2 2x(2\cos 2x\, dx)$$

$$= \frac{1}{2}\cdot\frac{\sin^3 2x}{3} + C$$

$$= \frac{\sin^3 2x}{6} + C$$

5. $u = 2x;\; du = 2\, dx;\; u = \cos 2x;\; du = -2\sin 2x\, dx$

$$\int \sin^3 2x\, dx = \int \sin^2 2x \sin 2x\, dx = \int (1 - \cos^2 2x)\sin 2x\, dx = \int \sin 2x\, dx - \int \cos^2 2x \sin 2x\, dx$$

$$= \frac{1}{2}\int \sin 2x(2\, dx) + \frac{1}{2}\int \cos^2 2x(-2\sin 2x\, dx) = -\frac{1}{2}\cos 2x + \frac{1}{2}\frac{\cos^3 2x}{3} + C$$

$$= -\frac{1}{2}\cos 2x + \frac{1}{6}\cos^3 2x + C$$

9. $u = \cos x;\; du = -\sin x\, dx$

$$\int_0^{\pi/4} 5\sin^5 x\, dx = 5\int_0^{\pi/4} \sin x \sin^4 x\, dx = 5\int_0^{\pi/4} \sin x(1 - \cos^2 x)^2 dx = 5\int_0^{\pi/4} \sin x(1 - 2\cos^2 x + \cos^4 x)dx$$

$$= 5\left(\int \sin x\, dx - 2\int \cos^2 x \sin x\, dx + \int \cos^4 x \sin x\, dx\right) = 5(-\cos x) + 10\frac{\cos^3 x}{3} - 5\frac{\cos^5 x}{5}\Big|_0^{\pi/4}$$

$$= 5\left(-\frac{1}{\sqrt{2}}\right) + \frac{10}{3}\left(\frac{1}{\sqrt{2}}\right)^3 - \left(\frac{1}{\sqrt{2}}\right)^5 - \left[5(-1) + \frac{10}{3}(1) - 1\right]$$

$$= -\frac{5}{\sqrt{2}} + \frac{10}{3}\frac{1}{2\sqrt{2}} - \frac{1}{4\sqrt{2}} - \left(-5 + \frac{10}{3} - 1\right) = \frac{-60 + 10(2) - 3}{12\sqrt{2}} - \left[\frac{-15 + 10 - 3}{3}\right]$$

$$= \frac{-43 + 32\sqrt{2}}{12\sqrt{2}} = \frac{64 - 43\sqrt{2}}{24} = 0.1329$$

13. $\displaystyle\int 2(1 + \cos 3\phi)^2 d\phi = \int 2(1 + 2\cos 3\phi + \cos^2 3\phi)d\phi = \int (2 + 4\cos 3\phi + 2\cos^2 3\phi)d\phi$

$$= 2\phi + \frac{4\sin 3\phi}{3} + \int (1 + \cos 6\phi)d\phi = 2\phi + \frac{4\sin 3\phi}{3} + \phi + \frac{1}{6}\int \cos 6\phi(6\, d\phi)$$

$$= 3\phi + \frac{4\sin 3\phi}{3} + \frac{\sin 6\phi}{6} + C = \frac{1}{3}(9\phi + 4\sin 3\phi + \sin 3\phi \cos 3\phi) + C$$

17. $u = \tan x;\ du = \sec^2 x\, dx$

$$\int_0^{\pi/4} \frac{\tan x}{\cos^4 x}\, dx = \int_0^{\pi/4} \tan x \sec^4 x\, dx = \int_0^{\pi/4} \tan x \sec^2 x \left(1 + \tan^2 x\right) dx = \int_0^{\pi/4} \left(\tan x\right)^1 \sec x\, dx + \int_0^{\pi/4} \tan^3 x \sec^2 x\, dx$$

$$= \frac{1}{2}\left(\tan x\right)^2 + \frac{1}{4}\left(\tan x\right)^4 \bigg|_0^{\pi/4} = \frac{1}{2}\left(1\right)^2 + \frac{1}{4}\left(1\right)^4 = \frac{1}{2} + \frac{1}{4} = \frac{3}{4}$$

21. $\displaystyle\int 0.5 \sin s \sin 2s\, ds = \int 0.5 \sin s \cdot 2 \sin s \cos s\, ds = \int \sin^2 s \cos s\, ds = \frac{\sin^3 s}{3} + C$

25. $\displaystyle\int \frac{1 - \cot x}{\sin^4 x}\, dx$

$$= \int (1 - \cot x) \csc^4 x\, dx = \int (1 - \cot x) \csc^2 x (1 + \cot^2 x)\, dx = \int (1 - \cot x + \cot^2 x - \cot^3 x \csc^2 x\, dx)$$

$$= \int \csc^2 x\, dx - \int \cot x \csc^2 x\, dx + \int \cot^2 x \csc^2 x\, dx - \int \cot^3 x \csc^2 x\, dx$$

$$= \int \csc^2 x\, dx + \int \cot x(-\csc^2 x\, dx) - \int \cot^2 x(-\csc^2 x\, dx) + \int \cot^3 x(-\csc^2 x\, dx)$$

$$= -\cot x + \frac{\cot^2 x}{2} - \frac{\cot^3 x}{3} + \frac{\cot^4 x}{4} + C = \frac{1}{4}\cot^4 x - \frac{1}{3}\cot^3 x + \frac{1}{2}\cot^2 x - \cot x + C$$

29. $\displaystyle\int \sec^6 x\, dx = \int \sec^4 x(1 + \tan^2 x)\, dx = \int \sec^4 x\, dx + \int \sec^4 x \tan^2 x\, dx$

$$= \int \sec^2 x(1 + \tan^2 x)\, dx + \int \sec^2 x(1 + \tan^2 x)x\, dx$$

$$= \int \sec^2 x\, dx + \int \tan^2 x \sec^2 x\, dx + \int \tan^2 x \sec^2 x\, dx + \int \tan^4 x \sec^2 x\, dx$$

$$= \tan x + \frac{1}{3}\tan^3 x + \frac{1}{3}\tan^3 x + \frac{1}{5}\tan^5 + C = \frac{1}{5}\tan^5 x + \frac{2}{3}\tan^3 x + \tan x + C$$

33. $u = e^{-x} \Rightarrow du = -e^{-x} \Rightarrow -du = \dfrac{dx}{e^x}$

$$\int \frac{\sec e^{-x}}{e^x}\, dx = -\int \sec u\, du = -\ln\left|\sec u + \tan u\right| + C$$

$$= -\ln\left|\sec e^{-x} + \tan e^{-x}\right| + C$$

37. Rotate about x-axis, disks

$$V = \pi \int_0^\pi y^2 dx = \pi \int_0^\pi \sin^2 x \, dx = \pi \int_0^\pi \frac{1}{2}(1 - \cos 2x) \, dx = \frac{\pi}{2} \int_0^\pi dx - \frac{\pi}{2} \int_0^\pi \cos 2x \, dx$$

$$= \frac{\pi}{2} x - \frac{\pi}{2} \times \frac{1}{2} \sin 2x \Big|_0^\pi = \frac{\pi^2}{2} - \frac{\pi}{4} \sin 2\pi - 0 + \frac{\pi}{4} \sin 0 = \frac{1}{2}\pi^2 = 4.935$$

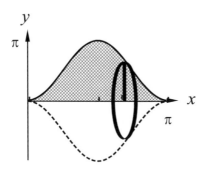

41. $\displaystyle\int \sin x \cos x \, dx; \ u = \sin x; \ du = \cos x \, dx$

$$\int u \, du = \frac{1}{2}u^2 + C = \frac{1}{2}\sin^2 x + C_1$$

Let $u = \cos x; \ du = -\sin x \, dx$

$$-\int \cos x (-\sin x) \, dx = -\frac{1}{2}\cos^2 x + C_2$$

$$\frac{1}{2}\sin^2 x + C_1 = \frac{1}{2}(1 - \cos^2 x + C_1)$$

$$= \frac{1}{2} - \frac{1}{2}\cos^2 x + C_1$$

$$-\frac{1}{2}\cos^2 x + C_2 = \frac{1}{2} - \frac{1}{2}\cos^2 x + C_1$$

$$C_2 = C_1 + \frac{1}{2}$$

45. $a = \sin^2 t \cos t$

$$v = \int \sin^2 t \cos t \, dt = \frac{\sin^3 t}{3} + v_0$$

$$v(0) = 6 = \frac{\sin^3 0}{3} + v_0 \Rightarrow v_0 = 6$$

$$v = \frac{\sin^3 t}{3} + 6$$

$$s = \int \frac{\sin^3 t}{3}\,dt + \int 6dt$$

$$s = \frac{1}{3}\int \sin^2 t \sin t\,dt + \int 6dt$$

$$s = \frac{1}{3}\int \left(1 - \cos^2 t\right)\sin t\,dt + 6\int dt$$

$$s = \frac{1}{3}\int \sin t\,dt - \frac{1}{3}\int \cos^2 t \sin t\,dt + 6\int dt$$

$$s = \frac{-1}{3}\cos t + \frac{1}{9}\cos^3 t + 6t + s_0$$

$$s(0) = 0 = -\frac{1}{3}\cos 0 + \frac{1}{9}\cos^3 0 + 6(0) + s_0 \Rightarrow s_0 = \frac{2}{9}$$

$$s = -\frac{1}{3}\cos t + \frac{1}{9}\cos^3 t + 6t + \frac{2}{9}$$

49. $V_{rms} = \sqrt{\dfrac{1}{1/60.0}\displaystyle\int_0^{1/60.0}(340\sin 120\pi t)^2 dt} = \sqrt{60}\sqrt{\displaystyle\int_0^{1/60.0} 340^2\frac{1-\cos 240\pi t}{2}\,dt}$

$$= \sqrt{60}\sqrt{\frac{340^2}{2}\left(t - \frac{1}{240\pi}\sin 240\pi t\right)\Bigg|_0^{1/60.0}} = 240\ \text{V}$$

28.6   Inverse Trigonometric Forms

1. Change dx to $-x\,dx$, then

$$\int \frac{-x\,dx}{\sqrt{9-x^2}} = \frac{1}{2}\int (9-x^2)^{-1/2}(-2x\,dx) = \frac{1}{2}\frac{(9-x^2)^{1/2}}{\frac{1}{2}} + C = \sqrt{9-x^2} + C$$

5. $a = 8;\ u = x;\ du = dx;$

$$\int \frac{12\,dx}{64+x^2} = \frac{3}{2}\tan^{-1}\frac{x}{8} + C$$

9. $\displaystyle\int_0^2 \frac{3e^{-t}dt}{1+9e^{-2t}}$

$$= -\int_0^2 \frac{-3e^{-t}dt}{1+(3e^{-t})^2} \text{ which has form } \int\frac{dx}{1+x^2}$$

$$= -\tan^{-1}(3e^{-t})\Big|_0^2$$

$$= -\tan^{-1}\frac{3}{e^2} + \tan^{-1}3$$

$$= 0.8634$$

13. $u = 9x^2 + 16;\ du = 18x\,dx$

$$\int \frac{8x\,dx}{9x^2+16} = \frac{8}{18}\int \frac{18x\,dx}{9x^2+16}$$

$$= \frac{4}{9}\ln|9x^2+16| + C$$

17. $a = 1;\ u = e^x;\ du = e^x\,dx$

$$\int \frac{2e^x\,dx}{\sqrt{1-e^{2x}}} = 2\sin^{-1}e^x + C$$

21. $a = 2;\ u = x+2;\ du = dx$

$$\int \frac{4\,dx}{\sqrt{-4x-x^2}} = \int \frac{4\,dx}{\sqrt{4-(x+2)^2}}$$

$$= 4\int \frac{dx}{\sqrt{4-(x+2)^2}}$$

$$= 4\sin^{-1}\left(\frac{x+2}{2}\right) + C$$

25. $a = 2;\ u = x;\ du = dx;\ n = -\dfrac{1}{2};\ u = 4-x^2;$

$$du = -2x\,dx$$

$$\int \frac{2-x}{\sqrt{4-x^2}}dx = \int \frac{2\,dx}{\sqrt{4-x^2}} - \int \frac{x\,dx}{\sqrt{4-x^2}}$$

$$= 2\int \frac{dx}{\sqrt{4-x^2}} + \frac{1}{2}\int \frac{(-2x\,dx)}{(4-x^2)^{1/2}}$$

$$= 2\sin^{-1}\frac{x}{2} + \frac{1}{2}(4-x^2)^{1/2}\cdot 2 + C$$

$$= 2\sin^{-1}\frac{x}{2} + \sqrt{4-x^2} + C$$

29. $\displaystyle\int \frac{x^2+3x^5}{1+x^6}dx = \int \frac{x^2}{1+x^6}dx + \int \frac{3x^5}{1+x^6}dx$

$$\int \frac{x^2}{1+x^6}dx = \frac{1}{3}\int \frac{3x^2}{1+\left(x^3\right)^2}dx = \frac{1}{3}\tan^{-1}x^3$$

$$\int \frac{3x^5}{1+x^6}dx = \frac{1}{2}\int \frac{6x^5\,dx}{1+x^6} = \frac{1}{2}\ln\left(1+x^6\right)$$

$$\int \frac{x^2+3x^5}{1+x^6}dx = \frac{1}{3}\tan^{-1}x^3 + \frac{1}{2}\ln\left(1+x^6\right) + c$$

33. (a) General power, $\displaystyle\int u^{-1/2}du$ where $u = 4-9x^2$.

     $du = -18x\,dx$; numerator can fit du of denominator.
     Square root becomes $-1/2$ power.
     Does not fit inverse sine form.

     (b) Inverse sine; $a = 2;\ u = 3x;\ du = 3\,dx$

     (c) Logarithmic; $u = 4-9x;\ du = -9\,dx$

37. $y = \dfrac{1}{1+x^2};\ A = \displaystyle\int_0^2 \frac{1}{1+x^2}dx;$

     $a = 1;\ u = x;\ du = dx$

$$A = \frac{1}{1}\tan^{-1}\frac{x}{1}\Big|_0^2 = \tan^{-1}2 - \tan^{-1}0 = 1.11$$

41. $a = d; u = x; du = dx$

$$kd \int \frac{dx}{d^2 + x^2} = kd \cdot \frac{1}{d} \tan^{-1} \frac{x}{d} + C$$

$$= k \tan^{-1} \left(\frac{x}{d} \right) + C$$

45. $y = \dfrac{1}{1 + x^6}$; x-axis, $x = 1$; $x = 2$, WRT y-axis

$$I_y = k \int_1^2 x^2 y \, dx = k \int_1^2 x^2 \frac{1}{1 + x^6} dx = k \int_1^2 \frac{x^2 dx}{1 + x^6}$$

$$a = 1; u = x^3; du = 3x^2 dx$$

$$I_y = \frac{k}{3} \int_1^2 \frac{3x^2 dx}{1 + x^6} = \frac{k}{3} \cdot \frac{1}{1} \tan^{-1} x^3 \Big|_1^2$$

$$= \frac{k}{3} \tan^{-1} x^3 \Big|_1^2$$

$$I_y = \frac{k}{3} (\tan^{-1} 8 - \tan^{-1} 1) = 0.22k$$

28.7 Integrations by Parts

1. $u = \sqrt{1 - x}, dv = x \, dx$

$$du = \frac{-1}{2\sqrt{1 - x}} dx, v = \frac{x^2}{2}$$

$$\int x\sqrt{1 - x} \, dx = \frac{x^2}{2} \sqrt{1 - x} - \int \frac{x^2}{2} \frac{-dx}{2\sqrt{1 - x}}$$

$$= \frac{x^2}{2} \sqrt{1 - x} + \frac{1}{4} \int \frac{x^2 dx}{\sqrt{1 - x}}$$

No, the substitution $u = \sqrt{1 - x}, dv = x \, dx$ does not work since $\int \dfrac{x^2 dx}{\sqrt{1 - x}}$ is more complex than $\int x\sqrt{1 - x} \, dx$.

5. $\int 4xe^{2x} dx$; $u = x$; $du = dx$; $dv = e^{2x} dx$

$$v = \frac{1}{2} \int e^{2x} (2dx) = \frac{1}{2} e^{2x}$$

$$\int 4e^{2x} dx = \frac{4}{2} xe^{2x} - \frac{4}{2} \int e^{2x} dx$$

$$= 2xe^{2x} - \int e^{2x} (2 \, dx)$$

$$= 2xe^{2x} - e^{2x} + C$$

9. $\int 2 \tan^{-1} x \, dx$; $u = \tan^{-1} x$; $du = \dfrac{1}{1 + x^2} dx$;

$dv = dx$;

$v = x$

$$2 \int \tan^{-1} x \, dx$$

$$= 2 \left(x \tan^{-1} x - \int \frac{x \, dx}{1 + x^2} \right)$$

$u = x^2; du = 2x \, dx$

$$= 2 \left(x \tan^{-1} x - \frac{1}{2} \int \frac{2x \, dx}{1 + x^2} \right)$$

$$= 2x \tan^{-1} x - \ln |1 + x^2| + C$$

OR

$$= 2x \tan^{-1} x - 2 \ln \sqrt{1 + x^2} + C$$

13. $\int x \ln x \, dx$; $u = \ln x$; $du = \dfrac{dx}{x}$; $dv = x \, dx$; $v = \dfrac{x^2}{2}$

$$\int x \ln x \, dx = \frac{1}{2} x^2 \ln x - \frac{1}{2} \int x^2 \frac{dx}{x}$$

$$= \frac{1}{2} x^2 \ln x - \frac{1}{2} \int x \, dx$$

$$= \frac{1}{2} x^2 \ln x - \frac{1}{2} \cdot \frac{x^2}{2} + C$$

$$= \frac{1}{2} x^2 \ln x - \frac{1}{4} x^2 + C$$

17. $\displaystyle\int_0^{\pi/2} e^x \cos x \, dx$; $u = e^x$; $du = e^x dx$; $dv = \cos x \, dx$;

$v = \sin x$

$$\int_0^{\pi/2} e^x \cos x \, dx = e^x \sin x - \int \sin x e^x dx$$

$u = e^x$; $du = e^x dx$; $dv = \sin x \, dx$; $v = -\cos x$

$$\int_0^{\pi/2} e^x \cos x \, dx = e^x \sin x - \left(-e^x \cos x + \int e^x \cos x \, dx \right)$$

$$\int_0^{\pi/2} e^x \cos x \, dx = e^x \sin x + e^x \cos x - \int e^x \cos x \, dx$$

$$2 \int e^x \cos x \, dx = e^x \sin x + e^x \cos x$$

$$\int_0^{\pi/2} e^x \cos x \, dx = \frac{1}{2} e^x (\sin x + \cos x) \Big|_0^{\pi/2} = \frac{1}{2} e^{\pi/2} (1 + 0) - \frac{1}{2} e^0 (0 + 1)$$

$$= \frac{1}{2} e^{\pi/2} - \frac{1}{2} = \frac{1}{2} (e^{\pi/2} - 1) = 1.91$$

21. $\displaystyle\int \frac{3x^3}{\sqrt{1-x^2}} \, dx = \int 3x^2 \left(1-x^2\right)^{-1/2} (x \, dx)$

$u = 3x^2$, $dv = -\dfrac{1}{2} \left(1-x^2\right)^{-1/2} (-2x \, dx)$

$du = 6x \, dx$ $v = -\left(1-x^2\right)^{1/2}$

$$\int \frac{3x^3}{\sqrt{1-x^2}} \, dx = -3x^2 \left(1-x^2\right)^{1/2} - \int -\left(1-x^2\right)^{1/2} 6x \, dx$$

$$= -3x^2 \left(1-x^2\right)^{1/2} - \int 3\left(1-x^2\right)^{1/2} (-2x \, dx)$$

$$= -3x^2 \left(1-x^2\right)^{1/2} - 2\left(1-x^2\right)^{3/2} + C$$

$$= -\sqrt{1-x^2} \left(x^2 + 2\right) + C$$

25. $u = \cos(\ln x)$, $dv = dx$

$v = x$

$du = -\sin(\ln x) \cdot \dfrac{1}{x} dx$

$$\int \cos(\ln x) \, dx = x \cos(\ln x) + \int x \sin(\ln x) \frac{1}{x} dx$$

$u = \sin(\ln x)$, $dv = dx$

$v = x$

$du = \cos(\ln x) \cdot \dfrac{1}{x} dx$

$$\int \cos(\ln x) \, dx = x \cos(\ln x) + x \sin(\ln x) - \int x \cos$$

$$(\ln x) \cdot \frac{1}{x} dx$$

$$2 \int \cos(\ln x) \, dx = x \cos(\ln x) + x \sin(\ln x)$$

$$\int \cos(\ln x) \, dx = \frac{x}{2} \left(\cos(\ln x) + \sin(\ln x) \right) + C$$

29. $A = \displaystyle\int_0^2 xe^{-x} dx$; $u = x$; $du = dx$; $dv = e^{-x} dx$;

$v = \displaystyle\int e^{-x} dx = -e^{-x}$

$$A = -xe^{-x} \Big|_0^2 = \int_0^2 -e^{-x} dx = -xe^{-x} - e^{-x} \Big|_0^2$$

$$= 2e^{-2} - e^{-2} - (0-1) = 1 - \frac{3}{e^2} = 0.594$$

33. $\bar{x} = \dfrac{\displaystyle\int_0^{\pi/2} x(\cos x) \, dx}{\displaystyle\int_0^{\pi/2} \cos x \, dx}$

Let $u = x$; $du = dx$; $dv = \cos x$; $v = \sin x$

$$\bar{x} = \frac{x \sin x \Big|_0^{\pi/2} - \displaystyle\int_0^{\pi/2} \sin x \, dx}{\sin x \Big|_0^{\pi/2}}$$

$$= \frac{x \sin x \Big|_0^{\pi/2} - (-\cos x)_0^{\pi/2}}{1}$$

$$= x \sin x + \cos x \Big|_0^{\pi/2} = \frac{\pi}{2} - 1 = 0.571$$

37. a) $A = \int_0^\pi x \sin x = \sin x - x \cos x \big|_0^\pi = \pi$

b) $A = -\int_\pi^{2\pi} x \sin x = -\sin x - x \cos x \big|_\pi^{2\pi} = 3\pi$

c) $A = \int_\pi^{3\pi} x \sin x = \sin x - x \cos x \big|_{2\pi}^{3\pi} = 5\pi$

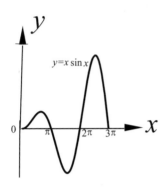

41. $i = e^{-2t} \cos t = \dfrac{dy}{dt}; t = 0, q = q_0 = 0$

$q = \int e^{-2t} \cos t \, dt; u = e^{-2t}; du = -2e^{-2t} dt;$

$dv = \cos x \, dt; v = \sin t$

$q = \int e^{-2t} \cos t \, dt = e^{-2t} \sin t + 2 \int e^{-2t} \sin t \, dt;$

$u = e^{-2t}; du = -2e^{-2t} dt; dv = \sin t \, dt;$

$v = -\cos t$

$\int e^{-2t} \cos t \, dt = e^{-2t} \sin t + 2 \left(-e^{-2t} \cos t - \int 2e^{-2t} \cos t \, dt \right)$

$\int e^{-2t} \cos t \, dt = e^{-2t} \sin t - 2e^{-2t} \cos t - 4 \int e^{-2t} \cos t \, dt$

$5 \int e^{-2t} \cos t \, dt = e^{-2t}(\sin t - 2 \cos t)$

$\int e^{-2t} \cos t \, dt = \dfrac{1}{5} e^{-2t}(\sin t - 2 \cos t) + C; t = 0; q = 0; q$

$0 = \dfrac{1}{5}(1)(\sin 0 - 2 \cos 0) + C;$ therefore, $C = \dfrac{2}{5}$

$q = \dfrac{1}{5} e^{-2t}(\sin t - 2 \cos t) + \dfrac{2}{5} = \dfrac{1}{5}[e^{-2t}(\sin t - 2 \cos t) + 2]$

28.8 Integration by Trigonometric Substitution

1. Delete the x^2 before the radical in the denominator.

$$\int \frac{dx}{\sqrt{1-x^2}} = \sin^{-1} x + C$$

5. $\int \frac{dx}{x^2 \sqrt{x^2+1}}$. Let $x = \tan\theta$, $dx = \sec^2\theta\,d\theta$

$$\int \frac{dx}{x^2\sqrt{x^2+1}} = \int \frac{\sec^2\theta\,d\theta}{\tan^2\theta\sqrt{\tan^2\theta+1}} = \int \frac{\sec^2\theta\,d\theta}{\tan^2\theta\sqrt{\sec^2\theta}} = \int \frac{\sec\theta^2\,d\theta}{\tan^2\theta\sec\theta}$$

$$= \int \frac{d\theta}{\tan\theta} = \int \cot\theta\,d\theta = \int \frac{\cos^2\theta}{\sin^2\theta}\cdot\frac{1}{\cos\theta}\,d\theta = \int \frac{1}{\sin\theta}\cdot\frac{\cos\theta}{\sin\theta}\,d\theta$$

$$= \int \csc\theta\cot\theta\,d\theta$$

9. Let $x = \sin\theta$; $dx = \cos\theta\,d\theta$

$$\int \frac{\sqrt{1-x^2}}{x^2}\,dx = \int \frac{\sqrt{1-\sin^2\theta}}{\sin^2\theta}\cos\theta\,d\theta = \int \frac{\cos^2\theta}{\sin^2\theta}\,d\theta$$

$$= \int \cot^2\theta\,d\theta = \int \left(\csc^2\theta - 1\right)d\theta$$

$$= \int \csc^2\theta\,d\theta - \int d\theta = -\cot\theta - \theta + C$$

$$= \frac{-\sqrt{1-x^2}}{x} - \sin^{-1} x + C$$

13. Let $z = 3\tan\theta$; $dz = 3\sec^2\theta\,d\theta$

$$\int \frac{6\,dz}{z^2\sqrt{z^2+9}} = 6\int \frac{3\sec^2\theta\,d\theta}{9\tan^2\theta\sqrt{9\tan^2\theta+9}} = 6\int \frac{3\sec^2\theta\,d\theta}{27\tan^2\theta\sqrt{\tan^2\theta+1}} = \frac{6}{9}\int \frac{\sec\theta\,d\theta}{\tan^2\theta} = \frac{6}{9}\int \frac{\cos\theta\,d\theta}{\sin^2\theta}$$

$$= \frac{6}{9}\int \csc\theta\cot\theta\,d\theta = -\frac{6}{9}\csc\theta + C = \frac{-6}{9\sin\theta} + C$$

$$\tan\theta = \frac{z}{3};\ \sin\theta = \frac{z}{\sqrt{9+z^2}}$$

$$\frac{-6}{9\sin\theta} + C = \frac{-6}{\dfrac{9z}{\sqrt{9+z^2}}} + C = -\frac{2\sqrt{z^2+9}}{3z} + C$$

17. $\displaystyle\int_0^{0.5} \frac{x^3\,dx}{\sqrt{1-x^2}}$, $x = \sin\theta$; $dx = \cos\theta\,d\theta$

$$\int \frac{\sin^3\theta\cos\theta\,d\theta}{\sqrt{1-\sin^2\theta}} = \int \sin^3\theta\,d\theta = \int \sin\theta\sin^2\theta\,d\theta = \int \sin\theta(1-\cos^2\theta)\,d\theta = \int \sin\theta\,d\theta - \int \cos^2\theta\sin\theta\,d\theta$$

$$= -\cos\theta + \frac{\cos^3\theta}{3}$$

$$\cos\theta = \sqrt{1-x^2};\ -\sqrt{1-x^2} + \frac{1}{3}\left(\sqrt{1-x^2}\right)^3 \Big|_0^{0.5} = -\sqrt{1-0.5^2} + \frac{1}{3}\left(\sqrt{1-0.5^2}\right)^3 + \sqrt{1} - \frac{1}{3}\sqrt{1} = 0.017$$

21. $\displaystyle\int \frac{dy}{y\sqrt{4y^2-9}}$; $2y = 3\sec\theta$; $y = \frac{3}{2}\sec\theta$; $dy = \frac{3}{2}\sec\theta\tan\theta\,d\theta$

$$\int \frac{\frac{3}{2}\sec\theta\tan\theta\,d\theta}{\frac{3}{2}\sec\theta\sqrt{4\left(\frac{3}{2}\sec\theta\right)^2-9}} = \int \frac{\tan\theta\,d\theta}{\sqrt{9\sec^2\theta-9}} = \int \frac{\tan\theta\,d\theta}{3\sqrt{\sec^2\theta-1}} = \int \frac{\tan\theta\,d\theta}{3\tan\theta} = \frac{1}{3}\int d\theta$$

$$= \frac{1}{3}\theta + C = \frac{1}{3}\sec^{-1}\frac{2}{3}y + C$$

$$\int_{2.5}^{3} \frac{dy}{y\sqrt{4y^2-9}} = \frac{1}{3}\sec^{-1}\left(\frac{2y}{3}\right)\Big|_{2.5}^{3} = \frac{1}{3}\cos^{-1}\left(\frac{3}{2y}\right)\Big|_{2.5}^{3} = 0.03997$$

25. $\displaystyle\int \frac{2\,dx}{\sqrt{e^{2x}-1}}$; $a = 1$; $e^x = \sec\theta$; $\theta = \sec^{-1}e^x$; $e^x\dfrac{dx}{d\theta} = \sec\theta\tan\theta$

$$dx = \frac{\sec\theta\tan\theta\,d\theta}{e^x} = \frac{\sec\theta\tan\theta\,d\theta}{\sec\theta} = \tan\theta\,d\theta = 2\int\frac{\tan\theta\,d\theta}{\sqrt{\sec^2\theta-1}} = 2\int\frac{\tan\theta\,d\theta}{\tan\theta} = 2\theta + C = 2\sec^{-1}e^x + C$$

29.

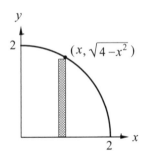

$$A = \int_0^2 \sqrt{4-x^2}\,dx, \qquad x = 2\sin\theta$$
$$dx = 2\cos\theta\,d\theta$$

$$\int\sqrt{4-x^2}\,dx = 2\int\sqrt{4-4\sin^2\theta}\cos\theta\,d\theta = 4\int\cos^2\theta\,d\theta$$

$$= 4\int\frac{1+\cos 2\theta}{2}\,d\theta = 2\theta + \int\cos 2\theta\,(2\,d\theta)$$

$$= 2\theta + \sin 2\theta = 2\theta + 2\sin\theta\cos\theta$$

$$= 2\sin^{-1}\frac{x}{2} + 2\left(\frac{x}{2}\right)\left(\frac{\sqrt{4-x^2}}{2}\right)$$

$$A = \int_0^2 \sqrt{4-x^2}\,dx = 2\sin^{-1}\frac{x}{2} + \frac{1}{2}x\sqrt{4-x^2}\,\Big|_0^2$$

$$A = \pi$$

33. Third runner: $x = \int \dfrac{48\, dt}{0.25t^2 + 4}$

$x = 48\tan^{-1}\dfrac{t}{4}$

time	1st runner	2nd runner	3rd runner
3.0	28.2	28.3	30.9
4.0	35.3	35.0	37.7

a) Third runner ahead after 3.0 h

b) After 4.0 h third runner is in first place, first runner is in second place and the second runner is in third place.

37. $V = kQ \displaystyle\int_{-a}^{a} \dfrac{dx}{\sqrt{b^2 + x^2}}$

$x = b\tan\theta; \ dx = b\sec^2\theta\, d\theta$

$V = kQ \displaystyle\int \dfrac{b\sec^2\theta\, d\theta}{\sqrt{b^2 + b^2\tan^2\theta}} = kQ \int \dfrac{b\sec^2\theta\, d\theta}{b\sec\theta} = kQ \int \sec\theta\, d\theta = kQ \ln|\sec\theta + \tan\theta|$

$\quad = kQ \ln\left| \dfrac{\sqrt{x^2 + b^2}}{b} + \dfrac{x}{b} \right| \Bigg|_{-a}^{a} = kQ \ln\left| \dfrac{\sqrt{x^2 + b^2} + x}{x} \right| \Bigg|_{-a}^{a}$

$\quad = kQ \left(\ln\left| \dfrac{\sqrt{x^2 + b^2} + a}{a} \right| - \ln\left| \dfrac{\sqrt{x^2 + b^2} - a}{-a} \right| \right)$

$\quad = kQ(\ln|\sqrt{a^2 + b^2} + a| - \ln|a| - \ln|\sqrt{a^2 + b^2}| + \ln|a|)$

$V = kQ \left(\ln\left| \dfrac{\sqrt{a^2 + b^2} + a}{\sqrt{a^2 + b^2} - a} \right| \right)$

41. $u = (x - 4)^{2/3} \Rightarrow x = u^{3/2} - 4$

$du = \dfrac{2}{3}(x - 4)^{-1/3}\, dx$

$\dfrac{3}{2}u^{1/2}\, du = dx$

$\displaystyle\int x(x - 4)^{2/3}\, dx = \int (u^{3/2} - 4)(u)\left(\dfrac{3}{2}u^{1/2}\, du \right)$

$\quad = \dfrac{3}{2} \displaystyle\int (u^3 - 4u^{3/2})\, du$

$\quad = \dfrac{3}{2}\left[\dfrac{u^4}{4} - \dfrac{4u^{5/2}}{5/2} \right] + C$

$\quad = \dfrac{3}{8}(x - 4)^{8/3} + \dfrac{12}{5}(x - 4)^{5/3} + C$

28.9   Integration by Partial Fractions: Nonrepeated Linear Factors

1. $\dfrac{10-x}{x^2+x-2} = \dfrac{10-x}{(x-1)(x+2)} = \dfrac{A}{x-1} + \dfrac{B}{x+2}$

$10 - x = A(x+2) + B(x-1)$

for $x = -2$:

$10 - (-2) = A(-2+2) + B(-2-1)$

$12 = -3B$

$B = -4$

for $x = 1$:

$10 - 1 = A(1+2) + B(1-1)$

$9 = 3A$

$A = 3$

$\dfrac{10-x}{x^2+x-2} = \dfrac{3}{x-1} + \dfrac{-4}{x+2}$

5. $\dfrac{x^2-6x-8}{x^3-4x} = \dfrac{x^2-6x-8}{x(x^2-4)}$

$\phantom{\dfrac{x^2-6x-8}{x^3-4x}} = \dfrac{x^2-6x-8}{x(x+2)(x-2)}$

$\phantom{\dfrac{x^2-6x-8}{x^3-4x}} = \dfrac{A}{x} + \dfrac{B}{x+2} + \dfrac{C}{x-2}$

9. $\displaystyle\int \dfrac{8\,dx}{x^2-4} = \int \dfrac{dx}{(x+2)(x-2)}$

$\phantom{\int \dfrac{8dx}{x^2-4}} = 8\displaystyle\int \dfrac{-\frac{1}{4}}{x+2}\,dx + 8\int \dfrac{\frac{1}{4}}{x-2}\,dx$

$\phantom{\int \dfrac{8dx}{x^2-4}} = \dfrac{-8}{4}\ln|x+2| + \dfrac{8}{4}\ln|x-2| + C$

$\phantom{\int \dfrac{8dx}{x^2-4}} = 2\ln\left|\dfrac{x-2}{x+2}\right| + C$

13. $\displaystyle\int_0^1 \dfrac{2t+4}{3t^2+5t+2}\,dt$

$ = \displaystyle\int_0^1 \dfrac{8}{3t+2}\,dt - \int_0^1 \dfrac{2}{t+1}\,dt$

$ = \dfrac{8\ln(3t+2)}{3}\Big|_0^1 - 2\ln(t+1)\Big|_0^1$

$ = \dfrac{8\ln 5}{3} - \dfrac{8\ln 2}{3} - 2\ln 2 + 2\ln 1$

$ = 1.057$

17. $\displaystyle\int \dfrac{12x^2-4x-2}{4x^3-x}\,dx = 2\int \dfrac{6x^2-2x-1}{4x^3-x}\,dx$

$2\displaystyle\int \dfrac{6x^2-2x-1}{x(4x^2-1)}\,dx = 2\int \dfrac{6x^2-2x-1}{x(2x+1)(2x-1)}\,dx$

$= 2\displaystyle\int \dfrac{dx}{x} + 2\int \dfrac{\frac{3}{2}}{2x+1}\,dx - 2\int \dfrac{\frac{1}{2}}{2x-1}\,dx$

$= 2\ln|x| + \dfrac{3\ln|2x+1|}{2} - \dfrac{\ln|2x-1|}{2} + C$

$= \dfrac{4\ln|x|}{2} + \dfrac{3\ln|2x+1|}{2} - \dfrac{\ln|2x-1|}{2} + C$

$= \dfrac{1}{2}\ln\left|\dfrac{x^4(2x+1)^3}{2x-1}\right| + C$

21. $\displaystyle\int \dfrac{dV}{(V^2-4)(V^2-9)}$

$= \displaystyle\int \dfrac{dV}{(V-2)(V+2)(V+3)(V-3)}$

$= \displaystyle\int \dfrac{\frac{1}{30}}{V-3}\,dV - \int \dfrac{\frac{1}{30}\,dV}{V+3} - \int \dfrac{\frac{1}{20}\,dV}{V-2} + \int \dfrac{\frac{1}{20}\,dV}{V+2}$

$= \dfrac{1}{30}\ln|V-3| - \dfrac{1}{30}\ln|V+3|$

$\quad - \dfrac{1}{20}\ln|V-2| + \dfrac{1}{20}\ln|V+2| + C$

$= \dfrac{2}{60}\ln|V-3| - \dfrac{2}{60}\ln|V+3| - \dfrac{3}{60}\ln|V-2|$

$\quad + \dfrac{3}{60}\ln|V+2| + C$

$= \dfrac{1}{60}\ln\left|\dfrac{(V+2)^3(V-3)^2}{(V-2)^3(V+3)^2}\right| + C$

25. $\dfrac{1}{u(a+bu)} = \dfrac{A}{u} + \dfrac{B}{a+bu}$

$1 = A(a+bu) + Bu$

for $u = 0$: $1 = aA$

$A = \dfrac{1}{a}$

for $u = -\dfrac{a}{b}$: $1 = B\left(-\dfrac{a}{b}\right)$

$\phantom{for u = -\dfrac{a}{b}:}B = -\dfrac{b}{a}$

$$\int \frac{du}{u(a+bu)}$$

$$= \int \frac{\frac{1}{a}}{u} \, du + \int \frac{-\frac{b}{a}}{a+bu} \, du$$

$$= \frac{1}{a} \ln |u| - \frac{1}{a} \int \frac{b \, du}{a+bu}$$

$$= \frac{1}{a} \ln |u| - \frac{1}{a} \ln |a+bu| + C$$

$$= -\frac{1}{a}(-\ln |u| + \ln |a+bu|) + C$$

$$= -\frac{1}{a} \ln \frac{|a+bu|}{|u|} + C$$

$$= -\frac{1}{a} \ln \left| \frac{a+bu}{u} \right| + C$$

29. $u = \sin\theta$, $du = \cos\theta \, d\theta$

$$\int \frac{\cos\theta \, d\theta}{\sin^2\theta + 2\sin\theta - 3} = \int \frac{du}{u^2 + 2u - 3} = \int \frac{du}{(u+3)(u-1)}$$

$$1 = A(u-1) + B(u+3)$$

$$u = 1, \quad 1 = B(1+3) \Rightarrow B = \frac{1}{4}$$

$$u = -3, \quad 1 = A(-3-1) \Rightarrow A = -\frac{1}{4}$$

$$\int \frac{\cos\theta \, d\theta}{\sin^2\theta + 2\sin\theta - 3} = \int \frac{-\frac{1}{4} du}{u+3} + \int \frac{\frac{1}{4} du}{u-1}$$

$$= -\frac{1}{4} \ln |u+3| + \frac{1}{4} \ln |u-1| + C$$

$$= -\frac{1}{4} \ln |\sin\theta + 3| + \frac{1}{4} \ln |\sin\theta - 1| + C$$

33. $\dfrac{dy}{dx} = \dfrac{3x+5}{x^2+5x} \Rightarrow y = \int \dfrac{(3x+5) \, dx}{x^2+5x}$

$$= \ln |x| + 2 \ln |x+5| + C$$

$$0 = \ln |1| + 2 \ln |6| + C \Rightarrow C = -\ln 36$$

$$y = \ln |x| + \ln (x+5)^2 - \ln 36$$

$$y = \ln \left| \frac{x(x+5)^2}{36} \right|$$

28.10 Integration by Partial Fractions: Other Cases

1. $\dfrac{2}{x(x+3)^2} = \dfrac{A}{x} + \dfrac{B}{x+3} + \dfrac{C}{(x+3)^2}$

5. $\displaystyle\int \frac{x-8}{x^3 - 4x^2 + 4x} = \int \frac{-2}{x} \, dx + \int \frac{2}{x-2} \, dx - \int \frac{3}{(x-2)^2} \, dx$

$$= -2 \ln |x| + 2 \ln |x-2| + \frac{3}{x-2} + C$$

$$= 2 \ln \left| \frac{x-2}{x} \right| + \frac{3}{x-2} + C$$

9. $\dfrac{2s}{(s-3)^3} = \dfrac{A}{s-3} + \dfrac{B}{(s-3)^2} + \dfrac{C}{(s-3)^3}$

$$= \frac{A(s-3)^2 + B(s-3) + C}{(s-3)^3}$$

$$2s = A(s-3)^2 + B(s-3) + C$$

$$s = 3, \ 6 = C$$

$$2s = A(s-3)^2 + B(s-3) + 6$$

$$\left. \begin{array}{l} s = 1, \quad 2 = 4A - 2B + 6 \\ s = 2, \quad 4 = A - B + 6 \end{array} \right\} A = 0, \ = 2$$

$$\int_1^2 \frac{2s}{(s-3)} \, ds = \int_1^2 \frac{2}{(s-3)^3} \, ds + \int_1^2 \frac{6}{(s-3)^3} \, dx$$

$$= 2 \cdot \frac{(s-3)^{-2+1}}{-2+1} + 6 \cdot \frac{(s-3)^{-3+1}}{-3+1} \bigg|_1^2$$

$$= \frac{-2}{(s-3)} - \frac{3}{(s-3)^2} \bigg|_1^2 = -\frac{5}{4}$$

13. $\dfrac{x^2+x+5}{(x+1)(x^2+4)} = \dfrac{A}{x+1} + \dfrac{Bx+C}{x^2+4} = \dfrac{A(x^2+4)+(Bx+C)(x+1)}{(x+1)(x^2+4)}$

$\qquad x^2+x+5 = A(x^2+4)+(Bx+C)(x+1)$

$\left. \begin{array}{lr} x=0, & 5 = 4A + C \\ x=1, & 7 = 5A+2B+2C \\ x=2, & 11 = 8A+6B+3C \end{array} \right\} A=1,\ B=0,\ C=1$

$\displaystyle\int_0^2 \frac{x^2+x+5}{(x+1)(x^2+4)}\,dx = \int_0^2 \frac{dx}{x+1} + \int_1^2 \frac{1}{x^2+4}\,dx \ln|x+1| + \left.\frac{\tan^{-1}\frac{x}{2}}{2}\right|_0^2 = 1.491$

17. $\displaystyle\int \frac{5x^2+8x+16}{x^2\left(x^2+4x+8\right)}\,dx = \int \frac{3\,dx}{x^2+4x+8} + \int \frac{2\,dx}{x^2} = \frac{3}{2}\tan^{-1}\left(\frac{x+2}{2}\right) - \frac{2}{x} + C$

21. $\displaystyle\int \frac{-x^3+x^2+x+3}{(x+1)(x^2+1)^2}\,dx = \int \frac{2\,dx}{(x^2+1)^2} - \int \frac{x\,dx}{x^2+1} + \int \frac{dx}{x+1} = \tan^{-1}x + \frac{x}{x^2+1} - \frac{1}{2}\ln(x^2+1) + \ln|x+1| + C$

25.
$\qquad A = -\displaystyle\int_1^3 \frac{x-3}{x^3+x^2}\,dx = -\int_1^3 \frac{x-3}{x^2(x+1)}\,dx$

$\dfrac{x-3}{x^2(x+1)} = \dfrac{Ax+B}{x^2} + \dfrac{C}{x+1} = \dfrac{(Ax+B)(x+1)+Cx^2}{x^2(x+1)} = \dfrac{Ax^2+Ax+Bx+B+Cx^2}{x^2(x+1)}$

$\qquad x-3 = (A+C)x^2 + (A+B)x + B$

(1) $A+C=0,\ C=-4$
(2) $A+B=1,\ A=4$
(3) $B=-3$

$\qquad \dfrac{x-3}{x^2(x+1)} = \dfrac{4x-3}{x^2} - \dfrac{4}{x+1} = \dfrac{4}{x} - \dfrac{3}{x^2} - \dfrac{4}{x+1}$

$\displaystyle\int_1^3 \frac{x-3}{x^3+x^2}\,dx = \int_1^3 \frac{4}{x}\,dx - 3\int_1^3 \frac{dx}{x^2} - 4\int_1^3 \frac{dx}{x+1} = 4\ln|x|\Big|_1^3 + \frac{3}{x}\Big|_1^3 - 4\ln(x+1)\Big|_1^3$

$\qquad\qquad = 4(\ln 3 - \ln 1) + \frac{3}{3} - \frac{3}{1} - 4\ln(4) + 4\ln 2 = \ln 3^4 + 1 - 3 - \ln 4^4 + \ln 2^4$

$\qquad\qquad\qquad = -2 + \ln 81 - \ln 256 + \ln 16 = -2 + \ln\frac{81\cdot 16}{256} = -2 + \ln\frac{81}{16} = -0.3781$

$A = 2 - \ln\dfrac{81}{16} = 0.3781$

29. $\dfrac{t^2 + 14t + 27}{(2t+1)(t+5)^2} = \dfrac{A}{2t+1} + \dfrac{B}{t+5} + \dfrac{C}{(t+5)^2}$

$\dfrac{t^2 + 14t + 27}{(2t+1)(t+5)^2} = \dfrac{A(t+5)^2 + B(t+5)(2t+1) + C(2t+1)}{(2t+1)(t+5)^2}$

$t^2 + 14t + 27 = At^2 + 10At + 25A + 2Bt^2 + 11Bt + 5B + 2Ct + C$

$t^2 + 14t + 27 = (A + 2B)t^2 + (10A + 11B + 2C)t + 25A + 5B + C$

(1) $\quad\quad A + 2B = 1$
(2) $\quad 10A + 11B + 2C = 14$ $\left.\right\} A = 1,\ B = 0,\ C = 2$
(3) $\quad\quad 25A + 5B + C = 27$

$\dfrac{ds}{dt} = \dfrac{t^2 + 14t + 27}{(2t+1)(t+5)^2} = \dfrac{1}{2t+1} + \dfrac{2}{(t+5)^2}$

$s = \dfrac{1}{2}\displaystyle\int_0^{2.00} \dfrac{2}{2t+1}\,dt + \int_0^{2.00} \dfrac{2}{(t+5)^2}\,dt;\ s = \dfrac{1}{2}\ln|2t+1|\Big|_0^{2.00} - 2\cdot \dfrac{1}{(t+5)}\Big|_0^{2.00}$

$s = \dfrac{1}{2}\ln 5.00 - \dfrac{1}{2}\ln 1 - \dfrac{2}{7.00} + \dfrac{2}{5.00} = 0.919\text{ m}$

28.11 Integration by Use of Tables

1. $\displaystyle\int \dfrac{x\,dx}{\left(2+3x\right)^2}$ is formula 3 with $u = x$, $du = dx$, $a = 2$, $b = 3$.

5. $u = x^2$, $du = 2x\,dx$

$\displaystyle\int \dfrac{x\,dx}{\left(4 - x^4\right)^{3/2}} = \int \dfrac{x\,dx}{\left(2^2 - \left(x^2\right)^2\right)^{3/2}} = \int \dfrac{\frac{1}{2}\,du}{\left(2^2 - u^2\right)^{3/2}}$

which is formula 25 with $a = 2$.

9. Formula #1; $u = x$; $a = 2$; $b = 5$; $du = dx$

$\displaystyle\int \dfrac{3x\,dx}{2+5x} = 3\int \dfrac{3x\,dx}{2+5x}$

$= 3\left\{ \dfrac{1}{25}(2+5x) - 2\ln|2+5x| \right\} + C$

$= \dfrac{3}{25}\left[2+5x - 2\ln|2+5x| \right] + C$

13. Formula #24; $u = y$, $a = 2$

$\displaystyle\int \dfrac{8\,dy}{\left(y^2+4\right)^{3/2}} = \dfrac{8y}{4\sqrt{y^2+4}} + C$

$= \dfrac{2y}{\sqrt{y^2+4}} + C$

17. Formula #17; $u = 2x$; $du = 2dx$; $a = 3$

$\displaystyle\int \dfrac{\sqrt{4x^2-9}}{x}\,dx = \int \dfrac{\sqrt{(2x)^2 - 3^2}}{2x}\,dx$

$= \sqrt{4x^2-9} - 3\sec^{-1}\left(\dfrac{2x}{3}\right) + C$

21. Formula #52; $u = r^2$; $du = 2r\,dr$

$$6 \int \tan^{-1} r^2 (r\,dr)$$

$$= 3 \int \tan^{-1} r^2 (2r\,dr)$$

$$= 3 \left[r^2 \tan^{-1} r^2 - \frac{1}{2} \ln(1 + r^4) \right] + C$$

$$= 3r^2 \tan^{-1} r^2 - \frac{3}{2} \ln(1 + r^4) + C$$

25. $\displaystyle \int \frac{dx}{2x\sqrt{x^2 + \frac{1}{4}}} = \int \frac{dx}{2x\sqrt{\frac{4x^2+1}{4}}} = \int \frac{dx}{x\sqrt{4x^2 + 1}}$

Formula #11; $u = 2x$; $du = 2\,dx$; $a = 1$

$$\int \frac{dx}{x\sqrt{4x^2 + 1}} = \int \frac{2\,dx}{2x\sqrt{(2x)^2 + 1^2}}$$

$$= -\ln\left(\frac{1 + \sqrt{4x^2 + 1}}{2x} \right) + C$$

29. Formula #40; $a = 1$; $u = x$; $du = dx$; $b = 5$

$$\int_0^{\pi/12} \sin\theta \cos 5\theta\, d\theta = -\frac{\cos(-4\theta)}{2(-4)} - \frac{\cos 6\theta}{12}$$

$$= \frac{1}{8} \cos 4\theta - \frac{1}{12} \cos 6\theta \Big|_0^{\pi/12}$$

$$= 0.0208$$

33. let $u = x^2$, $du = 2x\,dx$
$\qquad\quad u^2 = x^4$

$$\int \frac{2x\,dx}{(1 - x^4)^{3/2}} = \int \frac{du}{(1 - u^2)^{3/2}}$$

Formula #25: $a = 1$

$$\int \frac{2x\,dx}{(1 - x^4)^{3/2}} = \frac{u}{\sqrt{1 - u^2}} + C;$$
$$\int \frac{2x\,dx}{(1 - x^4)^{3/2}} = \frac{x^2}{\sqrt{1 - x^4}} + C$$

37. Formula #46; $u = x^2$; $du = 2x\,dx$; $n = 1$

$$\int x^3 \ln x^2\, dx = \frac{1}{2} \int x^2 \ln x^2 (2x\,dx)$$

$$= \frac{1}{2} \left[(x^2)^2 \left(\frac{\ln x^2}{2} - \frac{1}{4} \right) \right]$$

$$= \frac{1}{2} \left[\frac{x^4}{2} \left(\ln x^2 - \frac{1}{2} \right) \right]$$

$$= \frac{1}{4} x^4 \left(\ln x^2 - \frac{1}{2} \right) + C$$

41. Let $u = t^3$, $du = 3t^2\,dt$

$$\int t^2 \left(t^6 + 1 \right)^{3/2} dt = \int \frac{1}{3} \left(u^2 + 1 \right)^{3/2} du,\ \#19,\ a = 1$$

$$= \frac{1}{3} \left[\frac{u}{4} \left(u^2 + 1 \right)^{3/2} + \frac{3u}{8} \sqrt{u^2 + 1} + \frac{3}{8} \ln\left(u + \sqrt{u^2 + 1} \right) \right]$$

$$= \frac{1}{12} t^3 \left(t^6 + 1 \right)^{3/2} + \frac{t^3}{8} \sqrt{t^6 + 1} + \frac{1}{8} \ln\left(t^3 + \sqrt{t^6 + 1} \right) + C$$

$$= \frac{1}{4} \left[\frac{\sin^4 u \cos^2 u}{6} + \frac{\sin^4 u}{12} \right] + C$$

$$= \frac{1}{24} \sin^4 4x \cos^2 4x + \frac{1}{48} \sin^4 4x + C$$

45. $y = 0.00037x^2$, $\dfrac{dy}{dx} = 0.00074x$

$$s = \int_a^b \sqrt{1 + \left(\frac{dy}{dx} \right)^2}\ dx$$

$$s = 2 \int_0^{1140} \sqrt{1 + \left(0.00074x \right)^2}\ dx = 2527.08$$

The cables are 2530 m long.

49. $F = w \int_0^3 lhdh = w \int_0^3 x(3-y)\,dy = w \int_0^3 \frac{3-y}{\sqrt{1+y}}\,dy$

Formula #6

$\int \frac{3-y}{\sqrt{1+y}}\,dy = 3\int \frac{dy}{\sqrt{1+y}} - \int \frac{y\,dy}{\sqrt{1+y}}$

$= 3\frac{(1+y)^{1/2}}{\frac{1}{2}} - \left[\frac{-2(2-y)\sqrt{1+y}}{3(1)^2}\right] + C$

$F = w\int_0^3 \frac{3-y}{\sqrt{1+y}}\,dy$

$= w\left[6(1+y)^{1/2} + \frac{2}{3}(2-y)(1+y)^{1/2}\right]\Big|_0^3$

$= w\left[6(2) + \frac{2}{3}(-1)(2) - 6(1) - \frac{2}{3}(2)(1)\right]$

$F = w\left(12 - \frac{4}{3} - 6 - \frac{4}{3}\right) = \frac{10w}{3} = \frac{10(62.4)}{3} = 208 \text{ lb}$

Chapter 28   Review Exercises

1. $u = -8x,\ du = -8dx$

$\int e^{-8x}\,dx = -\frac{1}{8}\int e^{-8x}(-8dx) = -\frac{1}{8}e^{-8x} + C$

5. $\int_0^{\pi/2} \frac{4\cos\theta\,d\theta}{1+\sin\theta} = 4\int_0^{\pi/2} \frac{\cos\theta\,d\theta}{1+\sin\theta};\ u$

$= 1+\sin\theta,$

$du = \cos\theta$

$= 4\ln(1+\sin\theta)\Big|_0^{\pi/2}$

$= 2.77$

9. $\int_0^{\pi/2} \cos^3 2\theta\,d\theta = \int_0^{\pi/2} \cos^2 2\theta\cos 2\theta\,d\theta$

$= \int_0^{\pi/2}\left(-\sin^2 2\theta\right)\cos 2\theta\,d\theta$

$= \int_0^{\pi/2} \cos 2\theta\,d\theta - \int_0^{\pi/2} \sin^2 2\theta\cos 2\,d\theta$

$= \frac{1}{2}\int_0^{\pi/2}\cos 2\theta(2d\theta) - \frac{1}{2}\int_0^{\pi/2}\sin^2 2\theta\cos 2\theta(2d\theta)$

$= \frac{1}{2}\left[\sin 2\theta - \frac{1}{3}\sin^3 2\theta\right]\Big|_0^{\pi/2}$

$= \frac{1}{2}\left[\left(\sin\pi - \frac{1}{3}\sin^3\pi\right) = \left(\sin 0 - \frac{1}{3}\sin^3 0\right)\right]$

$= \frac{1}{2}(0) = 0$

13. $\int(\sin t + \cos t)^2\cdot \sin t\,dt$

$= \int\left(\sin^2 t + 2\sin t\cos t + \cos^2 t\right)\cdot\sin t\,dt$

$= \int(1+2\sin t\cos t)\cdot\sin t\,dt$

$= \int\left(\sin t + 2\sin^2 t\cos t\right)dt$

$= \int\sin t\,dt + 2\int\sin^2 t(\cos t\,dt)$

$= -\cos t + \frac{2\sin^3 t}{3} + C$

17. $\int 6\sec^4 3x\,dx = 6\int \sec^2 3x \sec^2 3x\,dx$

$= 6\int \left(1 + \tan^2 3x\right)\sec^2 3x\,dx$

$= 2\int \sec^2 3x\left(3dx\right) + 2\int \tan^2 3x \sec^2 3x\left(3dx\right)$

$= 2\tan 3x + 2\dfrac{\tan^3 3x}{3} + C$

$= \dfrac{2}{3}\tan^3 3x + 2\tan 3x + C$

21. $\int \dfrac{3x\,dx}{4 + x^4} = 3\int \dfrac{x\,dx}{4 + x^4} = 3\int \dfrac{1}{2^2 + \left(x^2\right)^2}x\,dx$

$= \dfrac{3}{2}\int \dfrac{1}{2^2 + \left(x^2\right)^2}2x\,dx$

$= \dfrac{3}{2}\left(\dfrac{1}{2}\tan^{-1}\dfrac{x^2}{2} + C_1\right)$

$= \dfrac{3}{4}\tan^{-1}\dfrac{x^2}{2} + C$ where $C = \dfrac{3}{2}C_1$.

25. $u = e^{2x}, \; du = e^{2x}\left(2dx\right)$

$\int \dfrac{e^{2x}\,dx}{\sqrt{e^{2x} + 1}} = \dfrac{1}{2}\int \left(e^{2x} + 1\right)^{-1/2}e^{2x}\left(2dx\right)$

$= \dfrac{1}{2}\left(e^{2x} + 1\right)^{1/2}\left(2\right) + C = \sqrt{e^{2x} + 1} + C$

29. $\displaystyle\int_0^{\pi/6} 3\sin^2 3\phi\,d\phi = \int_0^{\pi/6} 3\cdot\dfrac{\left(1 - \cos 6\phi\right)}{2}d\phi$

$= \displaystyle\int_0^{\pi/6} \dfrac{3}{2}d\phi - \dfrac{1}{4}\int_0^{\pi/6}\cos 6\phi\left(6d\phi\right)$

$= \dfrac{3}{2}\phi\Big|_0^{\pi/6} - \dfrac{1}{4}\sin 6\phi\Big|_0^{\pi/6}$

$= \dfrac{3}{2}\left[\dfrac{\pi}{6} - 0\right] - \dfrac{1}{4}\left[\sin \pi - \sin 0\right] = \dfrac{\pi}{4}$

33. $\dfrac{3u^2 - 6u - 2}{u^2\left(3u + 1\right)} = \dfrac{Au + B}{u^2} + \dfrac{C}{3u + 1}$

$= \dfrac{\left(Au + B\right)\left(3u + 1\right) + Cu^2}{u^2\left(3u + 1\right)}$

$3u^2 - 6u - 2 = 3Au^2 + Au + 3Bu + B + Cu^2$

$3u^2 - 6u - 2 = \left(3A + C\right)u^2 + \left(A + 3B\right)u + B$

(1) $3A + C = 3, \; 3\left(0\right) + C = 3, \; C = 3$

(2) $A + 3B = -6; \; A + 3\left(-2\right) = -6, \; A = 0$

(3) $\qquad B = -2$

$\int \dfrac{3u^2 - 6u - 2}{u^2\left(3u + 1\right)}du = \int \dfrac{-2}{u^2}du + \int \dfrac{3}{3u + 1}du$

$= \dfrac{2}{u} + \ln\left|3u + 1\right| + C$

37. $\displaystyle\int_1^e 3\cos\left(\ln x\right)\cdot\dfrac{dx}{x} = 3\sin\left(\ln x\right)\Big|_1^e$

$= 3\sin\left(\ln e\right) - 3\sin\left(\ln 1\right)$

$= 3\sin\left(1\right) - 3\sin\left(0\right)$

$= 3\sin 1 - 3\cdot 0 = 3\sin 1 \approx 2.52$

41. $u = \cos x, \; du = -\sin x\,dx$

$\int \dfrac{\sin x\cos^2 s}{5 + \cos^2 x}dx = -\int \dfrac{u^2\,du}{5 + u^2} = -\int\left(1 - \dfrac{5}{5 + u^2}\right)du$

$= -u + \dfrac{5}{\sqrt{5}}\tan^{-1}\dfrac{u}{\sqrt{5}} + C$

$= -\cos x + \dfrac{5}{\sqrt{5}}\tan^{-1}\dfrac{\cos x}{\sqrt{5}} + C$

45. $\int e^{\ln 4x}\,dx = \int 4x\,dx = 2x^2 + C$

$\int \ln e^{4x}\,dx = \int 4x\cdot\ln e\,dx = \int 4x\,dx = 2x^2 + C$

49. Use the general power formula. Let $u = e^x + 1$,
    $du = e^x\,dx, \; n = 2$.

$\int e^x\left(e^x + 1\right)^2 dx = \int e^x\left(e^{2x} + 2e^x + 1\right)dx$

$= \int\left(e^{3x} + 2e^{2x} + e^x\right)dx$

$= \int e^{3x}\,dx + \int 2e^{2x}\,dx + \int e^x\,dx$

$= \dfrac{1}{3}\int e^{3x}\left(3dx\right) + 2\left(\dfrac{1}{2}\right)\int e^{2x}\left(2dx\right) + \int e^x\,dx$

$= \dfrac{1}{3}e^{3x} + e^{2x} + e^x + C_2; \; C_2 = C_1 + \dfrac{1}{3}$

53. (a) $u = x^2 + 4$
        $du = 2x\,dx$

$\int \dfrac{x}{\sqrt{x^2 + 4}}dx = \int \dfrac{\frac{1}{2}du}{\sqrt{u}} = \dfrac{1}{2}\dfrac{u^{-1/2+1}}{2 - \frac{1}{2} + 1} + C = u^{1/2} + C$

$= \sqrt{x^2 + 4} + C$

(b) $x = 2\tan\theta$
    $dx = 2\sec^2\theta\,d\theta$

$$\int \frac{x\,dx}{\sqrt{x^2+4}} = \int \frac{2\tan\theta\left(2\sec^2\theta\,d\theta\right)}{\sqrt{4\tan^2\theta+4}}$$

$$= \int \frac{2\tan\theta\sec^2\theta\,d\theta}{\sec\theta}$$

$$= \int 2\tan\theta\sec\theta\,d\theta$$

$$= 2\sec\theta + C$$

$$= 2\frac{\sqrt{x^2+4}}{2} + C$$

$$= \sqrt{x^2+4} + C$$

(a) is simpler

57. $\int \sin 2x\,dx = \frac{1}{2}\int \sin(2x)(2\,dx)$

$$= 2\int \sin x(\cos x\,dx)$$

$$= -2\int \cos x(-\sin x\,dx)$$

61. $x^2 + y^2 = 5^2$; $y = \sqrt{25-x^2}$; $A = 2\int_3^5 \sqrt{25-x^2}\,dx$

$$A = 2\left[\frac{x}{2}\sqrt{25-x^2} + \frac{25}{2}\sin^{-1}\frac{x}{5}\right]_3^5 \quad \text{Formula 15 in}$$

table of integrals.

$$A = 2\left[\frac{5}{2}\sqrt{0} + \frac{25}{2}\sin^{-1}1\right] - 2\left[\frac{3}{2}\sqrt{16} + \frac{25}{2}\sin^{-1}\frac{3}{5}\right]$$

$$= 2\left[\frac{25}{2}\left(\frac{\pi}{2}\right)\right] - 2\left[6 + \frac{25}{2}(0.6435)\right]$$

$$= 2[19.63] - 2[14.04] = 11.18$$

65. $y = xe^x$, $y = 0$, $x = 2$

shells: $V = \int_0^2 2\pi xy\,dx = 2\pi\int_0^2 x\left(xe^x\right)dx$

$$V = 2\pi\int_0^2 x^2 e^x\,dx, \text{ Formula 45}$$

$$V = 2\pi\left[e^x\left(x^2 - 2x + 2\right)\right]_0^2$$

$$V = 2\pi\left[e^2\left(4 - 4 + 2\right) - e^0\left(0 - 0 + 2\right)\right]$$

$$V = 2\pi\left[2e^2 - 2\right] = 4\pi\left[e^2 - 1\right]$$

$$V = 80.29$$

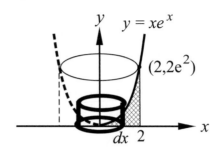

69. $y = \ln\sin x$, $x = 0.5$ to $x = 2.5$

$$\frac{dy}{dx} = \frac{1}{\sin x}(\cos x) = \cot x$$

$$\sqrt{1 + \left(\frac{dy}{dx}\right)^2} = \sqrt{+\cot^2 x}$$

$$= \sqrt{\csc^2 x}$$

$$= \csc x$$

$$L = \int_{0.5}^{2.5} \csc x\,dx = \ln\left|\frac{\sin x}{\cos x + 1}\right|_{0.5}^{2.5}$$

$$L = \ln\frac{\sin 2.5}{\cos 2.5 + 1} - \ln\frac{\sin 0.5}{\cos 0.5 + 1} = 2.47 \text{ m}$$

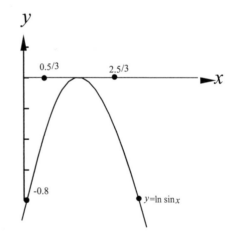

73. $\Delta S = \int (c_v/T)\,dT = \int \frac{a + bT + cT^2}{T}\,dT$

$$= \int \frac{a}{T}\,dT + \int b\,dT + \int cT\,dT$$

$$= a\ln T + bT + \frac{1}{2}cT^2 + C$$

77.
$$\int \frac{dv}{32 - 0.5v} = \int dt$$

$$-\frac{1}{0.5} \ln|32 - 0.5v| = t + C$$

$$-\frac{1}{0.5} \ln|32 - 0.5(0)| = 0 + C$$

$$C = -\frac{1}{0.5} \ln|32|$$

81. $V = \pi \int_{2.00}^{4.00} y^2 \, dx = \pi \int_{2.00}^{4.00} e^{-0.2x} \, dx;$ and

$u = 0.2x \, du = -0.2dx$

$$= -\frac{\pi}{0.2} \int_{2.00}^{4.00} e^{-0.2x} (-0.2) \, dx = \frac{-\pi}{0.2} e^{-0.2x} \Big|_{2.00}^{4.00}$$

$$= -\frac{\pi}{2} \left[e^{-0.8} - e^{-0.4} \right] = 3.47 \text{ cm}^3$$

85.

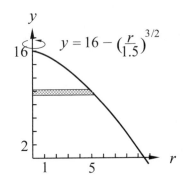

$$r = 1.5x^{2/3} \Rightarrow x = \left(\frac{r}{1.5} \right)^{3/2}$$

$$\frac{d}{dr} \left(16 - \left(\frac{r}{1.5} \right)^{3/2} \right) = -\left(\frac{r}{1.5} \right)^{1/2}$$

$$S = \int_0^{1.5\left(16^{2/3}\right)} 2\pi r \sqrt{1 + \frac{r}{1.5}} \, dr$$

$$S = 644 \text{ ft}^2$$

Chapter 29

PARTIAL DERIVATIVES AND DOUBLE INTEGRALS

29.1 Functions of Two Variables

1. $f(x, y) = 3x^2 + 2xy - y^3$

$f(-2, 1) = 3(-2)^2 + 2(-2)(1) - (1)^3$

$= 7$

5. From geometry:

$A = 2\pi rh + 2\pi r^2;\ V = \pi r^2 h,\ h = \dfrac{V}{\pi r^2}$

$A = 2\pi r \left(\dfrac{V}{\pi r^2}\right) + 2\pi r^2 = \dfrac{2V}{r} + 2\pi r^2$

9. $f(x, y) = 2x - 6y$

$f(0, -4) = 2(0) - 6(-4) = 24$

$f(-3, 2) = 2(-3) - 6(2) = -18$

13. $Y(y, t) = \dfrac{2 - 3y}{t - 1} + 2y^2 t$

$Y(3, 2) = \dfrac{2 - 3(3)}{2 - 1} + 2(3)^2 (1) = 29$

$Y(y, 2) = \dfrac{2 - 3y}{2 - 1} + 2y^2 (2)$

$= 2 - 3y + 4y^2$

17. $H(p, q) = p - \dfrac{p - 2q^2 - 5q}{p + q}$

$H(p, q + k)$

$= p - \dfrac{p - 2(q + k)^2 - 5(q + k)}{p + q + k}$

$= \dfrac{p(p + q + k) - p + 2(q^2 + 2kq + k^2) + 5(q + k)}{p + q + k}$

$= \dfrac{p^2 + pq + pk - p + 2q^2 + 4kq + 2k^2 + 5q + 5k}{p + q + k}$

21. $f(x, y) = xy + x^2 - y^2$

$f(x, x) - f(x, 0)$

$= x(x) + x^2 - x^2 - [x(0) + x^2 - 0^2]$

$= x^2 + x^2 - x^2 - x^2$

$= 0$

25. $f(x, y) = \frac{\sqrt{y}}{2x}$; considering $\sqrt{y}$, $y \geq 0$ for real values of $f(x, y)$; considering $2x, x \neq 0$ to avoid division by zero. Thus, $y \geq 0$ and $x \neq 0$.

29. $M = mv = 0.160(45)$

$M = 7.2\ \text{kg} \cdot \text{m/s}$

33. $p = \dfrac{nRT}{V}$; $n = 3$ mol, $R = 8.31$ J/mol·K

$T = 300$ K, $V = 50$ m^3

$p = \dfrac{3(8.31)(300)}{50} = 150$ Pa

37. $i = \dfrac{6.0 \sin(0.01t)}{R + 0.12} = \dfrac{6.0 \sin(0.01(0.75))}{1.50 + 0.12}$

$i = 0.028$A

41. $p = 2l + 2w;\ l = \dfrac{p - 2w}{2}$

$A = lw = \dfrac{p - 2w}{2} w = \dfrac{pw - 2w^2}{2}$

$p = 250$ cm, $w = 55$ cm

$A = \dfrac{250(55) - 2(55)^2}{2} = 3850$ cm^2

29.2   Curves and Surfaces in Three Dimensions

1. $3x - y + 2z + 6 = 0$

intercepts: $(0, 0, -3), (0, 6, 0), (-2, 0, 0)$

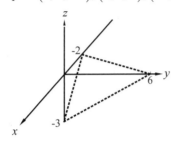

5. $x + y + 2z - 4 = 0$; plane

Intercepts: $(4, 0, 0), (0, 4, 0), (0, 0, 2)$

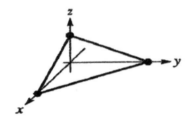

9. $z = y - 2x - 2$; plane

Intercepts: $(-1, 0, 0), (0, 2, 0), (0, 0, -2)$

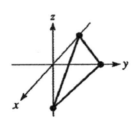

13. $x^2 + y^2 + z^2 = 4$

Intercepts: $(\pm 2, 0, 0), (0, \pm 2, 0), (0, 0, \pm 2)$

Traces:
yz-plane: $y^2 + z^2 = 4$, circle, $r = 2$
xz-plane: $x^2 + z^2 = 4$, circle, $r = 2$
xy-plane: $x^2 + y^2 = 4$, circle, $r = 2$

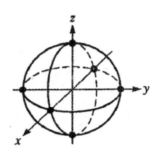

17. $z = 2x^2 + y^2 + 2$

Intercepts: No x-int., no y-int., $(0,0,2)$

Traces:

yz-plane:: $z = y^2 + 2$; parabola, $V(0,0,2)$
xz-plane: $z = 2x^2 + 2$, parabola, $V(0,0,2)$
xy-plane: No trace, $(2x^2 + y^2 + 2 \neq 0)$

Section: For $z = 4$, $2x^2 + y^2 = 2$, ellipse

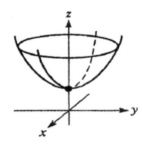

21. $x^2 + y^2 = 16$

Intercepts: $(\pm 4, 0, 0), (0, \pm 4, 0)$, no z-int.

Traces and Sections:

Since z is not present in the equation, the trace and sections are circles $x^2 + y^2 = 16$, with $r = 4$, for all z. This is a cylindrical surface.

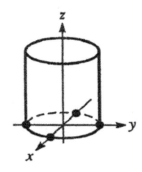

25. $z = \dfrac{1}{x^2 + y^2} \Rightarrow z > 0$

in xz-plane: $z = \dfrac{1}{x^2}$

in xz-plane: $z = \dfrac{1}{y^2}$

$z = \text{constant} \Rightarrow \text{circular sections}$

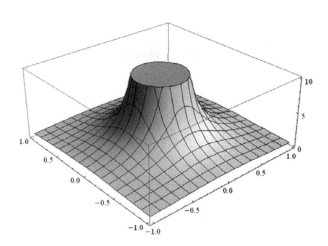

29. $z = y^4 - 4y^2 - 2x^2$

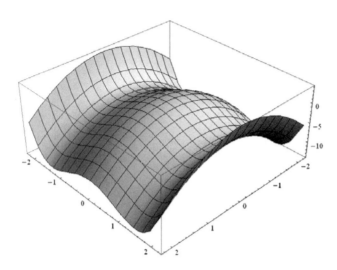

41. $p = \dfrac{T}{2V}$

<u>Sections:</u>

for constant p: $T = \text{constant} \cdot V$, a line

for constant V: $p = \text{constant} \cdot T$, a line

for constant T:

$p = \dfrac{\text{constant}}{V}$, one branch of a rotated hyperbola.

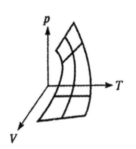

33. (a) $r = 2$ describes a cylinder with z-axis as axis and radius 2.

(b) $\theta = 2$ describes a plane with $\theta = 2$ for all r and z.

(c) $z = 2$ describes a plane 2 units above xy-plane.

29.3   Partial Derivatives

1. $z = \dfrac{x \ln y}{y^2 + 1}$

$\dfrac{\partial z}{\partial x} = \dfrac{\ln y}{y^2 + 1}$

$\dfrac{\partial z}{\partial y} = \dfrac{x(y^2 + 1)1/y - 2xy \ln y}{(y^2 + 1)^2}$

$\dfrac{\partial z}{\partial y} = \dfrac{x(y^2 + 1) - 2xy^2 \ln y}{y(y^2 + 1)^2}$

37. $t = 4x - y^2$

$t = -4;\ y^2 = 4(x + 1)$, parabola, $V(-1, 0)$

$t = 0;\ y^2 = 4x$, parabola, $V(0, 0)$

$t = 8;\ y^2 = 4(x - 2)$, parabola, $V(2, 0)$

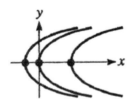

5. $f(x, y) = xe^{-2y}$

$\dfrac{\partial f}{\partial x} = (1)e^{-2y} = e^{-2y}$

$\dfrac{\partial f}{\partial y} = xe^{-2y}(-2) = -2xe^{-2y}$

9. $\phi = r\sqrt{1 + 2rs}$

$$\frac{\partial\phi}{\partial r} = r\left(\frac{1}{2}\right)(1 + 2rs)^{-1/2}(2s)$$

$$+ (1 + 2rs)^{1/2}$$

$$= \frac{rs}{(1 + 2rs)^{1/2}} + (1 + 2rs)^{1/2}$$

$$= \frac{1 + 3rs}{\sqrt{1 + 2rs}}$$

$$\frac{\partial\phi}{\partial s} = r\left(\frac{1}{2}\right)(1 + 2rs)^{-1/2}(2r)$$

$$= \frac{r^2}{\sqrt{1 + 2rs}}$$

13. $z = \sin x^2 y$

$$\frac{\partial z}{\partial x} = \left(\cos x^2 y\right)(2xy)$$

$$= 2xy \cos x^2 y$$

$$\frac{\partial z}{\partial y} = \left(\cos x^2 y\right)\left(x^2\right)$$

$$= x^2 \cos x^2 y$$

17. $f(x, y) = \dfrac{2\sin^3 2x}{1 - 3y}$

$$\frac{\partial f}{\partial x} = \frac{2(3)(\sin^2 2x)(\cos 2x)(2)}{1 - 3y}$$

$$= \frac{12 \sin^2 2x \cos 2x}{1 - 3y}$$

$$\frac{\partial f}{\partial y} = -(2\sin^3 2x)(1 - 3y)^{-2}(-3)$$

$$= \frac{6 \sin^3 2x}{(1 - 3y)^2}$$

21. $z = \sin x + \cos xy - \cos y$

$$\frac{\partial z}{\partial x} = \cos x - (\sin xy)(y)$$

$$= \cos x - y \sin xy$$

$$\frac{\partial z}{\partial y} = -(\sin xy)(x) + \sin y$$

$$= -x \sin xy + \sin y$$

25. $z = 3xy - x^2$

$$\frac{\partial z}{\partial x} = 3y - 2x$$

$$\left.\frac{\partial z}{\partial y}\right|_{(1,-2,-7)} = 3(-2) - 2(1) = -8$$

29. $z = 2xy^3 - 3x^2y$

$$\frac{\partial z}{\partial x} = 2y^3 - 6xy$$

$$\frac{\partial z}{\partial y} = 6xy^2 - 3x^2$$

$$\frac{\partial^2 z}{\partial x^2} = -6y, \frac{\partial^2 z}{\partial x^2} = 12xy$$

$$\frac{\partial^2 z}{\partial x \partial y} = \frac{\partial^2 z}{\partial x \partial y} = 6y^2 - 6x$$

33. $A = \pi r^2 + \pi r\sqrt{r^2 + h^2}$

$$\frac{\partial A}{\partial r} = 2\pi r + \pi r \cdot \frac{2r}{2\sqrt{r^2 + h^2}} + \pi\sqrt{r^2 + h^2}$$

$$= 2\pi r + \frac{\pi r^2}{\sqrt{r^2 + h^2}} + \pi\sqrt{r^2 + h^2}$$

$$\frac{\partial A}{\partial h} = \pi r \cdot \frac{2h}{2\sqrt{r^2 + h^2}} = \frac{\pi rh}{\sqrt{r^2 + h^2}}$$

37. $z = 9 - x^2 - y^2$

$$\frac{\partial z}{\partial y} = -2y$$

$$\left.\frac{\partial z}{\partial y}\right|_{(1, 2, 4)} = -4$$

$$\left.\frac{\partial z}{\partial y}\right|_{(2, 2, 1)} = -4$$

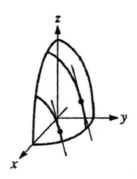

41. $V = \pi r^2 h + \dfrac{1}{2}\left(\dfrac{4}{3}\pi r^3\right)$

$\dfrac{\partial V}{\partial R} = 2\pi rh + 2\pi r^2 \Big|_{r=2.65,\ h=4.20}$

$\dfrac{\partial V}{\partial R} = 2\pi(2.65)(4.20) + 2\pi(2.65)^2$

$\dfrac{\partial V}{\partial R} = 114 \text{ cm}^2$

45. $i_b = 50\left(e_b + 5e_c\right)^{1.5}$

$\dfrac{\partial i_b}{\partial e_c} = 50(1.5)\left(e_b + 5e_c\right)^{0.5}(5)$

$\qquad = 375\left(e_b + 5e_c\right)^{0.5}$

For $e_b = 200$ V and $e_c = -20$ V

$\dfrac{\partial i_b}{\partial e_c} = 375(200 - 100)^{0.5}$

$\qquad = 3750\mu\text{A/V} = 3.75\ 10^{-3} 1/\Omega$

29.4   Double Integrals

1. $\displaystyle\int_0^1 \int_{x^2}^{x}(x+y)\,dy\,dx = \int_0^1 xy + \dfrac{y^2}{2}\bigg|_{x^2}^{x}\,dx$

$\qquad = \displaystyle\int_0^1\left(x^2 + \dfrac{x^2}{2} - \left(x^3 + \dfrac{x^4}{2}\right)\right)dx$

$\qquad = \displaystyle\int_0^1\left(-\dfrac{x^4}{2} - x^3 + \dfrac{3x^2}{2}\right)dx$

$\qquad = -\dfrac{x^5}{10} - \dfrac{x^4}{4} + \dfrac{3x^2}{6}\bigg|_0^1 = \dfrac{3}{20}$

5. $\displaystyle\int_1^2 \int_0^{y^2} xy^2\,dx\,dy = \int_1^2 y^2\left(\dfrac{1}{2}x^2\right)\bigg|_0^{y^2}\,dy$

$\qquad = \displaystyle\int_1^2 y^2\left(\dfrac{1}{2}y^4\right)dy$

$\qquad = \dfrac{1}{2}\displaystyle\int_1^2 y^6\,dy$

$\qquad = \dfrac{1}{14}y^7\bigg|_1^2$

$\qquad = \dfrac{127}{14}$

9. $\displaystyle\int_0^{\pi/6} \int_{\pi/3}^{y} \sin x\,dx\,dy$

$\qquad = \displaystyle\int_0^{\pi/6} (-\cos x)\bigg|_{\pi/3}^{y}\,dy$

$\qquad = -\displaystyle\int_0^{\pi/6}\left(\cos y - \cos\dfrac{\pi}{3}\right)dy$

$\qquad = -\displaystyle\int_0^{\pi/6}\left(\cos y - \dfrac{1}{2}\right)dy$

$\qquad = -\sin y + \dfrac{1}{2}y\bigg|_0^{\pi/6}$

$\qquad = -\sin\dfrac{\pi}{6} + \dfrac{\pi}{12}$

$\qquad = \dfrac{\pi}{12} - \dfrac{1}{2}$

$\qquad = \dfrac{\pi - 6}{12}$

13. $\displaystyle\int_1^2\int_0^x yx^3 e^{xy^2}\,dy\,dx = \frac{1}{2}\int_1^2\int_0^x x^2\left(2xye^{xy^2}\right)dx$

$\displaystyle\qquad = \frac{1}{2}\int_1^2 x^2\left(e^{xy^2}\right)\Big|_0^x dx$

$\displaystyle\qquad = \frac{1}{2}\int_1^2 x^2\left(e^{x^3}-1\right)dx$

$\displaystyle\qquad = \frac{1}{6}\int_1^2 3x^2 e^{x^3}\,dx - \frac{1}{2}\int_1^2 x^2\,dx$

$\displaystyle\qquad = \frac{1}{6}e^{x^3} - \frac{1}{6}x^3\Big|_1^2$

$\displaystyle\qquad = \frac{1}{6}\left[e^8 - 8 - (e-1)\right]$

$\displaystyle\qquad = 495.2$

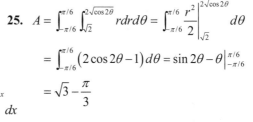

$z = 2 + x^2 + y^2$

17.

$V = \displaystyle\int_0^4\int_x^{4-x} z\,dy\,dx = \int_0^4\int_x^{4-x}(4-x-y)\,dy\,dx = \int_0^4\left(4y - xy - \frac{1}{2}y^2\right)\Big|_x^{4-x} dx$

$= \displaystyle\int_0^4\left[4(4-x) - x(4-x) - \frac{1}{2}(4-x)^2\right]dx = \int_0^4\left(8 - 4x + \frac{1}{2}x^2\right)dx$

$= 8x - 2x^2 + \dfrac{1}{6}x^3\Big|_0^4 = 32 - 32 + \dfrac{64}{6} = \dfrac{32}{3}$

25. $A = \displaystyle\int_{-\pi/6}^{\pi/6}\int_{\sqrt2}^{2\sqrt{\cos 2\theta}} r\,dr\,d\theta = \int_{-\pi/6}^{\pi/6}\frac{r^2}{2}\Big|_{\sqrt2}^{2\sqrt{\cos 2\theta}} d\theta$

$= \displaystyle\int_{-\pi/6}^{\pi/6}(2\cos 2\theta - 1)\,d\theta = \sin 2\theta - \theta\Big|_{-\pi/6}^{\pi/6}$

$= \sqrt3 - \dfrac{\pi}{3}$

29. $z = 4 - x - 2y$ from $y = x^2$ to $y = 1$, from

$x = 0$ to $x = \dfrac{1}{2}$

Intercepts of plane: $(4,0,0),(0,2,0),(0,0,4)$

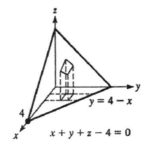

$y = 4 - x$

$x + y + z - 4 = 0$

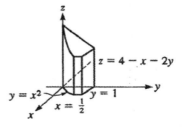

$z = 4 - x - 2y$

$y = x^2$ $y = 1$

$x = \frac{1}{2}$

21.

$V = \displaystyle\int_0^2\int_x^2 z\,dy\,dx\int_0^2\int_x^2(2 + x^2 + y^2)\,dy\,dx = \int_x^2\left(2y + x^2 y + \frac{1}{3}y^3\right)\Big|_x^2 dx$

$= \displaystyle\int_x^2\left(4 + 2x^2 + \frac{8}{3} - 2x - x^3 - \frac{1}{3}x^3\right)dx = \int_x^2\left(\frac{20}{3} - 2x + 2x^2 - \frac{4}{3}x^3\right)dx$

$= \dfrac{20}{3}x - x^3 + \dfrac{2}{3}x^3 - \dfrac{1}{3}x^4\Big|_0^2 = \dfrac{40}{3} - 4 + \dfrac{16}{3} - \dfrac{16}{3} = \dfrac{28}{3}$

Chapter 29 Review Exercises

1. $f(x, y) = 3x^2 y - y^3$

$f(-1, 4) = 3(-1)^2 (4) - 4^3 = -52$

$f(2, -3) = 3(2)^2 (-3) - (-3)^3 = -9$

5. $x - y + 2z - 4 = 0$, plane

intercepts: $(0, 0, 2), (0, -4, 0), (4, 0, 0)$

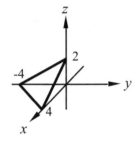

9. $z = 5x^3 y^2 - 2xy^4$

$\dfrac{\partial z}{\partial x} = 5y^2 (3x^2) - 2y^4 (1)$

$\quad = 15x^2 y^2 - 2y^4$

$\dfrac{\partial z}{\partial y} = 5x^3 (2y) - 2x(4y^3)$

$\quad = 10x^3 y - 8xy^3$

13. $z = \dfrac{2x - 3e^y}{x^2 y + 1}$

$\dfrac{\partial z}{\partial x} = \dfrac{(x^2 y + 1)(2) - (2x - 3e^y)(2xy)}{(x^2 y + 1)^2}$

$\quad = \dfrac{2 - 2x^2 y + 6xye^y}{(x^2 y + 1)^2}$

$\dfrac{\partial z}{\partial y} = \dfrac{(x^2 y + 1)(-3e^y) - (2x - 3e^y)(x^2)}{(x^2 y + 1)^2}$

$\quad = \dfrac{-(3e^y - 3x^2 e^y + 2x^3 + 3x^2 ye^y)}{(x^2 y + 1)^2}$

17. $z = \cos^{-1} \sqrt{x + y}$

$\dfrac{\partial z}{\partial x} = \dfrac{-1}{\sqrt{1 - (x + y)}} \left(\dfrac{1}{2}\right)(x + y)^{-1/2} (1)$

$\quad = \dfrac{-1}{2\sqrt{(x + y)(1 - x - y)}}$

$\dfrac{\partial z}{\partial y} = \dfrac{-1}{\sqrt{1 - (x + y)}} \left(\dfrac{1}{2}\right)(x + y)^{-1/2} (1)$

$\quad = \dfrac{-1}{2\sqrt{(x + y)(1 - x - y)}}$

21. $r = 4e^s \cos - 2te^{-s}$

$\dfrac{\partial r}{\partial s} = 4e^s \cos 2t + 2te^{-s}$

$\dfrac{\partial^2 r}{\partial s^2} = 4e^s \cos 2t - 2te^{-s}$

$\dfrac{\partial r}{\partial t} = -8e^s \sin 2t - 2e^{-s}$

$\dfrac{\partial^2 r}{\partial t^2} = -16e^s \cos 2t$

$\dfrac{\partial^2 r}{\partial r \partial t} = \dfrac{\partial^2 r}{\partial t \partial r} = -8e^s \sin 2t + 2e^{-s}$

25. $\displaystyle\int_0^3 \int_1^x (x + 2y)\, dy\, dx = \int_0^3 (xy + y^2)\Big|_1^x dx$

$\quad = \displaystyle\int_0^3 (x^2 + x^2 - x - 1)\, dx = \int_0^3 (2x^2 - x - 1)\, dx$

$\quad = \dfrac{2}{3}x^3 - \dfrac{1}{2}x^2 - x\Big|_0^3 = \dfrac{2}{3}(27) - \dfrac{1}{2}(9) - 3 - 0 = \dfrac{21}{2}$

29. $\displaystyle\int_1^e \int_1^x \dfrac{\ln y}{xy}\, dy\, dx = \int_1^e \int_1^x \dfrac{1}{x}\left[\ln y \left(\dfrac{dy}{y}\right)\right] dx$

$\quad = \displaystyle\int_1^e \dfrac{1}{x}\left(\dfrac{\ln^2 y}{2}\right)\Big|_1^x dx = \int_1^e \dfrac{\ln^2 x}{2x}\, dx$

$\quad = \dfrac{1}{6}\ln^3 x\Big|_1^e = \dfrac{1}{6}\ln^3 e - 0 = \dfrac{1}{6}$

33. $z = e^{x+y}$

Intercepts: $(0, 0, 1)$

Traces:

$x > 0$ for all x and y since z is defined by an exponential.

In xy-plane: $z = e^x$

In yz-plane: $z = e^y$

Section: $z = 2 = e^{x+y}$

$\ln 2 = x + y$ (a line)

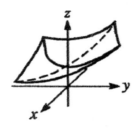

37. $v = \dfrac{rE}{r + R}$

$\dfrac{\partial v}{\partial r} = \dfrac{(r + R)E - rE(1)}{(r + R)^2} = \dfrac{ER}{(r + R)^2}$

$\dfrac{\partial v}{\partial R} = -rE(r + R)^{-2}(1) = \dfrac{-rE}{(r + R)^2}$

41. $T = 2\pi\sqrt{\dfrac{\ell}{g}}$

$\dfrac{\partial T}{\partial \ell} = 2\pi\sqrt{\dfrac{1}{g}}\left(\dfrac{1}{2}\ell^{-1/2}\right) = \dfrac{\pi}{\sqrt{g\ell}}$

$\dfrac{T}{2\ell} = \dfrac{2\pi\sqrt{\ell/g}}{2\ell} = \dfrac{\pi}{\sqrt{g\ell}}$

$\dfrac{\partial T}{\partial \ell} = \dfrac{T}{2\ell}$

45. $M = \dfrac{k}{\pi\left(b^2 - a^2\right)}\displaystyle\int_0^{2\pi}\int_a^b r^2\,dr\,d\theta$

$= \dfrac{k}{\pi\left(b^2 - a^2\right)}\displaystyle\int_0^{2\pi}\dfrac{r^3}{3}\Big|_c^b\,d\theta$

$= \dfrac{k}{\pi\left(b^2 - a^2\right)}\cdot\dfrac{b^3 - a^3}{3}\cdot\displaystyle\int_0^{2\pi}d\theta$

$= \dfrac{k}{\pi(b+a)(b-a)}\cdot\dfrac{(b-a)\left(b^2 + ab + a^2\right)}{3}\cdot 2\pi$

$= \dfrac{2k\left(b^2 + ab + a^2\right)}{3(b+a)}$

49. $V = \displaystyle\int_0^4\int_0^{\sqrt{16-x^2}} z\,dy\,dz = \int_0^4\int_0^{\sqrt{16-x^2}}(8-x)\,dy\,dz$

$= \displaystyle\int_0^4 (8-x)\,y\Big|_0^{\sqrt{16-x^2}}\,dx = \int_0^4 (8-x)\sqrt{16-x^2}\,dx$

$= 8\displaystyle\int_0^4\sqrt{16-x^2}\,dx - \int_0^4 x\sqrt{16-x^2}\,dx$

$= 8\left[\dfrac{x}{2}\sqrt{16-x^2} + \dfrac{16}{2}\sin^{-1}\dfrac{x}{4}\right] + \dfrac{1}{3}\left(16-x^2\right)^{3/2}\Big|_0^4$

$= 8\left[2(0) + 8\sin^{-1}1\right] + \dfrac{1}{3}(0) - 8\left[0 + 8\sin^{-1}0\right]$

$-\dfrac{1}{3}(16)^{3/2} = 64\sin^{-1}1 - \dfrac{64}{3}$

$= 64\left(\dfrac{\pi}{2}\right) - \dfrac{64}{3} = 32\left(\pi - \dfrac{2}{3}\right)$

$= 79.2$

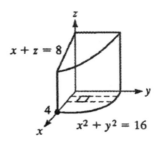

Chapter 30

EXPANSION OF FUNCTIONS IN SERIES

30.1 Infinite Series

1. $\displaystyle\sum_{u=1}^{\infty} 0.5^n = 0.5 + 0.5^2 + 0.5^3 + 0.5^4 + \cdots + 0.5^n + \cdots$

$s_1 = 0.5, s_2 = 0.75, s_3 = 0.875, s_4 = 0.9375.$
Converges

5. $a_n = \dfrac{1}{n+2}; \; n = 0, 1, 2, 3, \ldots$

$a_0 = \dfrac{1}{0+2} = \dfrac{1}{2}$ $\qquad a_2 = \dfrac{1}{2+2} = \dfrac{1}{4}$

$a_1 = \dfrac{1}{1+2} = \dfrac{1}{3}$ $\qquad a_3 = \dfrac{1}{3+2} = \dfrac{1}{5}$

9. $a_n = \cos\dfrac{n\pi}{2}, n = 0, 1, 2, 3, \ldots$

$a_0 = \cos\dfrac{0\cdot\pi}{2} = 1$

$a_1 = \cos\dfrac{1\cdot\pi}{2} = 0$

$a_2 = \cos\dfrac{2\cdot\pi}{2} = -1$

$a_3 = \cos\dfrac{3\cdot\pi}{2} = 0$

(a) $1, 0, -1, 0$

(b) $1 + 0 - 1 + 0 - \cdots$

13. $\dfrac{1}{2\times3} - \dfrac{1}{3\times4} + \dfrac{1}{4\times5} - \dfrac{1}{5\times6} + \cdots$

$n = 1, \; a_1 = \dfrac{1}{(1+1)(1+2)} = \dfrac{1}{2\times3}$

$n = 2, \; a_2 = \dfrac{-1}{(2+1)(2+2)} = \dfrac{-1}{3\times4}$

$a_n = \dfrac{(-1)^{n+1}}{(n+1)(n+2)}$

17. $1 + \dfrac{1}{2} + \dfrac{2}{3} + \dfrac{3}{4} + \dfrac{4}{5} + \cdots$

$S_0 = 1; \; S_1 = 1 + \dfrac{1}{2} = \dfrac{3}{2} = 1.5$

$S_2 = 1 + \dfrac{1}{2} + \dfrac{2}{3} = \dfrac{13}{6} = 2.1666667$

$S_3 = 1 + \dfrac{1}{2} + \dfrac{2}{3} + \dfrac{3}{4} = \dfrac{35}{12} = 2.9166667$

$S_4 = 1 + \dfrac{1}{2} + \dfrac{2}{3} + \dfrac{3}{4} + \dfrac{4}{5} = \dfrac{223}{60} = 3.7166667$

Divergent

21. $\displaystyle\sum_{n=1}^{\infty} \dfrac{2n+1}{n^2(n+1)^2}$

First five terms:

$a_1 = \dfrac{3}{4}; \; a_2 = \dfrac{5}{36}; \; a_3 = \dfrac{7}{144}; \; a_4 = \dfrac{9}{400}; \; a_5 = \dfrac{11}{900}$

First five partial sums:

$S_1 = 0.75;$

$S_2 = \dfrac{3}{4} + \dfrac{5}{36} = 0.8888889$

$S_3 = \dfrac{3}{4} + \dfrac{5}{36} + \dfrac{7}{144} = 0.9375000$

$S_4 = \dfrac{3}{4} + \dfrac{5}{36} + \dfrac{7}{144} + \dfrac{9}{400} = 0.9600000$

$S_5 = \dfrac{3}{4} + \dfrac{5}{36} + \dfrac{7}{144} + \dfrac{9}{400} + \dfrac{11}{900} = 0.9722222$

Convergent, converging to 1 (approx. sum)

25. $1 + 2 + 4 + \cdots + 2^n + \cdots;\ n = 0, 1, 2, 3, \cdots$

$S_0 = 1$
$S_1 = 3$
$S_2 = 7$
$S_3 = 15$
$\vdots$
$S_n = 2^{n+1} - 1$

$\lim\limits_{n \to \infty} S_n = \lim\limits_{n \to \infty} (2^{n+1} - 1) = \infty$, divergent

29. $10 + 9 + 8.1 + 7.29 + 6.561 + \cdots + 10(0.9)^n + \cdots;$
$n = 0, 1, 2, 3, \ldots S_n = 100 - 100(0.9)^{n+1};$
$n = 0, 1, 2, 3, \ldots$

$\lim\limits_{n \to \infty} S_n = \lim\limits_{n \to \infty} 100 - 100(0.9)^{n+1} = 100$, convergent

33. $\sum\limits_{n=0}^{\infty} (x-4)^n$ is a GS with $a_1 = 1$, $r = x - 4$ which

converges for $|x - 4| < 1 \Leftrightarrow -1 < x - 4 < 1$ or
$3 < x < 5$.

37. $\lim\limits_{x \to \infty} \dfrac{e^x}{x^3} \overset{LH}{=} \lim\limits_{x \to \infty} \dfrac{e^x}{3x^2} \overset{LH}{=} \lim\limits_{x \to \infty} \dfrac{e^x}{6x} \overset{LH}{=} \lim\limits_{x \to \infty} \dfrac{e^x}{6} = \infty$, divergent

$\dfrac{\infty}{\infty} \qquad \dfrac{\infty}{\infty} \qquad \dfrac{\infty}{\infty}$

41. $S_n = \dfrac{a_1(1 - r^n)}{(1 - r)};\ r \neq 1$; geometric series

Series: $\dfrac{1}{2} + \dfrac{1}{4} + \dfrac{1}{8} + \cdots;\ a_n = \dfrac{1}{2^n}, a = \dfrac{1}{2}, r = \dfrac{1}{2}$

$f(x) = \dfrac{a_1(1 - r^x)}{(1 - r)};\ f(x) = \dfrac{\frac{1}{2}(1 - r^x)}{\left(1 - \frac{1}{2}\right)} = (1 - r^x)$

x	y
0	0
1	$\frac{1}{2}$
2	$\frac{3}{4}$
3	$\frac{7}{8}$
4	$\frac{15}{16}$
5	$\frac{31}{32}$

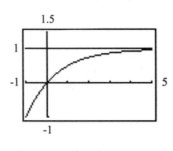

The infinite series approaches 1.

45. $\sum\limits_{n=0}^{\infty} x^n = 1 + x + x^2 + \cdots + x^n + \cdots$

For $|x| < 1, a_1 = 1, r = x,$

$S = \dfrac{1}{1 - x}$

$\sum\limits_{n=0}^{\infty} x^n = \dfrac{1}{1 - x}$

30.2 Maclaurin's Series

1. $f(x) = \dfrac{2}{2 + x},\ f(0) = 1$

$f'(x) = \dfrac{-2}{(2 + x)^2},\ f'(0) = -\dfrac{1}{2}$

$f''(x) = \dfrac{4}{(2 + x)^3},\ f''(0) = \dfrac{1}{2}$

$f'''(x) = \dfrac{-12}{(2 + x)^4},\ f'''(0) = -\dfrac{3}{4}$

$f(x) = \dfrac{2}{2 + x} = 1 - \dfrac{1}{2}x + \dfrac{1}{4}x^2 - \dfrac{1}{8}x^3 + \cdots$

5.
$f(x) = \cos x$	$f(0) = 1$
$f'(x) = -\sin x$	$f'(0) = 0$
$f''(x) = -\cos x$	$f''(0) = -1$
$f'''(x) = \sin x$	$f'''(0) = 0$
$f^{iv}(x) = \cos x$	$f^{iv}(0) = 1$

$f(x) = \cos x = f(0) + f''(0)\dfrac{x^2}{2!} + f^{iv}(0)\dfrac{x^4}{4!} - \cdots$

$\cos x = 1 - 1\dfrac{x^2}{2} + 1\dfrac{x^4}{24} - \cdots$

$\cos x = 1 - \dfrac{1}{2}x^2 + \dfrac{1}{24}x^4 - \cdots$

9. $f(x) = e^{-2x}$ $\qquad$ $f(0) = 1$
    $f'(x) = -2e^{-2x}$ $\qquad$ $f'(0) = -2$
    $f''(x) = 4e^{-2x}$ $\qquad$ $f''(0) = 4$

    $$e^{-2x} = 1 - 2x + 4\frac{x^2}{2} - \cdots = 1 - 2x + 2x^2 - \cdots$$

13. $f(x) = \dfrac{1}{(1-x)}$ $\qquad$ $f(0) = 1$

    $f'(x) = \dfrac{1}{(1-x)^2}$ $\qquad$ $f'(0) = 1$

    $f''(x) = \dfrac{2}{(1-x)^3}$ $\qquad$ $f''(0) = 2$

    $$\frac{1}{(1-x)} = 1 + x + \frac{2x^2}{2} + \cdots = 1 + x + x^2 + \cdots$$

17. $f(x) = \cos^2 x$ $\qquad$ $f(0) = 1$
    $f'(x) = -2\sin x \cos x$ $\qquad$ $f'(0) = 0$
    $f''(x) = 2 - 4\cos^2 x$ $\qquad$ $f''(0) = -2$
    $f'''(x) = 8\sin x \cos x$ $\qquad$ $f'''(0) = 0$
    $f^{iv}(x) = 16\cos^2 x - 8$ $\qquad$ $f^{iv}(0) = 8.$

    $$\cos^2 x = 1 - 2\frac{x^2}{2!} + 8\frac{x^4}{4!} - \cdots$$
    $$= 1 - x^2 + \frac{1}{3}x^4 - \cdots$$

21. $f(x) = \tan^{-1} x$

    $f'(x) = \dfrac{1}{1+x^2} = (1+x^2)^{-1}$

    $f''(x) = -(1+x^2)^{-2}2x = -2x(1+x^2)^{-2}$
    $f'''(x) = -2x[-2(1+x^2)^{-3}(2x)] + (1+x^2)^{-2}(-2)$
    $f(0) = 0$

    $f'(0) = 1$

    $f''(0) = 0$
    $f'''(0) = -2$

    $$f(x) = 0 + 1x + \frac{0x^2}{2!} - \frac{2x^3}{3!} + \cdots = x - \frac{1}{3}x^3 + \cdots$$

25. $f(x) = \ln\cos x$

    $f'(x) = -\dfrac{1}{\cos x}\sin x = -\tan x$

    $f''(x) = -\sec^2 x$
    $f'''(x) = -2\sec x \sec x \tan x = -2\sec^2 x \tan x$
    $f^{iv}(x) = -2\sec^2 x \sec^2 x - 2\tan x(2\sec x \sec x \tan x)$
    $f(0) = \ln 1 = 0$

    $f'(0) = 0$

    $f''(0) = -1$
    $f'''(0) = 0$
    $f^{iv}(0) = -2 - 0 = -2$

    $$f(x) = 0 + 0x - \frac{1x^2}{2!} + \frac{0x^3}{3!} - \frac{2x^4}{4!} + \cdots$$
    $$= -\frac{1}{2}x^2 - \frac{1}{12}x^4 - \cdots$$

29.

$$1-x \overline{)\begin{array}{l} 1 + x + x^2 + x^3 + \cdots \\ 1 \\ \underline{1-x} \\ x \\ \underline{x - x^2} \\ x^2 \\ \underline{x^2 - x^3} \\ x^3 \\ \underline{x^3 - x^4} \\ x^4 \end{array}}$$

33. (a) $f(x) = e^x$ $\qquad$ $f(0) = 1$
    $f'(x) = e^x$ $\qquad$ $f'(0) = 1$
    $f''(x) = e^x$ $\qquad$ $f''(0) = 1$

    $$e^x = 1 + x + \frac{1}{2}x^2 + \cdots$$

    (b) $f(x) = e^{x^2}$ $\qquad$ $f(0) = 1$
    $f'(x) = 2xe^{x^2}$ $\qquad$ $f'(0) = 0$
    $f''(x) = 2e^{x^2}(2x^2 + 1)$ $\qquad$ $f''(0) = 2$
    $f'''(x) = 4xe^{x^2}(2x^2 + 3)$ $\qquad$ $f'''(0) = 0$
    $f^{iv}(x) = 4e^{x^2}(4x^4 + 12x^2 + 3)$ $\qquad$ $f^{iv}(0) = 12$

    $$e^{x^2} = 1 + \frac{2x^2}{2} + \frac{12x^4}{4!} + \cdots = 1 + x^2 + \frac{1}{2}x^4 + \cdots$$

37. $y = \cosh x,$ $\qquad\qquad y(0) = 1$

$y = \sinh x,$ $\qquad\qquad y'(0) = 0$

$y'' = \cosh x,$ $\qquad\qquad y''(0) = 1$

$y''' = \sinh x,$ $\qquad\qquad y'''(0) = 0$

$y^{iv} = \cosh x$ $\qquad\qquad y^{iv}(0) = 1$

$$\cosh x = 1 + 0 \cdot x + 1\frac{x^2}{2!} + 0 \cdot \frac{x^3}{3!} + \cdots$$

$$= 1 + \frac{x^2}{2!} + \frac{x^4}{4!} + \cdots$$

41. $y = 4e^{-0.2t}\cos t,\ y(0) = 4$

$y' = -0.8e^{-0.2t}\cos t - 4.0e^{-0.2t}\sin t,\ y'(0) = -0.8$

$y'' = -3.84e^{-0.2t}\cos t + 1.6e^{-0.2t}\sin t,\ y''(0) = -3.84$

$$y = 4e^{-0.2t}\cos t = 4 - 0.8t - 3.84\frac{t^2}{2!} + \cdots$$

$$= 4 - 0.8t - 1.92t^2 + \cdots$$

30.3 Operations with Series

1. $e^x = 1 + x + \dfrac{x^2}{2!} + \dfrac{x^3}{3!} + \cdots$

$e^{2x^2} = 1 + 2x^2 + \dfrac{(2x^2)^2}{2!} + \dfrac{(2x^2)^3}{3!} + \cdots$

$e^{2x^2} = 1 + 2x^2 + 2x^4 + \dfrac{4}{3}x^6 + \cdots$

5. $f(x) = \sin\left(\dfrac{1}{2}x\right);\ \sin x = x - \dfrac{x^3}{3!} + \dfrac{x^5}{5!} - \dfrac{x^7}{7!} + \cdots$

$f(x) = \sin\left(\dfrac{1}{2}x\right)$

$= \dfrac{1}{2}x - \dfrac{\left(\frac{1}{2}x\right)^3}{6} + \dfrac{\left(\frac{1}{2}x\right)^5}{120} - \dfrac{\left(\frac{1}{2}x\right)^7}{7!} + \cdots$

$= \dfrac{1}{2}x - \dfrac{x^3}{2^3 3!} + \dfrac{x^5}{2^5 5!} - \dfrac{x^7}{2^7 7!} + \cdots$

9. $f(x) = \ln(1+x^2);\ \ln(1+x) = x - \dfrac{x^2}{2} + \dfrac{x^3}{3} - \dfrac{x^4}{4} + \cdots$

$\ln(1+x^2) = x^2 - \dfrac{(x^2)^2}{2} + \dfrac{(x^2)^3}{3} - \dfrac{(x^2)^2}{4} + \cdots$

$= x^2 - \dfrac{1}{2}x^4 + \dfrac{1}{3}x^6 - \dfrac{1}{4}x^8 + \cdots$

13. $\displaystyle\int_0^{0.5} e^{\sqrt{x}}\,dx = \int_0^{0.5}\left(1 - \sqrt{x} + \dfrac{x}{2}\right)dx = 0.327$

17. $f(x) = \dfrac{2}{1-x^2} = \dfrac{1}{1+x} + \dfrac{1}{1-x}$

$f(x) = 1 - x + x^2 - x^3 + \cdots + 1 + x + x^2 + x^3 + \cdots$

$f(x) = 1 - x + x^2 - x^3 + x^4 - x^5 + x^6 - x^7 + x^8 + \cdots$

$\qquad + 1 + x + x^2 + x^3 + x^4 + x^5 + x^6 + x^7 + x^8 + \cdots$

$f(x) = 2(1 + x^2 + x^4 + x^6 + \cdots)$

21. $x^2\ln(1-x)^2$

$= x^2(2\ln(1-x))$

$= 2x^2\ln(1 + (-x))$

$= 2x^2\left((-x) - \dfrac{(-x)^2}{2} + \dfrac{(-x)^3}{3} - \dfrac{(-x)^4}{4} + \cdots\right)$

$= 2x^2\left(-x - \dfrac{x^2}{2} - \dfrac{x^3}{3} - \dfrac{x^4}{4} - \cdots\right)$

$= -2x^3 - x^4 - \dfrac{2}{3}x^5 - \dfrac{1}{2}x^6 - \cdots$

25. $e^x = 1 + x + \dfrac{x^2}{2!} + \dfrac{x^3}{3!} + \dfrac{x^4}{4!} + \cdots$

$\dfrac{d}{dx}e^x = 0 + 1 + \dfrac{2x}{2!} + \dfrac{3x^2}{6} + \dfrac{4x^3}{24} + \cdots$

$= 1 + x + \dfrac{x^2}{2} + \dfrac{x^3}{6} + \cdots$

$= 1 + x + \dfrac{x^2}{2!} + \dfrac{x^3}{3!} + \cdots = e^x$

29. $y = 4e^{-0.2t}\cos t = 4e^{-t/5}\cos t$

$= 4\left[1 - \dfrac{t}{5} + \dfrac{t^2}{50} - \cdots\right]$

$\cdot\left[1 - \dfrac{t^2}{2} + \dfrac{t^4}{24} - \dfrac{t^6}{720} + \cdots\right]$

$= 4\left[1 - \dfrac{1}{2}t^2 + \dfrac{t^4}{24} - \dfrac{t^6}{720} - \dfrac{t}{5} + \dfrac{t^3}{10} - \dfrac{t^5}{120} + \dfrac{t^7}{3600}\right.$

$\left. + \dfrac{t^2}{50} - \dfrac{t^4}{100} + \dfrac{t^6}{1200} - \dfrac{t^8}{36{,}000} + \cdots\right]$

$= 4\left[1 - \dfrac{t}{5} - \dfrac{12t^2}{25}\right] = 4 - \dfrac{4t}{5} - \dfrac{48t^2}{25}$

33. $\lim\limits_{x\to 0}\dfrac{\sin x - x}{x^3} = \lim\limits_{x\to 0}\dfrac{x - \frac{1}{6}x^3 + \frac{1}{120}x^5 - \frac{1}{5040}x^7 + \cdots - x}{x^3}$

$\qquad = \lim\limits_{x\to 0}\left(-\dfrac{1}{6} + \dfrac{1}{120}x^2 - \dfrac{1}{5040}x^4 + \cdots\right)$

$\qquad = -\dfrac{1}{6}$

37. $\lim\limits_{x\to 0}\dfrac{\ln\left(1+x^2\right)}{1-\cos x} = \lim\limits_{x\to 0}\dfrac{x^2 - \frac{x^4}{2} + \frac{x^6}{3} - \frac{x^8}{4} + \cdots}{1 - \left(1 - \frac{x^2}{2!} + \frac{x^4}{4!} - \frac{x^6}{6!} + \frac{x^8}{8!} - \cdots\right)}$

$\qquad = \lim\limits_{x\to 0}\dfrac{x^2\left(1 - \frac{x^2}{2} + \frac{x^4}{3} - \frac{x^6}{4} + \cdots\right)}{x^2\left(\frac{1}{2!} - \frac{x^2}{4!} + \frac{x^4}{6!} - \frac{x^6}{8!} + \cdots\right)} = 2$

41. $K = \left[\left(1-\dfrac{v^2}{c^2}\right)^{-1/2} - 1\right]mc^2$

$\qquad = \left[1 - \dfrac{1}{2}\left(\dfrac{-v^2}{c^2}\right)\right.$

$\qquad \left.+ \dfrac{-\frac{1}{2}\left(-\frac{1}{2}-1\right)}{2!}\left(\dfrac{-v^2}{c^2}\right)^2 + \cdots - 1\right]mc^2$

$\qquad = \left[1 + \dfrac{1}{2}\dfrac{v^2}{c^2} + \dfrac{3}{8}\dfrac{v^4}{c^4} + \cdots\right]mc^2$

$\qquad = \dfrac{1}{2}mv^2 + \dfrac{3}{8}\dfrac{mv^4}{c^4}$

$\qquad = \dfrac{1}{2}mv^2$ for v much smaller than c.

45. $y_1 = \ln(1+x),$

$\quad y_2 = x,$

$\quad y_3 = x - \dfrac{1}{2}x^2,$

$\quad y_4 = x - \dfrac{1}{2}x^2 + \dfrac{1}{3}x^3$

30.4 Computations by Use of Series Expansions

1. $e^x = 1 + x + \dfrac{x^2}{2!} + \cdots$

$\qquad e^{-0.1} = 1 + (-0.1) + \dfrac{(-0.1)^2}{2!} + \cdots$

$\qquad e^{-0.1} = 0.905$

5. $\sin 0.1$, (2 terms); $\sin x = x - \dfrac{x^3}{3!}$

$\qquad \sin 0.1 = 0.1 - \dfrac{(0.1)^3}{6} = 0.09983333$

$\qquad (0.0998334$ calculator$)$

9. $\cos \pi°$, (2 terms); $\pi° = \dfrac{\pi^2}{180}$ radians

$\qquad \cos x = 1 - \dfrac{x^2}{2!}$

$\qquad \cos 3° = 1 - \dfrac{\left(\frac{\pi^2}{180}\right)^2}{2} = 0.9984967733$

$\qquad \left(0.9984791499 \text{ calculator}\right)$

13. $\sin 0.3625$, (3 terms): $\sin x = x - \dfrac{x^3}{3!} + \dfrac{x^5}{5!}$

$\qquad \sin 0.3625 = 0.3625 - \dfrac{(0.3625)^3}{6} + \dfrac{(0.3625)^5}{5!}$

$\qquad\qquad = 0.3546130$

$\qquad (0.3546129$ calculator$)$

17. $(1+x)^6 = 1 + 6x + 15x^2 + 20x^3 + 15x^4$

$\qquad\qquad + 6x^5 + x^6$

$\qquad (1.032)^6 = 1 + 6(0.032) + 15(0.032)^2$

$\qquad\qquad = 1.20736$

$\qquad (1.032)^6 = 1.20803$, calculator

21. $\sqrt{1.1076} = 1.1076^{1/2} = (1 + 0.1076)^{1/2}$

$(1 + x)^n = 1 + nx + \dfrac{n(n - 1)x^2}{2!} + \cdots$

$x = 0.1076$ and $n = \frac{1}{2}$

$\sqrt{1.1076}$

$\quad = 1 + \dfrac{1}{2}(0.1076) + \dfrac{\frac{1}{2}\left(-\frac{1}{2}\right)(0.1076)^2}{2} + \cdots$

$\quad = 1 + 0.0538000 - 0.0014472 + \cdots$

$\quad = 1.0523528$

25. From Exercise 5, $\sin(0.1) = 0.1 + \frac{0.1^3}{6} = 0.1001667$

The maximum possible error is the value of the first term omitted,

$\dfrac{x^5}{5!} = \left|\dfrac{0.1^5}{120}\right| = 8.3 \times 10^{-8}$

29. $\left(1 + x\right)^n = 1 + nx + \dfrac{n(n - 1)}{2!}x^2$

$\sqrt{3.92} = 2\left(1 + (-0.02)\right)^{1/2}$

$\quad = 2\left[1 + \dfrac{1}{2}(-0.02) + \dfrac{\frac{1}{2}\left(\frac{1}{2} - 1\right)}{2!}(-0.02)^2\right]$

$\quad = 1.9799$

33. $e^x = 1 + x + \dfrac{x^2}{2} + \dfrac{x^3}{3!} + \dfrac{x^4}{4!} + \cdots > 1 + x + \dfrac{x^2}{2}$

for $x > 0$ since the terms of the expansion for e^x after those on right hand side of the inequality have a positive value.

37. $f(t) = \dfrac{E}{R}(1 - e^{-Rt/L});\ e^x = 1 + x + \dfrac{x^2}{2} + \cdots$

$e^{-Rt/L} = 1 - \dfrac{Rt}{L} + \dfrac{R^2t^2}{2L^2} + \cdots$

$i = \dfrac{E}{R}\left[1 - \left(1 - \dfrac{Rt}{L} + \dfrac{R^2t^2}{2L^2}\right)\right] = \dfrac{E}{L}\left(t - \dfrac{Rt^2}{2L}\right)$

The approximation will be valid for small values of t.

30.5 Taylor Series

1. $f(x) = x^{1/2}, f(1) = 1$

$f'(x) = \dfrac{1}{2x^{1/2}}, f'(1) = \dfrac{1}{2}$

$f''(x) = -\dfrac{1}{4x^{3/2}}, f''(1) = -\dfrac{1}{4}$

$f'''(x) = \dfrac{3}{8x^{5/2}}, f'''(1) = \dfrac{3}{8}$

$\sqrt{x} = 1 + \dfrac{1}{2}(x-1) + \dfrac{-\frac{1}{4}(x-1)^2}{2!} + \dfrac{\frac{3}{8}(x-1)^3}{3!} + \cdots$

$\sqrt{x} = 1 + \dfrac{1}{2}(x-1) - \dfrac{1}{8}(x-1)^2 + \dfrac{1}{16}(x-1)^3 - \cdots$

5. $\sqrt{4.2}$; $\sqrt{x} = 2 + \dfrac{(x-4)}{4} - \dfrac{(x-4)^2}{64} + \dfrac{(x-3)^3}{512}$

$\sqrt{4.2} = 2 + \dfrac{(4.2-4)}{4} - \dfrac{(4.2-4)^2}{64} + \dfrac{(4.2-4)^3}{512}$

$= 2.049;\ (2.04939 \text{ calculator})$

9. $\sin x = \dfrac{1}{2} + \dfrac{\sqrt{3}}{2}\left(x - \dfrac{\pi}{6}\right) - \dfrac{1}{4}\left(x - \dfrac{\pi}{6}\right)^2$

$\sin\dfrac{59\pi}{360} = \sin\dfrac{29.5\pi}{180}$

$= \dfrac{1}{2} + \dfrac{\sqrt{3}}{2}\left(\dfrac{29.5\pi}{180} - \dfrac{\pi}{6}\right) - \dfrac{1}{4}\left(\dfrac{29.5\pi}{180} - \dfrac{\pi}{6}\right)^2$

$= 0.49242\quad (0.4924235601 \text{ calculator})$

13. $\sin x;\ a = \dfrac{\pi}{3}$

$f(x) = \sin x \qquad f\left(\dfrac{\pi}{3}\right) = \dfrac{\sqrt{3}}{2}$

$f'(x) = \cos x \qquad f'\left(\dfrac{\pi}{3}\right) = \dfrac{1}{2}$

$f''(x) = -\sin x \qquad f''\left(\dfrac{\pi}{3}\right) = -\dfrac{\sqrt{3}}{2}$

$\sin x = \dfrac{\sqrt{3}}{2} + \dfrac{1}{2}\left(x - \dfrac{\pi}{3}\right) - \dfrac{\sqrt{3}}{2!}\left(x - \dfrac{\pi}{3}\right)^2 - \cdots$

$= \dfrac{1}{2}\left[\sqrt{3} + \left(x - \dfrac{\pi}{3}\right) - \dfrac{\sqrt{3}}{2!}\left(x - \dfrac{\pi}{3}\right)^2 - \cdots\right]$

17. $\tan x;\ a = \dfrac{\pi}{4}$

$f(x) = \tan x \qquad f\left(\dfrac{\pi}{4}\right) = 1$

$f'(x) = \sec^2 x \qquad f'\left(\dfrac{\pi}{4}\right) = (\sqrt{2})^2 = 2$

$f''(x) = 2\sec x \sec x \tan x = 2\sec^2 x \tan x$

$f''(x) = 2(\sqrt{2})^2(1) = 4$

$\tan x = 1 + 2\left(x - \dfrac{\pi}{4}\right) + \dfrac{4\left(x - \frac{\pi}{4}\right)^2}{2!} + \cdots$

$= 1 + 2\left(x - \dfrac{\pi}{4}\right) + 2\left(x - \dfrac{\pi}{4}\right)^2 + \cdots$

21. $f(x) = \dfrac{1}{x+2},\ f(3) = \dfrac{1}{5}$

$f'(x) = -\dfrac{1}{(x+2)^2},\ f'(3) = -\dfrac{1}{25}$

$f''(x) = \dfrac{2}{(x+2)^3},\ f''(3) = \dfrac{2}{125}$

$\dfrac{1}{x+2} = \dfrac{1}{5} - \dfrac{1}{25}(x-3) + \dfrac{1}{125}(x-3)^2$

25. $\sqrt{9.3}$; $a = 9$

$f(x) = \sqrt{x} \qquad f(9) = 3$

$f'(x) = \dfrac{1}{2\sqrt{x}} \qquad f'(9) = \dfrac{1}{6}$

$f''(x) = -\dfrac{1}{4x^{3/2}} \qquad f''(9) = -\dfrac{1}{108}$

$\sqrt{x} = 3 + \dfrac{1}{6}(x-9) - \dfrac{1}{108}\dfrac{(x-9)^2}{2!}$

$\sqrt{9.3} = 3 + \dfrac{1}{6}(0.3) - \dfrac{1}{108}\dfrac{(0.3)^2}{2} = 3.0496$

29. $\sin x = \dfrac{1}{2}\left[\sqrt{3} + \left(x - \dfrac{\pi}{3}\right) - \dfrac{\sqrt{3}}{2}\left(x - \dfrac{\pi}{3}\right)^2\right];\ a = \dfrac{\pi}{3}$

$61° = 60° + 1° = \dfrac{\pi}{3} + \dfrac{\pi}{180}$

$\sin 61° = \dfrac{1}{2}\left[\sqrt{3} + \dfrac{\pi}{180} - \dfrac{\sqrt{3}}{2}\left(\dfrac{\pi}{180}\right)^2\right] = 0.87462$

33. Expand $f(x) = 2x^3 + x^2 - 3x + 5$ about $x = 1$

$$f'(x) = 6x^2 + 2x - 3$$

$$f''(x) = 12x + 1$$

$$f'''(x) = 12$$

$$f(x) = f(1) + f'(1)(x-1)\frac{f''(1)(x-1)^2}{2!} + \frac{f'''(1)(x-1)^2}{3!}$$

$$= 2(1)^3 + 1^2 - 3(1) + 5 + \left(6(1)^2 + 2(1) - 3\right)(x-1)$$

$$+ \frac{(12(1)+2)(x-1)^2}{2!} + \frac{12(x-1)^3}{3!}$$

$$= 5 + 5(x-1) + 7(x-1)^2 + 2(x-1)^3$$

37. $i = 6\sin \pi t,$ $\qquad i(\pi/2) = 6\sin \pi^2/2$

$i' = 6\pi \cos \pi t,$ $\qquad i'(\pi/2) = 6\pi \cos \pi^2/2$

$i'' = -6\pi^2 \sin \pi t,$ $\quad i''(\pi/2) = -6\pi^2 \sin \pi^2/2$

$$i = 6\sin \frac{\pi^2}{2} + 6\pi \cos \frac{\pi^2}{2}\left(t - \frac{\pi}{2}\right)$$

$$- 3\pi^2 \sin \frac{\pi^2}{2}\left(t - \frac{\pi}{2}\right)^2 + \cdots$$

41. $f(x) = \frac{1}{x}; x = 0$ to $x = 4$

(a) $y_1 = \frac{1}{x}$

(b) $y_2 = \frac{1}{2} - \frac{1}{4}(x-2)$

Graph of part (b) will fit the graph of part (a) well
for values of x close to $x = 2$.

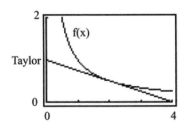

30.6 Introduction to Fourier Series

1. $f(x) = \begin{cases} -2, -\pi \le x < 0 \\ 2, 0 \le x < \pi \end{cases}$

$$a_0 = \frac{1}{2\pi}\int_{-\pi}^{0}(-2)dx + \frac{1}{2\pi}\int_{0}^{\pi}2\,dx = 0$$

$$a_n = \frac{1}{\pi}\int_{-\pi}^{0} -2\cos nx\,dx + \frac{1}{\pi}\int_{0}^{\pi}2\cos nx\,dx = 0$$

$$b_n = \frac{1}{\pi}\int_{-\pi}^{0} -2\sin nx\,dx + \frac{1}{\pi}\int_{0}^{\pi}2\sin nx\,dx$$

$$= \frac{4}{\pi}\left(\frac{1}{n} - \frac{\cos(\pi n)}{n}\right)$$

$$b_n = \frac{4}{\pi}(1 - \cos \pi n) = \begin{cases} \dfrac{8}{n\pi}, n \text{ odd} \\ \\ 0, n \text{ neven} \end{cases}$$

$$b_1 = \frac{8}{\pi}, b_3 = \frac{8}{3\pi}, b_5 = \frac{8}{5\pi}$$

$$f(x) = \frac{8}{\pi}\sin x + \frac{8}{3\pi}\sin 3x + \frac{8}{5\pi}\sin 5x + \cdots$$

$$f(x) = \frac{8}{\pi}\left(\sin x + \frac{1}{3}\sin 3x + \frac{1}{5}\sin 5x + \cdots\right)$$

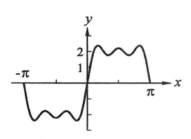

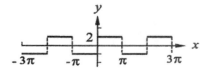

5. $f(x) = \begin{cases} 1 & -\pi \leq x < 0 \\ 2 & 0 \leq x < \pi \end{cases}$

$a_0 = \dfrac{1}{2\pi} \displaystyle\int_{-\pi}^{0} 1\, dx + \dfrac{1}{2\pi} \displaystyle\int_{0}^{\pi} 2\, dx$

$= \dfrac{x}{2\pi}\Big|_{-\pi}^{0} + \dfrac{2x}{2\pi}\Big|_{0}^{\pi}$

$= 0 + \dfrac{\pi}{2\pi} + \dfrac{2\pi}{2\pi} - 0 = \dfrac{1}{2} + 1 = \dfrac{3}{2}$

$a_1 = \dfrac{1}{\pi} \displaystyle\int_{-\pi}^{0} 1\cos x\, dx + \dfrac{1}{\pi} \displaystyle\int_{0}^{\pi} 2\cos x\, dx$

$= \dfrac{1}{\pi} \sin x\Big|_{-\pi}^{0} + \dfrac{2}{\pi} \sin x\Big|_{0}^{\pi}$

$= \dfrac{1}{\pi}(0-0) + \dfrac{2}{\pi}(0-0) = 0$

$a_n = 0$ since $\sin n\pi = 0$

$b_1 = \dfrac{1}{\pi} \displaystyle\int_{-\pi}^{0} 1\sin x\, dx + \dfrac{1}{\pi} \displaystyle\int_{0}^{\pi} 2\sin x\, dx$

$= -\dfrac{1}{\pi} \cos x\Big|_{-\pi}^{0} - \dfrac{2}{\pi} \cos x\Big|_{0}^{\pi}$

$= -\dfrac{1}{\pi}(1+1) - \dfrac{2}{\pi}(-1-1)$

$= -\dfrac{2}{\pi} + \dfrac{4}{\pi} = \dfrac{2}{\pi}$

$b_2 = \dfrac{1}{\pi} \displaystyle\int_{-\pi}^{0} 1\sin 2x\, dx + \dfrac{1}{\pi} \displaystyle\int_{0}^{\pi} 2\sin x\, dx$

$= -\dfrac{1}{2\pi} \cos 2x\Big|_{-\pi}^{0} - \dfrac{1}{\pi} \cos 2x\Big|_{0}^{\pi}$

$= -\dfrac{1}{2\pi}(1-1) - \dfrac{1}{\pi}(1-1) = 0$

$b_3 = \dfrac{1}{\pi} \displaystyle\int_{-\pi}^{0} \sin 3x\, dx + \dfrac{1}{\pi} \displaystyle\int_{0}^{\pi} 2\sin 3x\, dx$

$= -\dfrac{1}{3\pi} \cos 3x\Big|_{-\pi}^{0} - \dfrac{2}{3\pi} \cos 3x\Big|_{0}^{\pi}$

$= -\dfrac{1}{3\pi}(1+1) - \dfrac{2}{3\pi}(-1-1) = \dfrac{2}{3\pi}$

Therefore, $b_n = 0$ for n even; $b_n = \dfrac{2}{n\pi}$ for n odd.

Therefore, $f(x) = \dfrac{3}{2} + \dfrac{2}{\pi} \sin x + \dfrac{2}{3\pi} \sin 3x + \cdots$

9. $f(x) = \begin{cases} -1 & -\pi \leq x < 0 \\ 0 & 0 \leq x < \frac{\pi}{2} \\ 1 & \frac{\pi}{2} \leq x < \pi \end{cases}$

$a_0 = \dfrac{1}{2\pi} \displaystyle\int_{-\pi}^{0} -dx + \dfrac{1}{2\pi} \displaystyle\int_{\pi/2}^{\pi} dx$

$= \dfrac{1}{2\pi} x\Big|_{-\pi}^{0} + \dfrac{1}{2\pi} x\Big|_{\pi/2}^{\pi}$

$= -\dfrac{1}{2\pi} \left(x\Big|_{-\pi}^{0} - x\Big|_{\pi/2}^{\pi} \right)$

$= -\dfrac{1}{2\pi} \left[\pi - \left(\pi - \dfrac{\pi}{2}\right) \right] = -\dfrac{1}{4}$

$a_1 = \dfrac{1}{\pi} \displaystyle\int_{-\pi}^{0} -\cos x\, dx + \dfrac{1}{\pi} \displaystyle\int_{\pi/2}^{\pi} c\cos x\, dx$

$= -\dfrac{1}{\pi} \sin x\Big|_{-\pi}^{0} + \dfrac{1}{\pi} \sin x\Big|_{\pi/2}^{\pi}$

$= -\dfrac{1}{\pi} \left(\sin x\Big|_{-\pi}^{0} - \sin x\Big|_{\pi/2}^{\pi} \right) = -\dfrac{1}{\pi}$

$a_2 = \dfrac{1}{\pi} \displaystyle\int_{-\pi}^{0} -\cos 2x\, dx + \dfrac{1}{\pi} \displaystyle\int_{\pi/2}^{\pi} \cos 2x\, dx$

$= -\dfrac{1}{2\pi} \sin 2x\Big|_{-\pi}^{0} + \dfrac{1}{2\pi} \sin 2x\Big|_{\pi/2}^{\pi}$

$= -\dfrac{1}{2\pi} \left(\sin 2x\Big|_{-\pi}^{0} - \sin 2x\Big|_{\pi/2}^{\pi} \right) = 0$

$a_3 = \dfrac{1}{\pi} \displaystyle\int_{-\pi}^{0} -\cos 3x\, dx + \dfrac{1}{\pi} \displaystyle\int_{\pi/2}^{\pi} \cos 3x\, dx$

$= -\dfrac{1}{3\pi} \sin 3x\Big|_{-\pi}^{0} + \dfrac{1}{3\pi} \sin 3x\Big|_{\pi/2}^{\pi}$

$= -\dfrac{1}{3\pi} \left(\sin 3x\Big|_{-\pi}^{0} - \sin 3x\Big|_{\pi/2}^{\pi} \right) = \dfrac{1}{3\pi}$

Therefore, $a_n = \pm\dfrac{1}{n\pi}$ for n odd; $a_n = 0$ for n even.

$$b_1 = \frac{1}{\pi}\int_{-\pi}^{0} -\sin x \, dx + \frac{1}{\pi}\int_{\pi/2}^{\pi}\sin x \, dx$$

$$= \frac{1}{\pi}\cos x\Big|_{-\pi}^{0} - \frac{1}{\pi}\cos x\Big|_{\pi/2}^{\pi}$$

$$= \frac{1}{\pi}\left(\cos x\Big|_{-\pi}^{0} - \cos x\Big|_{\pi/2}^{\pi}\right) = \frac{3}{\pi}$$

$$b_2 = \frac{1}{\pi}\int_{-\pi}^{0} -\sin 2x \, dx + \frac{1}{\pi}\int_{\pi/2}^{\pi}\sin 2x \, dx$$

$$= \frac{1}{2\pi}\cos 2x\Big|_{-\pi}^{0} - \frac{1}{2\pi}\cos 2x\Big|_{\pi/2}^{\pi}$$

$$= \frac{1}{2\pi}\left(\cos x\Big|_{-\pi}^{0} - \cos 2x\Big|_{\pi/2}^{\pi}\right) = -\frac{1}{\pi}$$

$$b_3 = \frac{1}{\pi}\int_{-\pi}^{0} -\sin 3x \, dx + \frac{1}{\pi}\int_{\pi/2}^{\pi}\sin 3x \, dx$$

$$= \frac{1}{3\pi}\cos 3x\Big|_{-\pi}^{0} - \frac{1}{3\pi}\cos 3x\Big|_{\pi/2}^{\pi}$$

$$= \frac{1}{3\pi}\left(\cos 3x\Big|_{-\pi}^{0} - \cos 3x\Big|_{\pi/2}^{\pi}\right) = -\frac{1}{\pi}$$

$$b_n = \pm\frac{1}{\pi}\text{ for }n > 1$$

$$f(x) = -\frac{1}{4} - \frac{1}{\pi}\cos x + \frac{1}{3\pi}\cos 3x - \cdots$$
$$+ \frac{3}{\pi}\sin x - \frac{1}{\pi}\sin 2x + \frac{1}{\pi}\sin 3x - \cdots$$

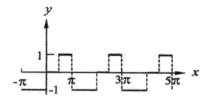

13. $a_0 = \dfrac{1}{2\pi}\int_{-\pi}^{\pi} e^x dx = \dfrac{e^{\pi} - e^{-\pi}}{2\pi}$

$a_1 = \dfrac{1}{\pi}\int_{-\pi}^{\pi} e^x \cos x \, dx = -\dfrac{e^{\pi} - e^{-\pi}}{2\pi}$

$a_2 = \dfrac{1}{\pi}\int_{-\pi}^{\pi} e^x \cos 2x \, dx = \dfrac{e^{\pi} - e^{-\pi}}{5\pi}$

$b_1 = \dfrac{1}{\pi}\int_{-\pi}^{\pi} e^x \sin x \, dx = -\dfrac{e^{\pi} - e^{-\pi}}{2\pi}$

$b_2 = \dfrac{1}{\pi}\int_{-\pi}^{\pi} e^x \sin 2x \, dx = -\dfrac{2(e^{\pi} - e^{-\pi})}{5\pi}$

$$e^x = \frac{e^{\pi} - e^{-\pi}}{2\pi} - \frac{e^{\pi} - e^{-\pi}}{2\pi}\cos x + \frac{e^{\pi} - e^{-\pi}}{5\pi}\cos 2x + \cdots$$
$$+ \frac{e^{\pi} - e^{-\pi}}{2\pi}\sin x - \frac{2(e^{\pi} - e^{-\pi})}{5\pi}\sin 2x + \cdots$$

$$e^x = \frac{e^{\pi} - e^{-\pi}}{\pi}\left(\frac{1}{2} - \frac{1}{2}\cos x + \frac{1}{5}\cos 2x + \cdots + \frac{1}{2}\sin x\right.$$
$$\left. - \frac{2}{5}\sin 2x + \cdots\right)$$

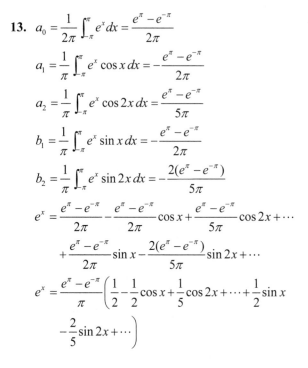

17. $f(x) = \begin{cases} 1, -\pi \le x < 0 \\ 2, 0 \le x < \pi \end{cases}$

Graph $y_1 = \dfrac{3}{2} + \dfrac{2}{\pi}\sin x + \dfrac{2}{3\pi}\sin 3x$

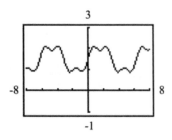

21. $F(t) = \begin{cases} 0, & -\pi \leq t < 0 \\ t^2 + t, & 0 < t < \pi \end{cases}$

$$a_0 = \frac{1}{2\pi} \int_0^\pi (t^2 + t) dt = \frac{\pi^2}{6} + \frac{\pi}{4}$$

$$a_1 = \frac{1}{\pi} \int_0^\pi (t^2 + t) \cos t \, dt = -\frac{2}{\pi} - 2$$

$$a_2 = \frac{1}{\pi} \int_0^\pi (t^2 + t) \cos 2t \, dt = \frac{1}{2}$$

$$a_3 = \frac{1}{\pi} \int_0^\pi (t^2 + t) \cos 3t \, dt = \frac{-2 - 2\pi}{9\pi}$$

$$b_1 = \frac{1}{\pi} \int_0^\pi (t^2 + t) \sin t \, dt = \pi - \frac{4}{\pi} + 1$$

$$b_2 = \frac{1}{\pi} \int_0^\pi (t^2 + t) \sin 2t \, dt = \frac{-\pi - 1}{2}$$

$$b_3 = \frac{1}{\pi} \int_0^\pi (t^2 + t) \sin 3t \, dt = \frac{\pi}{3} - \frac{4}{27\pi} + \frac{1}{3}$$

$$F(t) = \frac{\pi^2}{6} + \frac{\pi}{4} - \left(\frac{2}{\pi} + 2\right) \cos t + \frac{1}{2} \cos 2t$$

$$- \left(\frac{2 + 2\pi}{9\pi}\right) \cos 3t + \cdots + \left(\pi - \frac{4}{\pi} + 1\right) \sin t$$

$$- \left(\frac{\pi + 1}{2}\right) \sin 2t + \left(\frac{\pi}{3} - \frac{4}{27\pi} + \frac{1}{3}\right) \sin 3t + \cdots$$

30.7   More About Fourier Series

1. $f(x) = \begin{cases} 2 & -\pi \leq x < -\dfrac{\pi}{2}, \dfrac{\pi}{2} \leq x < \pi \\ 3 & -\dfrac{\pi}{2} \leq x < \dfrac{\pi}{2} \end{cases}$

From Example 2,

$$f(x) = \frac{1}{2} + \frac{2}{\pi}\left(\cos x - \frac{\cos 3x}{3} + \frac{\cos 5x}{5} - \cdots\right) + 2.$$

$$= \frac{5}{2} + \frac{2}{\pi}\left(\cos x - \frac{\cos 3x}{3} + \frac{\cos 5x}{5} - \cdots\right)$$

5. $f(x) = \begin{cases} 5 & -3 \leq x < 0 \\ 0 & 0 \leq x < 3 \end{cases}$

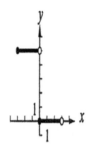

from the graph $f(x)$ is neither odd nor even.

9. $f(x) = |x| \quad -4 \leq x < 4$

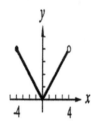

is even from the graph.

13. From the graph, $f(x) = 2 - x$, $-4 \leq x < 4$ is not odd or even. Fourier series may contain both sine and cosine terms. In fact, the expansion contains only sine terms.

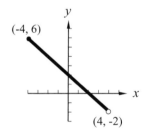

17. $f(x)\begin{cases} 5 & -3 \le x < 0 \\ 0 & 0 \le x < 3 \end{cases}$

period $= 6 = 2L$, $L = 3$

$a_0 = \dfrac{1}{2L} \displaystyle\int_{-L}^{L} f(x)\,dx$

$= \dfrac{1}{6} \displaystyle\int_{-3}^{0} 5\,dx + \dfrac{1}{6} \displaystyle\int_{0}^{3} 0 \cdot dx = \dfrac{5}{2}$

$a_n = \dfrac{1}{L} \displaystyle\int_{-L}^{L} f(x) \cos \dfrac{n\pi x}{L}\,dx$

$= \dfrac{1}{3} \displaystyle\int_{-3}^{0} 5 \cos \dfrac{n\pi x}{3}\,dx + \dfrac{1}{3} \displaystyle\int_{0}^{3} 0 \cdot \cos \dfrac{n\pi x}{3}\,dx$

$a_n = \dfrac{5\sin(n\pi)}{n\pi} = 0, n = 1,2,3\cdots$

$b_n = \dfrac{1}{L} \displaystyle\int_{-L}^{L} f(x) \sin \dfrac{n\pi x}{L}\,dx$

$= \dfrac{1}{3} \displaystyle\int_{-3}^{0} 5 \sin \dfrac{n\pi x}{3}\,dx + \dfrac{1}{3} \displaystyle\int_{0}^{3} 0 \cdot \sin \dfrac{2\pi x}{3}\,dx$

$b_n = \dfrac{5\cos(n\pi) - 5}{n\pi} = \dfrac{5}{\pi}\left(\dfrac{\cos(n\pi) - 1}{n} \right)$

n	b_n
1	$\dfrac{5}{\pi} \cdot (-2) = \dfrac{-10}{\pi}$
2	0
3	$\dfrac{5}{\pi}\left(-\dfrac{2}{3}\right) = \dfrac{-10}{3\pi}$
4	0
5	$\dfrac{5}{\pi}\left(-\dfrac{2}{5}\right) = \dfrac{-10}{5\pi}$

$f(x) = a_0 + a_1 \cos \dfrac{\pi x}{L} + a_2 \cos \dfrac{2\pi x}{L} + a_3 \cos \dfrac{3\pi x}{L} + \cdots$
$\quad + b_1 \sin \dfrac{\pi x}{L} + b_2 \sin \dfrac{2\pi x}{L} + b_3 \sin \dfrac{3\pi x}{L} + \cdots$

$f(x) = \dfrac{5}{2}$

$\quad - \dfrac{10}{\pi}\left(\sin \dfrac{\pi x}{3} + \dfrac{1}{3}\sin \dfrac{3\pi x}{3} + \dfrac{1}{5}\sin \dfrac{5\pi x}{3} + \cdots \right)$

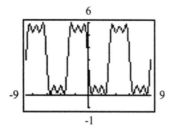

21. $f(x) = \begin{cases} -x & -4 \le x < 0 \\ x & 0 \le x < 4 \end{cases}$

$a_0 = \dfrac{1}{8} \displaystyle\int_{-4}^{0} -x\,dx + \dfrac{1}{8} \displaystyle\int_{0}^{4} x\,dx = -\dfrac{1}{16}x^2 \Big|_{-4}^{0} + \dfrac{1}{16}x^2 \Big|_{0}^{4} = 2$

$a_n = \dfrac{1}{4} \displaystyle\int_{-4}^{0} -x \cos \dfrac{n\pi x}{4}\,dx + \dfrac{1}{4} \displaystyle\int x \cos \dfrac{n\pi x}{4}\,dx$

$= -\dfrac{1}{4}\dfrac{16}{(n\pi)^2} \displaystyle\int_{-4}^{0} \dfrac{n\pi}{4}x \cos \dfrac{n\pi}{4}x \dfrac{n\pi}{4}\,dx + \dfrac{1}{4}\dfrac{16}{(n\pi)^2} \displaystyle\int_{0}^{4} \dfrac{n\pi}{4}x \cos \dfrac{n\pi}{4}x \dfrac{n\pi}{4}\,dx$

$= -\dfrac{4}{(n\pi)^2} \left(\cos \dfrac{n\pi x}{4} + \dfrac{n\pi x}{4} \sin \dfrac{n\pi x}{4} \right) \Big|_{-4}^{0} + \dfrac{4}{(n\pi)^2} \left(\cos \dfrac{n\pi x}{4} + \dfrac{n\pi x}{4} \sin \dfrac{n\pi x}{4} \right) \Big|_{0}^{4}$

$= -\dfrac{4}{(n\pi)^2}\left(\cos 0 - [\cos(-n\pi) + n\pi \sin n\pi] \right) + \dfrac{4}{(n\pi)^2}\left(\cos n\pi + n\pi \sin n\pi - [\cos 0] \right)$

$= -\dfrac{4}{(n\pi)^2}\left(1 - \cos n\pi - n\pi \sin n\pi \right) + \dfrac{4}{(n\pi)^2}\left(\cos n\pi + n\pi \sin n\pi - 1 \right)$

$= -\dfrac{4}{(n\pi)^2}\left(1 - \cos n\pi - n\pi \sin n\pi - \cos n\pi - n\pi \sin n\pi + 1 \right)$

$= -\dfrac{4}{(n\pi)^2}\left(2 - 2\cos n\pi - 2n\pi \sin n\pi \right)$

$a_1 = -\dfrac{16}{\pi^2};\ a_2 = 0;\ a_3 = -\dfrac{16}{9\pi^2}$

$b_n = \dfrac{1}{4} \displaystyle\int_{-4}^{0} -x \sin \dfrac{n\pi x}{4}\,dx + \dfrac{1}{4} \displaystyle\int_{0}^{4} x \sin \dfrac{n\pi x}{4}\,dx$

$= -\dfrac{1}{4}\dfrac{16}{(n\pi)^2} \displaystyle\int_{-4}^{0} \dfrac{n\pi x}{4} \sin \dfrac{n\pi x}{4} \left(\dfrac{n\pi}{4}\,dx \right) + \dfrac{1}{4}\dfrac{16}{(n\pi)^2} \displaystyle\int_{0}^{4} \dfrac{n\pi x}{4} \sin \dfrac{n\pi x}{4} \cdot \dfrac{n\pi\,dx}{4}$

$= -\dfrac{4}{(n\pi)^2} \left(\sin \dfrac{n\pi x}{4} - \dfrac{n\pi x}{4} \cos \dfrac{n\pi x}{4} \right) \Big|_{-4}^{0} + \dfrac{4}{(n\pi)^2} \left(\sin \dfrac{n\pi x}{4} - \dfrac{n\pi x}{4} \cos \dfrac{n\pi x}{4} \right) \Big|_{0}^{4}$

$$= -\frac{4}{(n\pi)^2}\{-[\sin(-n\pi) + n\pi\cos(-n\pi)]\} + \frac{4}{(n\pi)^2}(\sin n\pi - n\pi\cos n\pi)$$

$$= -\frac{4}{(n\pi)^2}(\sin n\pi - n\pi\cos n\pi) + \frac{4}{(n\pi)^2}(\sin n\pi - n\pi\cos n\pi) = 0, \text{ for all } n.$$

Therefore,

$$f(x) = 2 - \frac{16}{\pi^2}\cos\frac{\pi x}{4} - \frac{16}{9\pi^2}\cos\frac{3\pi x}{4} = 2 - \frac{16}{\pi^2}\left(\cos\frac{\pi x}{4} + \frac{1}{9}\cos\frac{3\pi x}{4} + \cdots\right)$$

and comparing with calculator,

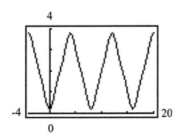

$$a_n = \frac{16}{n^2\pi^2}\cos n\pi$$

$$a_1 = \frac{16}{1^2\pi^2}\cos\pi = \frac{-16}{\pi^2}$$

$$a_2 = \frac{16}{2^2\pi^2}\cos 2\pi = \frac{4}{\pi^2}$$

$$a_3 = \frac{16}{3^2\pi^2}\cos 3\pi = \frac{-16}{9\pi^2}$$

$$f(x) = \frac{4}{3} - \frac{16}{\pi^2}\cos\frac{\pi x}{2} + \frac{4}{\pi^2}\cos\frac{2\pi x}{2}$$
$$-\frac{16}{9\pi^2}\cos\frac{3\pi x}{2} + \cdots$$

$$f(x) = \frac{4}{3} - \frac{16}{\pi^2}\left(\cos\frac{\pi x}{2} - \frac{1}{4}\cos\pi x + \frac{1}{9}\cos\frac{3\pi x}{2} - \cdots\right)$$

25. Expand $f(x) = x^2$ in a half-range cosine series for $0 \le x < 2$.

$$a_0 = \frac{1}{L}\int_0^L f(x)dx = \frac{1}{2}\int_0^2 x^2 dx$$

$$= \frac{1}{2}\left.\frac{x^3}{3}\right|_0^2 = \frac{1}{6}(2^3 - 0) = \frac{4}{3}$$

$$a_n = \frac{2}{L}\int_0^L f(x)\cos\frac{n\pi x}{L}dx, (n = 1, 2, 3, \ldots)$$

$$a_n = \frac{2}{2}\int_0^2 x^2 \frac{n\pi x}{2} dx$$

$$= \frac{2x}{\frac{n^2\pi^2}{4}}\cos\frac{n\pi x}{2} + \left(\frac{x^2}{\frac{n\pi}{2}} - \frac{2}{\frac{n^3\pi^3}{8}}\right)\left.\sin\frac{n\pi x}{2}\right|_0^2$$

$$a_n = \frac{8x}{n^2\pi^2}\cos\frac{n\pi x}{2} + \left(\frac{2x^2}{n\pi} - \frac{16}{n^3\pi^3}\right)\left.\sin\frac{n\pi x}{2}\right|_0^2$$

$$a_n = \frac{16}{n^2\pi^2}\cos n\pi + \left(\frac{8}{n\pi} - \frac{16}{n^3\pi 3}\right)\sin n\pi$$

Chapter 30   Review Exercises

1. $f(x) = \dfrac{1}{1+e^x} = (1+e^x)^{-1}$

$f'(x) = -(1+e^x)^{-2}(e^x) = e^x(1+e^x)^{-2}$

$f''(x) = -(1+e^x)^{-2}e^x + e^x(2)(1+e^x)^{-3}(e^x)$

$f'''(x) = -(1+e^x)^{-2}e^x + e^x(2)(1+e^x)^{-3}(e^x)$
$\qquad\quad + 2e^{2x}(-3)(1+e^x)^{-4}e^x$
$\qquad\quad + (1+e^x)^{-3}(2e^{2x})(2)$

$f(0) = \dfrac{1}{1+1} = \dfrac{1}{2}$

$f'(0) = -1(2^{-2}) = -\dfrac{1}{4}$

$f''(0) = -\dfrac{1}{4} + \dfrac{1}{4} = 0$

$f'''(0) = -\dfrac{1}{4} + \dfrac{1}{4} - \dfrac{3}{8} + \dfrac{1}{2} = \dfrac{1}{8}$

$f(x) = \dfrac{1}{2} - \dfrac{1}{4}x + \dfrac{0x^2}{2!} + \left(\dfrac{1}{8}\right)\dfrac{x^3}{3!} + \cdots$

$\qquad = \dfrac{1}{2} - \dfrac{1}{4}x + \dfrac{1}{48}x^3 - \cdots$

5. $f(x) = (x+1)^{1/3} \qquad f(0) = 1$

$f'(x) = \dfrac{1}{3}(x+1)^{-2/3} \qquad f'(0) = \dfrac{1}{3}$

$f''(x) = -\dfrac{2}{9}(x+1)^{-5/3} \quad f''(0) = -\dfrac{2}{9}$

$f(x) = 1 + \dfrac{1}{3}x - \dfrac{2x^2}{9(2)} + \cdots$

$\qquad = 1 + \dfrac{1}{3}x - \dfrac{1}{9}x^2 + \cdots$

9. $f(x) = \cos(a+x), \qquad f(0) = \cos a$

$f'(x) = -\sin(a+x), \qquad f'(0) = -\sin a$

$f''(x) = -\cos(a+x), \qquad f''(0) = -\cos a$

$f(x) = \cos(a+x) = \cos a - \sin a \cdot x - \dfrac{\cos a}{2!}x^2 + \cdots$

13.   See Exercise 5.

$\sqrt[3]{1+x} = 1 + \dfrac{1}{3}x - \dfrac{1}{9}x^2 + \cdots$

$\sqrt[3]{1+0.3} = 1 + \dfrac{1}{3}(0.3) - \dfrac{1}{9}(0.3)^2$

$\sqrt[3]{1.3} = 1.09$

17. $\ln(1+x) = x - \dfrac{x^2}{2} + \dfrac{x^3}{3} - \cdots$

$\ln\big[1 + (-0.1828)\big]$

$= -0.1828 - \dfrac{(-0.1828)^3}{2} + \dfrac{(-0.1828)^2}{3} - \cdots$

$= -0.2015$

$\ln 1.2237^{-1} = \ln 0.8172 = -0.2019$

21. $f(x) = \sqrt{x}, a = 144$

$f(x) = \sqrt{x}, f(144) = 12$

$f'(x) = \dfrac{1}{2}x^{-1/2}, f'(144) = \dfrac{1}{24}$

$f''(x) = -\dfrac{1}{4}x^{-3/2}, f''(144) = -\dfrac{1}{6912}$

$\sqrt{x} = 12 + \dfrac{1}{24}(x-144) - \dfrac{1}{6912}\dfrac{(x-144)^2}{2} + \cdots$

$\sqrt{148} = 12 + \dfrac{4}{24} - \dfrac{4^2}{13{,}824} = 12.1655$

25. $f(x) = \cos x, a = \dfrac{\pi}{3}$

$f(x) = \cos x, f\left(\dfrac{\pi}{3}\right) = \dfrac{1}{2}$

$f'(x) = -\sin x, f'\left(\dfrac{\pi}{3}\right) = \dfrac{1}{2}\sqrt{3}$

$f''(x) = -\cos x, f''\left(\dfrac{\pi}{3}\right) = -\dfrac{1}{2}$

$f(x) = \dfrac{1}{2} + \dfrac{1}{2}\sqrt{3}\left(x - \dfrac{\pi}{3}\right) - \dfrac{1}{4}\left(x - \dfrac{\pi}{3}\right)^2 + \cdots$

29. $f(x)\begin{cases} 0 & -\pi < x < 0 \\ x-1 & 0 \le x < \pi \end{cases}$

is Example 2 of section 30-6 shifted down 1 unit.
The Fourier series is

$f(x) = -1 + \dfrac{2+\pi}{4} - \dfrac{2}{\pi}\cos x - \dfrac{2}{9\pi}\cos 3x - \cdots$

$\qquad + \left(\dfrac{\pi-2}{\pi}\right)\sin x - \dfrac{1}{2}\sin 2x + \cdots$

33. $f(x) = \begin{cases} 0 & -\pi \le x < -\pi/2, \ \pi/2 < x < \pi \\ 1 & -\pi/2 \le x \le \pi/2 \end{cases}$

$a_0 = \dfrac{1}{2\pi}\int_{-\pi}^{-\pi/2} 0\,dx + \dfrac{1}{2\pi}\int_{-\pi/2}^{\pi/2} dx + \dfrac{1}{2\pi}\int_{\pi/2}^{\pi} 0\,dx$

$= \dfrac{x}{2\pi}\Big|_{-\pi/2}^{\pi/2} = \dfrac{1}{2\pi}\left(\dfrac{\pi}{2}+\dfrac{\pi}{2}\right) = \dfrac{1}{2}$

$a_n = \dfrac{1}{\pi}\int_{-\pi}^{-\pi/2} 0\cos nx\,dx + \dfrac{1}{\pi}\int_{-\pi/2}^{\pi/2}\cos nx\,dx$

$+ \dfrac{1}{\pi}\int_{\pi/2}^{\pi} 0\cos nx\,dx = \dfrac{1}{n\pi}\sin nx\Big|_{-\pi/2}^{\pi/2}$

$= \dfrac{1}{n\pi}\left[\sin\dfrac{n\pi}{2} - \sin\dfrac{(-n\pi)}{2}\right] = \dfrac{2}{n\pi}\sin\dfrac{n\pi}{2}$

$a_1 = \dfrac{2}{\pi}\sin\dfrac{\pi}{2} = \dfrac{2}{\pi}, \ a_2 = \dfrac{2}{2\pi}\sin\pi = 0, \ a_3 = \dfrac{2}{3\pi}\sin\dfrac{3\pi}{2}$

$= -\dfrac{2}{3\pi}$

$b_n = \dfrac{1}{\pi}\int_{-\pi}^{-\pi/2} 0\sin nx\,dx + \dfrac{1}{\pi}\int_{-\pi/2}^{\pi/2}\sin nx\,dx$

$+ \dfrac{1}{\pi}\int_{\pi/2}^{\pi} 0\sin nx\,dx = -\dfrac{1}{n\pi}\cos nx\Big|_{-\pi/2}^{\pi/2}$

$= -\dfrac{1}{n\pi}\left[\cos\dfrac{n\pi}{2} - \cos\dfrac{(-n\pi)}{2}\right] = 0;$

$\left(\text{for all } n \text{ since } \cos\theta = \cos(-\theta)\right)$

$f(x) = \dfrac{1}{2} + \dfrac{2}{\pi}\left(\cos x - \dfrac{1}{3}\cos 3x + \cdots\right)$

37. It is a geometric series for which $|r| < 1 = 0.80$.

Therefore the series converges.

$S = \dfrac{1000}{1 - 0.80} = 5000$

41. $\displaystyle\int \sin x^2\,dx = \int\left(x^2 - \dfrac{x^6}{3!} + \dfrac{x^{10}}{5!} - \cdots\right)dx$

$= \dfrac{x^3}{3} - \dfrac{x^7}{7\cdot 3!} + \dfrac{x^{11}}{11\cdot 5!} - \cdots$

45. $\dfrac{d}{dx}\tan x$

$= \dfrac{d}{dx}1 + 2\left(x - \dfrac{\pi}{4}\right) + 2\left(x - \dfrac{\pi}{4}\right)^2 + \dfrac{8}{3}\left(x - \dfrac{\pi}{4}\right)^3 + \cdots$

$\sec^2 x = 2 + 4\left(x - \dfrac{\pi}{4}\right) + 8\left(x - \dfrac{\pi}{4}\right)^2 + \cdots$

49. $\ln(1+x)^4 = 4\ln(1+x) = 4\left(x - \dfrac{x^2}{2} + \dfrac{x^3}{3} - \dfrac{x^4}{4} + \cdots\right)$

$= 4x - 2x^2 + \dfrac{4x^3}{4} - x^4 + \cdots$

53.
$$\begin{aligned}
f(x) &= \cos x & f(0) &= 1 \\
f'(x) &= -\sin x & f'(0) &= 0 \\
f''(x) &= -\cos x & f''(0) &= -1 \\
f'''(x) &= \sin x & f'''(0) &= 0 \\
f^{iv}(x) &= \cos x & f^{iv}(0) &= 1
\end{aligned}$$

$f(x) = 1 + 0x - \dfrac{1}{2}x^2 + \dfrac{0x^3}{3!} + \dfrac{1x^4}{4!} = 1 - \dfrac{1}{2}x^2 + \dfrac{1}{24}x^4 + \cdots$

$\sec x = \dfrac{1}{\cos x} = \dfrac{1}{1 - \frac{1}{2}x^2 - \frac{1}{24}x^4} = 1 + \dfrac{1}{2}x^2 + \dfrac{5}{24}x^4 + \cdots$

The long division is

$$
\begin{array}{r}
1 + \frac{1}{2}x^2 + \frac{5}{24}x^4 \\
1 - \frac{1}{2}x^2 + \frac{1}{24}x^4 \overline{)\,1 + 0 \quad\ + 0 \quad\ + 0 \quad\ + 0} \\
1 - \frac{1}{2}x^2 + \frac{1}{24}x^4 \\
\hline
\frac{1}{2}x^2 - \frac{1}{24}x^4 \\
\frac{1}{2}x^2 - \frac{1}{4}x^4 \\
\hline
\frac{5}{24}x^4 + 0 \quad\ + 0 \\
\frac{5}{24}x^4 - \frac{5}{48}x^2 + \frac{5}{576}x^4 \\
\hline
\end{array}
$$

57. $a_0 = \dfrac{1}{1}\int_0^1 x^2\,dx = \dfrac{1}{3}$

$a_1 = \dfrac{2}{1}\int_0^1 x^2\cos\dfrac{\pi x}{1}\,dx = \dfrac{-4}{\pi^2}$

$a_2 = \dfrac{2}{1}\int_0^1 x^2\cos\dfrac{2\pi x}{1}\,dx = \dfrac{1}{\pi^2}$

$a_3 = \dfrac{2}{1}\int_0^1 x^2\cos\dfrac{3\pi x}{1}\,dx = \dfrac{-4}{9\pi^2}$

$f(x) = \dfrac{1}{3} - \dfrac{4}{\pi^2}\cos\pi x + \dfrac{1}{\pi^2}\cos 2\pi x$

$\qquad\qquad - \dfrac{4}{9\pi^2}\cos 3\pi x + \cdots$

61. $A = \displaystyle\int_{0.1}^{0.2}\dfrac{x - \sin x}{x^2}\,dx = \int_{0.1}^{0.2}\dfrac{x - \left(x - \dfrac{x^3}{3!} + \dfrac{x^5}{5!}\right)}{x^2}\,dx$

$= \displaystyle\int_{0.1}^{0.2}\dfrac{\dfrac{x^3}{3!} + \dfrac{x^5}{5!}}{x^2}\,dx = \int_{0.1}^{0.2}\left(\dfrac{x}{6} - \dfrac{x^3}{120}\right)dx$

$= \left(\dfrac{x^2}{12} - \dfrac{x^4}{480}\right)\Big|_{0.1}^{0.2} = \left(\dfrac{0.4}{12} - \dfrac{0.0016}{480}\right) - \left(\dfrac{0.01}{12} - \dfrac{0.0001}{480}\right)$

$= 0.0025$

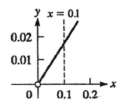

65. $\cos t = 1 - \dfrac{1}{2}t^2 + \dfrac{1}{24}t^4 - \dfrac{1}{720}t^6 + \cdots$

$y = 3.2\cos(880\pi t)$

$= 3.2\left(1 - \dfrac{1}{2}(880\pi t)^2 + \dfrac{1}{24}(880\pi t)^4 - \dfrac{1}{720}(880\pi t)^6 + \cdots\right)$

$y = 3.2 - 1.239\times10^6\,\pi^2 t^2 + 7.9959\times10^{10}\,\pi^4 t^4$

$\qquad - 2.064\times10^{15}\,\pi^6 t^6 + \cdots$

69. $\cos\theta = \dfrac{r}{r+h}$, $r = 6400$ km

$4.8 = r\theta$, $\theta = \dfrac{4.8}{r}$

$\cos\dfrac{4.8}{6400} = \dfrac{6400}{6400+h}$

$\qquad h = 0.0018$ km $= 1.8$ m

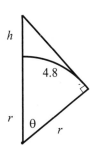

73. $\dfrac{N_0}{1 - e^{-k/T}} = N_0\left(\dfrac{1}{1 - e^{-k/T}}\right)$; Let $x = e^{-k/T}$

$N_0\left(\dfrac{N_0}{1 - e^{-k/T}}\right) = N_0\left(\dfrac{1}{1 - x}\right)$

The Maclaurin's expansion for $f(x) = \dfrac{1}{1 - x}$ is:

$f(x) = \dfrac{1}{1 - x}$; $\qquad\qquad f(0) = 1$

$f'(x) = \dfrac{1}{(1 - x)^2}$; $\qquad\qquad f'(0) = 1$

$f''(x) = \dfrac{2}{(1 - x)^3}$; $\qquad\qquad f''(0) = 2$

$f(x) = 1 + x + \dfrac{2x^2}{2!} + \cdots = 1 + x + x^2 + \cdots$

Substituting $e^{-k/T}$ for x; $f(x) = 1 + e^{-kT} + e^{-2k/T} + \cdots$;

$\therefore \dfrac{N_0}{1 - e^{-kT}} = N_0\left(1 + e^{-k/T} + e^{-2kT} + \cdots\right)$

77. $\sin x = x - \dfrac{1}{6}x^3 + \dfrac{1}{120}x^5 - \dfrac{1}{5040}x^7 + \dfrac{1}{362,880}x^9 + \cdots$

$\cos x = 1 - \dfrac{1}{2}x^2 + \dfrac{1}{24}x^4 - \dfrac{1}{720}x^6 + \dfrac{1}{40,320}x^8 + \cdots$

(a)

$\sin^2 x = \dfrac{1 - \cos 2x}{2}$

$= \dfrac{1}{2}\left(1 - 1 - \dfrac{1}{2}(2x)^2 + \dfrac{1}{24}(2x)^4 - \dfrac{1}{720}(2x)^6 + \dfrac{1}{40,320}(2x)^8\right) + \cdots$

$\sin^2 x = \dfrac{1 - \cos 2x}{2}$

$= \dfrac{1}{2}\left(1 - 1 - \dfrac{1}{2}(2x)^2 + \dfrac{1}{24}(2x)^4 - \dfrac{1}{720}(2x)^6 + \dfrac{1}{40,320}(2x)^8\right) + \cdots$

$\sin^2 x = x^2 - \dfrac{1}{3}x^4 + \dfrac{2}{45}x^6 - \dfrac{1}{315}x^8 + \cdots$

(b)

$\sin^2 x = 1 - \cos^2 x = 1 - \left(\dfrac{1 + \cos 2x}{2}\right) = 1 - \dfrac{1}{2}(1 + 1 - \dfrac{1}{2}(2x)^2$

$\qquad + \dfrac{1}{24}(2x)^4 - \dfrac{1}{720}(2x)^6 + \dfrac{1}{40,320}(2x)^8 + \cdots$

$\sin^2 x = 1 - \dfrac{1}{2}(2 - \dfrac{1}{2}(2x)^2 + \dfrac{1}{24}(2x)^4 - \dfrac{1}{720}(2x)^6$

$\qquad + \dfrac{1}{40,320}(2x)^8 + \cdots$

$\sin^2 x = x^2 - \dfrac{1}{3}x^4 + \dfrac{2}{45}x^6 - \dfrac{1}{315}x^8 + \cdots$

(c)

$\sin^2 x = \left(x - \dfrac{1}{6}x^3 + \dfrac{1}{120}x^5 - \dfrac{1}{5040}x^7 + \cdots\right)^2$

$\sin^2 x = x^2 - \dfrac{1}{3}x^4 + \dfrac{2}{45}x^6 - \dfrac{1}{315}x^8 + \cdots$

(d)

$\sin^2 x = 1 - \cos^2 x$

$= 1 - \left(1 - \dfrac{1}{2}x^2 + \dfrac{1}{24}x^4 - \dfrac{1}{720}x^6 + \dfrac{1}{40,320}x^8 + \cdots\right)^2$

$\sin^2 x = 1 - \cos^2 x$

$= 1 - \left(1 - \dfrac{1}{2}x^2 + \dfrac{1}{24}x^4 - \dfrac{1}{720}x^6 + \dfrac{1}{40,320}x^8 + \cdots\right)^2$

$\sin^2 x = 1 - \left(1 - x^2 + \dfrac{1}{3}x^4 - \dfrac{2}{45}x^6 + \dfrac{1}{320}x^8 + \cdots\right)$

$\sin^2 x = x^2 - \dfrac{1}{3}x^4 + \dfrac{2}{45}x^6 - \dfrac{1}{320}x^8 + \cdots$

Chapter 31

DIFFERENTIAL EQUATIONS

31.1 Solutions of Differential Equations

1. $y = c_1 e^{-x} + c_2 e^{2x}$

$$\frac{dy}{dx} = -c_1 e^{-x} + 2c_2 e^{2x}$$

$$\frac{d^2y}{dx^2} = c_1 e^{-x} + 4c_2 e^{2x}$$

$$\begin{aligned}
\frac{d^2y}{dx^2} - \frac{dy}{dx} &= c_1 e^{-x} + 4c_2 e^{2x} - (-c_1 e^{-x} + 2c_2 e^{2x}) \\
&= c_1 e^{-x} + 4c_2 e^{2x} + c_1 e^{-x} - 2c_2 e^{2x} \\
&= 2c_1 e^{-x} + 2c_2 e^{2x} = 2(c_1 e^{-x} + c_2 e^{2x}) \\
&= 2y
\end{aligned}$$

$$y = 4e^{-x}, \frac{dy}{dx} = -4e^{-x}, \frac{d^2y}{dx^2} = 4e^{-x}$$

$$\begin{aligned}
\frac{d^2y}{dx^2} - \frac{dy}{dx} &= 4e^{-x} - (-4e^{-x}) = 8e^{-x} \\
&= 2(4e^{-x}) = 2y
\end{aligned}$$

5. $y'' + 3y' - 4y = 3e^x$; $y = c_1 e^x + c_2 e^{-4x} + \frac{3}{5}xe^x$

$$\begin{aligned}
y' &= c_1 e^x - 4c_2 e^{-4x} + \frac{3}{5}(xe^x + e^x) \\
&= c_1 e^x - 4c_2 e^{-4x} + \frac{3}{5}xe^x + \frac{3}{5}e^x
\end{aligned}$$

$$\begin{aligned}
y'' &= c_1 e^x + 16c_2 e^{-4x} + \frac{3}{5}e^x + \frac{3}{5}(xe^x + e^x) \\
&= c_1 e^x + 16c_2 e^{-4x} + \frac{3}{5}e^x + \frac{3}{5}xe^x + \frac{3}{5}e^x \\
&= c_1 e^x + 16c_2 e^{-4x} + \frac{3}{5}xe^x + \frac{6}{5}e^x
\end{aligned}$$

Substitute y, y', y'' into differential equation.

$$c_1 e^x + 16c_2 e^{-4x} + \frac{3}{5}xe^x + \frac{6}{5}e^x$$

$$+ 3\left(c_1 e^x - 4c_2 e^{-4x} + \frac{3}{5}xe^x + \frac{3}{5}e^x\right)$$

$$- 4\left(c_1 e^x + c_2 e^{-4x} + \frac{3}{5}xe^x\right) = 3e^x$$

$$c_1 e^x + 16c_2 e^{-4x} + \frac{3}{5}xe^x + \frac{6}{5}e^x$$

$$+ 3c_1 e^x - 12c_2 e^{-4x} + \frac{9}{5}xe^x + \frac{9}{5}e^x$$

$$- 4c_1 e^x - 4c_2 e^{-4x} - \frac{12}{5}xe^x = 3e^x$$

$$\frac{15}{5}e^x = 3e^x; \ 3e^x = 3e^x \text{ identity}$$

general solution

9. $y = 3\cos 2x$; $y'' = -12\cos 2x$

$$y'' + 4y = -12\cos 2x + 4(3\cos 2x) = 0$$

$$y = c_1 \sin 2x + c_2 \cos 2x;$$

$$y'' = -4c_1 \sin 2x - 4c_2 \cos 2x$$

$$\begin{aligned}
y'' + 4y &= -4c_1 \sin 2x - 4c_2 \cos 2x \\
&\quad + 4(c_1 \sin 2x + c_2 \cos 2x) \\
&= 0
\end{aligned}$$

13. $y = 2 + x - x^3$

$$\frac{dy}{dx} = 1 - 3x^2$$

17. $y'' + 9y = 4\cos x$; $2y = \cos x$; $y = \frac{1}{2}\cos x$;

$$y' = -\frac{1}{2}\sin x; \ y'' = -\frac{1}{2}\cos x$$

Substitute y and y''.

$$-\frac{1}{2}\cos x + 9\left(\frac{1}{2}\cos x\right) = 4\cos x$$

$$-\frac{1}{2}\cos x + \frac{9}{2}\cos x = 4\cos x$$

$$\frac{8}{2}\cos x = 4\cos x; \ 4\cos x = 4\cos x \text{ identity}$$

21. $x\frac{d^2y}{dx^2} + \frac{dy}{dx} = 0$; $y = c_1 \ln x + c_2$

$$\frac{dy}{dx} = \frac{c_1}{x} = c_1 x^{-1}; \ \frac{d^2y}{dx^2} = -c_1 x^{-2} = -\frac{c_1}{x^2}$$

Substitute $\frac{dy}{dx}$ and $\frac{d^2y}{dx^2}$.

$$x\left(\frac{-c_1}{x^2}\right) + c_1 x^{-1} = 0; \ -\frac{c_1}{x} + \frac{c_1}{x} = 0$$

$$0 = 0 \text{ identity}$$

25. $y = c_1 e^x + c_2 e^{2x} + \dfrac{3}{2}$

$y' = c_1 e^x + 2c_2 e^{2x}$

$y'' = c_1 e^x + 4c_2 e^{2x}$

$y'' - 3y' + 2y = c_1 e^x + 4c_2 e^{2x} - 3\left(c_1 e^x + 2c_2 e^{2x}\right)$

$\qquad + 2\left(c_1 e^x + c_2 e^{2x} + \dfrac{3}{2}\right)$

$\qquad = c_1 e^x + 4c_2 e^{2x} - 3c_1 e^x - 6c_2 e^{2x}$

$\qquad + 2c_1 e^x + 2c_2 e^{2x} + 3$

$\qquad = 3$

29. $(y')^2 + xy' = y$

$y = cx + c^2; \; y' = c$

Substitute.

$(c)^2 + x(c) = y; \; c^2 + cx = y; \; y = y$ identity

33. $y = c_1 + c_2 x + c_3 e^x$

$\dfrac{dy}{dx} = c_2 + c_3 e^x$

$\dfrac{d^2 y}{dx^2} = c_3 e^x$

$\dfrac{d^3 y}{dx^3} = c_3 e^x = \dfrac{d^2 y}{dx^2}$

37. $N = N_0 k e^{kt}$

$\dfrac{dN}{dt} = N_0 k e^{kt} = kN$

31.2 Separation of Variables

1. $2xy\,dx + (x^2 + 1)dy = 0$

$\dfrac{2x\,dx}{x^2 + 1} + \dfrac{dy}{y} = 0$

$\ln(x^2 + 1) + \ln y = \ln c$

$\ln(y(x^2 + 1)) = \ln c$

$y(x^2 + 1) = c$

5. $y^2 dx + dy = 0$; divide by y^2;

$dx + \dfrac{dy}{y^2} = 0$; integrate

$x + \dfrac{y^{-1}}{-1} = c; \; x - \dfrac{1}{y} = c$

9. $x^2 + (x^3 + 5)y' = 0$; $(x^3 + 5)\dfrac{dy}{dx} = -x^2$

$(x^3 + 5)dy = -x^2 dx; \; dy = \dfrac{-x^2 dx}{x^3 + 5}$; integrate

$y = -\dfrac{1}{3}\ln(x^3 + 5) + c$

$3y + \ln(x^3 + 5) = 3c_1$; $3c_1$ is constant

$3y + \ln(x^3 + 5) = c$

13. $e^{x^2} dy = x\sqrt{1-y}\,dx$; $\dfrac{dy}{\sqrt{1-y}} = \dfrac{x\,dx}{e^{x^2}}$

$\dfrac{dy}{(1-y)^{1/2}} = e^{-x^2} x\,dx$; integrate

$-\dfrac{(1-y)^{1/2}}{\frac{1}{2}} = -\dfrac{1}{2}e^{-x^2} + c$

$-2\sqrt{1-y} = -\dfrac{1}{2}e^{-x^2} + c$; multiply by -2

$4\sqrt{1-y} = e^{-x^2} - 2c_1$; $-2c_1 = c$

$4\sqrt{1-y} = e^{-x^2} + c$

17. $y' - y = 4$; $\dfrac{dy}{dx} = 4 + y$; $\dfrac{dy}{4+y} = dx$; integrate

$\ln(4+y) = x + c$

21. $y \tan x \, dx + \cos^2 x \, dy = 0; \quad \dfrac{\tan x \, dx}{\cos^2 x} + \dfrac{dy}{y} = 0$

$(\tan x)^1 \sec^2 x \, dx + \dfrac{dy}{y} = 0; \text{ integrate}$

$\dfrac{1}{2} \tan^2 x + \ln y = c_1; \; 2c_1 = c; \; \tan^2 x + 2\ln y = c$

25. $e^{\cos \theta} \tan \theta \, d\theta + \sec \theta \, dy = 0$

$e^{\cos \theta} \sin \theta \, d\theta + dy = 0$

$-e^{\cos \theta} + y = c$

29. $\sqrt{1 + y^2} \, dx - y \, dy = x^2 y \, dy$

$\dfrac{dx}{x^2 + 1} - \dfrac{y}{\sqrt{1 + y^2}} \, dy = 0$

$\tan^{-1} x - \sqrt{1 + y^2} = c$

33. $2y(x^3 + 1)dy + 3x^2(y^2 - 1)dx = 0$

$\dfrac{2y \, dx}{y^2 - 1} + \dfrac{3x^2 \, dx}{x^3 + 1} = 0$

Integrate: $\ln(y^2 - 1) + \ln(x^3 + 1) = c_1$

$\ln(y^2 - 1)(x^3 + 1) = c_1; \; (y^2 - 1)(x^3 + 1) = e^{c_1}$

$(y^2 - 1)(x^3 + 1) = c$

37. $\dfrac{dy}{dx} = (1 - y)\cos x; \; x = \dfrac{\pi}{6}$ when $y = 0$

$dy = (1 - y)\cos x \, dx$

$\dfrac{1}{1 - y} \, dy = \cos x \, dx; \text{ integrate}$

$-\ln(1 - y) = \sin x + c; \; \sin x + \ln(1 - y) = c$

Substitute $x = \dfrac{\pi}{6}, \; y = 0; \; \sin \dfrac{\pi}{6} + \ln 1 = c; \; c = \dfrac{1}{2}$

$\sin x + \ln(1 - y) = \dfrac{1}{2}; \; 2\ln(1 - y) = 1 - 2\sin x$

41. $\dfrac{1}{2} \ln\left(x^2 + 1\right) + \ln y = \ln c$

$\dfrac{1}{2} \ln\left(0^2 + 1\right) + \ln e = \ln c$

$\dfrac{1}{2} \ln 1 + 1 = \ln c \Rightarrow \ln c = 1$

$\dfrac{1}{2} \ln\left(x^2 + 1\right) + \ln y = 1$ as in Example 6

31.3   Integrating Combinations

1. $x \, dy + y \, dx + 2xy^2 dy = 0$

$\dfrac{x \, dy + y \, dx}{xy} + 2y \, dy = 0$

$\dfrac{d(xy)}{xy} + 2y \, dy = 0$

$\ln xy + y^2 = c$

5. $y \, dx - x \, dy + x^3 dx = 2 \, dx;$

$x \, dy - y \, dx - x^3 dx = -2 \, dx$

$\dfrac{(x \, dy - y \, dx)}{x^2} - x \, dx = -\dfrac{2 \, dx}{x^2}$

$\dfrac{y}{x} - \dfrac{1}{2}x^2 = 2x^{-1} = \dfrac{2}{x} + c_1; \; y - \dfrac{1}{2}x^3 = 2 + c_1 x$

$2y - x^3 = 4 + 2c_1 x; \; x^3 - 2y = -2c_1 x - 4; \; -2c_1 = c$

$x^3 - 2y = cx - 4$

9. $\sin x \, dy = \left(1 - y \cos x\right) dx$

$\sin x \, dy + y \cos x \, dx = dx$

$d\left(y \sin x\right) = dx$

$y \sin x = x + c$

13. $\tan(x^2 + y^2)dy + x\,dx + y\,dy = 0$;

$$dy + \frac{x\,dx + y\,dy}{\tan(x^2 + y^2)} = 0$$

$$d(x^2 + y^2) = 2x\,dx + 2y\,dy = 2(x\,dx + y\,dy)$$
$$dy + \cot(x^2 + y^2)(x\,dx + y\,dy) = 0$$

$$dy + \frac{1}{2}\cot(x^2 + y^2)2d(x^2 + y^2)$$

$$y + \frac{1}{2}\ln\sin(x^2 + y^2) = c;\ y = c - \frac{1}{2}\ln\sin(x^2 + y^2)$$

17. $10x\,dy + 5y\,dx + 3y\,dy = 0$
$5(2x\,dy + y\,dx) + 3y\,dy = 0$; multiply by y
$5(2xy\,dy + y^2dx) + 3y^2dy = 0$; $5d(xy^2) + 3y^2dy = 0$
$5xy^2 + y^3 = c$

21. $y\,dx - x\,dy = y^3dx + y^2x\,dy$; $x = 2, y = 4$

$$\frac{y\,dx - x\,dy}{y^2} = y\,dx + x\,dy;\ d\left(\frac{x}{y}\right) = d(xy)$$

$$\frac{x}{y} = xy + c;\ x = 2, y = 4;\ \frac{2}{4} = 2(4) + c;\ c = -\frac{15}{2}$$

$$\frac{x}{y} = xy - \frac{15}{2};\ \text{multiply by } 2y$$

$$2x = 2xy^2 - 15y$$

25. $e^{-x}dy - 2y\,dy = ye^{-x}\,dx$
$e^{-x}dy - ye^{-x}\,dx = 2y\,dy$
$$d\left(ye^{-x}\right) = d\left(y^2\right)$$
$$ye^{-x} = y^2 + c$$

31.4 The Linear Differential Equation of the First Order

1. $dy + \left(\dfrac{2}{x}\right)y\,dx = 3\,dx$

$$ye^{\int \frac{2}{x}dx} = \int 3e^{\int \frac{2}{x}dx}dx + c$$

$$ye^{2\ln x} = \int 3e^{2\ln x}dx + c$$

$$ye^{\ln x^2} = \int 3e^{\ln x^2}dx + c$$

$$yx^2 = \int 3x^2dx + c$$

$$yx^2 = x^3 + c$$
$$y = x + cx^{-2}$$

5. $dy + 2y\,dx = 2e^{-4x}dx$; $P = 2,\ Q = 2e^{-4x}$;

$e^{\int 2\,dx} = e^{2x}$

$$ye^{2x} = \int 2e^{-4x}e^{2x}dx = -\int e^{-2x}\left(-2\,dx\right)$$

$$= -e^{-2x} + c$$

$$y = -e^{-2x}e^{-2x} + ce^{-2x} = -e^{-4x} + ce^{-2x}$$

9. $dy = 3x^2\left(2 - y\right)dx$

$$\frac{dy}{y - 2} = -3x^2dx$$

$$\ln\left(y - 2\right) = -x^3 + \ln c$$

$$\ln\frac{y - 2}{c} = -x^3$$

$$y = ce^{-x^3} + 2$$

13. $dr + r\cot\theta\,d\theta = d\theta$; $dr + \cot\theta\,r\,d\theta = d\theta$
$P = \cot\theta, Q = 1, e^{\int \cot\theta\,d\theta} = e^{\ln\sin\theta} = \sin\theta$

$$r\sin\theta = \int \sin\theta\,d\theta + c;\ r\sin\theta = -\cos\theta + c$$

$$r = -\frac{\cos\theta}{\sin\theta} + \frac{c}{\sin\theta} = -\cot\theta + c\,\csc\theta$$

17. $y' + y = x + e^x$

$dy + y\,dx = (x + e^x)\,dx, \qquad P(x) = 1$

$ye^{\int dx} = \int (x + e^x) e^{\int dx}\,dx$

$ye^x = \int (xe^x + e^{2x})\,dx$

$ye^x = xe^x - e^{-x} + \dfrac{1}{2}e^{2x} + c$

21. $y' = x^3(1 - 4y); \quad \dfrac{dy}{dx} = x^3 - 4x^3 y;$

$dy = x^3\,dx - 4x^3 y\,dx$

$dy + 4x^3 y\,dx = x^3\,dx; \quad P = 4x^3, Q = x^3;$

$e^{4\int x^3\,dx} = e^{x^4}$

$ye^{x^4} = \int x^3 e^{x^4}\,dx + c = \dfrac{1}{4}e^{x^4} 4x^3\,dx + c$

$\qquad = \dfrac{1}{4}e^{x^4} + c$

$y = \dfrac{1}{4} + ce^{-x^4}$

25. $\sqrt{1 + x^2}\,dy + x(1 + y)\,dx = 0$

$\sqrt{1 + x^2}\,dx + x\,dx + xy\,dx = 0$

$dy + \dfrac{x}{\sqrt{1 + x^2}} y\,dx = -\dfrac{x}{\sqrt{1 + x^2}}\,dx, \quad P(x) = \dfrac{x}{\sqrt{1 + x^2}}$

$ye^{\int \frac{x}{\sqrt{1+x^2}}\,dx} = \int \dfrac{-x}{\sqrt{1 + x^2}}\,e^{\int \frac{x}{\sqrt{1+x^2}}\,dx}\,dx$

$ye^{\sqrt{1+x^2}} = \int \dfrac{-x}{\sqrt{1 + x^2}}\,e^{\sqrt{1+x^2}}\,dx$

$ye^{\sqrt{1+x^2}} = -e^{\sqrt{1+x^2}} + c$

29. $y' = 2(1 - y)$; solve by separation of variables.

$\dfrac{dy}{1 - y} = 2\,dx; \ -\ln(1 - y) = 2x - \ln c;$

$\ln \dfrac{c}{1 - y} = 2x;$

$c = (1 - y)e^{2x};\ 1 - y = ce^{-2x};\ y = 1 - ce^{-2x}$

33. $\dfrac{dy}{dx} + 2y\cot x = 4\cos x;\ x = \dfrac{\pi}{2},\ y = \dfrac{1}{3};$

$dy + 2y\cot x\,dx = 4\cos x\,dx;\ P = 2\cot x;$

$Q = 4\cos x$

$e^{\int P\,dx} = e^{\int 2\cot x\,dx} = 2^{2\ln|\sin x|} = e^{\ln|\sin x|^2}$

$\qquad = |\sin x|^2$

$y(\sin x)^2 = \int 4\cos x(\sin x)^2\,dx + c = \dfrac{4(\sin x)^3}{3} + c$

$y = \dfrac{4}{3}\sin x + c(\csc^2 x)$

$x = \dfrac{\pi}{2}$ when $y = \dfrac{1}{3};\ \dfrac{1}{3} = \dfrac{4}{3}\sin \dfrac{\pi}{2} + c;\ c = -1$

$y = \dfrac{4}{3}\sin x - \csc^2 x$

37. $y' + P(x)y = Q(x)y^2$

$dy + P(x)y\,dx = Q(x)y^2\,dx$ which is not linear

because $Q(x)y^2$ is not a function of x only.

Let $u = \dfrac{1}{y} \Rightarrow dy = -y^2\,du$, then $dy + P(x)y\,dx$

$= Q(x)y^2\,dx$ is $-y^2\,du + P(x)y\,dx = Q(x)y^2\,dx$

$du - P(x)u\,dx = -Q(x)\,dx$ which is linear.

31.5 Numerical Solutions of First-Order Equations

1. $\dfrac{dy}{dx} = x + 1$

x	y	$x+1$	dy	y(correct)
0.0	1.00	1.0	0.20	1.00
0.2	1.20	1.2	0.24	1.22
0.4	1.44	1.4	0.28	1.48
0.6	1.72	1.6	0.32	1.78
0.8	2.04	1.8	0.36	2.12
1.0	2.40	2.0	0.40	2.50

$y = \dfrac{1}{2}x^2 + x + c$

$y = 1$ when $x = 0$

$c = 1$

$y = \dfrac{1}{2}x^2 + x + 1$

5.

x	y approximate	y exact
0.0	1	1
0.1	$1.0 + (0.0+1)(0.1) = 1.10$	1.105
0.2	$1.10 + (0.1+1)(0.1) = 1.21$	1.220
0.3	$1.21 + (0.2+1)(0.1) = 1.33$	1.345
0.4	$1.33 + (0.3+1)(0.1) = 1.46$	1.480
0.5	$1.46 + (0.4+1)(0.1) = 1.60$	1.625
0.6	$1.60 + (0.5+1)(0.1) = 1.75$	1.780
0.7	$1.75 + (0.6+1)(0.1) = 1.91$	1.945
0.8	$1.91 + (0.7+1)(0.1) = 2.08$	2.120
0.9	$2.08 + (0.8+1)(0.1) = 2.26$	2.305
1.0	$2.26 + (0.9+1)(0.1) = 2.45$	2.500

9. $\dfrac{dy}{dx} = xy + 1$, $x = 0$ to $x = 0.4$, $\Delta x = 0.1$, $(0, 0)$

x	y	y to 4 places
0	0	0
0.1	0.1003339594	0.1003
0.2	0.20268804	0.2027
0.3	0.3091639819	0.3092
0.4	0.42203172548	0.4220

13. $\dfrac{dy}{dx} = \cos(x+y)$, $x = 0$ to $x = 0.6$, $\Delta x = 0.1$, $\left(0, \dfrac{\pi}{2}\right)$

x	y
0	$\frac{\pi}{2} = 1.5708$
0.1	1.5660
0.2	1.5521
0.3	1.5302
0.4	1.5011
0.5	1.4656
0.6	1.4244

17. $\dfrac{di}{dt} = 2i = \sin t$

t	i	$\sin t - 2i$	di
0.0	0.0000	0.0000	0.0000
0.1	0.0000	0.0998	0.0100
0.2	0.0100	0.1787	0.0179
0.3	0.0279	0.2398	0.0240
0.4	0.0518	0.2857	0.0286
0.5	0.0804	0.3186	0.0319

$i = 0.0804$ A for $t = 0.5$ s

$di + 2i\,dt = \sin t\,dt$; $e^{\int 2\,dt} = e^{2t}$

$ie^{2t} = \int e^{2t} \sin t\,dt = \dfrac{e^{2t}(2\sin t - \cos t)}{4+1} + c$

$i = \dfrac{1}{5}(2\sin t - \cos t) + ce^{-2t}$ (Formula 49)

$i = 0$ for $t = 0$, $0 = \dfrac{1}{5}(0-1) + c$, $c = \dfrac{1}{5}$

$i = \dfrac{1}{5}(2\sin t - \cos t + e^{-2t})$

$i = 0.0898$ A for $t = 0.5t$

31.6   Elementary Applications

1. $y^2 = cx$

$$2y\frac{dy}{dx} = c = \frac{y^2}{x}$$

$\dfrac{dy}{dx} = \dfrac{y}{2x}$ for slope of any member of family.

$\dfrac{dy}{dx} = -\dfrac{2x}{y}$ for slope of orthogonal trajectories.

$$y\,dy = -2x\,dx$$

$$\frac{y^2}{2} = -x^2 + \frac{c}{2}$$

$$y^2 + 2x^2 = c$$

5. $\dfrac{dy}{dx} = \dfrac{2x}{y};\ y\,dy = 2x\,dx;\ \dfrac{1}{2}y^2 = x^2 + c$

Substitute $x = 2,\ y = 3;\ \dfrac{1}{2}(9) = 4 + c;\ c = 0.5$

$\dfrac{1}{2}y^2 = x^2 + 0.5;\ y^2 = 2x^2 + 1;\ y = \pm\sqrt{2x^2 + 1}$

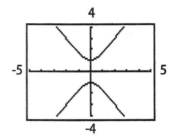

9. See Example 2; $\dfrac{dy}{dx} = ce^x;\ y = ce^x;\ c = \dfrac{y}{e^x}$

Substitute for c in the equation for the derivative.

$\dfrac{dy}{dx} = \dfrac{y}{e^x}e^x = y;\ \left.\dfrac{dy}{dx}\right|_{OT} = -\dfrac{1}{y};\ y\,dy = -dx$

Integrating, $\dfrac{y^2}{2} = -x + \dfrac{c}{2};\ y^2 = c - 2x$

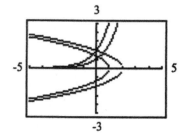

13. From Example 3, $N = N_0\,(0.5)^{t/5.27}$

$N = N_0\,(0.5)^{2.00/5.27} = N_0\,(0.769)$

76.9% of the initial amount remains

17. $\dfrac{dN}{dt} = r - kN$

$$\frac{dN}{r - kN} = dt$$

$$-\frac{1}{k}\ln(r - kN) = t + c$$

$N = 0$ for $t = 0$

$$c = -\frac{1}{k}\ln r$$

$$-\frac{1}{k}\ln(r - kN) = t - \frac{1}{k}\ln r$$

$$\ln\frac{r - kN}{r} = -kt$$

$$r - kN = re^{-kt}$$

$$N = \frac{r}{k}(1 - e^{-kt})$$

21.
$$\frac{dP}{dt} = kP + I$$

$$\frac{k\,dP}{kP + I} = k\,dt$$

$$\ln(kP + I) = kt + c$$

$$\ln(0.010(34.2) + 0.1) = 0.010(0) + c \Rightarrow$$

$$c = \ln 0.442$$

$$\ln(0.010P + 0.1) = 0.010t + \ln 0.442$$

$$\ln(0.010P + 0.1) = 0.010(14) + \ln 0.442$$

$$P = 38.8\text{ million}$$

25. $\dfrac{dM}{dt} = KM$

$\dfrac{dM}{M} = K\,dt;\ \ln M = Kt + c;\ M = 5250,\ t = 0$

$\ln 5250 = c;\ \ln M = Kt + \ln 5250;$

$M = 5460,\ t = 2.00$

$$\ln 5460 = 2.00K + \ln 5250; \; K = \ln \frac{\frac{5460}{5250}}{2} = 0.0196$$

$$\ln M = 0.0196t + \ln 5250; \; \ln \frac{M}{5250} = 0.0196t$$

$$\frac{M}{5250} = e^{0.0196t}; \; M = 5250e^{0.0196t}$$

29. $\frac{dP}{dt} = kP; \; \frac{dP}{P} = k\,dt; \; k = \text{interest rate}$

$\ln P = kt + c; \; t = 0, \; P = P_0, \text{ the original investment.}$

$\ln P_0 = c; \; \ln P = kt + \ln P_0; \; \ln \frac{P}{P_0} = kt; \; \frac{P}{P_0} = e^{kt}$

$P = P_0 e^{0.04t}; \; \$1000 \text{ invested for } t = 1 \text{ year at}$

$k = 4\%$

$P = 1000e^{0.04(1)} = \$1040.81$

33. See Example 4; $i = \frac{E}{R}\left(1 - e^{-(R/L)t}\right)$

$$\lim_{i \to \infty}\left(\frac{E}{R} - \frac{E}{R}e^{-(R/L)t}\right) = \frac{E}{R} - 0 = \frac{E}{R}$$

37. $Ri + \frac{q}{C} = 0; \; i = \frac{dq}{dt}; \; R\frac{dq}{dt} + \frac{q}{C} = 0;$

$\frac{dq}{q} + \frac{1}{RC}\,dt = 0$

Integrating, $\ln q = -\frac{1}{RC}t + c$

Let $q = q_0$ when $t = 0; \; \ln q_0 = c;$

$\ln q = -\frac{1}{RC}t + \ln q_0; \; \ln q - \ln q_0 = -\frac{1}{RC}t;$

$\ln \frac{q}{q_0} = -\frac{1}{RC}t; \; \frac{q}{q_0} = e^{(-1/RC)t}; \; q = q_0 e^{-t/RC}$

41. Mass of boat $= 10$ slugs $= 10(32) = 320$ lb

Force driving boat $= 20$ lb; water slows down by $2v$

$q = 32$ ft/s^2

$320\frac{dv}{dt} = 20 - 2v; \; dv = \frac{20}{320}dt - \frac{2}{320}v\,dt$

$dv + \frac{1}{160}v\,dt = \frac{1}{16}dt$

$P = \frac{1}{160}, Q = \frac{1}{16}; \; e^{\int(dt/160)} = e^{t/160}$

$ve^{t/160} = \frac{1}{16}\int e^{t/160}dt = \frac{1}{16}(160)\int e^{t/160}\frac{1}{160}dt$

$ve^{t/160} = 10e^{t/160} + c$

$t = 0, v = 8.0 \text{ mi/h} = \frac{8.0(5280 \text{ ft})}{3600 \text{ s}} = 11.733 \text{ ft/s}$

$11.733e^0 = 10e^0 + c, c = 1.733; \; v = 10 + 1.733e^{-t/160}$
$t = 3.0 \times 60 \text{ s} = 180 \text{ s};$
$v = 10 + 1.733e^{-180/160} = 11 \text{ ft/s}$

45. $\frac{dp}{dh} = kp, \; h = 0, \; p = 15/\text{in.}^2$

$h = 9800 \text{ ft}, \; p = 10.0 \text{ lb/in.}^2$

$\int\frac{dp}{p} = \int k\,dh; \; \ln p = kh + c; \; \ln 15 = 0 + c$

Therefore $c = \ln 15$

$\ln p = kh + \ln 15; \; \ln p - \ln 15 = kh;$

$\ln \frac{p}{15} = kh; \; \frac{p}{15} = e^{kh}$

$p = 15e^{kh}; \; 10.0 = 15e^{k9800}; \; \frac{10.0}{15} = e^{k9800} = 0.667$

$e^k = (0.667)^{1/9800} = (0.667)^{10^{-4}}; \; p = 15(0.667)^{10^{-4}h}$

49. $\frac{dx}{dt} = 1 - 0.25x$

$\frac{dx}{0.25x - 1} = -dt$

$-4\ln(0.25x - 1) = -t + c$

$x = 12 \text{ ft}^3 \text{ when } t = 0$

$c = 4\ln 2$

$4\ln(0.25x-1) = -t + 4\ln 2$

$4\ln\dfrac{0.25x-1}{2} = -t$

$0.25x - 1 = 2e^{-0.25t}$

$x = 4\left(1 + 2e^{-0.25t}\right)$

31.7 Higher-Order Homogeneous Equations

1. $D^2 y - 5Dy = 0$

$m^2 - 5m = 0$

$m(m-5) = 0$

$m = 0,\ m = 5$

$y = c_1 + c_2 e^{5x}$

5. $3\dfrac{d^2 y}{dx^2} + 4\dfrac{dy}{dx} + y = 0$

$3D^2 y + 4Dy + y = 0;;\ 3m^2 + 4m + 1 = 0$

$(3m+1)(m+1) = 0;\ m_1 = -\dfrac{1}{3},\ m_2 = -1$

$y = c_1 e^{-(1/3)x} + c_2 e^{-x}$

9. $\qquad 2D^2 y - 3y = Dy$

$2D^2 y - Dy - 3y = 0$

$2m^2 - m - 3 = 0$

$(2m-3)(m+1) = 0$

$m = \dfrac{3}{2},\ m = -1$

$y = c_1 e^{\frac{3x}{2}} + c_2 e^{-x}$

13. $3D^2 y + 8Dy - 3y = 0;\ 3m^2 + 8m - 3 = 0;$

$(3m-1)(m+3) = 0;\ m_1 = \frac{1}{3}$ and $m_2 = -3;$

$y = c_1 e^{x/3} + c_2 e^{-3x}$

17. $2\dfrac{d^2 y}{dx^2} - 4\dfrac{dy}{dx} + y = 0;\ 2D^2 y - 4Dy + y = 0;$

$2m^2 - 4m + 1 = 0$

Quadratic formula: $m = \dfrac{4 \pm \sqrt{16-8}}{4};$

$m_1 = 1 + \dfrac{\sqrt{2}}{2},$

$m_2 = 1 - \dfrac{\sqrt{2}}{2}$

$y = c_1 e^{(1+(\sqrt{2}/2))x} + c_2 e^{(1-(\sqrt{2}/2))x};$

$y = c_1 e^x e^{(\sqrt{2}/2)x} + c_2 e^x e^{-(\sqrt{2}/2)x}$

$\qquad = e^x\big(c_1 e^{x(\sqrt{2}/2)} + c_2 e^{-x(\sqrt{2}/2)}\big)$

21. $y'' = 3y' + y;\ D^2 y - 3Dy - y = 0;\ m^2 - 3m - 1 = 0$

Quadratic formula:

$m = \dfrac{3 \pm \sqrt{9+4}}{2};\ m_1 = \dfrac{3}{2} + \dfrac{\sqrt{13}}{2};\ m_2 = \dfrac{3}{2} - \dfrac{\sqrt{13}}{2}$

$y = c_1 e^{((3/2)+(\sqrt{13}/2))x} + c_2 e^{((3/2)-(\sqrt{13}/2))x};$

$y = e^{3x/2}\big(c_1 e^{x(\sqrt{13}/2)} + c_2 e^{-x(\sqrt{13}/2)}\big)$

25. $2D^2 y + 5aDy - 12a^2 = 0,\ a > 0$

$2m^2 + 5am - 12a^2 = 0$

$(2m - 3a)(m + 4a) = 0$

$m = \dfrac{3a}{2}, \qquad m = -4a$

$y = c_1 e^{\frac{3a}{2}x} + c_2 e^{-4ax}$

29. $D^2 y - Dy = 12y$

$D^2 y - Dy - 12y = 0;\ y = 0$ when $x = 0;\ y = 1$
when $x = 1$

$m^2 - m - 12 = 0;\ (m-4)(m+3) = 0;\ m_1 = 4;$

$m_2 = -3$

$y = c_1 e^{4x} + c_2 e^{-3x}$

Substituting given values: $0 = c_1 + c_2;$
therefore, $c_1 = -c_2;$

$1 = c_1 e^4 + c_2 e^{-3}$

$1 = -c_2 e^4 + c_2 e^{-3} = -c_2 e^4 + \dfrac{c_2}{e^3} = \dfrac{-c_2 e^7 + c_2}{e^3}$

$e^3 = c_1(1 - e^7);$ therefore,

$$c_2 = \frac{e^3}{(1-e^7)}, c_1 - \frac{e^3}{(1-e^7)}$$

$$y = -\frac{e^3}{(1-e^7)}e^{4x} + \frac{e^3}{(1-e^7)}e^{-3x}$$

$$= \frac{e^3}{e^7-1}e^{4x} - \frac{e^3}{e^7-1}e^{-3x}$$

$$y = \frac{e^3}{e^7-1}(e^{4x} - e^{-3x})$$

33. $D^4y - 5D^2y + 4y = 0, m^4 - 5m^2 + 4 = 0$

   $m = -2, -1, 1, 2$

   $y = c_1e^x + c_2e^{-x} + c_3e^{2x} + c_4e^{-2x}$

37. $D^2y + 4Dy = 0 \Rightarrow D(D+4)y = 0 \Rightarrow \frac{dz}{dx} = 0,$

   $z = (D+4)y$

   $z = c_0 \Rightarrow (D+4)y = c_0 \Rightarrow \frac{dy}{dx} + 4y = c_0$

   $dy + 4y\,dx = c_0\,dx \Rightarrow ye^{\int 4\,dx} = \int c_0 e^{\int 4\,dx}\,dx + c_2$

   $ye^{4x} = \int c_0 e^{4x}\,dx + c_2$

   $ye^{4x} = \frac{c_0}{4}e^{4x} + c_2$

   $y = \frac{c_0}{4} + c_2e^{-4x}, \ c_1 = \frac{c_0}{4}$

   $y = c_1 + c_2e^{-4x}$

31.8 Auxiliary Equations with Repeated or Complex Roots

1. $\frac{d^2y}{dx^2} + 10\frac{dy}{dx} + 25y = 0$

   $D^2y + 10Dy + 25y = 0$

   $m^2 + 10m + 25 = 0$

   $(m+5)^2 = 0$

   $m = -5, \ -5$

   $y = e^{-5x}(c_1 + c_2x)$

5. $D^2y - 2Dy + y = 0; m^2 - 2m + 1 = 0;$
   $(m-1)^2 = 0; m = 1, 1$

   $y = e^x(c_1 + c_2x); y = (c_1 + c_2x)e^x$

9. $D^2y + 9y = 0; m^2 + 9 = 0; m_1 = 3j$
   and $m_2 = -3j, \alpha = 0, \beta = 3$

   $y = e^{0x}(c_1 \sin 3x + c_2 \cos 3x);$
   $y = c_1 \sin 3x + c_2 \cos 3x$

13. $D^4y - y = 0$

   $m^4 - 1 = 0$

   $(m^2 - 1)(m^2 + 1) = 0$

   $(m-1)(m+1)(m^2+1) = 0$

   $m = \pm 1, \ m = \pm j$

   $y = c_1e^x + c_2e^{-x} + c_3 \sin x + c_4 \cos(-x)$

   $y = c_1e^x + c_2e^{-x} + c_3 \sin x + c_4 \cos x$

17. $16D^2y - 24Dy + 9y = 0; 16m^2 - 24m + 9 = 0;$

   $(4m-3)^2 = 0; m = \frac{3}{4}, \frac{3}{4}$

   $y = e^{3x/4}(c_1 + c_2x)$

21. $2D^2y + 5y = 4Dy; 2D^2y + 5y - 4Dy = 0;$
   $2m^2 - 4m + 5 = 0$

   Quadratic formula:

   $$m = \frac{4 \pm \sqrt{16 - 40}}{4} = \frac{4 \pm 2\sqrt{-6}}{4}$$

   $m_1 = 1 + \frac{\sqrt{6}}{2}j; m_2 = 1 - \frac{\sqrt{6}}{2}j; \alpha = 1, \beta = \frac{1}{2}\sqrt{6}$

   $$y = e^x\left(c_1 \cos \frac{1}{2}\sqrt{6}x + c_2 \sin \frac{1}{2}\sqrt{6}x\right)$$

25. $2D^2y - 3Dy - y = 0$; $2m^2 - 3m - 1 = 0$

By the quadratic formula, $m = \dfrac{3 \pm \sqrt{9+8}}{4}$

$m_1 = \dfrac{3}{4} + \dfrac{\sqrt{17}}{4}$, $m_2 = \dfrac{3}{4} - \dfrac{\sqrt{17}}{4}$

$y = c_1 e^{((3/4)+(\sqrt{17}/4))x} + c_2 e^{((3/4)+(\sqrt{17}/4))x}$;

$y = e^{(3/4)x}(c_1 e^{x(\sqrt{17}/4)} + c_2 e^{-x(\sqrt{17}/4)})$

29. $D^3y - 6D^2y + 12Dy - 8y = 0$

$\quad m^3 - 6m^2 + 12m - 8 = 0$

$\quad (m-2)(m^2 - 4m + 4) = 0$

$\quad (m-2)(m-2)(m-2) = 0$

$m = 2, 2, 2$ repeated root

$y = e^{2x}(c_1 + c_2 x + c_3 x^2)$

33. $D^2y + 2Dy + 10y = 0$; $m^2 + 2m + 10 = 0$

By the quadratic formula, $m = \dfrac{-2 \pm \sqrt{4-40}}{2}$

$m_1 = -1 + 3j$; $m_2 = -1 - 3j$, $\alpha = -1$, $\beta = 3$;

$y = e^{-x}(c_1 \sin 3x + c_2 \cos 3x)$

Substituting $y = 0$ when $x = 0$;

$0 = e^0(c_1 \sin 0 + c_2 \cos 0)$; $c_2 = 0$

Substituting $y = e^{-\pi/6}$, $x = \dfrac{\pi}{6}$;

$e^{-\pi/6} = e^{-\pi/6}\left(c_1 \sin \dfrac{\pi}{2}\right)$; $e^{-\pi/6} = e^{-\pi/6}c_1$;

$c_1 = 1$

$y = e^{-x} \sin 3x$

37. $y = c_1 e^{3x} + c_2 e^{-3x}$

$(m-3)(m+3) = 0$

$m^2 - 9 = 0$; $(D^2 - 9)y = 0$

41. $D^2y + 4Dy + 4y = 0 \Rightarrow m^2 + 4m + 4 = 0 \Rightarrow m = \pm 2$

$y = f(t) = e^{-2t}(c_1 + c_2 t)$

$y = f(0) = 0 = e^{-0}(c_1 + c_2 \cdot 0) \Rightarrow c_1 = 0$

$y = f(t) = e^{-2t} \cdot c_2 \cdot t$

$y = f(1) = e^{-2} \cdot c_2 \cdot 1 = 0.50 \Rightarrow c_2 = 0.50e^2 = 3.7$

$y = 0.5e^2 \cdot te^{-2t} = 3.7te^{-2t}$

31.9 Solutions of Nonhomogeneous Equations

1. b: $x^2 + 2x + e^{-x}$

$y_p = A + Bx + Cx^2 + Ee^{-x}$

5. $D^2y - Dy - 2y = 4$; $m^2 - m - 2 = 0$

$(m-2)(m+1) = 0$; $m_1 = 2$, $m_2 = -1$

$y_c = c_1 e^{2x} + c_2 e^{-x}$

$y_p = A$; $Dy_p = 0$; $D^2y_p = 0$

Substituting in diff. equation, $0 - 0 - 2A = 4$;

$A = -2$

Therefore, $y_p = -2$; $y = c_1 e^{2x} + c_2 e^{-x} - 2$

9. $y'' - 3y' = 2e^x + xe^x$; $D^2y - 3Dy = 2e^x + xe^x$

$m^2 - 3m = 0$; $m(m-3) = 0$; $m_1 = 0$, $m_2 = 3$

$y_c = c_1 e^0 + c_2 e^{3x} = c_1 + c_2 e^{3x}$; $y_p = Ae^x + Bxe^x$

$Dy_p = Ae^x + B(xe^x + e^x) = Ae^x + Bxe^x + Be^x$

$D^2y_p = Ae^x + Be^x + B(xe^x + e^x)$

$\quad = Ae^x + Be^x + Bxe^x + Be^x$

$D^2y_p = Ae^x + 2Be^x + Bxe^x$

Substituting in diff. equation:

$Ae^x + 2Be^x + Bxe^x - 3(Ae^x + Bxe^x + Be^x)$

$\quad = 2e^x + xe^x$

$Ae^x + 2Be^x + Bxe^x - 3Ae^x - 3Bxe^x - 3Be^x$

$\quad = 2e^x + xe^x$

$$-2Ae^x - Be^x - 2Bxe^x$$
$$= 2e^x + xe^x$$
$$e^x(-2A - B) + xe^x(-2B)$$
$$2e^x + xe^x$$
$$-2A - B = 2; \quad -2A + \frac{1}{2} = 2; \quad -2A = \frac{3}{2};$$
$$A = -\frac{3}{4}; \quad -2B = 1; \quad B = -\frac{1}{2}$$
$$y = c_1 + c_2 e^{3x} - \frac{3}{4}e^x - \frac{1}{2}xe^x$$

13. $\dfrac{d^2y}{dx^2} - 2\dfrac{dy}{dx} + y = 2x + x^2 + \sin 3x$

$D^2y - 2Dy + y = 2x + x^2 + \sin 3x$

$m^2 - 2m + 1 = 0; \quad (m-1)^2 = 0; \quad m = 1, 1$
$y_c = e^x(c_1 + c_2 x);$
$y_p = A + Bx + Cx^2 + E\sin 3x + F\cos 3x$
$Dy_p = B + 3Cx + 3E\cos 3x - 3F\sin 3x$
$D^2 y_p = 2c - 9E\sin 3x - 9F\cos 3x$

Substituting in diff. equation:

$2C - 9E\sin 3x - 9F\cos 3x$
$\quad -2(B + 2Cx + 3E\cos 3x - 3F\sin 3x)$
$\quad + A + Bx + Cx^2 + E\sin 3x + F\cos 3x$
$\quad = 2x + x^2 + \sin 3x$
$2C - 9E\sin 3x - 9F\cos 3x - 2B - 4Cx - 6E\cos 3x$
$\quad + 6F\sin 3x + A + Bx + Cx^2 + E\sin 3x + F\cos 3x$
$\quad = 2x + x^2 + \sin 3x$
$2C - 2B + A - 8E\sin 3x + 6F\sin 3x - 8F\cos 3x$
$\quad - 6E\cos 3x - 4Cx + Bx + Cx^2$
$\quad = 2x + x^2 + \sin 3x$
$(2C - 2B + A) + \sin 3x(6F - 8E) + \cos 3x(-8F - 6E)$
$\quad + x(B - 4) + Cx^2$
$\quad = 2x + x^2 + \sin 3x$
$2C - 2B + A = 0; \quad A = 10$

$6F - 8E = 1; \quad -8F - 6E = 0; \quad E = -\dfrac{2}{25};$

$F = \dfrac{3}{50}; \quad B - 4 = 2; \quad B = 6; \quad C = 1$

$y = e^x(c_1 + c_2 x) + 10 + 6x + x^2$
$\quad - \dfrac{2}{25}\sin 3x + \dfrac{3}{50}\cos 3x$

17. $D^2y - Dy - 30y = 10; \quad m^2 - m - 30 = 0;$
$(m - 6)(m + 5) = 0; \quad m_1 = -5; \quad m_2 = 6$
$y_c = c_1 e^{-5x} + c_2 e^{6x}; \quad y_p = A; \quad Dy_p = 0;$

$D^2 y_p = 0; \quad 0 = 0 - 30A = 10; \quad A = -\dfrac{1}{3};$

$y_p = -\dfrac{1}{3}; \quad y = c_1 e^{-5x} + c_2 e^{6x} - \dfrac{1}{3}$

21. $D^2y - 4y = \sin x + 2\cos x; \quad m^2 - 4 = 0;$
$m_1 = 2; \quad m_2 = -2$
$y_c = c_1 e^{2x} + c_2 e^{-2x}$
$y_p = A\sin x + B\cos x; \quad Dy_p = A\cos x - B\sin x$
$D^2 y_p = -A\sin x - B\cos x$
$-A\sin x - B\cos x - 4A\sin x - 4B\cos x$
$\quad = \sin x + 2\cos x$
$-5A\sin x - 5B\cos x$
$\quad = \sin x + 2\cos x$

$-5A = 1; \quad A = -\dfrac{1}{5}; \quad -5B = 2; \quad B = -\dfrac{2}{5}$

$y_p = -\dfrac{1}{5}\sin x - \dfrac{2}{5}\cos x$

$y = c_1 e^{2x} + c_2 e^{-2x} - \dfrac{1}{5}\sin x - \dfrac{2}{5}\cos x$

25. $D^2y + 5Dy + 4y = xe^x + 4; \quad m^2 + 5m + 4 = 0;$
$(m + 1)(m + 4) = 0; \quad m_1 = -1, \quad m_2 = -4$
$y_c = c_1 e^{-x} + c_2 e^{-4x}; \quad y_p = Ae^x + Bxe^x + C$
$Dy_p = Ae^x + B(xe^x + e^x) = Ae^x + Bxe^x + Be^x$
$D^2 yp = Ae^x - B(xe^x + e^x) + Be^x$
$\quad = Ae^x + 2Be^x + Bxe^x$
$Ae^x + 2Be^x + Bxe^x + 5(Ae^x + Bxe^x + Be^x)$
$\quad + 4(Ae^x + Bxe^x + C) = xe^x + 4$

$(10A + 7B)e^x + 10Bxe^x + 4C = xe^x + 4$

$10A + 7B = 0; \quad 10B = 1; \quad B = \dfrac{1}{10}$

$4C = 4; \quad C = 1; \quad 10A + 7\left(\dfrac{1}{10}\right) = 0$

$A = -\dfrac{7}{100}; \quad y_p = -\dfrac{7}{100}e^x + \dfrac{1}{10}e^x + 1$

$y = c_1 e^{-x} + c_2 e^{-4x} - \dfrac{7}{100}e^x + \dfrac{1}{10}xe^x + 1$

29. $D^2 y + y = \cos x$

$m^2 + 1 = 0$

$m = \pm i$

$y_c = c_1 \sin x + c_2 \cos x$

let $y_p = x(A \sin x + B \cos x)$

$Dy_p = x(A \cos x - B \sin x) + A \sin x + B \cos x$

$D^2 y_p = x(-A \sin x - B \cos x) + A \cos x$
$\quad - B \sin x + A \cos x - B \sin x$

$D^2 y_p + y_p = \cos x$

$-x(A \sin x + B \cos x) + 2A \cos x - 2B \sin x$
$+ x(A \sin x + B \cos x) = \cos x$

$2A \cos x - 2B \sin x = \cos x$

$2A = 1, B = 0$

$A = \dfrac{1}{2}$

$y_p = \dfrac{1}{2} x \sin x$

$y = y_c + y_p$

$y = c_1 \sin x + c_2 \cos x + \dfrac{1}{2} x \sin x$

33. $D^2 y - Dy - 6y = 5 - e^x;\ m^2 - m - 6 = 0;$

$(m - 3)(m + 2) = 0;\ m_1 = 3,\ m_2 = -2$

$y_c = c_1 e^{3x} + c_2 e^{-2x}$

$y_p = A + Be^x;\ Dy_p = Be^x;\ D^2 y_p = Be^x$

$Be^x - Be^x - 6(A + Be^x) = 5 - e^x$

$-6A - 6Be^x = 5 - e^x;\ -6A = 5;\ A = -\dfrac{5}{6};$

$-6B = -1;\ B = \dfrac{1}{6}$

$y_p = -\dfrac{5}{6} + \dfrac{1}{6} e^x;\ y = c_1 e^{3x} + c_2 e^{-2x} + \dfrac{1}{6} e^x - \dfrac{5}{6}$

Substituting $x = 0$ when $y = 2;\ 3c_1 + 3c_2 = 8$

$Dy = 3c_1 e^{3x} - 2c_2 e^{-2x} + \dfrac{1}{6} e^x$

Substituting, $Dy = 4$ when $x = 0,\ 18c_1 = 12c_2 = 23$

Solving the two linear equations simultaneously,

$c_1 = \dfrac{11}{6},\ c_2 = \dfrac{5}{6}$

$y = \dfrac{11}{6} e^{3x} + \dfrac{5}{6} e^{-2x} + \dfrac{1}{6} e^x - \dfrac{5}{6}$

$= \dfrac{1}{6}\left(11 e^{3x} + 5 e^{-2x} + e^x - 5\right)$

37. $Dy - y = x^2$

$m - 1 = 0 \Rightarrow m = 1$

$y_c = c_1 e^x$

$y_p = A + Bx + Cx^2$

$Dy_p = B + 2Cx$

$B + 2Cx - A - Bx - Cx^2 = x^2$

$-C = 1 \Rightarrow C = -1$

$2C - B = 0 \Rightarrow 2(-1) - B = 0 \Rightarrow B = -2$

$A - B = 0 \Rightarrow A = B = -2$

$y = c_1 e^x - 2 - 2x - x^2$

31.10 Applications of Higher-Order Equations

1. $x = c_1 \sin 4t + c_2 \cos 4t;\ x = 0,\ Dx = 2$ for $t = 0$

$0 = c_1 \sin(4(0)) + c_2 \cos(4(0)) \Rightarrow c_2 = 0$

$x = c_1 \sin 4t$

$Dx = 4c_1 \cos 4t$

$4c_1 \cos(4(0)) = 2 \Rightarrow c_1 = \dfrac{1}{2}$

$x = \dfrac{1}{2} \sin 4t$

5. a) $b^2 < 4h^2 = 400$

$b < 20$, underdamped

b) $b^2 > 4k^2 = 400$

$b > 20$, overdamped

9. $D^2 y + bDy + 25y = 0;\ m^2 + bm + 25 = 0$

$m = \dfrac{-b \pm \sqrt{b^2 - 4(25)}}{2} = \dfrac{-b \pm \sqrt{b^2 - 100}}{2}$

For critical damping: $b^2 - 100 = 0;\ b = 10\,(b > 0)$

13. To get spring constant: $F = kx$; $4.00 = k(0.125)$;
$k = 32.0$ lb/ft
To get mass of object: $F = ma$; $4.00 = m(32.0)$;

$$m = 0.125 \frac{\text{lb} \cdot \text{s}^2}{\text{ft}} \text{ (a slug)}$$

Using Newton's Second Law:
mass × accel. = restoring force

$$0.125 \frac{d^2x}{dt^2} = -32.0x; \quad D^2x + 256x = 0;$$

$m^2 + 256 = 0$;
$m = \pm 16.0j$; $x = c_1 \sin 16.0t + c_2 \cos 16.0t$
$D_x = 16.0c_1 \cos 16.0t - 16.0c_2 \sin 16.0t$

Let $x = 0.250$ ft when $t = 0$;
$0.250 = c_1 \sin 0 + c_2 \cos 0$;
$c_2 = 0.250$
Let $Dx = 0$ when $t = 0$,
$0 = 16.0c_1 \cos 0 - 16.0c_2 \sin 0$;
$c_1 = 0$; $x = 0.250 \cos 16.0t$

17. $$L\frac{d^2q}{dt^2} + R\frac{dq}{dt} + \frac{q}{C} = E$$

$$0.200 \frac{d^2\theta}{dt^2} + 8.00 \frac{dq}{dt} + 10^6 q = 0$$

$$0.200m^2 + 8.00m + 10^6 = 0$$

$$m = \frac{-8.00 \pm \sqrt{64.0 - 0.800 \times 10^6}}{0.400};$$

$m = -20.0 \pm 2240j$; $\alpha = -20.0$, $\beta = 2240$

$q = e^{-20.0t}(c_1 \sin 2240t + c_2 \cos 2240t)$; $t = 0$, $q = 0$

Therefore, $c_2 = 0$

$$\frac{dq}{dt} = e^{-20.0t}(2240c_1 \cos 2240t - 2240c_2 \sin 2240t) + (c_1 \sin 2240t + c_2 \cos 2240t)(-20.0e^{-20.0t})$$

$t = 0$, $i = 0.500$; therefore, $c_1 = 2.24 \times 10^{-4}$;
$q = e^{-20.0t}(2.24 \times 10^{-4}) \sin 2240t$
$q = 2.24 \times 10^{-4} e^{-20.0t} \sin 2240t$

21. $$0.500D^2q + 10.0Dq + \frac{q}{200 \times 10^6} = 120 \sin 120\pi t$$

$$1.00D^2q + 20.0Dq + 10^4 q = 240 \sin 120\pi t$$
$$1.00m^2 + 20.0m + 10^4 = 0$$

$$m = \frac{-20.0 \pm \sqrt{400 - 4 \times 10^4}}{200} = -10.0 \pm 99.5j;$$
$q_c = e^{-10.0t}(c_1 \sin 99.5t + c_2 \cos 99.5t)$;
$q_p = A \sin 120\pi t + B \cos 120\pi t$

$Dq_p = 120\pi A \cos 120\pi t - 120\pi B \sin 120\pi t$
$D^2q_p = -142\,000 A \sin 120\pi t - 142\,000B \cos 120\pi t$
$\qquad - 142\,000\,A120\pi t - 142\,000\,B \cos 120\pi t$
$\qquad + 7540A \cos 120\pi t - 7540B \sin 120\pi t$
$\qquad + 10\,000A \sin 120\pi t + 10\,000B \cos 120\pi t$
$\qquad = 240 \sin 120\pi t$
$-132\,000A - 7540B = 240$
$-132\,000B + 7540A = 0$
$A = -1.81 \times 10^{-3}, B = -1.03 \times 10^{-4}$
$q = e^{-10.0t}(c_1 \sin 99.5t + c_2 \cos 99.5t)$
$\qquad -1.81 \times 10^{-3} \sin 120\pi t - 1.03 \times 10^{-4} \cos 120\pi t$

25. $$1.00D^2q + 5.00Dq + \frac{q}{150 \times 10^{-6}} = 120 \sin 100t$$

$$1.00D^2q + 5.00Dq + 6670q = 120 \sin 100t$$

$q_p = A \sin 100t + B \cos 100t$
$Dq_p = 100A \cos 100t - 100B \sin 100t$
$D^2q_p = -10^4 A \sin 100t - 10^4 B \cos 100t$

$-10^4 A \sin 100t - 10^4 B \cos 100t$
$\quad + 5.00(100A \cos 100t - 100B \sin 100t)$
$\quad + 6670(A \sin 100t + B \cos 100t)$
$\quad = 120 \sin 100t$

$(-10^4 A - 500B + 6670A) \sin 100t$
$\quad + (-10^4 B + 500A + 6670B) \cos 100t$
$\quad = 120 \sin 100t$
$3330A + 500B = -120$
$500A - 3330B = 0$
$A = -0.0352$; $B = 0.005\,28$;
$q_p = -0.0352 \sin 100t + 0.005\,28 \cos 100t$;
$i_p = -3.52 \cos 100t + 0.528 \sin 100t$

31.11 Laplace Transforms

1. $f(t) = 1, t > 0$

$$L(f) = \int_0^\infty e^{-st} \cdot 1 \, dt$$

$$= \lim_{c\to\infty} \frac{-1}{s} \int_0^c e^{-st}(-s \, dt)$$

$$L(f) = -\frac{1}{s} \lim_{c\to\infty} e^{-st}\Big|_0^c$$

$$= -\frac{1}{s}\left[\lim_{c\to\infty}(e^{-sc} - e^{-s(0)})\right]$$

$$= -\frac{1}{s}\lim_{c\to\infty}(0-1)$$

$$L(f) = \frac{1}{s}$$

5. $f(t) = e^{3t}$; from transform (3) of the table,

$a = -3; \ L(3t) = \dfrac{1}{s-3}$

9. $f(t) = \cos 2t - \sin 2t; \ L(f) = L(\cos 2t) - L(\sin 2t)$

By transforms (5) and (6),

$$L(f) = \frac{s}{s^2+4} - \frac{2}{s^2+4}; \ L(f) = \frac{s-2}{s^2+4}$$

13. $y'' + y'; \ f(0) = 0; \ f'(0) = 0$
$L[f''(y) + f'(y)]$
$= L(f'') + L(f')$
$= s^2 L(f) - sf(0) - f'(0) + sL(f) - f(0)$
$= s^2 L(f) - s(0) - 0 + sL(f) - 0$
$= s^2 L(f) + sL(f)$

17. $L^{-1}(F) = L^{-1}\left(\dfrac{2}{s^3}\right) = 2L^{-1}\left(\dfrac{1}{s^3}\right);$

$L^{-1}(F) = \dfrac{2t^2}{2} = t^2;$ transform (2)

21. $L^{-1}(F) = L^{-1}\dfrac{1}{(s+1)^3} = L^{-1}\dfrac{1}{2}\left[\dfrac{2}{(s+1)^3}\right]$

$= \dfrac{1}{2}t^2 e^{-t};$

transform (12)

25. $F(s) = \dfrac{4s^2 - 8}{(s+1)(s-2)(s-3)} = \dfrac{-\frac{1}{3}}{s+1} + \dfrac{-\frac{8}{3}}{s-2} + \dfrac{7}{s-3}$

$$L^{-1}(F) = -\frac{1}{3}L^{-1}\left(\frac{1}{s+1}\right) - \frac{8}{3}L^{-1}\left(\frac{1}{s-2}\right)$$

$$+ 7L^{-1}\frac{1}{s-3}$$

$$f(t) = -\frac{1}{3}e^{-t} - \frac{8}{3}e^{2t} + 7e^{3t}$$

29. $L\{tf(t)\} = L\left(te^{-at}\right) = -\dfrac{d}{ds}\left(\dfrac{1}{s+a}\right)$

$$= -\left(\frac{-1}{(s+a)^2}\right)$$

$$= \frac{1}{(s+a)^2}$$

31.12 Solving Differential Equations by Laplace Transforms

1. $2y' - y = 0; \ y(0) = 2$
$L(2y') - L(y) = L(0)$
$2L(y') - L(y) = 0$
$2sL(y) - 2(2) - L(y) = 0$

$$L(y) = \frac{4}{2s-1} = 2\frac{1}{s-\frac{1}{2}}, \text{ from transform 3,}$$

$$y = 2e^{\frac{1}{2}t}$$

5. $y' + y = 0; \ y(0) = 1; \ L(y') + L(y) = L(0);$
$L(y') + L(y) = 0; \ sL(y) - y(0) + L(y) = 0$
$sL(y) - 1 + L(y) = 0; \ (s+1)L(y) = 1;$

$$L(y) = \frac{1}{s+1}; \ a = -1, \text{ transforms (3)}; \ y = e^{-t}$$

9. $y' + 3y = e^{-3t}$; $y(0) = 1$; $L(y') + L(3y) = L(e^{-3t})$;

$L(y') + 3L(y) = L(3^{-3t})$;

$[sL(y-1] + 3L(y) = \dfrac{1}{s+3}$

$(s+3)L(y) = \dfrac{1}{s+3} + 1$; $L(y) = \dfrac{1}{(s+3)^2} + \dfrac{1}{s+3}$

The inverse is found from transforms (11) and (3).

$y = te^{-3t} + e^{-3t} = (1+t)e^{-3t}$

13. $4y'' + 4y' + 5y = 0$, $y(0) = 1$, $y'(0) = -\dfrac{1}{2}$

$4L(y'') + 4L(y') + 5L(y) = 0$

$4\left(s^2 L(y) - sy(0) - y'(0)\right) + 4\left(sL(y) - y(0)\right)$

$\quad + 5L(y) = 0$

$4s^2 L(y) - 4s + 2 + 4sL(y) - 4 + 5L(y) = 0$

$\left(4s^2 + 4s + 5\right)L(y) = 4s + 2$

$L(y) = \dfrac{4s+2}{4s^2 + 4s + 5} = \dfrac{4\left(s + \frac{1}{2}\right)}{4\left(s^2 + s + \frac{1}{4}\right) + 4}$

$L(y) = \dfrac{s + \frac{1}{2}}{\left(s + \frac{1}{2}\right)^2 + 1}$,   #20, $a = \dfrac{1}{2}$, $b = 1$

$\quad y = e^{-1/2t} \cos t$

17. $y'' + y = 1$; $y(0) = 1$; $y'(0) = 1$;
$L(y'') + L(y) = L(1)$;

$s^2 L(y) - s - 1 + L(y) = \dfrac{1}{s}$

$(s^2 + 1)L(y) = \dfrac{1}{s} + s + 1$;

$L(y) = \dfrac{1}{s(s^2 + 1)} + \dfrac{s}{s^2 + 1} + \dfrac{1}{s^2 + 1}$

By transforms (7), (5), and (6),
$y = 1 - \cos t + \cos t + \sin t$; $y = 1 + \sin t$

21. $y'' - 4y = 10e^{3t}$, $y(0) = 5$, $y'(0) = 0$;
$L(y'') - 4L(y) = 10 \cdot L(e^{3t})$

$s^2 L(y) - s \cdot y(0) - y'(0) - 4L(y) = \dfrac{10}{s-3}$;

$s^2 L(y) - 5s - 0 - 4L(y) = \dfrac{10}{s-3}$

$(s^2 - 4)L(y) = 5s + \dfrac{10}{s-3}$

$L(y) = \dfrac{5s}{(s+2)(s-2)} + \dfrac{10}{(s+2)(s-2)(s-3)}$

$L(y) = \dfrac{\frac{5}{2}}{s+2} + \dfrac{\frac{5}{2}}{s-2} + \dfrac{\frac{1}{2}}{s+2} + \dfrac{-\frac{5}{2}}{s-2} + \dfrac{2}{s-3}$

$L(y) = \dfrac{3}{s+2} + \dfrac{2}{s-3}$

$y = 3e^{-2t} + 2e^{3t}$

25. $2v' = 6 - v$; since the object starts from rest,
$f(0) = 0$, $f'(0) = 0$

$2L(v') + L(v) = 6L(1)$; $2sL(v) - 0 + L(v) = \dfrac{6}{s}$;

$(2s + 1)L(v) = \dfrac{6}{s}$

$L(v) = \dfrac{6}{s(2s+1)} = 6\left[\dfrac{\frac{1}{2}}{s\left(s + \frac{1}{2}\right)}\right]$

By transforms (4), $v = 6(1 - e^{-t/2})$

$\left((s + 0.2)^2 + 80^2\right)L(y) = 4(s + 0.1) + 4(0.1)$

$L(y) = 4\dfrac{s + 0.1}{(s+0.1)^2 \, 80^2} + \dfrac{0.4}{80}\dfrac{80}{(s+0.1)^2 + 80^2}$

#20, $a = 0.1$, $b = 80$; #19, $a = 0.1$, $b = 80$

$y = 4e^{-0.1t} \cos 80t + 0.005e^{-0.1t} \sin 80t$

29. $L\dfrac{d^2 q}{dt^2} + R\dfrac{dq}{dt} + \dfrac{q}{i} = E$

$0(q'') + 50q' + \dfrac{q}{4 \times 10^{-6}} = 40$

$50q' + \dfrac{10^6}{4}q = 40$

$50L(q') + \dfrac{10^6}{4}L(q) = L(40)$

$50[sL(q) - q(0)] + \dfrac{10^6}{4}L(q) = \dfrac{40}{5}$

$$L(q)\left(50s + \frac{10^6}{4}\right) = \frac{40}{s}$$

$$L(q) = \frac{40}{s\left(50s + \frac{10^6}{4}\right)}$$

$$L(q) = F(s) = \frac{40}{50s(s + 5000)}$$

$$= \frac{40}{50(5000)} \cdot \frac{5000}{s(s + 5000)}$$

$$= 0.00016 \cdot \frac{5000}{s(s + 5000)}$$

$$L^{-1}[F(s)] = f(t) = \frac{1.60}{10^4}(1 - e^{-5000t}) = 9; \text{ from}$$

transform (4), $a = 5000$

$$q = 1.60 \times 10^{-4}(1 - e^{-5000t})$$

@ $t = 0, i = 0$

$$0 = 226c_1 + 66.7CE \Rightarrow c_1 = \frac{-66.7CE}{226}$$

$$i = e^{-66.7t}[(-66.7CE\cos(226t) + 226E\sin(226t)$$

$$+ \frac{66.7^2CE}{226}\sin(226t) + 66.7CE\cos(226t)]$$

$$i = \left(\frac{66.7^2CE}{226} + 226CE\right)e^{-66.7t}\sin(226t)$$

$$i = \left(\frac{66.7^2(300 \times 10^{-6})(60)}{226}\right.$$

$$\left. + 226(300 \times 10^{-6})(60)\right)e^{-66.7t}\sin(226t)$$

$$i = 4.42e^{-66.7t}\sin(226t)$$

33. $D^2y + 9y = 18\sin 3t;\ y = 0, Dy = 0, t = 0$

$y'' + 9y = 18\sin 3t;\ L(y'') + 9L(y) = 18L(\sin 3t)$

$$s^2L(y) - sy(0) - y'(0) + 9L(y) = 18\frac{3}{s^2 + 9} = \frac{54}{s^2 + 9}$$

$$L(y)(s^2 + 9) = \frac{54}{s^2 + 9}$$

$$L(y) = \frac{54}{(s^2 + 9)^2} = F(s); \text{ from transform (15)},$$
$$a = 3;$$

$$2a^3 = 54$$

$$y = L^{-1}\left[\frac{54}{(s^2 + 9)^2}\right] = L^{-1}\left[\frac{2(27)}{(s^2 + 9)^2}\right]$$

$$y = \sin 3t - 3t\cos 3t$$

37. $L\frac{d^2q}{dt^2} + R\frac{dq}{dt} + \frac{q}{C} = E;\ q(0) = 0, q'(0) = 0$

$$Lm^2 + Rm + \frac{1}{C} = 0$$

$$m = \frac{-R}{2L} \pm \sqrt{\frac{1}{LC} - \frac{R^2}{4L^2}}j = -66.7 \pm 226j$$

$$q = e^{-66.7t}(c_1\sin(226t) + c_2\cos(226t) + CE$$
@ $t = 0, q = 0 \Rightarrow c_2 = -CE$
$$q = e^{-66.7t}(c_1\sin(226t) - CE\cos(226t) + CE$$

$$i = \frac{dq}{dt}$$
$$= e^{-66.7t}(226c_1\cos(226t) + 226CE\sin(226t)$$
$$- 66.7(c_1\sin(226t) - CE\cos(226t))e^{-66.7t}$$

Chapter 31   Review Exercises

1. $4xy^3 dx + \left(x^2 +1\right)dy = 0$; divide by y^3 and $x^2 +1$

$$\frac{4x}{x^2 +1}dx + \frac{dy}{y^3} = 0; \text{ integrating,}$$

$$2\ln\left(x^2 +1\right) - \frac{1}{2y^2} = c$$

5. $2D^2 y + Dy = 0$. The auxiliary equation is

$$2m^2 + m = 0$$

$$m\left(2m+1\right) = 0; \; m_1 = 0 \text{ and } m_2 = -\frac{1}{2}$$

$$y = c_1 e^0 + c_2 e^{-1/2}; \; y = c_1 + c_2 e^{-x/2}$$

9. $\left(x+y\right)dx + \left(x+y^3\right)dy = 0; \; x\,dx + y\,dx + x\,dy +$

$y^3 \, dy = 0; \; x\,dx + d\left(xy\right) + y^3 \, dy = 0$

Integrating, $\frac{1}{2}x^2 + xy + \frac{1}{4}y^4 = c_1;$

$$2x^2 + 4xy + y^4 = c$$

13. $dy = \left(2y + y^2\right)dx; \; \frac{dy}{\left(2y + y^2\right)} = dx$

$$-\frac{1}{2}\ln\left(\frac{2+y}{y}\right) = x + \ln \; c_1; \; \ln\frac{2+y}{y}$$

$$= -2x - \ln c_1^2 \left(\frac{2+y}{y}\right) = -2x; \; \frac{2+y}{y} = \frac{e^{-2x}}{c_1^2};$$

$$y = c_1^2 \left(y+2\right)e^{2x}; \; y = c\left(y+2\right)e^{2x}$$

17. $y' + 4y = 2e^{-2x}; \; \frac{dy}{dx} + 4y = 2e^{-2x}; \; dy + 4\cdot y\,dx$

$$= 2e^{-2x}dx$$

$$e^{\int 4dx} = e^{4x}; \; e^{4x}dy + 4ye^{4x}dx = 2e^{-2x}\cdot e^{4x}dx;$$

$$d\left(ye^{4x}\right) = 2e^{2x}dx; \; \int d\left(ye^{4x}\right) = \int e^{2x}\cdot\left(2dx\right);$$

$$ye^{4x} = e^{2x} + c;$$

$$y = e^{-2x} + ce^{-4x}$$

21. $2D^2 s + Ds - 3s = 6; \; 2m^2 + m - 3 = 0;$

$$\left(m-1\right)\left(2m+3\right) = 0; \; m_1 = 1, \; m_2 = -\frac{3}{2}$$

$s_c = c_1 e^t + c_2 e^{-3t/2}; \; s_p = A; \; s_p' = 0; \; s_p'' = 0$

Substituting into the differential equation,

$2\left(0\right) + 0 - 3A = 6; \; A = -2; \; s_p = -2.$

$$s = c_1 e^t + c_2 e^{-3t/2} - 2$$

25. $9D^2 y - 18Dy + 8y = 16 + 4x$

$$D^2 y - 2Dy + \frac{8}{9}y = \frac{16}{9} + \frac{4}{9}x$$

$$m^2 - 2m + \frac{8}{9} = 0; \; m = \frac{2 \pm \sqrt{4 - 4\left(\frac{8}{9}\right)}}{2}; \; m_1 = \frac{2}{3}; \; m_2 = \frac{4}{3}$$

$$y_c = c_1 e^{2x/3} + c_2 e^{4x/3}; \; y_p = A + Bx; \; y_p' = B; \; y_p'' = 0$$

Substituting into the differential equation.

$$0 - 2B + \frac{8}{9}\left(A + Bx\right) = \frac{16}{9} + \frac{4}{9}x$$

$$\left(-2B + \frac{8}{9}A\right) + \frac{8}{9}Bx = \frac{16}{9} + \frac{4}{9}x; \; -2B + \frac{8}{9}A = \frac{16}{9};$$

$$\frac{8}{9}B = \frac{4}{9}; \; B = \frac{1}{2}; \; -2\left(\frac{1}{2}\right) + \frac{8}{9}A = \frac{16}{9};$$

$$A = \frac{25}{8}; \; y_p = \frac{1}{2}x + \frac{25}{8}$$

$$y = c_1 e^{2x/3} + c_2 e^{4x/3} + \frac{1}{2}x + \frac{25}{8}$$

29. $y'' - 7y' - 8y = 2e^{-x}$

$$D^2 y_c - 7Dy_c - 8y_c = 0$$

$$m^2 - 7m - 8 = 0$$

$$\left(m+1\right)\left(m-8\right) = 0$$

$$m = -1, \; m = 8$$

$$y_c = c_1 e^{-x} + c_2 e^{8x}$$

Let $y_p = Axe^{-x}$, $y_p' = Ae^{-x}(1-x)$, $y_p'' = Ae^{-x}(x-2)$

$Ae^{-x}(x-2) - 7Ae^{-x}(1-x) - 8Axe^{-x} = 2e^{-x}$

$$A(x-2) - 7A(1-x) - 8Ax = 2$$

$$-9A = 2$$

$$A = \frac{-2}{9}$$

$$y_p = \frac{-2}{9}xe^{-x}$$

$$y = y_c + y_p$$

$$y = c_1 e^{-x} + c_2 e^{8x} + \frac{-2}{9}xe^{-x}$$

33. $3y' = 2y \cot x;$ $\dfrac{dy}{dx} = \dfrac{2}{3}y \cot x;$ $\dfrac{dy}{y} = \dfrac{2}{3}\cot x\, dx;$

$\ln y = \dfrac{2}{3}\ln \sin x + \ln c;$ $\ln y - \ln \sin^{2/3} x = \ln c;$

$\ln \dfrac{y}{\sin^{2/3} x} = \ln c;$ $\dfrac{y}{\sin^{2/3} x} = c;$ $y = c \sin^{2/3} x$

Substituting $y = 2$ when $x = \dfrac{\pi}{2}$, $2 = c \sin^{2/3} \dfrac{\pi}{2};$ $c = 2$

Therefore $y = 2\sin^{2/3} x = 2\sqrt[3]{\sin^2 x};$ $y^3 = 8\sin^2 x$

37. $D^2 v + Dv + 4v = 0.$ $m^2 + m + 4 = 0;$ $m = \dfrac{-1 \pm \sqrt{-15}}{2}$

$m_1 = \dfrac{1}{2} + \dfrac{\sqrt{15}}{2} j;$ $m_2 = -\dfrac{1}{2} - \dfrac{\sqrt{15}}{2} j;$ $\alpha = -\dfrac{1}{2},$ $\beta = \dfrac{\sqrt{15}}{2};$

by Eq. (30-17)

$v = e^{-t/2}\left(c_1 \sin \dfrac{\sqrt{15}}{2} t + c_2 \cos \dfrac{\sqrt{15}}{2} t \right)$

$Dv = e^{-t/2}\left(\dfrac{\sqrt{15}}{2} c_1 \cos \dfrac{\sqrt{15}}{2} t - c_2 \sin \dfrac{\sqrt{15}}{2} t \right)$

$+\left(c_1 \sin \dfrac{\sqrt{15}t}{2} c_2 \cos \dfrac{\sqrt{15}}{2} t \right)\left(-\dfrac{1}{2} e^{-t} \right)$

Substituting $Dv = \sqrt{15}$, $v = 0$ when $t = 0;$ $c_2 = 0;$ $c_1 = 2.$

$v = 2e^{-t/2} \sin\left(\dfrac{1}{2}\sqrt{15}t \right)$

41. $L(4y') - L(y) = 0;$ $4L(y') - L(y)$

$= 0;$ $4sL(y) - 1 - L(y)$

by Eq. (30-24)

$(4s - 1)L(y) = 1;$ $L(y) = \dfrac{1}{4s - 1} = \dfrac{\frac{1}{4}}{s - \frac{1}{4}}$

By transform (3), $y = e^{t/4}.$

45. $y'' - 6y' + 9y = t,$ $y(0) = 0,$ $y'(0) = 1$

$L(y'') - 6L(y') + 9L(y) = L(t)$

$s^2 L(y) - sy(0) - y'(0) - 6(sL(y) - y(0))$

$+9L(y) = \dfrac{1}{s^2}$

$s^2 L(y) - 1 - 6sL(y) + 9L(y) = \dfrac{1}{s^2}$

$(s-3)^2 L(y) = 1 + \dfrac{1}{s^2} = \dfrac{s^2 + 1}{s^2}$

$L(y) = \dfrac{s^2 + 1}{s^2(s-3)^2}$

$= \dfrac{10}{9(s-3)^2} - \dfrac{2}{27(s-3)} + \dfrac{2}{27s} + \dfrac{1}{9s^2}$

$y = \dfrac{10}{9}te^{3t} - \dfrac{2}{27}e^{3t} + \dfrac{2}{27} + \dfrac{t}{9}$

49. $16y'' + 9y = 3e^x,$ $y(0) = 0,$ $y'(0) = 0$

$16L(y'') + 9L(y) = 3L(e^x)$

$16(s^2 L(y) - sy(0)) + 9L(y) = \dfrac{3}{s-1}$

$16\left(s^2 + \dfrac{9}{16} \right)L(y) = \dfrac{3}{s-1}$

$L(y) = \dfrac{1}{16}\dfrac{3}{\left(s^2 + \frac{9}{16}\right)(s-1)}$

$L(y) = \dfrac{1}{16}\left(\dfrac{48}{25(s-1)} - \dfrac{768}{25(16)} \dfrac{s}{\left(s^2 + \frac{9}{16}\right)} - \dfrac{768}{25(16)}\dfrac{1}{s^2 + \frac{9}{16}} \right)$

$y = \dfrac{3}{25}\left(e^x - \cos\dfrac{3x}{4} - \dfrac{4}{3}\sin\dfrac{3x}{4} \right)$

53. (a) $dy - 2y\,dx = e^{3x}\,dx$

$ye^{\int -2\,dx} = \int e^{\int -2\,dx} e^{3x}\,dx$

$ye^{-2x} = \int e^{x}\,dx = e^{x} + c_1$

$y = e^{3x} + c_1 e^{2x}$, $y = 1$ when $x = 0$

$1 = 1 + c_1 \Rightarrow c_1 = 0$

$y = e^{3x}$

(b) $L(y') - 2L(y) = L(e^{3x})$

$sL(y) - y(0) - 2L(y) = \dfrac{2}{s-3}$

$sL(y) - 1 - 2L(y) = \dfrac{2}{s-3}$

$L(y) = \dfrac{2}{(s-3)(s-2)} + \dfrac{1}{s-2}$

#9, $a = -3$, $b = -2$; #4, $a = -2$

$y = e^{3x} - e^{-2x} + e^{2x} = e^{3x}$

57. $\dfrac{dx}{dt} = 2t$

$x = t^2 + c$

$1 = 0^2 + c \Rightarrow c = 1$

$x = t^2 + 1$

$xy = 1$

$x\dfrac{dy}{dt} + y\dfrac{dx}{dt} = 0$

$x\dfrac{dy}{dt} + y(2t) = 0$

$\dfrac{x}{y}\,dy + 2t\,dt = 0$

$\dfrac{1}{y^2}\,dy + 2t\,dt = 0$

$-\dfrac{1}{y} + t^2 = c$

$-\dfrac{1}{1} + 0^2 = c \Rightarrow = -1$

$-\dfrac{1}{y} + t^2 = -1$

$-1 + yt^2 = -y$

$y(t^2 + 1) = 1$

$y = \dfrac{1}{t^2 + 1}$

In terms of t, $x = t^2 + 1$ and $y = \dfrac{1}{t^2 + 1}$.

61. $\dfrac{dm}{dt} = km$, $k < 0$

$\dfrac{dm}{m} = k\,dt$

$\ln m = kt + c$

$\ln m_0 = k(0) + c \Rightarrow c = \ln m_0$

$\ln m = kt + \ln m_0$

$\ln \dfrac{m}{m_0} = kt$

$\dfrac{m}{m_0} = e^{kt}$

$m = m_0 e^{kt}$

65. $\dfrac{dy}{dx} = \dfrac{y}{y-x}$, $(-1, 2)$, $y > 0$

$y\,dy = y\,dx + x\,dy = d(xy)$

$\dfrac{y^2}{2} = xy + C$

$\dfrac{2^2}{2} = (-1)(2) + C \Rightarrow C = 4$

$\dfrac{y^2}{2} = xy + 4$

$y^2 = 2xy + 8 \Rightarrow y = x \pm \sqrt{x^2 - 8}$ from which the path consists of two functions

$f_1(x) = x + \sqrt{x^2 - 8}$, $x \ge \sqrt{8}$ and

$f_2(x) = x - \sqrt{x^2 - 8}$, $x \ge \sqrt{8}$

69. $N = N_0 e^{-kt}$

$\dfrac{N_0}{2} = N_0 e^{-kt(1.28 \times 10^9)}$

$e^{k(1.28 \times 10^9)} = 2$

$k(1.28 \times 10^9) = \ln 2$

$k = \dfrac{\ln 2}{1.28 \times 10^9}$

$N = N_0 e^{-\frac{\ln 2}{1.28 \times 10^9} t}$

$0.75 N_0 = N_0 e^{\frac{-\ln 2}{1.28 \times 10^9} t}$

$$e^{\frac{\ln 2}{1.28\times10^9}\cdot t} = \frac{4}{3}$$

$$\frac{\ln 2}{1.28\times10^9}\cdot t = \ln\frac{4}{3}$$

$$t = \frac{\ln\frac{4}{3}}{\frac{\ln 2}{1.28\times10^9}}$$

$$t = 5.31\times10^8 \text{ years}$$

73. See Example 2, Section 31.6

$$y = cx^5, \ c = \frac{y}{x^5}$$

$$y' = 5cx^4 = 5\left(\frac{y}{x^5}\right)x^4 = \frac{5y}{x}$$

$$y'|OT = -\frac{x}{5y}; \ 5y\,dy = -x\,dx$$

$$\frac{5}{2}y^2 = -\frac{1}{2}x^2 + \frac{1}{2}c$$

$$5y^2 + x^2 = c$$

77. $L\dfrac{di}{dt} + Ri = E$

$$2\frac{di}{dt} + 40i = 20$$

$$\frac{di}{dt} + 20i = 10$$

$$di + 20i\,dt = 10\,dt$$

$$ie^{\int 20\,dt} = \int 10e^{\int 20\,dt} + c$$

$$ie^{20t} = 0.5e^{20t} + c$$

$$(0)e^{20(0)} = 0.5e^{20(0)} + c$$

$$c = -0.5$$

$$ie^{20t} = 0.5e^{20t} - 0.5 = 0.5\left(e^{20t} - 1\right)$$

$$i = 0.5\left(1 - e^{-20t}\right)$$

81. $LD^2q + RDq + \dfrac{q}{C} = E \quad \left(D = \dfrac{d}{dt}\right)$

$L = 0.5$ H, $R = 6\,\Omega$, $C = 20$ mF, $E = 24\sin 10t$

$$0.5D^2q + 6Dq + 50q = 24\sin 10t, \ D^2q + 12Dq + 100q$$
$$= 48\sin 10t$$

$$m^2 + 12m + 100 = 0, \ m = \frac{-12\pm\sqrt{144-400}}{2} = -6\pm 8j$$

$$q_e = e^{-6t}\left(c_1\sin 8t + c_2\cos 8t\right), \ q_p = A\sin 10t + B\cos 10t$$

$$Dq_p = 10A\cos 10t - 10B\sin 10t, \ D^2q_p = -100A\sin 10t$$
$$-100B\cos 10t$$

$$-100A\sin 10t - 100B\cos 10t + 12\left(10A\cos 10t - 10B\sin 10t\right)$$
$$+100\left(A\sin 10t + B\cos 10t\right) = 48\sin 10t$$

$$-120B = 48, \ B = -0.4; \ 120A = 0, \ A = 0$$

$$q = e^{-6t}\left(c_1\sin 8t + c_2\cos 8t\right) - 0.4\cos 10t$$

$$q = 0 \text{ when } t = 0, \ 0 = c_2 - 0.4, \ c_2 = 0.4$$

$$q = e^{-6t}\left(c_1\sin 8t + 0.4\cos 8t\right) - 0.4\cos 10t$$

$$Dq = e^{-6t}\left(-6c_1\sin 8t - 2.4\cos 8t + 8c_1\cos 8t - 3.2\sin 8t\right)$$
$$+4\sin 10t$$

$$Dq = 0 \text{ when } t = 0; \ 0 = -2.4 + 8c_1, \ c_1 = 0.3$$

$$q = e^{-6t}\left(0.3\sin 8t + 0.4\cos 8t\right) - 0.4\cos 10t$$

85.
$$2\frac{di}{dt} + i = 12, \ i(0)0$$

$$2L(i') + L(i) = L(12)$$

$$2\left[sL(i) - 0\right] + L(i) = \frac{12}{s}$$

$$L(i) = \frac{12}{s(2s+1)} = \frac{12(1/2)}{s(s+1/2)}$$

$$i = 12\left(1 - e^{-t/2}\right)$$

$$i(0.3) = 12\left(1 - e^{-0.3/2}\right)$$

$$i = 1.67 \text{ A}$$

89. R = rate in L/t at which mixtures flow in/out of container

$V(t)$ = volume of O_2 in L in container at time t

$V(0) = 5.00$ L. When 5.00 L of air passed into the container, $Rt = 5$, $t = \dfrac{R}{5}$. Find $V\left(\dfrac{R}{5}\right)$.

$$\frac{dV}{dt} = 0.20R - \frac{V}{5.00}R = -0.20R(V-1)$$

$$\frac{dV}{V-1} = -0.20 \ R \ dt$$

$$\ln(V-1) = -0.20 \ Rt + C$$

$$\ln(5.00-1) = -0.20R(0) + C \Rightarrow C = \ln 4.00$$

$$\ln(V-1) = -0.20 \ Rt + \ln 4.00$$

$$\ln(V-1) = -0.20R\left(\frac{5}{R}\right) + 4.00$$

$$\ln\frac{V-1}{4.00} = -0.20R\left(\frac{5}{R}\right)$$

$$e^{\ln\frac{V-1}{4.00}} = e^{-0.20R\left(\frac{5}{R}\right)} \Rightarrow \frac{V-1}{4.00} = e^{-0.20R\left(\frac{5}{R}\right)}$$

$$V = 4.00e^{-0.20R\left(\frac{5}{R}\right)} + 1$$

$$V = 2.47 \text{ L}$$

93. $EI\dfrac{d^2 y}{dx^2} = M, \ M = 2000x - 40x^2$

$$EID^2 y = 2000x - 40x^2$$

$$D^2 y = \frac{1}{EI}\left(2000x - 40x^2\right)$$

$$Dy = \frac{1}{EI}\left(1000x^2 - \frac{40}{3}x^3\right) + c_1$$

$$y = \frac{1}{EI}\left(\frac{1000}{3}x^3 - \frac{10}{3}x^4\right) + c_1 x + c_2$$

$$y = 0 \text{ for } x = 0 \text{ and } x = L$$

$$0 = \frac{1}{EI}(0-0) + 0 + c_2, \ c_2 = 0$$

$$0 = \frac{1}{EI}\left(\frac{1000}{3}L^3 - \frac{10}{3}L^4\right) + c_1 L$$

$$c_1 = \frac{1}{EI}\left(\frac{10}{3}L^3 - \frac{1000}{3}L^2\right)$$

$$y = \frac{1}{EI}\left(\frac{1000}{3}x^3 - \frac{10}{3}x^4\right) + \frac{1}{EI}\left(\frac{1000}{3}L^3 - \frac{1000}{3}L^2\right)x$$

$$= \frac{10}{3EI}\left(100x^3 - x^4 + L^3 x - 100L^2 x\right)$$